Grundlehren Text Editions

The *Grundlehren der mathematischen Wissenschaften*, Springer's first book series in higher mathematics, was founded by Richard Courant in 1920 as a series of modern textbooks. Its objective was to lead students to current research questions through basic, comprehensive books. Today, it is often no longer possible to start from the basics and, in a single book, reach the frontiers of current research. As a result, volumes in the series have become increasingly specialised and advanced.

The *Text Editions* of selected *Grundlehren* volumes have been specially adapted by their authors for use in graduate-level teaching and study. The most relevant chapters have been selected, rewritten for better comprehension, and exercises of varying difficulty have been added. Volumes in this series maintain a consistent level throughout, enabling students to build on their knowledge in preparation for research.

Herbert Lange

Abelian Varieties over the Complex Numbers

A Graduate Course

Springer

Herbert Lange
Department Mathematik
University of Erlangen-Nuremberg
Erlangen, Germany

ISSN 1618-2685 ISSN 2627-5260 (electronic)
Grundlehren Text Editions
ISBN 978-3-031-25569-4 ISBN 978-3-031-25570-0 (eBook)
https://doi.org/10.1007/978-3-031-25570-0

Mathematics Subject Classification (2020): 14-01, 14Kxx, 14D20, 14H40, 14H42

This Springer imprint is published by the registered company Springer Nature Switzerland AG
The registered company address is: Gewerbestrasse 11, 6330 Cham, Switzerland

Preface

The purpose of the Springer series *Grundlehren Text Editions* is to publish textbooks derived from *Grundlehren* books by their authors. They may serve as an introduction to the subject of the corresponding volume. When I was asked to write such a manuscript, I agreed because, being retired, I had the time and, moreover, because I had already forgotten most of the details of our volume.

The present text consists of seven chapters, which in addition to parts of [24] also contains some new subjects, which I think belong here. I tried to simplify the presentation whenever possible and moreover corrected some errors and numerous typos.

Each chapter contains several sections, all ending with a set of exercises. In fact, there are more exercises than in the original volume, including among them some easier ones which are recommended in particular for beginners.

Erlangen, April 2022 *Herbert Lange*

Contents

Introduction

The abelian varieties over the complex numbers is probably the most widely investigated class of algebraic varieties. There are several reasons for this fact. Firstly, they are closely related to theta functions, which pop up throughout mathematics and even in physics. Secondly, to every smooth projective complete algebraic variety one can associate several abelian varieties in a natural way (see Chapters 4 and 5) which often help to study the original variety. Finally, abelian varieties occur in several other mathematical branches such as number theory and integrable systems. So the theory is interesting not only for algebraic geometers, but also for mathematicians in related disciplines.

Apart from some additions which will be mentioned below, the present volume consists of about half of the theory given in [24], namely that which may be considered the most important at the current time. However, ultimately the choice of topics reflects my personal taste. Due to the textbook nature of this book, I have tried to simplify proofs whenever possible.

Abelian varieties over the complex numbers are special complex tori, that is, quotients of finite-dimensional complex vector spaces modulo a lattice of maximal rank. Chapter 1 gives an introduction to the theory of complex tori and their line bundles. The main point is that the sections of a line bundle can be interpreted as theta functions on the underlying complex vector space.

In Chapter 2 abelian varieties are defined as algebraic complex tori, algebraic meaning that the complex torus admits an embedding into some projective space. Polarizations of an abelian variety are introduced and the decomposition of polarized abelian varieties is investigated. Finally, some special results are given, such as the dual polarization, the Pontryagin product and the endomorphism algebra of a simple abelian variety.

Chapter 3 deals with moduli spaces of polarized abelian varieties. For this they are given by elements of the Siegel upper half space, which implies that isomorphisms can be described by actions of some symplectic group. The classical theta transformation formula implies that the moduli spaces with level structure can be embedded into some projective space, which implies the same for most moduli spaces.

Chapters 4 and 5 deal with examples of abelian varieties. To every smooth projective curve one can associate an abelian variety, its Jacobian. In Chapter 4 the main results about them are proved. The theta divisor of a Jacobian is investigated, a proof of Torelli's Theorem given and a universal property of the Jacobian using the Poincaré bundle outlined. Finally, a proof of the criterion of Matsusaka–Ran for a principally polarized abelian variety to be a product of Jacobians is given.

In Chapter 5 we study some other special abelian varieties. For abelian surfaces one can say more about projective embeddings. To every smooth projective variety one can associate two abelian varieties, its Albanese and Picard varieties, which are both a generalization of the Jacobian varieties. To every étale double cover of smooth projective curves one can associate its Prym variety. Finally the last section deals with Intermediate Jacobians.

Chapters 6 and 7 contain two more special topics. In Chapter 6 the Fourier–Mukai transform for certain sheaves and the Fourier transform on the Chow and cohomology rings of an abelian variety are studied. This implies some results on the Chow ring of an abelian variety. Finally, Chapter 7 contains an introduction to the Hodge conjecture for abelian varieties. The Hodge group is introduced and proofs of the conjecture in the two easiest cases of general polarized abelian varieties and general Jacobians are given.

Each chapter is divided into sections and each section into subsections. The last subsection of each section contains exercises of varying degrees of difficulty and sometimes also quotes important results which should encourage the reader to further studies. The first exercises are in general very easy and should be worked out in particular by beginners. They are often just the proof of an equation or a check of the commutativity of a diagram, which in both cases helps to understand the definitions. The later exercises are often just standard results, included in order to both challenge the reader as well as broaden their knowledge.

As for the prerequisites, we use the basic language of algebraic geometry and complex analysis. Since abelian varieties are some of the most easily accessible algebraic varieties, we do not need the whole theory. For example Chapters 1 to 6 of Griffiths–Harris [55] or the first chapters of Hartshorne [61] in the algebraic case are more than sufficient. Any other good introductory book (e.g. Harris [60] or Shafarevich [124] for algebraic varieties and Wells [142] or Fischer [43] in the analytic case) will also do. Whenever some deeper result is applied, a precise reference is given.

The chapters can be read in the following order:

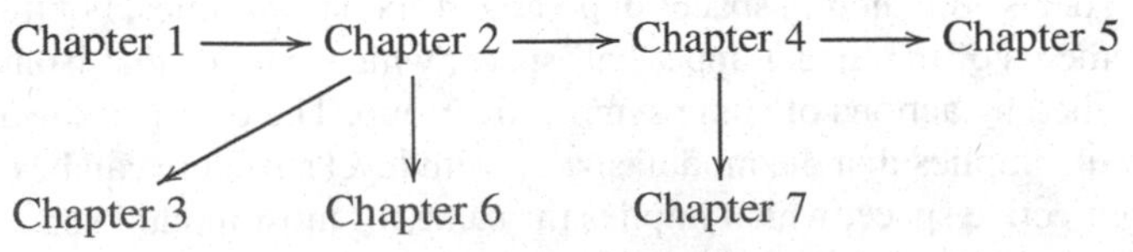

Sometimes one does not need the whole chapter to understand the next one. For example one could read large parts of Chapter 5 right after Chapter 2.

Finally, a word on the intended audience of this book. According to the guidelines of the series *Grundlehren Text Editions* it is mainly directed at graduate students who want to learn the subject. By this I do not mean that the book is easy to read, certainly for some parts one has to work. Besides students specializing algebraic geometry, those of other branches, such as number theory and cryptography, mathematical physics and integral systems, where abelian varieties play a role, will hopefully find this book helpful.

It is a pleasure to acknowledge the help of some colleagues: First I thank Christina Birkenhake for her work on the original *Grundlehren* volume, which went into this manuscript. Moreover I thank Hans-Joachim Schmid for his support. Finally, I would like to thank the referees and the editor for valuable suggestions.

Notation

$H^i(E)$	We use the same symbol to denote a vector bundle and its corresponding locally free sheaf. If E is a locally free sheaf and there is no ambiguity about the base space X of E, we write $H^i(E)$ instead of $H^i(X, E)$.
$h^i(E)$	dimension of the vector space $H^i(E)$.
$x \in D$	for a divisor D and a point x on a variety, $x \in D$ means that x is contained in the support of D.
$\mathrm{M}(g \times g', \mathcal{R})$	module of $(g \times g')$-matrices over a ring $\mathcal{R}$.
$\mathrm{M}_g(\mathcal{R})$	algebra of $(g \times g)$-matrices over a ring $\mathcal{R}$.
$\mathrm{diag}(x_1, \dots, x_g)$	diagonal matrix with entries $x_1, \dots, x_g$.
$\mathrm{d}x_1 \wedge \cdots \wedge \check{\mathrm{d}x_\nu} \wedge \cdots \wedge \mathrm{d}x_g$	$\mathrm{d}x_1 \wedge \cdots \wedge \mathrm{d}x_{\nu-1} \wedge \mathrm{d}x_{\nu+1} \wedge \cdots \wedge \mathrm{d}x_g$, the differential $(g-1)$-form with $\mathrm{d}x_\nu$ omitted.
$\#S$ or $\lvert S \rvert$	cardinality of a set S.
$\mathbf{e}(\cdot)$	exponential function $z \mapsto \mathrm{e}^z$.
$\mathcal{S}_n$	symmetric group of degree n.
$\mathbb{C}_1$	the circle group $\{z \in \mathbb{C} \mid \lvert z \rvert = 1\}$.
$\overline{v}$	image of $v \in V \simeq \mathbb{C}^g$ under a projection map $\pi\colon V \to X = V/\Lambda$.
$\sim$	linear equivalence of divisors.
$\equiv$	algebraic equivalence of line bundles and divisors.
L^n	n-th tensor power of a line bundle L.
(L^g)	self-intersection number of a line bundle L on a g-dimensional complex torus.
$\langle S \rangle$	vector space or group generated by a set S.
δ_{IJ}	Kronecker symbol for subsets $I, J \subset \{1, \dots, n\}$.
Im	imaginary part and also the image of a map. It is clear in every case what is meant.
$\mathbf{1}_X$	the identity map on a variety X or a bundle X.

For more notation see the notation index at the end of the book.

Chapter 1
Line Bundles on Complex Tori

This first chapter deals with complex tori, that is quotients $X = V/\Lambda$ of a finite-dimensional complex vector space V by a lattice Λ of maximal rank. Complex abelian varieties are special complex tori, so everything proved here is in particular valid for abelian varieties.

In his fundamental paper [84] Lefschetz derived, among other results, the most important topological properties of the first part of Section 1.1. In the second part we explain the Hodge decomposition of the complex cohomology groups of X. The rest of this chapter deals with line bundles on complex tori. In Section 1.2 they are introduced and described by factors of automorphy, which are by definition 1-cocycles of the lattice Λ with values in $H^0(\mathcal{O}_X^*)$.

In Section 1.3 a proof of the Appell–Humbert Theorem is given. It was first proved by Humbert in [66] for dimension 2 applying a result of Appell [6] and by Lefschetz in general in [84]. The present formulation is due to Weil [140] and Mumford [97]. It describes each line bundle uniquely by its first Chern class, which can be considered as a hermitian form on V, and a character on Λ with values in the circle group $\mathbb{C}_1$. In Section 1.4 the dual complex torus $\widehat{X}$ of X is introduced. This allows us to define the Poincaré bundle $\mathcal{P}$ on $X \times \widehat{X}$ and its universal property.

In Section 1.4 the global sections of a line bundle L on X are described by certain theta functions on the vector space V, which allow us to compute the dimension $h^0(L)$ explicitly in terms of $c_1(L)$. In the next section this is used to compute the dimension of every cohomology group $h^i(L)$ by changing the complex structure of V and comparing it with h^0 of a line bundle M on the new complex torus. In Section 1.7 finally the Riemann–Roch Theorem for any line bundle on X is derived by just taking the alternating sum of the dimensions of its cohomology groups.

H. Lange, *Abelian Varieties over the Complex Numbers*, Grundlehren Text Editions,
https://doi.org/10.1007/978-3-031-25570-0_1

1.1 Complex Tori

1.1.1 Definition of Complex Tori

Let V denote a complex vector space of dimension g. A *lattice* in V is by definition a discrete subgroup of maximal rank in V. It is a free abelian group of rank $2g$. A lattice Λ in V acts in a natural way on the vector space V and the quotient

$$X = V/\Lambda$$

is called a *complex torus*. It is easy to see that X is a complex manifold of dimension g. A meromorphic function on $\mathbb{C}^g$, periodic with respect to Λ, may be considered as a meromorphic function on X. Conversely, every meromorphic function on X is of this form. An *abelian variety* is a complex torus admitting sufficiently many meromorphic functions. Although the topic of this book is abelian varieties, the first chapter deals more generally with arbitrary complex tori.

Remark 1.1.1 In the theory of linear algebraic groups there is the notion of a *torus*. Such a torus is an affine group, whereas a complex torus is compact. So one has to distinguish both notions. The affine tori do not occur in this book. See also Exercise 1.1.6 (6), where real tori are defined.

The addition in V induces the structure of an abelian complex Lie group on X. We write its group operation additively and call the map

$$\mu \colon X \times X \to X, \qquad (x_1, x_2) \mapsto x_1 + x_2$$

the *addition map* of X.

Lemma 1.1.2 *Any connected compact complex Lie group X of dimension g is a complex torus.*

Proof Let X denote a connected compact complex Lie group of dimension g. First we claim that X is abelian: For this consider the commutator map

$$\Phi : X \times X \to X, \qquad (x, y) \mapsto xyx^{-1}y^{-1}$$

and let U be a coordinate neighbourhood of the unit element 1 in X. For every $x \in X$ there exist open neighbourhoods V_x of x and W_x of 1 in X with $\Phi(V_x, W_x) \subseteq U$, since $\Phi(x, 1) = 1 \in U$ and Φ is continuous. As X is compact, finitely many V_x cover X. Denoting by W the intersection of the corresponding finitely many open sets W_x, we get $\Phi(X, W) \subseteq U$. This implies $\Phi(X, W) = 1$, since holomorphic functions on compact manifolds are constant and $\Phi(1, y) = 1$ for every $y \in W$. As W is open and non-empty, this implies the assertion.

Let $\pi: V \to X$ be the universal covering map. The Lie group structure of X induces the structure of a simply connected complex Lie group on V, such that π is a homomorphism. Moreover V is abelian, since X is. Hence V is isomorphic to the vector space $\mathbb{C}^g$ (see Hochschild [63, Theorem 17.4.1]). Finally, the compactness of X implies that $\ker \pi$ is a lattice in V. □

Lemma 1.1.3 *Let $X = V/\Lambda$ be a complex torus.*

(a) *The natural map $\pi : V \to X$ may be considered as the universal covering map.*
(b) $\Lambda = \operatorname{Ker} \pi$ *is canonically isomorphic to the first homology group $H_1(X, \mathbb{Z})$.*
(c) *the vector space V may be considered as the tangent space T_0X of X at 0. From the Lie theoretical point of view, $\pi : V = T_0X \to X$ is just the exponential map.*

Proof (a) is clear. As for (b), the kernel Λ of π is the fundamental group $\pi_1(X)$. Since Λ is abelian, $\pi_1(X)$ is canonically isomorphic to $H_1(X, \mathbb{Z})$.

Finally, the complex torus X is locally isomorphic to V. Since T_0V can be identified with V in a natural way, this implies the first assertion and then also the second assertion of (c). □

Example 1.1.4 $g = \dim V = 1$.
Choosing a basis of V, we may identify V with the field of complex numbers $\mathbb{C}$. A lattice in $\mathbb{C}$ is generated by two complex numbers λ_1 and λ_2 which are linearly independent over $\mathbb{R}$. So we have the following picture:

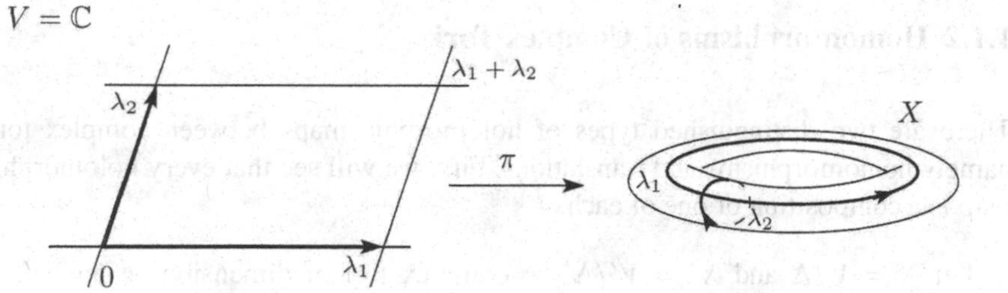

Identifying opposite sides of the parallelogram $0, \lambda_1, \lambda_1 + \lambda_2, \lambda_2$, we obtain the complex torus X. The images of the lines $\overline{0\lambda_1}$ and $\overline{0\lambda_2}$ are cycles on X, also denoted by λ_1, and λ_2. Obviously λ_1 and λ_2 generate the group $H_1(X, \mathbb{Z})$. A 1-dimensional complex torus is called an *elliptic curve*.

Let $X = V/\Lambda$ be an arbitrary complex torus again. In order to describe X, choose bases $e_1, \cdots, e_g$ of V and $\lambda_1, \cdots, \lambda_{2g}$ of the lattice Λ. Write λ_i in terms of the basis $e_1, \ldots, e_g$:

$$\lambda_i = \sum_{j=1}^{g} \lambda_{ji} e_j.$$

The matrix

$$\Pi = \begin{pmatrix} \lambda_{11} & \cdots & \cdots & \lambda_{1,2g} \\ \vdots & & & \vdots \\ \lambda_{g1} & \cdots & \cdots & \lambda_{g,2g} \end{pmatrix}$$

in $M(g \times 2g, \mathbb{C})$ is called a *period matrix* for X. The period matrix Π determines the complex torus X completely, but certainly it depends on the choice of the bases for V and Λ. Conversely, given a matrix $\Pi \in M(g \times 2g, \mathbb{C})$, one may ask: Is Π a period matrix for some complex torus? The following proposition gives an answer to this question.

Proposition 1.1.5 *$\Pi \in M(g \times 2g, \mathbb{C})$ is the period matrix of a complex torus if and only if the matrix $P = \binom{\Pi}{\overline{\Pi}} \in M_{2g}(\mathbb{C})$ is nonsingular, where $\overline{\Pi}$ denotes the complex conjugate matrix.*

Proof Π is a period matrix if and only if the column vectors of Π span a lattice in $\mathbb{C}^g$, in other words, if and only if the columns are linearly independent over $\mathbb{R}$.

Suppose first that the columns of Π are linearly dependent over $\mathbb{R}$. Then there is an $x \in \mathbb{R}^{2g}$, $x \neq 0$, with $\Pi x = 0$, and we get $Px = 0$. This implies $\det P = 0$.

Conversely, if P is singular, there are vectors $x, y \in \mathbb{R}^{2g}$, not both zero, such that $P(x+iy) = 0$. But $\Pi(x+iy) = 0$ and $\Pi(x-iy) = \overline{\overline{\Pi}(x+iy)} = 0$ imply $\Pi x = \Pi y = 0$. Hence the columns of Π are linearly dependent over $\mathbb{R}$. □

1.1.2 Homomorphisms of Complex Tori

There are two distinguished types of holomorphic maps between complex tori, namely homomorphisms and translations. First we will see that every holomorphic map is a composition of one of each.

Let $X = V/\Lambda$ and $X' = V'/\Lambda'$ be complex tori of dimensions g and g'. A *homomorphism of X to X'* is a holomorphic map $f\colon X \to X'$, compatible with the group structures. The *translation by an element* $x_0 \in X$ is defined to be the holomorphic map

$$t_{x_0}\colon X \to X, \qquad x \mapsto x + x_0.$$

Proposition 1.1.6 *Let $h\colon X \to X'$ be a holomorphic map.*

(a) *There is a unique homomorphism $f\colon X \to X'$ such that $h = t_{h(0)} f$, that is, $h(x) = f(x) + h(0)$ for all $x \in X$.*
(b) *There is a unique $\mathbb{C}$-linear map $F\colon V \to V'$ with $F(\Lambda) \subset \Lambda'$ inducing the homomorphism f.*

Proof Define $f = t_{-h(0)}h$. We can lift the composed map $V \xrightarrow{\pi} X \xrightarrow{f} X'$ to a holomorphic map F into the universal cover V' of X'

$$\begin{array}{ccc} V & \xrightarrow{\quad F \quad} & V' \\ & {\scriptstyle f\pi} \searrow \quad \swarrow {\scriptstyle \pi'} & \\ & X' & \end{array} \tag{1.1}$$

in such a way that $F(0) = 0$. The diagram implies that for all $\lambda \in \Lambda$ and $v \in V$ we have $F(v+\lambda) - F(v) \in \Lambda'$. Thus the continuous map $v \mapsto F(v+\lambda) - F(v)$ is constant and we get $F(v+\lambda) = F(v) + F(\lambda)$ for all $\lambda \in \Lambda$ and $v \in V$. Hence the partial derivatives of F are $2g$-fold periodic and thus constant by Liouville's theorem. It follows that F is $\mathbb{C}$-linear and f is a homomorphism. The uniqueness of F and f is obvious. □

Corollary 1.1.7 *Given a homomorphism $f : X = V/\Lambda \to X' = V'/\Lambda'$ and identifying $V = T_0X$ and $V' = T_0X'$, the $\mathbb{C}$-linear map $F : V \to V'$ above coincides with the differential* $\mathrm{d}f|_0 : T_0X \to T_0X'$.

Proof Identifying the projections π and π' with the corresponding exponential maps, the corollary is a consequence of the uniqueness property of the differential $\mathrm{d}f|_0$ with respect to diagram (1.1). □

Under addition the set of homomorphisms of X to X' forms an abelian group denoted by $\mathrm{Hom}(X, X')$. Proposition 1.1.6 gives an injective homomorphism of abelian groups

$$\rho_a : \mathrm{Hom}(X, X') \longrightarrow \mathrm{Hom}_{\mathbb{C}}(V, V'), \qquad f \mapsto F,$$

the *analytic representation of* $\mathrm{Hom}(X, X')$. The restriction F_Λ of F to the lattice Λ is $\mathbb{Z}$-linear. F_Λ determines F and f completely. Thus we get an injective homomorphism

$$\rho_r : \mathrm{Hom}(X, X') \to \mathrm{Hom}_{\mathbb{Z}}(\Lambda, \Lambda'), \qquad f \mapsto F_\Lambda,$$

the *rational representation of* $\mathrm{Hom}(X, X')$. We denote the extensions of ρ_a and ρ_r to

$$\mathrm{Hom}_{\mathbb{Q}}(X, X') := \mathrm{Hom}(X, X') \otimes_{\mathbb{Z}} \mathbb{Q}$$

by the same letters. These will also be referred to as the analytic and rational representations.

Since any subgroup of $\mathrm{Hom}_{\mathbb{Z}}(\Lambda, \Lambda') \simeq \mathbb{Z}^{4gg'}$ is isomorphic to some $\mathbb{Z}^m$, the injectivity of ρ_r implies:

Proposition 1.1.8 $\mathrm{Hom}(X, X') \simeq \mathbb{Z}^m$ *for some* $m \leq 4gg'$.

Let $X'' = V''/\Lambda''$ be a third complex torus. For $f \in \mathrm{Hom}(X, X')$ and $f' \in \mathrm{Hom}(X', X'')$ we have

$$\rho_a(f'f) = \rho_a(f')\rho_a(f)$$

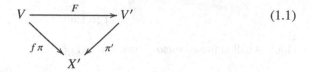

and similarly for ρ_r. This follows immediately from the uniqueness statement in Proposition 1.1.6 (b). In particular, if $X = X'$, ρ_a and ρ_r are representations of the ring $\mathrm{End}(X)$ of endomorphisms of X, respectively its extension

$$\mathrm{End}_{\mathbb{Q}}(X) := \mathrm{End}(X) \otimes_{\mathbb{Z}} \mathbb{Q},$$

which is called the *endomorphism algebra of* X.

Suppose $\Pi \in \mathrm{M}(g \times 2g, \mathbb{C})$ and $\Pi' \in \mathrm{M}(g' \times 2g', \mathbb{C})$ are period matrices for X and X' with respect to some bases of V, Λ and V', Λ' respectively. Let $f : X \to X'$ be a homomorphism. With respect to the chosen bases the representation $\rho_a(f)$ (respectively $\rho_r(f)$) is given by a matrix $A \in \mathrm{M}(g' \times g, \mathbb{C})$ (respectively $R \in \mathrm{M}(2g' \times 2g, \mathbb{Z})$). In terms of matrices the condition $\rho_a(f)(\Lambda) \subset \Lambda'$ means

$$A\Pi = \Pi' R. \tag{1.2}$$

Conversely, any two matrices $A \in \mathrm{M}(g' \times g, \mathbb{C})$ and $R \in \mathrm{M}(2g' \times 2g, \mathbb{Z})$ satisfying equation (1.2) define a homomorphism $X \to X'$. We apply this equation to prove the next proposition, which shows how ρ_a and ρ_r are related.

Proposition 1.1.9 *Let $X = V/\Lambda$ be a complex torus. The extended rational representation*

$$\rho_r \otimes 1 : \mathrm{End}_{\mathbb{Q}}(X) \otimes \mathbb{C} \to \mathrm{End}_{\mathbb{C}}(\Lambda \otimes \mathbb{C}) \simeq \mathrm{End}_{\mathbb{C}}(V \times V)$$

is equivalent to the direct sum of the analytic representation and its complex conjugate:

$$\rho_r \otimes 1 \simeq \rho_a \oplus \overline{\rho_a}.$$

Proof Let Π denote the period matrix of X with respect to some bases of V and Λ. Suppose $f \in \mathrm{End}(X)$. If A and R are the matrices of $\rho_a(f)$ and $\rho_r(f)$ with respect to the chosen bases, we have by equation (1.2),

$$\begin{pmatrix} A & 0 \\ 0 & \overline{A} \end{pmatrix} \begin{pmatrix} \Pi \\ \overline{\Pi} \end{pmatrix} = \begin{pmatrix} \Pi \\ \overline{\Pi} \end{pmatrix} R.$$

This implies the assertion, since $\begin{pmatrix} \Pi \\ \overline{\Pi} \end{pmatrix}$ is nonsingular by Proposition 1.1.5. □

Proposition 1.1.10 *Let $f : X \to X'$ be a homomorphism of complex tori.*

(a) $\mathrm{Im}\, f$ *is a complex subtorus of* X'.
(b) $\mathrm{Ker}\, f$ *is a closed subgroup of* X. *The connected component* $(\mathrm{Ker}\, f)^0$ *of* $\mathrm{Ker}\, f$ *containing* 0 *is a subtorus of* X *of finite index in* $\mathrm{Ker}\, f$.

Proof This is a consequence of Lemma 1.1.2. However, we will give a direct proof. Let $F = \rho_a(f)$.

(a): Since $\mathrm{Im}\, f = F(V)/(F(V) \cap \Lambda')$, we have to show that $F(V) \cap \Lambda'$ is a lattice in $F(V)$. But $F(V) \cap \Lambda'$ is discrete in $F(V)$ and generates $F(V)$ as an $\mathbb{R}$-vector space, since it contains $F(\Lambda)$.

(b): We have only to show that $(\operatorname{Ker} f)^0$ is a complex torus, since as a compact space, $\operatorname{Ker} f$ has only a finite number of connected components. F is a linear map, hence the connected component $(F^{-1}(\Lambda'))^0$ of $F^{-1}(\Lambda')$ containing 0 is a subvector space of V and $(\operatorname{Ker} f)^0 = (F^{-1}(\Lambda'))^0/\big((F^{-1}(\Lambda'))^0 \cap \Lambda\big)$. Finally, $(\operatorname{Ker} f)^0$ being compact, the group $(F^{-1}(\Lambda'))^0 \cap \Lambda$ is a lattice in $(F^{-1}(\Lambda'))^0$. □

As an example consider the product $X \times X'$ of the complex tori $X = V/\Lambda$ and $X' = V'/\Lambda'$. It is again a complex torus: $X \times X' = V \times V'/\Lambda \times \Lambda'$. The projections of $X \times X'$ onto its factors and the natural embeddings of X (respectively X') into $X \times X'$ are homomorphisms of complex tori. Obviously the analytic (respectively rational) representation of these homomorphisms are just the projections and natural embeddings of the corresponding vector spaces (respectively lattices).

Next we define a special class of homomorphisms of complex tori, the isogenies. They will be of particular importance in the sequel. An *isogeny* of a complex torus X to a complex torus X' is by definition a surjective homomorphism $X \to X'$ with finite kernel. The following lemma is easy to check.

Lemma 1.1.11 *For a homomorphism $f : X \to X'$ of complex tori the following conditions are equivalent:*

(i) *f is an isogeny;*
(ii) *f is surjective and* $\dim X = \dim X'$*;*
(iii) $df|_0 : T_0X \to T_0X'$ *is an isomorphism.*

If $\Gamma \subseteq X$ is a finite subgroup, the quotient X/Γ is a complex torus and the natural projection $p\colon X \to X/\Gamma$ is an isogeny. To see this, note that $\pi^{-1}(\Gamma) \subset V$ is a lattice containing Λ and $X/\Gamma = V/\pi^{-1}(\Gamma)$. Conversely it is clear that up to isomorphisms every isogeny is of this type.

The following proposition is an immediate consequence of Proposition 1.1.10.

Proposition 1.1.12 *Every homomorphism $f : X \to X'$ of complex tori factorizes canonically as follows:*

$$\begin{array}{ccccc} & & X/(\ker f)^0 & & \\ & \overset{g}{\nearrow} & & \overset{h}{\searrow} & \\ X & & \xrightarrow{\quad f \quad} & & \operatorname{Im} f \hookrightarrow X', \end{array}$$

with g surjective and h an isogeny.

This is the *Stein factorization* of the homomorphism f.

We define the *degree* $\deg f$ of a homomorphism $f : X \to X'$ to be the order of the group $\operatorname{Ker} f$, if it is finite, and 0 otherwise. In formulas

$$\deg f := \begin{cases} |\operatorname{Ker} f| & \text{if } \operatorname{Ker} f \text{ is finite;} \\ 0 & \text{otherwise.} \end{cases}$$

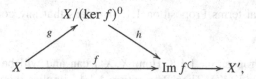

Proposition 1.1.13 *Let $f : X = V/\Lambda \to X' = V'/\Lambda'$ be a homomorphism.*

(a) *If f is an isogeny, then*

$$\deg f = |\Lambda' : \rho_r(f)(\Lambda)|.$$

(b) *If f is surjective and $g : X' \to X''$ a second homomorphism, then*

$$\deg(gf) = \deg f \cdot \deg g.$$

(c) *If $f \in \mathrm{End}(X)$, that is $\Lambda = \Lambda'$, then*

$$\deg f = \det \rho_r(f).$$

Proof (a) is a consequence of the group-theoretical homomorphism theorem. (b) is obvious. (c) follows from (a) and the fact that the index of $\rho_r(f)(\Lambda)$ in Λ is given by $\det \rho_r(f)$. Note that $\det \rho_r(f)$ is positive by Proposition 1.1.9. □

For any integer n define the *multiplication of X by n on X* as the homomorphism

$$n_X \colon X \to X, \quad x \mapsto nx.$$

If $n \neq 0$, its kernel $X_n := \mathrm{Ker}\, n_X$ is called the *group of n-division points* of X.

Proposition 1.1.14 *If X is of dimension g, then*

$$X_n \simeq (\mathbb{Z}/n\mathbb{Z})^{2g}.$$

In particular, n_X is an isogeny of degree n^{2g} for any integer $n \neq 0$.

Proof Suppose $X = V/\Lambda$. Then $X_n = \mathrm{Ker}\, n_X = \frac{1}{n}\Lambda/\Lambda \simeq \Lambda/n\Lambda \simeq (\mathbb{Z}/n\mathbb{Z})^{2g}$. □

In group-theoretical terms Proposition 1.1.14 means that any complex torus is a divisible group.

According to Proposition 1.1.8, $\mathrm{Hom}(X, X')$ can and will be considered as a subgroup of $\mathrm{Hom}_{\mathbb{Q}}(X, X')$. Then Proposition 1.1.14 implies that the definition of the degree of a homomorphism extends to $\mathrm{Hom}_{\mathbb{Q}}(X, X')$ by

$$\deg(rf) := r^{2g} \deg f$$

for any $r \in \mathbb{Q}$ and $f \in \mathrm{Hom}(X, X')$.

We will see now that isogenies are "almost" isomorphisms. Define the *exponent* $e = e(f)$ of an isogeny f to be the exponent of the finite group $\mathrm{Ker}\, f$. In other words $e(f)$ is the smallest positive integer n with $nx = 0$ for all x in $\mathrm{Ker}\, f$.

Proposition 1.1.15 *For any isogeny $f: X \to X'$ of exponent e there exists an isogeny $g: X' \to X$, unique up to isomorphisms, such that*

$$gf = e_X \qquad \text{and} \qquad fg = e_{X'}.$$

Proof As $\operatorname{Ker} f \subseteq \operatorname{Ker} e_X = X_e$, there is a unique map $g: X' \to X$ such that $gf = e_X$. With e_X and f also g is an isogeny. The kernel of g is contained in the kernel X'_e of $e_{X'}$, since for every $x' \in \operatorname{Ker} g$ there is an $x \in \operatorname{Ker} e_X$ with $f(x) = x'$ and $ex' = ef(x) = f(ex) = 0$. Thus $e_{X'} = f'g$ for some isogeny $f': X \to X'$ and we get

$$f'e_X = f'gf = e_{X'}f = fe_X.$$

This implies $f = f'$, since e_X is surjective. □

Corollary 1.1.16

(a) *Isogenies define an equivalence relation on the set of all complex tori.*
(b) *An element in* $\operatorname{End}(X)$ *is an isogeny if and only if it is invertible in* $\operatorname{End}_{\mathbb{Q}}(X)$.

Hence it makes sense to call two complex tori *isogenous* if there is an isogeny between them.

1.1.3 Cohomology of Complex Tori

The aim of this section is to compute the singular cohomology groups of complex tori $H^n(X,\mathbb{Z})$ with values in $\mathbb{Z}$. We only give the proof in the cases $n = 1$ and 2, where they are particularly easy. For arbitrary n consider Exercise 1.1.6 (7).

Let $X = V/\Lambda$ be a complex torus of dimension g. As a real manifold, X is isomorphic to the product of the $2g$ circles $S_1 \simeq S_1^{(i)} = \lambda_i\mathbb{R}/\lambda_i\mathbb{Z}$, where $\lambda_1, \dots, \lambda_{2g}$ denotes a basis of the lattice Λ. By the Künneth formula this implies that $H_n(X,\mathbb{Z})$ and $H^n(X,\mathbb{Z})$ are free abelian groups of finite rank for all $n = 1, \dots, 2g$. According to Lemma 1.1.3 we have identifications

$$\pi_1(X) = H_1(X,\mathbb{Z}) = \Lambda. \tag{1.3}$$

So by the universal coefficient theorem there is a natural isomorphism

$$H^1(X,\mathbb{Z}) \simeq \operatorname{Hom}(\pi_1(X),\mathbb{Z}) \tag{1.4}$$

(see Greenberg–Harper [53, 23.28]). This proves the first assertion of the following lemma. The second assertion is a consequence of the compatibility of the Künneth formula with the cup product.

Lemma 1.1.17 *Let $X = V/\Lambda$ be a complex torus. Then*

(a) *There is a canonical isomorphism*

$$H^1(X,\mathbb{Z}) \to \mathrm{Hom}(\Lambda,\mathbb{Z}).$$

(b) *The canonical map*

$$\wedge^2 H^1(X,\mathbb{Z}) \longrightarrow H^2(X,\mathbb{Z})$$

induced by the cup product is an isomorphism.

So we can identify $H^1(X,\mathbb{Z}) = \mathrm{Hom}(\Lambda,\mathbb{Z})$. Denoting by

$$\mathrm{Alt}^2(\Lambda,\mathbb{Z}) := \bigwedge^2 \mathrm{Hom}(\Lambda,\mathbb{Z})$$

the group of $\mathbb{Z}$-valued alternating 2-forms on Λ, we get as a consequence:

Corollary 1.1.18 *There is a canonical isomorphism*

$$H^2(X,\mathbb{Z}) \simeq \mathrm{Alt}^2(\Lambda,\mathbb{Z}).$$

The universal coefficient theorem yields $H^2(X,\mathbb{C}) = H^2(X,\mathbb{Z}) \otimes \mathbb{C}$. Denoting by $\mathrm{Alt}^2_{\mathbb{R}}(V,\mathbb{C})$ the group of $\mathbb{R}$-linear alternating 2-forms on V with values in $\mathbb{C}$ and applying the canonical isomorphism $\mathrm{Alt}^2(\Lambda,\mathbb{Z}) \otimes \mathbb{C} = \mathrm{Alt}^2_{\mathbb{R}}(V,\mathbb{C})$, we get the following corollary.

Corollary 1.1.19 *For $n = 1$ and 2 there are canonical isomorphisms*

$$H^n(X,\mathbb{C}) \simeq \mathrm{Alt}^n_{\mathbb{R}}(V,\mathbb{C}) = \bigwedge^n \mathrm{Hom}_{\mathbb{R}}(V,\mathbb{C}) \simeq \bigwedge^n H^1(X,\mathbb{C}).$$

1.1.4 The de Rham Theorem

The de Rham theorem states that integration of complex-valued C^∞-forms induces an isomorphism

$$H^n(X,\mathbb{C}) \simeq H^n_{\mathrm{dR}}(X) := \frac{\{d\text{-closed } n\text{-forms on } X\}}{d\{(n-1)\text{-forms on } X\}}. \tag{1.5}$$

We will explain how, in the case of a complex torus, in every class of n-forms in $H^n_{\mathrm{dR}}(X)$ one can distinguish a representative depending only on a basis of the lattice.

Fix a basis $\lambda_1,\dots,\lambda_{2g}$ of $\Lambda = H_1(X,\mathbb{Z})$ and denote by $x_1,\dots,x_{2g}$ the corresponding real coordinate functions of V. A complex-valued C^∞-form ω on X (respectively V) is called an *invariant n-form* if $t_x^*\omega = \omega$ for all $x \in X$ (respectively $t_v^*\omega = \omega$ for all

$v \in V$). Obviously the differentials $dx_1, \dots, dx_{2g}$ are invariant 1-forms on V. In particular, they are invariant with respect to translation by elements of Λ. Hence every dx_i is the pullback of a uniquely determined invariant 1-form on X via $\pi : V \longrightarrow X$. By abuse of notation we denote this 1-form on X also by dx_i.

Under the de Rham isomorphism the cohomology classes of $dx_1, \dots, dx_{2g}$ on X correspond to a basis of $H^1(X, \mathbb{C})$, since we have by construction

$$\int_{\lambda_i} dx_j = \delta_{ij}.$$

In particular the bases $dx_1, \dots, dx_{2g}$ and $\lambda_1, \dots, \lambda_{2g}$ are dual to each other. The cup product corresponds under the de Rham isomorphism to the exterior product of forms. Together with Exercise 1.1.6 (7) this implies that the classes of the n-forms $dx_{i_1} \wedge \cdots \wedge dx_{i_n}$, $i_1 < \cdots < i_n$, form a basis of $H^n(X, \mathbb{C})$.

Conversely it is obvious that every invariant differential form is a linear combination of these. In other words, they span the complex vector space of invariant n-forms on X. Denoting by $\mathrm{IF}^n(X)$ the vector space of invariant n-forms on X we obtain

Proposition 1.1.20 *The de Rham isomorphism induces an isomorphism*

$$H^n(X, \mathbb{C}) \simeq \mathrm{IF}^n(X).$$

1.1.5 The Hodge Decomposition

In the last subsection we used the real structure of the complex torus $X = V/\Lambda$ to compute the cohomology groups $H^n(X, \mathbb{C})$. Here we want to show that the complex structure of X yields a direct sum decomposition of these vector spaces, the *Hodge decomposition*. We include only part of the proof. For a complete proof we refer to [24]. But note that in the case of a complex torus it is considerably easier than for a general compact Kähler manifold (see [55]).

Theorem 1.1.21

(a) *For every integer $n \geq 0$ the de Rham and the Dolbeault isomorphisms induce an isomorphism*

$$H^n(X, \mathbb{C}) \simeq \bigoplus_{p+q=n} H^q(\Omega_X^p)$$

with Ω_X^p the sheaf of holomorphic p-forms on X.

(b) *For every pair (p, q) there is a natural isomorphism*

$$H^q(\Omega_X^p) \simeq \bigwedge^p \Omega \otimes \bigwedge^q \overline{\Omega}$$

with $\Omega := \mathrm{Hom}_{\mathbb{C}}(V, \mathbb{C})$ *and* $\overline{\Omega} := \mathrm{Hom}_{\overline{\mathbb{C}}}(V, \mathbb{C})$ *the group of $\mathbb{C}$-antilinear forms on V. In particular $h^p(\Omega_X^q) = h^q(\Omega_X^p)$.*

We start by describing the sheaf Ω^p_X. Identifying, as in Section 1.1.1, the vector space V with the complex tangent space $T_{X,0}$ of X at 0, the complex cotangent space $\Omega^1_{X,0}$ of X at 0 is

$$\Omega := \mathrm{Hom}_{\mathbb{C}}(V, \mathbb{C}).$$

For any $x \in X$ the translation t_{-x} induces a vector space isomorphism $\mathrm{d}t_{-x}\colon T_{X,x} \to T_{X,0}$. Using the dual isomorphism $(\mathrm{d}t_{-x})^*\colon \Omega = \Omega^1_{X,0} \to \Omega^1_{X,x}$, every p-covector $\varphi \in \bigwedge^p \Omega$ extends by

$$(\omega_\varphi)_x := (\wedge^p(\mathrm{d}t_{-x})^*)\varphi$$

to a translation-invariant holomorphic p-form ω_φ on X and the map $\varphi \mapsto \omega_\varphi$ defines a homomorphism of sheaves

$$\bigwedge^p \Omega \otimes_{\mathbb{C}} \mathcal{O}_X \longrightarrow \Omega^p_X. \tag{1.6}$$

Since the holomorphic forms, coming from a basis of $\bigwedge^p \Omega$, generate every fibre of Ω^p_X, this homomorphism is in fact an isomorphism. This proves

Lemma 1.1.22 *The sheaf Ω^p_X is a free $\mathcal{O}_X$-module of rank $\binom{g}{p}$.*

Choose a basis $e_1, \dots, e_g$ for V and denote by $v_1, \dots, v_g$ the corresponding complex coordinate functions on V. The differentials $\mathrm{d}v_1, \dots, \mathrm{d}v_g, \mathrm{d}\bar{v}_1, \dots, \mathrm{d}\bar{v}_g$ are linearly independent over $\mathbb{R}$. Hence by Proposition 1.1.20 they form a basis of the vector space $\mathrm{IF}^1(X)$ of invariant 1-forms on X. Here again, as in the previous subsection, we denote the invariant forms on V and the corresponding forms on X by the same letter. For a multi-index $I = (i_1 < \cdots < i_p)$ we write for short

$$\mathrm{d}v_I = \mathrm{d}v_{i_1} \wedge \ldots \wedge \mathrm{d}v_{i_p} \quad \text{and} \quad \mathrm{d}\bar{v}_I = \mathrm{d}\bar{v}_{i_1} \wedge \ldots \wedge \mathrm{d}\bar{v}_{i_p}.$$

An element $\varphi \in \mathrm{IF}^n(X)$ of the form

$$\varphi = \sum_{\#I=p, \#J=q} \alpha_{IJ} \mathrm{d}v_I \wedge \mathrm{d}\bar{v}_J$$

with $\alpha_{IJ} \in \mathbb{C}$ and $p + q = n$, is called an *invariant form of type* (p, q). This definition does not depend on the choice of the basis for V. Denoting by $\mathrm{IF}^{p,q}(X)$ the vector space of all invariant forms of type (p, q) in $\mathrm{IF}^n(X)$ we get the direct sum decomposition

$$\mathrm{IF}^n(X) = \bigoplus_{p+q=n} \mathrm{IF}^{p,q}(X).$$

Note that $\mathrm{IF}^{p,0}(X)$ coincides with the space $H^0(\Omega^p_X)$ of global sections of the sheaf Ω^p_X. So $\mathrm{IF}^{p,0}(X) \simeq \bigwedge^p \Omega$ by (1.6). Similarly there is an isomorphism $\bigwedge^q \overline{\Omega} \simeq \mathrm{IF}^{0,q}(X)$ given by $\varphi \mapsto \omega_\varphi$ with $(\omega_\varphi)_x := ((\wedge^q(\mathrm{d}t_{-x})^*))\varphi$ for all $\varphi \in \bigwedge^q \overline{\Omega}$ and $x \in X$. So we get an isomorphism

$$\bigwedge^p \Omega \otimes \bigwedge^q \overline{\Omega} \to \mathrm{IF}^{p,q}(X), \qquad \varphi_1 \otimes \varphi_2 \in \bigwedge^p \Omega \otimes \bigwedge^q \overline{\Omega} \mapsto \omega_{\varphi_1 \otimes \varphi_2} := \omega_{\varphi_1} \wedge \omega_{\varphi_2}.$$

Combining this with the isomorphism of Proposition 1.1.20 we obtain:

Proposition 1.1.23 *There are natural isomorphisms*

$$H^n(X,\mathbb{C}) \simeq \bigoplus_{p+q=n} \mathrm{IF}^{p,q}(X) \simeq \bigoplus_{p+q=n} \bigwedge^p \Omega \otimes \bigwedge^q \overline{\Omega}.$$

To complete the proof of Theorem 1.1.21, it remains to show that there is a canonical isomorphism

$$H^q(\Omega_X^p) \to \bigwedge^p \Omega \otimes \bigwedge^q \overline{\Omega},$$

which is called the *Dolbeault isomorphism*. For its proof we refer to [24].

Remark 1.1.24 Note that the spaces IF^n and $\mathrm{IF}^{p,q}$ are harmonic forms with respect to a suitable (i.e. flat) Kähler metric. In this form Proposition 1.1.23 generalizes to arbitrary compact Kähler manifolds (see [55]).

1.1.6 Exercises

(1) Let $X = V/\Lambda$ be a complex torus of dimension g. Show that

(a) there exist bases of V and Λ with respect to which the period matrix of X is of the form $(Z, \mathbf{1_g})$ with $Z \in \mathrm{M}_g(\mathbb{C})$ and $\det \operatorname{Im} Z \neq 0$. Conversely, every such matrix is the period matrix of a complex torus.

(b) $\det \begin{pmatrix} Z & 1 \\ \overline{Z} & 1 \end{pmatrix} = \det(2i \operatorname{Im} Z)$.

(2) Let $X = V/\Lambda$ be a complex torus.

(a) Show that X admits a complex subtorus of dimension g' if and only if there exists a subgroup $\Lambda' \subset \Lambda$ of rank $2g'$ such that the image of the canonical map $\Lambda' \otimes \mathbb{R} \to V$ is a $\mathbb{C}$-subvector space of V.

(b) Conclude from (a) that any complex torus admits at most countably many complex subtori.

(c) Give an example of a complex torus of dimension ≥ 2 not admitting any non-trivial complex subtorus.

(3) Let $\iota : Y \to X$ be an injective holomorphic map of complex tori such that the tangent map is injective everywhere. Show that $\iota(Y)$ is a complex subtorus of X and ι is an isomorphism onto it.

(4) Let X be a complex torus and $f : X \to G$ a holomorphic map into a complex Lie group G. Show that the map $X \to G$, $x \mapsto f(0)^{-1} f(x)$ is a homomorphism of complex Lie groups.

(5) (a) There is a bijection between the set of complex structures on the vector space $\mathbb{R}^{2g}$ and $\mathrm{GL}_g(\mathbb{C})\backslash \mathrm{GL}_{2g}(\mathbb{R})$.
(b) This induces a bijection between the set of isomorphism classes of complex tori of dimension g and the set of orbits in $\mathrm{GL}_g(\mathbb{C})\backslash \mathrm{GL}_{2g}(\mathbb{R})$ under the natural action of $\mathrm{GL}_{2g}(\mathbb{Z})$.

(6) (*Real Tori*) Let V be an $\mathbb{R}$-vector space of dimension n and Λ a lattice in V. The quotient group $T = V/\Lambda$ has the unique structure of a real analytic manifold such that the canonical projection $p\colon V \to T$ is real analytic. T is a connected compact abelian real Lie group, called the *real torus of dimension* n. This notion should not be mixed up with the notion of a *torus* in the theory of algebraic groups (see Remark 1.1.1). Show that

(a) Any two real tori of dimension n are isomorphic as real Lie groups.
(b) For any connected abelian real Lie group G of dimension n, there is an integer $m \le n$ such that $G \simeq T \times \mathbb{R}^{n-m}$ with T a real torus of dimension m. In particular, any connected compact abelian real Lie group is a real torus and any simply connected abelian real Lie group is an $\mathbb{R}$-vector space.
(c) Let S be a connected closed subgroup of a real torus T. Then S and T/S are real tori and $T \simeq S \times T/S$.

(7) Generalize Lemma 1.1.17 (b) to show that for any $n \ge 1$ the canonical map

$$\wedge^n H^1(X,\mathbb{Z}) \to H^n(X,\mathbb{Z}),$$

induced by the cup product, is an isomorphism. Conclude that Corollary 1.1.19 is valid for any positive integer n, where $\mathrm{Alt}_{\mathbb{R}}(V,\mathbb{C})$ is defined as $\wedge^n \mathrm{Hom}_{\mathbb{R}}(V,\mathbb{C})$.

(8) Show that $H_n(X,\mathbb{Z})$ and $H^n(X,\mathbb{Z})$ are free $\mathbb{Z}$-modules of rank $\binom{2g}{n}$ for all $n \ge 1$. Note that Exercise (12) below gives an explicit basis of $H_n(X,\mathbb{Z})$.

(9) For a complex torus X of dimension g: $\quad h^{1,1}(X) = \dim H^{1,1}(X) = g^2$.

(10) Let X be a complex torus, $\mu\colon X\times X \to X$ the addition map and $p_i\colon X\times X \to X$, $i = 1,2$, the natural projections. Show that a C^∞-one-form ω on X is translation-invariant if and only if $\mu^*\omega = p_1^*\omega + p_2^*\omega$.

(11) (*Pontryagin Product*) Let G be a real Lie group of dimension g. Let $\sigma\colon \Delta_p \to G$ and $\tau\colon \Delta_q \to G$ be singular p- respectively q-simplices. Here Δ_p denotes the standard p-simplex. If we divide the product $\Delta_p \times \Delta_q$ into $(p+q)$-simplices, then the map

$$\sigma * \tau\colon \Delta_p \times \Delta_q \to G\,, \quad \sigma * \tau(s,t) = \sigma(s)\tau(t)$$

is a singular $(p+q)$-chain. For singular p- and q-chains $\sigma = \sum m_i\sigma_i$ and $\tau = \sum n_j\tau_j$ define $\sigma * \tau = \sum m_i n_j \sigma_i * \tau_j$.

(a) Show that the boundary operator ∂ satisfies

$$\partial(\sigma * \tau) = (-1)^{\epsilon_1}\partial(\sigma) * \tau + (-1)^{\epsilon_2}\sigma * \partial(\tau),$$

where ϵ_1 and ϵ_2 are integers depending on p and q. Hence $*$ induces a bilinear map

$$*\colon H_p(G,\mathbb{Z}) \times H_q(G,\mathbb{Z}) \to H_{p+q}(G,\mathbb{Z}), \quad [\sigma] * [\tau] = [\sigma * \tau],$$

called the *Pontryagin product*. Moreover, show that the Pontryagin product coincides with the composition

$$H_p(G,\mathbb{Z}) \times H_q(G,\mathbb{Z}) \xrightarrow{\times} H_{p+q}(G \times G,\mathbb{Z}) \xrightarrow{\mu_*} H_{p+q}(G,\mathbb{Z}),$$

where $\times$ denotes the exterior homology product and $\mu\colon G \times G \to G$ is the multiplication map of G.

(b) For p-, q- and r-cycles s, τ and λ, and the unit element $1 \in G$ show

(i) $[1] * [\sigma] = [\sigma] * [1]$,

(ii) $([\sigma] * [\tau]) * [\lambda] = [\sigma] * ([\tau] * [\lambda])$,

(iii) $[\sigma] * [\tau] = (-1)^{pq}[\tau] * [\sigma]$, if G is commutative.

(c) Let $\iota\colon G' \to G$ be a Lie subgroup of dimension g'. Show that for any $[\sigma] \in H_p(G',\mathbb{Z})$ and $[\tau] \in H_{g'-p}(G',\mathbb{Z})$ and $[\lambda] \in H_{g-g'}(G,\mathbb{Z})$

$$(\iota_*[\sigma] \cdot (\iota_*[\tau] * [\lambda]))_G = (-1)^{\epsilon}([\sigma] \cdot [\tau])_{G'}([G'] \cdot [\lambda])_G,$$

where ϵ depends on g, g' and p and $(\,\cdot\,)_G$ and $(\,\cdot\,)_{G'}$ denote the intersection numbers in G and G' respectively.

(d) Let G' and $[\lambda]$ be as above. Use (c) to show that, if $[\sigma_1], \dots, [\sigma_n]$ are linearly independent elements in $H_\bullet(G',\mathbb{Z})$, then the elements

$$\iota_*[\sigma_1], \dots, \iota_*[\sigma_n], \iota_*[\sigma_1] * [\lambda], \dots, \iota_*[\sigma_n] * [\lambda]$$

are linearly independent in $H_\bullet(G,\mathbb{Z})$ (see Pontryagin [106]).

(12) Let $X = V/\Lambda$ be a complex torus of dimension g and $\lambda_1, \dots, \lambda_{2g}$ a basis of Λ. Via the identification $\Lambda = H_1(X,\mathbb{Z})$ of Section 1.1.3 the λ_i's can be considered as elements of $H_1(X,\mathbb{Z})$. Show that $\{\lambda_{i_1} * \cdots * \lambda_{i_p} \mid 1 \le i_1 < \cdots < i_p \le 2g\}$ is a basis of $H_p(X,\mathbb{Z})$ for any $1 \le p \le 2g$. (Hint: apply induction on p and use Exercise (11) (d) and Exercise (8) above. A different proof will be given in Lemma 2.5.12 below.)

(13) Let $X = V/\Lambda$ be a complex torus. Show that there is a canonical isomorphism

$$\phi_2 : H^2(\Lambda, \mathbb{Z}) \to H^2(X, \mathbb{Z}).$$

Here $H^2(\Lambda, \mathbb{Z})$ denotes the group cohomology. Actually there are canonical isomorphisms $\phi_n : H^n(\Lambda, \mathbb{Z}) \to H^n(X, \mathbb{Z})$ for all n.
(Hint: For the proof, use a suitable covering of V and explicit 2-cocycles.)

(14) Give an example of a complex torus whose group of automorphisms is not finite.
(Hint: Use a self product $E \times E$ of an elliptic curve.)

1.2 Line Bundles

In this section we introduce line bundles on a complex torus X, compute its first Chern class by means of a factor of automorphy and determine the Néron–Severi group of X.

1.2.1 Factors of Automorphy

Here it is explained how line bundles can be described by a factor of automorphy. Before we define these factors formally, let us explain how they arise: Let $X = V/\Lambda$ be a complex torus. According to Liouville's theorem there are no non-trivial holomorphic functions which are invariant with respect to Λ. However there are interesting almost invariant functions θ on V, meaning functions which satisfy an equation

$$\theta(v + \lambda) = f(\lambda, v)\theta(v).$$

The factor f is classically called a factor of automorphy. Its defining property arises by expanding the function $\theta((\lambda + \mu) + v) = \theta(\lambda + (\mu + v))$ in two ways. It turns out that f satisfies the relation

$$f(\lambda + \mu, v) = f(\lambda, \mu + v)f(\mu, v). \tag{1.7}$$

So f is just a 1-cocycle of Λ with values in $H^0(\mathcal{O}_V^*)$. For historical reasons we call them *factors of automorphy* or simply *factors*.

Under multiplication these 1-cocycles form an abelian group $Z^1(\Lambda, H^0(\mathcal{O}_V^*))$. The factors of the form

$$(\lambda, v) \mapsto h(\lambda + v)h(v)^{-1}$$

for some $h \in H^0(\mathcal{O}_V^*)$ are called *coboundaries*. They form the subgroup $B^1(\Lambda, H^0(\mathcal{O}_V^*))$ of $Z^1(V, H^0(\mathcal{O}_V^*))$. The *cohomology group* $H^1(\Lambda, H^0(\mathcal{O}_V^*))$ *of* Λ

with values in $H^0(\mathcal{O}_V^*)$ is defined to be the quotient

$$H^1(\Lambda, H^0(\mathcal{O}_V^*)) = Z^1(\Lambda, H^0(\mathcal{O}_V^*))/B^1(\Lambda, H^0(\mathcal{O}_V^*)).$$

This is the cohomology group of the group Λ with values in the multiplicative group $H^0(\mathcal{O}_V^*)$ considered as a Λ-module. The definition generalizes in an obvious way to any group cohomology. Let us note only that equation 1.7 reads in multiplicative form: $f(\lambda\mu, v) = f(\lambda, \mu v) f(\mu, v)$.

In order to describe any line bundle on X by a factor of automorphy, we need the following lemma.

Lemma 1.2.1 *Every holomorphic line bundle of a finite-dimensional complex vector space* V *is trivial.*

Suppose $X = V/\Lambda$ is a complex torus. As we saw in Section 1.1.3, the lattice Λ can be considered as the fundamental group $\pi_1(X)$ with respect to the base-point 0.

Proof From the exponential sequence $0 \to \mathbb{Z} \to \mathcal{O}_V \overset{e(2\pi i\cdot)}{\to} \mathcal{O}_V^* \to 1$ we obtain the exact sequence

$$H^1(\mathcal{O}_V) \to H^1(\mathcal{O}_V^*) \to H^2(V, \mathbb{Z}).$$

But $H^1(\mathcal{O}_V) = 0$ by the $\bar{\partial}$-Poincaré lemma (see Griffiths–Harris [55, p. 46]), whereas one knows from Algebraic Topology that $H^2(V, \mathbb{Z}) = 0$. This implies the assertion.□

Any factor f in $Z^1(\Lambda, H^0(\mathcal{O}_V^*))$ defines a line bundle on X as follows: According to Lemma 1.2.1 any line bundle on V is trivial. So we can start with the trivial line bundle $V \times \mathbb{C} \to V$ on V and consider the holomorphic action of Λ on it given by

$$\lambda(v, t) = (\lambda + v, f(\lambda, v)t)$$

for all $\lambda \in \Lambda$. This action is free and properly discontinuous, so the quotient

$$L = (V \times \mathbb{C})/\Lambda$$

is a complex manifold (see Section 2.3.5 below). Considering the projection $p : L \to X$ induced by the projection $V \times \mathbb{C} \to V$, one easily checks that L is a holomorphic line bundle on X.

By definition the group $\mathrm{Pic}(X)$ of line bundles on X is canonically isomorphic to the group $H^1(X, \mathcal{O}_X^*)$. Hence the construction above gives a map $Z^1(\Lambda, H^0(\mathcal{O}_{\bar{X}}^*)) \to H^1(X, \mathcal{O}_X^*)$. The following proposition shows that this map induces an isomorphism. For the proof we refer to [24, Proposition B1].

Proposition 1.2.2 *There is a canonical isomorphism*

$$\phi_1 : H^1(\Lambda, H^0(\mathcal{O}_V^*)) \to H^1(X, \mathcal{O}_X^*) = \mathrm{Pic}(X).$$

In particular, any line bundle L *on* X *can be described by a factor of automorphy.*

Let us describe the group of global sections of a line bundle L on X in terms of a factor of automorphy. For any line bundle L on X there is a natural isomorphism

$$H^0(X, L) \xrightarrow{\sim} H^0(V, \pi^* L)^\Lambda.$$

A trivialization $\alpha : \pi^* L \to V \times \mathbb{C}$ induces an isomorphism $H^0(V, \pi^* L)^\Lambda \xrightarrow{\sim} H^0(V, V \times \mathbb{C})^\Lambda$. If $f \in Z^1(V, H^0(\mathcal{O}_V^*))$ is the factor of automorphy associated to L with respect to the trivialization α, the elements in $H^0(V, V \times \mathbb{C})^\Lambda$ are just the holomorphic functions $\theta : V \to \mathbb{C}$ satisfying

$$\theta(\lambda + v) = f(\lambda, v)\theta(v) \tag{1.8}$$

for all $v \in V$ and $\lambda \in \Lambda$. The following proposition is obvious.

Proposition 1.2.3 *Via the composed isomorphism*

$$H^0(X, L) \xrightarrow{\sim} H^0(V, \pi^* L)^\Lambda \xrightarrow{\sim} H^0(V, V \times \mathbb{C})^\Lambda$$

the sections of L over X may be considered as holomorphic functions on V, satisfying the equation 1.8.

Note that the isomorphism depends on the trivialization: choosing another trivialization exactly means identifying the elements of $H^0(X, L)$ with holomorphic functions satisfying (1.8) with respect to an equivalent factor of automorphy. We will see later that choosing appropriate trivializations, the sections of L can be considered as theta-functions.

1.2.2 The First Chern Class of a Line Bundle

For a complex torus $X = V/\Lambda$ consider the exponential sequence $0 \to \mathbb{Z} \to \mathcal{O}_X \to \mathcal{O}_X^* \to 1$ and its long cohomology sequence

$$\cdots \to H^1(X, \mathbb{Z}) \to H^1(X, \mathcal{O}_X) \to H^1(X, \mathcal{O}_X^*) \to H^2(X, \mathbb{Z}) \to \cdots.$$

For any line bundle L on X the image of $L \in H^1(X, \mathcal{O}_X^*)$ in $H^2(X, \mathbb{Z})$ is called the *first Chern class* $c_1(L)$ of L. According to Corollary 1.1.18 the groups $H^2(X, \mathbb{Z})$ and $\mathrm{Alt}^2(\Lambda, \mathbb{Z})$ are canonically isomorphic. The following theorem shows how to compute $c_1(L)$ in terms of a factor of automorphy $f : \Lambda \times V \to \mathbb{C}^*$. Note that f can be written in the form $f = \mathbf{e}(2\pi i g)$ with a map $g : \Lambda \times V \to \mathbb{C}$, holomorphic in the second variable.

Theorem 1.2.4 *There is a canonical isomorphism $H^2(X, \mathbb{Z}) \to \mathrm{Alt}^2(\Lambda, \mathbb{Z})$ which maps the first Chern class $c_1(L)$ of a line bundle L on X with factor of automorphy $f = \mathbf{e}(2\pi i g)$ to the alternating form*

$$E_L(\lambda, \mu) := g(\mu, v + \lambda) + g(\lambda, v) - g(\lambda, v + \mu) - g(\mu, v)$$

for all $\lambda, \mu \in \Lambda$ and $v \in V$.

For the proof we need the following two lemmas. As usual let $Z^2(\Lambda, \mathbb{Z})$ denote the group of 2-cocycles of Λ with values in $\mathbb{Z}$.

Lemma 1.2.5 *The map* $\alpha : Z^2(\Lambda, \mathbb{Z}) \longrightarrow \mathrm{Alt}^2(\Lambda, \mathbb{Z})$ *defined by*

$$\alpha F(\lambda, \mu) := F(\lambda, \mu) - F(\mu, \lambda)$$

induces a canonical isomorphism, denoted by the same letter

$$\alpha : H^2(\Lambda, \mathbb{Z}) \xrightarrow{\simeq} \mathrm{Alt}^2(\Lambda, \mathbb{Z}).$$

Proof A 2-cocycle $F \in Z^2(\Lambda, \mathbb{Z})$ is a map $F : \Lambda \times \Lambda \to \mathbb{Z}$ satisfying

$$\partial F(\lambda, \mu, \nu) := F(\mu, \nu) - F(\lambda + \mu, \nu) + F(\lambda, \mu + \nu) - F(\lambda, \mu) = 0$$

for all $\lambda, \mu, \nu \in \Lambda$. One checks

$$\alpha F(\lambda + \mu, \nu) - \alpha F(\lambda, \nu) - \alpha F(\mu, \nu) = \partial F(\lambda, \nu, \mu) - \partial F(\nu, \lambda, \mu) - \partial F(\lambda, \mu, \nu) = 0.$$

Hence αF is an alternating bilinear form. Obviously $\alpha : Z^2(\Lambda, \mathbb{Z}) \longrightarrow \mathrm{Alt}^2(\Lambda, \mathbb{Z})$ is a homomorphism of groups.

Moreover, for the group $B^2(\Lambda, \mathbb{Z})$ of 2-coboundaries of Λ with values in $\mathbb{Z}$ we have $\alpha(B^2(\Lambda, \mathbb{Z})) = 0$, since the elements $\partial h(\lambda, \mu) = h(\mu) - h(\lambda + \mu) + h(\lambda)$ are symmetric in λ and μ. It follows that α descends to a homomorphism, denoted by the same letter,

$$\alpha : H^2(\Lambda, \mathbb{Z}) \longrightarrow \mathrm{Alt}^2(\Lambda, \mathbb{Z}).$$

To see that α is surjective, note that for all $f, g \in \mathrm{Hom}(\Lambda, \mathbb{Z})$ the map $f \otimes g$ is in $Z^2(\Lambda, \mathbb{Z})$ and thus

$$\alpha(f \otimes g)(\lambda, \mu) = f \otimes g(\lambda, \mu) - f \otimes g(\mu, \lambda) = f \wedge g(\lambda, \mu).$$

Since the elements $f \wedge g$ generate $\mathrm{Alt}^2(\Lambda, \mathbb{Z})$, this shows the surjectivity of α.

Suppose $F \in \operatorname{Ker} \alpha$, which means that F is symmetric. One checks that the multiplication law on $\mathbb{Z} \times \Lambda$,

$$(\ell, \lambda) \cdot (m, \mu) := (\ell + m + F(\lambda, \mu), \lambda + \mu),$$

defines the structure of a commutative group. Since the lattice Λ is a free group, the exact sequence

$$0 \to \mathbb{Z} \xrightarrow{i} \mathbb{Z} \times \Lambda \xrightarrow{p} \Lambda \to 0$$

with $i(\ell) = (\ell, 0)$ and $p(\ell, \lambda) = \lambda$ splits. Hence there is a section $s : \Lambda \to \mathbb{Z} \times \Lambda$, $s(\lambda) = (f(\lambda), \lambda)$ and the multiplication law yields

$$F(\lambda, \mu) = f(\lambda + \mu) - f(\lambda) - f(\mu).$$

So F is a boundary. This completes the proof of the lemma. □

Since V is contractible, $H^1(V,\mathbb{Z}) = 0$. So the following sequence is exact

$$0 \longrightarrow \mathbb{Z} = H^0(V,\mathbb{Z}) \longrightarrow H^0(\mathcal{O}_V) \stackrel{\mathbf{e}(2\pi i\cdot)}{\longrightarrow} H^0(\mathcal{O}_V^*) \longrightarrow 1. \tag{1.9}$$

The lattice Λ acts on each of these groups in a compatible way, so that (1.9) induces a long exact cohomology sequence. In particular, we get the connecting homomorphism

$$\delta : H^1(\Lambda, H^0(\mathcal{O}_V^*)) \longrightarrow H^2(\Lambda,\mathbb{Z}).$$

By definition δ maps the cocycle $f = \mathbf{e}(2\pi i g) \in Z^1(\Lambda, H^0(\mathcal{O}_V^*))$ to the 2-cocycle

$$\delta f(\lambda,\mu) = g(\mu, v+\lambda) - g(\lambda+\mu, v) + g(\lambda, v)$$

in $Z^2(\Lambda,\mathbb{Z})$, where $\lambda,\mu \in \Lambda$ and $v \in V$. Note that δf does not depend on the variable v, since f satisfies the cocycle relation $f(\lambda+\mu, v) = f(\mu+\lambda, v) = f(\mu, v+\lambda) f(\lambda, v)$.

The following lemma means that δ is compatible with the first Chern class map. Let $\phi_1 : H^1(\Lambda, H^0(\mathcal{O}_V^*)) \to H^1(X, \mathcal{O}_X^*)$ be the canonical isomorphism of Proposition 1.2.2 and $\phi_2 : H^2(\Lambda,\mathbb{Z}) \to H^2(X,\mathbb{Z})$ the canonical isomorphism of Exercise 1.1.6 (13). Then we have:

Lemma 1.2.6 *The following diagram is commutative*

$$\begin{array}{ccc} H^1(\Lambda, H^0(\mathcal{O}_V^*)) & \xrightarrow{\ \delta\ } & H^2(\Lambda,\mathbb{Z}) \\ \downarrow{\scriptstyle \phi_1} & & \downarrow{\scriptstyle \phi_2} \\ H^1(\mathcal{O}_X^*) & \xrightarrow[c_1]{} & H^2(X,\mathbb{Z}). \end{array}$$

Proof As always, let $\pi : V \to X$ be the canonical map. Let $\{U_i\}_I$ be an open covering of X, such that for every $i \in I$ there is a connected $W_i \subset \pi^{-1}(U_i)$ with $\pi_i := \pi|_{W_i} : W_i \to U_i$ biholomorphic. Then for every pair $(i, j) \in I^2$ there is a unique $\lambda_{ij} \in \Lambda$ such that

$$\pi_j^{-1}(x) = \lambda_{ij}\pi_i^{-1}(x)$$

for all $x \in U_i \cap U_j$. Recall that the homomorphisms ϕ_1, ϕ_2 and c_1 are defined as follows

$$\begin{aligned} (\phi_1 f)_{ij} &= f(\lambda_{ij}, \pi_i^{-1}) && \text{for} && f \in Z^1(\Lambda, H^0(\mathcal{O}_V^*)), \\ (\phi_2 F)_{ijk} &= F(\lambda_{ij}, \lambda_{jk}, \pi_i^{-1}) && \text{for} && F \in Z^2(\Lambda, H^0(V,\mathbb{Z})) = Z^2(\Lambda,\mathbb{Z}) \quad \text{and} \\ (c_1 h)_{ijk} &= g_{jk} - g_{ik} + g_{ij} && \text{for} && h = \{h_{ij} = \mathbf{e}(2\pi i g_{ij}\}_I \in Z^1(X, \mathcal{O}_X^*) \end{aligned}$$

for all $i, j, k \in I$. Using these definitions and the definition of δ above, it is easy to check that the diagram commutes (see Exercise 1.2.3 (2)). □

Proof (of Theorem 1.2.4) The canonical isomorphism of the theorem is the composed map

$$\alpha \circ \phi_2^{-1} : H^2(X, \mathbb{Z}) \to \operatorname{Alt}^2(\Lambda, \mathbb{Z}).$$

In fact, according to Lemma 1.2.6 the cocycle δf represents the element $\phi_2^{-1} c_1(L)$ of $H^2(\Lambda, \mathbb{Z})$. An immediate computation gives $\alpha \phi_2^{-1} c_1(L) = \alpha \delta f = E_L$. □

Recall the notations $\Omega = \operatorname{Hom}_{\mathbb{C}}(V, \mathbb{C})$ and $\overline{\Omega} = \operatorname{Hom}_{\overline{\mathbb{C}}}(V, \mathbb{C})$. The following lemma expresses the fact that for H^1 the de Rham isomorphism is compatible with the Dolbeault isomorphism.

Lemma 1.2.7 *The following diagram commutes*

$$\begin{array}{ccccc}
H^1(\Lambda, \mathbb{C}) & \xrightarrow{\simeq} & \operatorname{Hom}_{\mathbb{R}}(V, \mathbb{C}) & \xrightarrow{\simeq} & \Omega \oplus \overline{\Omega} \\
\downarrow{\scriptstyle \phi_1} & & & & \downarrow{\scriptstyle \gamma_1} \\
H^1(X, \mathbb{C}) & & \xleftarrow{\quad\rho\quad} & & H^{1,0}(X) \oplus H^{0,1}(X),
\end{array}$$

where ϕ_1 is given by Lemma 1.1.17 (a) combined with (1.3) *and* (1.4), *and γ_1 and ρ are given by Theorem 1.1.21.*

Proof For a proof see Exercise 1.2.3 (3). □

The canonical isomorphism $\operatorname{Alt}^2(\Lambda, \mathbb{Z}) \to H^2(X, \mathbb{Z})$ of Theorem 1.2.4 extends to an isomorphism

$$\beta_2 : \operatorname{Alt}^2(\Lambda, \mathbb{C}) \to H^2(X, \mathbb{C}).$$

On the one hand $\operatorname{Alt}^2(\Lambda, \mathbb{C}) = \operatorname{Alt}^2_{\mathbb{R}}(V, \mathbb{C})$ can be identified with $\wedge^2\Omega \oplus (\Omega \otimes \overline{\Omega}) \oplus \wedge^2\overline{\Omega}$. On the other hand we have the Hodge decomposition for $H^2(X, \mathbb{C})$.

Proposition 1.2.8 *The isomorphism β_2 respects both decompositions; that is, the following diagram is commutative, where γ_2 is given by Theorem 1.1.21.*

$$\begin{array}{ccccc}
\operatorname{Alt}^2(\Lambda, \mathbb{C}) & \xrightarrow{\simeq} & \operatorname{Alt}^2_{\mathbb{R}}(V, \mathbb{C}) & \xrightarrow{\simeq} & \wedge^2\Omega \oplus (\Omega \otimes \overline{\Omega}) \oplus \wedge^2\overline{\Omega} \\
\downarrow{\scriptstyle \beta_2} & & & & \downarrow{\scriptstyle \gamma_2} \\
H^2(X, \mathbb{C}) & \xrightarrow{\simeq} & H^2_{\mathrm{dR}}(X) & \xrightarrow{\simeq} & H^{2,0}(X) \oplus H^{1,1}(X) \oplus H^{0,2}(X).
\end{array}$$

Proof Recall that

$$\operatorname{Alt}^2_{\mathbb{R}}(V, \mathbb{C}) = \wedge^2 \operatorname{Hom}_{\mathbb{R}}(V, \mathbb{C}) = \wedge^2(\Omega \oplus \overline{\Omega}) \qquad \text{and}$$

$$H^2(X, \mathbb{C}) \simeq \wedge^2 H^1(X, \mathbb{C}) \simeq \wedge^2 H^1_{\mathrm{dR}}(X) = \wedge^2\big(H^{1,0}(X) \oplus H^{01}(X)\big).$$

So the assertion follows from the previous lemma. □

Theorem 1.2.4 associated to every line bundle $L \in \operatorname{Pic}(X)$ a $\mathbb{Z}$ valued alternating form on Λ and thus an alternating form $V \times V \to \mathbb{R}$. Conversely, the next proposition shows which alternating forms come from line bundles in this way.

Proposition 1.2.9 *For an alternating form $E : V \times V \to \mathbb{R}$ the following conditions are equivalent:*

(i) *There is an $L \in \operatorname{Pic}(X)$ such that E represents $c_1(L)$.*
(ii) *$E(\Lambda, \Lambda) \subseteq \mathbb{Z}$ and $E(iv, iw) = E(v, w)$ for all $v, w \in V$.*

Proof Consider the following diagram, where the upper line is part of the exact cohomology sequence of $0 \to \mathbb{Z} \to \mathcal{O}_X \to \mathcal{O}_X^* \to 1$, the map ι is the canonical one, β_2 and γ_2 are the isomorphisms of Proposition 1.2.8 and p the projection maps.

$$\begin{array}{ccccc}
H^1(\mathcal{O}_X^*) & \xrightarrow{c_1} & H^2(X,\mathbb{Z}) & \longrightarrow & H^2(\mathcal{O}_X) \\
 & & \downarrow \iota & & \| \\
 & & H^2(X,\mathbb{C}) & \xrightarrow{p} & H^{0,2}(X) \\
 & & \simeq \downarrow \beta_2^{-1} & & \simeq \downarrow \gamma_2^{-1} \\
 & & \wedge^2\Omega \oplus (\Omega \otimes \overline{\Omega}) \oplus \wedge^2\overline{\Omega} & \xrightarrow{p} & \wedge^2\overline{\Omega}.
\end{array}$$

By the previous proposition the diagram is commutative.

Now suppose $L \in H^1(\mathcal{O}_X^*)$ with $\beta_2^{-1}\iota c_1(L) = E = E_1 + E_2 + E_3$ with $E_1 \in \wedge^2\Omega$ etc. Then $E_1 = \overline{E_3}$, since E has values in $\mathbb{R}$, whereas by the diagram $E_3 = 0$, taking the exactness of the upper row into account. Hence $E = E_2$, which means that E satisfies (ii). The converse implication also follows from the diagram. □

The following lemma shows that the alternating forms of condition (ii) are just the imaginary parts of hermitian forms. Recall that a *hermitian form on V* is a map $H : V \times V \to \mathbb{C}$ which is linear in the first argument and satisfies $H(v, w) = \overline{H(w, v)}$ for all $v, w \in V$.

Lemma 1.2.10 *There is a $1-1$ correspondence between the set of hermitian forms H on V and the set of real-valued alternating forms E on V satisfying $E(iv, iw) = E(v, w)$ for all $v, w \in V$, given by*

$$E(v, w) = \operatorname{Im} H(v, w) \qquad \textit{and} \qquad H(v, w) = E(iv, w) + iE(v, w).$$

The proof is an easy exercise (see Exercise 1.2.3 (4)). In the sequel we consider the first Chern class of a line bundle either as an alternating form or as a hermitian form.

1.2.3 Exercises

(1) Let $X = V/\Lambda$ be a complex torus with projection $\pi : V \to X$. Construct for any $L \in \mathrm{Pic}(X)$ a functorial isomorphism $H^0(X, L) \to H^0(V, \pi^* L)^\Lambda$.

(2) Check the commutativity of the diagram in Lemma 1.2.6.

(3) Check the commutativity of the diagram in Lemma 1.2.7.

(4) Show that the map in Lemma 1.2.10 gives a bijection between the set of hermitian forms on V and the set of real-valued alternating forms on V satisfying $E(iv, iw) = E(v, w)$ for all $v, w \in V$.

(5) Let $X = V/\Lambda$ be a complex torus and $E : \Lambda \times \Lambda \to \mathbb{Z}$ an alternating form such that $E(iv, iw) = E(v, w)$ for the extension of E to V. Let $\Lambda_0 \subset \Lambda$ denote the kernel of E. Show that $V_0 = \mathbb{R}\Lambda_0$ is a complex subspace of V.

(6) Show that the map $\mathrm{Hom}_{\mathbb{C}}(V, \mathbb{C}) \to \mathrm{Hom}_{\mathbb{R}}(V, \mathbb{R})$, $\ell \mapsto \mathrm{Im}\,\ell$ is an isomorphism of $\mathbb{R}$-vector spaces.

(7) Let $X = V/\Lambda$ be a complex torus and $x_1, \dots, x_{2g}$ be real coordinate functions with respect to an $\mathbb{R}$-basis $\lambda_1, \dots, \lambda_{2g}$ of V.

(a) Show that the following diagram commutes

$$\begin{array}{ccc} H^2(\Lambda, \mathbb{C}) & \xrightarrow{\alpha} & \mathrm{Alt}^2_{\mathbb{R}}(V, \mathbb{C}) \\ \downarrow{\scriptstyle \phi_2} & & \downarrow{\scriptstyle \gamma_2} \\ H^2(X, \mathbb{C}) & \xrightarrow[\rho_2]{} & H^2_{\mathrm{dR}}(X), \end{array}$$

where α is an equivalent version of the canonical isomorphism of Lemma 1.2.5, ϕ_2 is the canonical isomorphism of Exercise 1.1.6 (13), ρ_2 is the de Rham isomorphism and γ_2 the isomorphism which sends the form E to $\sum_{1 \le \nu < \mu \le 2g} E(\lambda_\mu, \lambda_\nu) dx_\nu \wedge dx_\mu$.

(b) Conclude that the isomorphism $\alpha\phi_2^{-1} : H^2(X, \mathbb{C}) \to \mathrm{Alt}^2(\Lambda, \mathbb{Z})$ coincides with the isomorphism of Corollary 1.1.18.

(8) Let $X = V/\Lambda$ be a complex torus. In terms of sheaves the canonical isomorphism $\phi_1 : H^1(\Lambda, H^0(\mathcal{O}_V^*)) \xrightarrow{\sim} \mathrm{Pic}(X)$ can be described as follows: Let $f : \Lambda \times V \to \mathbb{C}^*$ denote a factor of automorphy and $[f]$ its class in $H^1(\Lambda, H^0(\mathcal{O}_V^*))$. Then $\phi_1[f]$ is the line bundle associated to the sheaf $\mathcal{L}$ given by

$$\mathcal{L}(U) = \left\{ \vartheta : \pi^{-1}(U) \to \mathbb{C} \;\middle|\; \begin{array}{c} \vartheta \text{ holomorphic with } \vartheta(v + \lambda) = f(\lambda, v)\vartheta(v) \\ \text{for all } (\lambda, v) \in \Lambda \times \pi^{-1}(U) \end{array} \right\}$$

for any open $U \subset X$.

(9) Given an isogeny $f : Y \to X$ of complex tori and a line bundle $M_0 \in \operatorname{Pic} X$, show that

(a) The set $\{M \in \operatorname{Pic}(X) \mid f^*(M) \simeq f^*(M_0)\}$ has a natural group structure with M_0 as identity.

(b) There is an isomorphism of groups

$$\{M \in \operatorname{Pic}(X) \mid f^*(M) \simeq f^*(M_0)\} \xrightarrow{\sim} \Lambda(L)/\rho_a(f)^{-1}\Lambda(M).$$

1.3 The Appell–Humbert Theorem and Canonical Factors

In the last section we saw that any line bundle on a complex torus $X = V/\Lambda$ can be given by a factor of automorphy. In this section we show that there is a canonical way to distinguish a factor in every class of $H^1(\Lambda, H^0(O_V^*))$. As a consequence we obtain the Appell–Humbert Theorem.

1.3.1 Preliminaries

Let $X = V/\Lambda$ be a complex torus. The image of the homomorphism $c_1 : H^1(O_X^*) \to H^2(X, \mathbb{Z})$ is called the *Néron–Severi group* $\operatorname{NS}(X)$ *of* X, so that we have the exact sequence

$$0 \to \operatorname{Pic}^0(X) \to \operatorname{Pic}(X) \xrightarrow{c_1} \operatorname{NS}(X) \to 0,$$

where $\operatorname{Pic}^0(X)$ denotes the group of line bundles with first Chern class 0. For every $H \in \operatorname{NS}(X)$ we define

$$\operatorname{Pic}^H(X) := \{L \in \operatorname{Pic}(X) \mid c_1(H) = H\}.$$

Consider the Néron–Severi group $\operatorname{NS}(X)$ as the group of hermitian forms $H : V \times V \to \mathbb{C}$ with $\operatorname{Im} H(\Lambda, \Lambda) \subseteq \mathbb{Z}$ and denote by $\mathbb{C}_1$ the circle group $\mathbb{C}_1 := \{z \in \mathbb{C} \mid |z| = 1\}$.

A *semicharacter for* H is a map $\chi : \Lambda \to \mathbb{C}_1$ satisfying

$$\chi(\lambda + \mu) = \chi(\lambda)\chi(\mu)\,\mathbf{e}(2\pi i \operatorname{Im} H(\lambda, \mu)) \qquad \text{for all} \quad \lambda, \mu \in \Lambda.$$

By definition the characters of the group Λ are exactly the semicharacters for $0 \in \operatorname{NS}(X)$. Define

$$\mathcal{P}(\Lambda) := \{(H, \chi) \mid H \in \operatorname{NS}(X), \chi \text{ a semicharacter for } H\}. \qquad (1.10)$$

Clearly $\mathcal{P}(\Lambda)$ is a group with respect to the composition $(H_1, \chi_1) \cdot (H_2, \chi_2) = (H_1 + H_2, \chi_1 \chi_2)$ and the following sequence is exact

$$1 \to \operatorname{Hom}(\Lambda, \mathbb{C}_1) \xrightarrow{\iota} \mathcal{P}(\Lambda) \xrightarrow{p} \operatorname{NS}(X),$$

where $\iota(\chi) = (\chi, 0)$ and $p(\chi, H) = H$ (see Exercise 1.3.4 (1)).

For any pair $(H, \chi) \in \mathcal{P}(\Lambda)$ define a map

$$a = a_{(H,\chi)} := \begin{cases} \Lambda \times V \to \mathbb{C}^* \\ (\lambda, v) \mapsto \chi(\lambda)\, \mathbf{e}\left(\pi H(v, \lambda) + \frac{\pi}{2} H(\lambda, \lambda)\right). \end{cases} \tag{1.11}$$

Lemma 1.3.1

(i) *The map $a_{(H,\chi)}$ induces a line bundle*

$$L(H, \chi) := V \times \mathbb{C}/\Lambda,$$

where Λ acts on $V \times \mathbb{C}$ by $\lambda(v, t) = (v + \lambda, a_{(H,\chi)}(\lambda, v)t)$.

(ii) *The map $\mathcal{P}(\Lambda) \to \operatorname{Pic}(X)$, $(H, \lambda) \mapsto L(H, \chi)$ is a surjective homomorphism of groups such that the following diagram commutes*

$$\begin{array}{ccc} \mathcal{P}(\Lambda) & & \\ \downarrow & \searrow^{p} & \\ \operatorname{Pic}(X) & \xrightarrow[c_1]{} & \operatorname{NS}(X). \end{array}$$

Proof For (i) it suffices to show that the map $a_{(H,\lambda)}$ is an element of $Z^1(\Lambda, H^0(\mathcal{O}_V^*))$, since this defines a line bundle by Proposition 1.2.2. For the easy proof of the cocycle condition see Exercise 1.3.4 (2). The action of Λ on $V \times \mathbb{C}$ follows from the definitions.

(ii): That the map $\mathcal{P}(\Lambda) \to \operatorname{NS}(X)$ is a homomorphism is an easy computation. It remains to show that $c_1(L(H, \lambda)) = H$. Writing $\chi(\lambda) = \mathbf{e}(2\pi i \varphi(\lambda))$, we have

$$a_{(H,\lambda)}(\lambda, v) = \mathbf{e}(2\pi i g(\lambda, v)) \quad \text{with} \quad g(\lambda, v) = \varphi(\lambda) - \frac{i}{2} H(\lambda, v) - \frac{i}{4} H(\lambda, \lambda).$$

According to Theorem 1.2.4, the imaginary part of the hermitian form $c_1(L(\lambda, v))$ is the alternating form

$$\begin{aligned} E_{L(H,\lambda)}(\lambda, \mu) &= g(\mu, v + \lambda) + g(\lambda, v) - g(\lambda, v + \mu) - g(\mu, v) \\ &= \frac{1}{2i}(H(\lambda, \mu) - H(\mu, \lambda)) = \operatorname{Im} H(\lambda, \mu). \end{aligned}$$

Now Lemma 1.2.10 gives the assertion. It proves in particular the surjectivity of the map. □

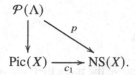

Lemma 1.3.2 *Any class in* $H^1(\Lambda, H^0(\mathcal{O}_V^*))$ *giving an element of* $\mathrm{Pic}^0(X)$ *is represented by a factor of automorphy with constant coefficients.*

Proof The exponential sequence on X induces the exact upper line of the commutative diagram

$$\begin{array}{ccccccc} H^1(X,\mathbb{Z}) & \longrightarrow & H^1(\mathcal{O}_X) & \longrightarrow & H^1(\mathcal{O}_X^*) & \xrightarrow{c_1} & H^1(X,\mathbb{Z}) \\ & & \uparrow{\scriptstyle p} & & \| & & \\ & & H^1(X,\mathbb{C}) & \xrightarrow{\epsilon} & H^1(\mathcal{O}_X^*), & & \end{array}$$

where p is the projection of the Hodge decomposition $H^1(X,\mathbb{C}) \simeq H^1(\mathcal{O}_X) \oplus H^0(\Omega_X)$ and ϵ the map induced by $\mathbf{e}(2\pi i\cdot) : \mathbb{C} \to \mathbb{C}^* \subseteq \mathcal{O}_X^*$. It follows that $\mathrm{Pic}^0(X) = \mathrm{Im}\,\epsilon$, that is, any line bundle of $\mathrm{Pic}^0(X)$ can be given by a cocycle in $Z^1(X, \mathcal{O}_X^*)$ with constant coefficients.

By the definition of the canonical isomorphisms $H^1(\Lambda, H^0(\mathcal{O}_V^*)) \to H^1(\mathcal{O}_X^*)$ of Proposition 1.2.2 one checks that also the corresponding element of $H^1(\Lambda, H^0(\mathcal{O}_V^*))$ can be represented by a factor with constant coefficients. □

1.3.2 The Theorem

Now we are in a position to prove the main theorem of this section.

Theorem 1.3.3 (Appell–Humbert Theorem) *For any complex torus* $X = V/\Lambda$ *there is a canonical isomorphism of exact sequences*

$$\begin{array}{ccccccccc} 1 & \longrightarrow & \mathrm{Hom}(\Lambda, \mathbb{C}_1) & \longrightarrow & \mathcal{P}(\Lambda) & \xrightarrow{p} & \mathrm{NS}(X) & \longrightarrow & 0 \\ & & \downarrow{\simeq} & & \downarrow{\simeq} & & \| & & \\ 1 & \longrightarrow & \mathrm{Pic}^0(X) & \longrightarrow & \mathrm{Pic}(X) & \xrightarrow[c_1]{} & \mathrm{NS}(X) & \longrightarrow & 0. \end{array}$$

Proof Clearly the sequences are exact and according to Lemmas 1.3.1 and 1.3.2 the diagram commutes. Hence it remains to show that the map

$$\psi : \mathrm{Hom}(\Lambda, \mathbb{C}_1) \to \mathrm{Pic}^0(X)$$

induced by the canonical map $\mathcal{P}(\Lambda) \to \mathrm{Pic}(X)$ is an isomorphism.

First we claim that ψ is surjective. To see this suppose $L \in \mathrm{Pic}^0(X)$. By Lemma 1.3.2 L admits a factor $f \in Z^1(\Lambda, H^0(\mathcal{O}_V^*))$ with constant coefficients. So $f : \Lambda \to \mathbb{C}^*$ is a homomorphism. It remains to show that f admits an equivalent cocycle with values in $\mathbb{C}_1$.

Writing $f(\lambda) = \mathbf{e}(2\pi i g(\lambda))$, we get for all $\lambda, \mu \in \Lambda$,

$$g(\lambda + \mu) \equiv g(\lambda) + g(\mu) \mod \mathbb{Z}.$$

Hence the imaginary part $\operatorname{Im} g : \Lambda \to \mathbb{R}$ is a homomorphism extending linearly to a function $V \to \mathbb{R}$, denoted by the same symbol. Define a $\mathbb{C}$-linear form

$$\ell : V \to \mathbb{C}, \quad v \mapsto \operatorname{Im} g(iv) + i \operatorname{Im} g(v).$$

Since $\mathbf{e}(2\pi i \ell) \in H^0(\mathcal{O}_V^*)$,

$$\chi_L(\lambda, \mu) := f(\lambda)\, \mathbf{e}(2\pi i \ell(v) - 2\pi i \ell(v + \lambda))$$

is a cocycle in $Z^1(\Lambda, H^0(\mathcal{O}_V^*))$ which is equivalent to f. Moreover, χ_L has constant values in $\mathbb{C}_1$, since

$$\chi_L(\lambda, \mu) = \mathbf{e}[2\pi i (g(\lambda) - \ell(\lambda))] = \mathbf{e}(2\pi i [\operatorname{Re} g(\lambda) - \operatorname{Im} g(i\lambda)]$$

and $\operatorname{Re} g(\lambda) - \operatorname{Im} g(i\lambda) \in \mathbb{R}$. Hence $L \simeq L(0, \chi_L)$ and ψ is surjective.

To see that ψ is injective, suppose there is another homomorphism $\chi : V \to \mathbb{C}_1$ with $L \simeq L(0, \chi)$. So χ_L and χ are equivalent cocycles in $Z^1(\Lambda, H^0(\mathcal{O}_V^*))$, that is, there is an $h \in H^0(\mathcal{O}_V^*)$ with

$$\chi_L(\lambda) = \chi(\lambda) h(v + \lambda) h(v)^{-1} \quad \text{for all } v \in V, \lambda \in \Lambda.$$

Now $|\chi_L| = |\chi| = 1$ implies $|h(v + \lambda)| = |h(v)|$ for all $v \in V, \lambda \in \Lambda$, so that h is bounded and periodic in V. By the theorem of Liouville h is constant, which gives $\chi = \chi_L$. □

1.3.3 Canonical Factors

Let $X = V/\Lambda$ be a complex torus. The Appell–Humbert Theorem tells us that there is a canonical way to associate a factor of automorphy to any line bundle on X. Let $L \in \operatorname{Pic}(X)$ with $c_1(L) = H$ and a semicharacter $\chi : \Lambda \to \mathbb{C}_1$ such that $L = L(H, \chi)$. The cocycle $a_L = a_{L(H,\chi)} \in Z^1(\Lambda, H^0(\mathcal{O}_V^*))$ defined by

$$a_L(\lambda, v) = \chi(\lambda)\, \mathbf{e}(\pi H(v, \lambda) + \frac{\pi}{2} H(\lambda, \lambda)) \quad \text{for all } (\lambda, v) \in \Lambda \times V \tag{1.12}$$

depends only on L (given the lattice $\Lambda \subset V$) and is therefore called the *canonical factor (of automorphy) of L.* In this section we compile some properties of line bundles which are most conveniently proved with canonical factors.

According to Proposition 1.1.6 every holomorphic map between complex tori is the composition of a homomorphism and a translation. We will study the behaviour of line bundles in both cases separately.

Lemma 1.3.4 *For any $L = L(H, \chi) \in \operatorname{Pic}(X)$ and $\overline{v} \in X$ with representative $v \in V$,*

$$t_{\overline{v}}^* L\,(H, \chi) = L(H, \chi \cdot \mathbf{e}(2\pi i \operatorname{Im} H(v, \cdot)))\,.$$

Proof Since t_v on V induces $t_{\overline{v}}$ on X, we have $t_{\overline{\lambda}} = \mathbf{1}_X$ for all $\lambda \in \Lambda$. Hence, if a_L is the canonical factor of L, then $(\mathbf{1}_\Lambda \times t_v)^* a_L$ is a factor for $t_{\overline{v}}^* L$, but not yet the canonical one. However, one immediately checks that for $g(w) = \mathbf{e}(-\pi H(w, v))$ the equivalent factor satisfies

$$(\mathbf{1}_\Lambda \times t_v)^* a_L(\lambda, w) g(w+\lambda) g(w)^{-1} = \chi(\lambda)\,\mathbf{e}(2\pi i \operatorname{Im} H(v, \lambda))\,\mathbf{e}(\pi H(w, \lambda) + \frac{\pi}{2} H(\lambda, \lambda))$$

and hence is the canonical factor of $t_{\overline{v}}^* L$, since $\chi\,\mathbf{e}(2\pi i \operatorname{Im} H(v, \cdot))$ is a semicharacter for H. □

Comparing hermitian forms and semicharacters, we get as an immediate consequence the following theorem, for the proof of which we refer to Exercise 1.3.4 (4).

Theorem 1.3.5 (Theorem of the Square) *For all $L \in \operatorname{Pic}(X)$ and $\overline{v}, \overline{w} \in X$,*

$$t_{\overline{v}+\overline{w}}^* L = t_{\overline{v}}^* L \otimes t_{\overline{w}}^* L \otimes L^{-1}.$$

Let $X' = V'/\Lambda'$ be a second complex torus and $f : X' \to X$ a homomorphism with analytic representation $F : V' \to V$ and rational representation $F_{\Lambda'} = F|_{\Lambda'} : \Lambda' \to \Lambda$.

Lemma 1.3.6 *For any $L(H, \chi) \in \operatorname{Pic}(X)$,*

$$f^* L(H, \chi) = L(F^* H, F_{\Lambda'}^* \chi).$$

Proof This follows from the Appell–Humbert Theorem using that $F^* H$ is the hermitian form of $f^* L$ and $F_{\Lambda'}^* \chi$ is a semicharacter for $F^* H$. □

As an example we apply the lemma to the *multiplication by n* on X for any integer n,

$$n_X : X \to X.$$

Here and in the sequel we use for the nth power the notation: $L^n := L^{\otimes n}$.

Proposition 1.3.7 *For every $L \in \operatorname{Pic}(X)$ and $n \in \mathbb{Z}$,*

$$n_X^* L \simeq L^{\frac{n^2+n}{2}} \otimes (-1)_X^* L^{\frac{n^2-n}{2}}\,.$$

Proof For $L = L(H, \chi)$ we get

$$\begin{aligned} L^{\frac{n^2+n}{2}} \otimes (-1)_X^* L^{\frac{n^2-n}{2}} &= L\left(\frac{n^2+n}{2} H + \frac{n^2-n}{2}(-1)_V^* H, \chi^{\frac{n^2+n}{2}} \cdot (-1)_V^* \chi^{\frac{n^2-n}{2}}\right) \\ &= L(n^2 H, \chi(n\cdot)) \qquad \text{(by Exercise 1.3.4 (3) (a))} \\ &= L(n_V^* H, n_\Lambda^* \chi) \\ &= n_X^* L(H, \chi) \qquad \text{(by Lemma 1.3.6).} \end{aligned}$$

So the Appell–Humbert Theorem gives the assertion. □

A line bundle $L \in \mathrm{Pic}(X)$ is called *symmetric* if $(-1)_X^* L \simeq L$. The following Corollary is a direct consequence of Proposition 1.3.7.

Corollary 1.3.8 *For every symmetric $L \in \mathrm{Pic}(X)$ and $n \in \mathbb{Z}$,*

$$n_X^* L \simeq L^{n^2}.$$

1.3.4 Exercises

(1) Show that the set $\mathcal{P}(\Lambda)$ defined in (1.10) is a group with respect to the composition given there such that the following sequence is exact:

$$1 \to \mathrm{Hom}(\Lambda, \mathbb{C}_1) \xrightarrow{\iota} \mathcal{P}(\Lambda) \xrightarrow{p} \mathrm{NS}(X).$$

(2) Show that the map $a = a_{(H,\chi)}$ of (1.11) satisfies the cocycle condition for H.

(3) Let a_L be the canonical factor of $L = L(H, \chi) \in \mathrm{Pic}(X)$. Show that for all $\lambda \in \Lambda$, $v, w \in V$ and $n \in \mathbb{Z}$,

(a) $\chi(n\lambda) = \chi(\lambda)^n$,
(b) $a_L(\lambda, v + w) = a_L(\lambda, v)\,\mathbf{e}(\pi H(w, \lambda))$,
(c) $a_L(\lambda, v)^{-1} = a_L(-\lambda, v)\,\mathbf{e}(-\pi H(\lambda, \lambda))$.

(4) Give a proof of the Theorem of the Square 1.3.5 by comparing the canonical factors of both sides.

(5) Show that a line bundle $L(H, \chi) \in \mathrm{Pic}(X)$ is symmetric if and only if $\chi(\Lambda) \subseteq \{\pm 1\}$.

(6) Let $\widetilde{E} : \Lambda \times \Lambda \to \mathbb{Z}/2\mathbb{Z}$ be an alternating bilinear form modulo 2 (alternating meaning that $\widetilde{E}(\lambda, \lambda) = 0$ for every $\lambda \in \Lambda$). Show that there is a map $f : \Lambda \to \mathbb{Z}/2\mathbb{Z}$ such that

$$\widetilde{E}(\lambda_1, \lambda_2) = f(\lambda_1 + \lambda_2) + f(\lambda_1) + f(\lambda_2) \quad \text{for all} \quad \lambda_1, \lambda_2 \in \Lambda.$$

Deduce from this and Exercise (3) that for every hermitian form H on V with corresponding alternating form $E : \Lambda\times\Lambda \to \mathbb{Z}$ there is a symmetric line bundle L with $c_1(L) = H$.

(7) Let $L = L(H,\chi)$ be a line bundle on $X = V/\Lambda$. Let $V_0 \subset V$ be the kernel of $E = \operatorname{Im} H$ and $\Lambda_0 = V_0 \cap \Lambda$. If $\chi|_{\Lambda_0} \equiv \mathbf{1}$, show that L is the pullback of a line bundle on the quotient $X' = V/(V_0 + \Lambda)$ under the natural projection $X \to X'$.

(8) (*First Chern class*) For a complex torus $X = V/\Lambda$ the composed map

$$c_1 : H^1(X, \mathcal{O}_X^*) \to H^2(X.\mathbb{Z}) \hookrightarrow H^2(X,\mathbb{C}) \xrightarrow{\sim} H^2_{\mathrm{dR}}(X)$$

is defined as follows: Suppose $L \in \operatorname{Pic}(X)$ is given by the cocycle $\{f_{\mu,\nu}\}_I$ with respect to a covering $\{U_\mu\}_I$ of X. Choose C^∞-one-forms φ_μ on U_μ with $\frac{1}{2\pi i}\mathrm{d}\log f_{\mu\nu} = \varphi_\nu - \varphi_\mu$. Then $c_1(L)$ is given by the global 2-form $\mathrm{d}\varphi_{\mu\nu}$.

(a) Let $L = L(H,\chi)$ and $v_1,\dots,v_g$ be complex coordinate functions with respect to a basis $e_1,\dots,e_g$ of V. Show that

$$c_1(L) = \frac{i}{2}\sum_{\mu,\nu=1}^{g} H(e_\nu, e_\mu)\mathrm{d}v_\nu \wedge \mathrm{d}\bar{v}_\mu.$$

(b) Suppose L is of type $(d_1,\dots,d_g)$ and $x_1,\dots,x_g,y_1,\dots,y_g$ are real coordinate functions of V with respect to a symplectic basis for L (see Section 1.5.1 below). Then

$$c_1(L) = -\sum_{\nu=1}^{g} d_\nu \mathrm{d}x_\nu \wedge \mathrm{d}y_\nu.$$

(Hint: use the cocycle $\phi_1 a_L$, where a_L is the canonical factor of L and ϕ_1 the isomorphism of Lemma 1.2.6.)

(9) (*The Néron–Severi group*) Let $(Z, \mathbf{1}_g)$ be the period matrix of a complex torus X. The group $\mathrm{NS}(X)$ may be considered as the group of alternating forms on $\mathbb{R}^{2g}$, integer-valued on $\mathbb{Z}^{2g}$, which induce a hermitian form on $\mathbb{C}^g$ via the $\mathbb{R}$-linear isomorphism $\mathbb{R}^{2g} \to \mathbb{C}^g$, $x \mapsto (Z, \mathbf{1}_g)x$. Let E be an alternating form on $\mathbb{R}^{2g}$ with matrix $\begin{pmatrix} A & B \\ -{}^tB & C \end{pmatrix} \in \mathrm{M}_{2g}(\mathbb{R})$. Show that the following conditions are equivalent:

(i) $E \in \mathrm{NS}(X)$;
(ii) $A, B, C \in \mathrm{M}_g(\mathbb{Z})$ with $A - BZ + {}^tZ{}^tB + {}^tZCZ = 0$.

Hence the Néron–Severi group $\mathrm{NS}(X)$ is isomorphic to the group of matrices $\begin{pmatrix} A & B \\ -{}^tB & C \end{pmatrix} \in \mathrm{M}_{2g}(\mathbb{Z})$ satisfying condition (ii).

In particular, if $g = 2$, $A = \begin{pmatrix} 0 & a \\ -a & 0 \end{pmatrix}$, $B = \begin{pmatrix} b & d \\ e & f \end{pmatrix}$ and $C = \begin{pmatrix} 0 & c \\ -c & 0 \end{pmatrix}$, condition (ii) reads: $a, b, c, d, e, f \in \mathbb{Z}$ with $a + ez_{11} - bz_{12} + fz_{21} - dz_{22} + c\det Z = 0$.

(10) Let X be a complex torus of dimension g.

(a) Show that the Néron–Severi group $\mathrm{NS}(X)$ is a free abelian group of finite rank.
(Hint: use that $\mathrm{NS}(X)$ is a subgroup of $H^2(X,\mathbb{Z})$.)

(b) The *Picard number* $\rho(X)$ of a complex torus X is by definition the rank of the Néron–Severi group $\mathrm{NS}(X)$. Show that

$$\rho(X) \le h^{1,1}(X) = g^2.$$

(11) (*Theorem of the Cube*)

(a) Let $X_\nu,\ \nu = 1,2,3$ be complex tori and $L \in \mathrm{Pic}(X_1 \times X_2 \times X_3)$ such that its restrictions to $X_1 \times X_2 \times \{0\},\ X_1 \times \{0\} \times X_3$ and $\{0\} \times X_2 \times X_3$ are trivial. Then L is trivial.

(b) Generalize this to products of $n > 3$ complex tori.

(Hint: for (a) use canonical factors.)

(12) Let X be a complex torus, Y any complex manifold and $\varphi_\nu : Y \to X$ holomorphic maps for $\nu = 1,2,3$. Show that for any $L \in \mathrm{Pic}(X)$:

$$\begin{aligned}&(\varphi_1+\varphi_2+\varphi_3)^*L\\ &\simeq (\varphi_1+\varphi_2)^*L \otimes (\varphi_1+\varphi_3)^*L \otimes (\varphi_2+\varphi_3)^*L \otimes \varphi_1^*L^{-1} \otimes \varphi_2^*L^{-1} \otimes \varphi_2^*L^{-1}.\end{aligned}$$

(Hint: use canonical factors.)

1.4 The Dual Complex Torus and the Poincaré Bundle

The Poincaré bundle of a complex torus X is a line bundle on the product $X \times \widehat{X}$, where $\widehat{X}$ denotes the dual complex torus of X, which we have to define first.

1.4.1 The Dual Complex Torus

Let $X = V/\Lambda$ be a complex torus of dimension g. According to the Appell–Humbert Theorem the map $\mathrm{Hom}(\Lambda, \mathbb{C}_1) \to \mathrm{Pic}^0(X),\ \chi \mapsto L(0,\chi)$ is an isomorphism of groups. The fact that $\mathrm{Hom}(\Lambda,\mathbb{C}_1) \simeq (\mathbb{R}/\mathbb{Z})^{2g}$ is a real torus suggests that $\mathrm{Pic}^0(X)$ can be given the structure of a complex torus.

Consider the complex vector space $\overline{\Omega} := \mathrm{Hom}_{\overline{\mathbb{C}}}(V,\mathbb{C})$ of $\mathbb{C}$-antilinear forms $\ell : V \to \mathbb{C}$. The underlying real vector space of $\overline{\Omega}$ is canonically isomorphic to $\mathrm{Hom}_{\mathbb{R}}(V,\mathbb{R})$ (see the analogue of Exercise 1.2.3 (6) for $\overline{\Omega}$). Hence the canonical

$\mathbb{R}$-bilinear form

$$\langle\ ,\ \rangle : \overline{\Omega} \longrightarrow \mathbb{R}, \quad (\ell, v) \mapsto \operatorname{Im} \ell(v)$$

is non-degenerate. This implies

$$\widehat{\Lambda} := \{\ell \in \overline{\Omega} \mid \langle \ell, v\rangle \in \mathbb{Z}\}$$

is a lattice in $\overline{\Omega}$, called the *dual lattice* of Λ. The quotient

$$\widehat{X} := \overline{\Omega}/\widehat{\Lambda}$$

is a complex torus of dimension g, the *dual complex torus* of X. Identifying V with the space of $\mathbb{C}$-antilinear forms $\overline{\Omega} \to \mathbb{C}$ by double antiduality, the nondegeneracy of the form $\langle\ ,\ \rangle$ implies that Λ is the lattice in V dual to $\overline{\Lambda}$. So we get

$$\widehat{\widehat{X}} = X.$$

Proposition 1.4.1 *The canonical homomorphism*

$$\overline{\Omega} \longrightarrow \operatorname{Hom}(\Lambda, \mathbb{C}_1), \quad \ell \mapsto e(2\pi i \langle \ell, \cdot\rangle)$$

induces an isomorphism $\widehat{X} \xrightarrow{\simeq} \operatorname{Pic}^0(X)$.

We often identify $\operatorname{Pic}^0(X)$ with the complex torus $\widehat{X}$.

Proof The nondegeneracy of the form $\langle\ ,\ \rangle$ implies that the map $\overline{\Omega} \longrightarrow \operatorname{Hom}(\Lambda, \mathbb{C}_1)$ is surjective. By definition $\widehat{\Lambda}$ is precisely its kernel. □

Let $X' = V'/\Lambda'$ be a second complex torus and $f : X' \longrightarrow X$ a homomorphism with analytic representation $F : V' \longrightarrow V$. The (anti-)dual map $F^* : \overline{\Omega} \longrightarrow \overline{\Omega'}$, $\ell \mapsto \ell \circ F$ induces a homomorphism $\widehat{f} : \widehat{X} \longrightarrow \widehat{X'}$, since $F^*\widehat{\Lambda} \subseteq \widehat{\Lambda'}$. It is called the *dual homomorphism of* f. According to Proposition 1.4.1 the following diagram commutes

$$\begin{array}{ccc} \widehat{X} & \xrightarrow{\simeq} & \operatorname{Pic}^0(X) \\ {\scriptstyle \widehat{f}}\downarrow & & \downarrow{\scriptstyle f^*} \\ \widehat{X'} & \xrightarrow{\simeq} & \operatorname{Pic}^0(X'). \end{array} \tag{1.13}$$

If $g : X \longrightarrow X''$ is a second homomorphism of complex tori, then

$$\widehat{gf} = \widehat{f}\widehat{g}.$$

To be more precise, "$\widehat{\ }$" is a functor from the category of complex tori into itself. The following proposition says that this functor is exact.

Proposition 1.4.2 *If* $0 \longrightarrow X' \longrightarrow X \longrightarrow X'' \longrightarrow 0$ *is an exact sequence of complex tori, the dual sequence* $0 \longrightarrow \widehat{X''} \longrightarrow \widehat{X} \longrightarrow \widehat{X'} \longrightarrow 0$ *is also exact.*

Proof Suppose $X'' = V''/\Lambda''$ and X, X' are as above. Applying the serpent lemma, the induced sequence of lattices $0 \to \Lambda' \to \Lambda \to \Lambda'' \to 0$ is exact. As a sequence of free abelian groups it splits, so that

$$0 \to \mathrm{Hom}(\Lambda'', \mathbb{C}_1) \to \mathrm{Hom}(\Lambda, \mathbb{C}_1) \to \mathrm{Hom}(\Lambda', \mathbb{C}_1) \to 0$$

is also exact. Using Proposition 1.4.1 and the Appell–Humbert Theorem this give the assertion. □

Proposition 1.4.3 *If $f : X' \to X$ is an isogeny, then $\widehat{f} : \widehat{X} \to \widehat{X'}$ is also an isogeny and*

$$\mathrm{Ker}\,\widehat{f} \simeq \mathrm{Hom}(\mathrm{Ker}\,f, \mathbb{C}_1).$$

In particular $\deg \widehat{f} = \deg f$.

Proof According to Lemma 1.1.11 we may assume that $\rho_a(f) = \mathbf{1}_V$. Then $\rho_a(\widehat{f}) = \mathbf{1}_{\overline{\Omega}}$. So $\widehat{f}$ is an isogeny. According to diagram (1.13) and the Appell–Humbert Theorem

$$\mathrm{Ker}\,\widehat{f} \simeq \mathrm{Ker}\left(\mathrm{Hom}(\Lambda, \mathbb{C}_1) \to \mathrm{Hom}(\Lambda', \mathbb{C}_1)\right) \simeq \mathrm{Hom}(\Lambda/\Lambda', \mathbb{C}_1).$$

Since $\mathrm{Ker}\,f \simeq \Lambda/\Lambda'$, this implies the last assertions. □

A consequence of the proposition is the following criterion for a line bundle to descend under an isogeny.

Corollary 1.4.4 *Let $f : X_1 \to X_2$ an isogeny of complex tori $X_i = V_i/\Lambda_i$ with analytic representation F. For any $L = L(H, \chi) \in \mathrm{Pic}(X_1)$ the following conditions are equivalent*

(i) $L = f^*M$ *for an* $M \in \mathrm{Pic}(X_2)$*;*
(ii) $\mathrm{Im}\,H(F^{-1}\Lambda_2, F^{-1}\Lambda_2) \subseteq \mathbb{Z}$.

Proof Suppose (ii). Since f is an isogeny, F is an isomorphism and (ii) implies ${F^{-1}}^*H \in \mathrm{NS}(X_2)$. Choose an $\widetilde{M} \in \mathrm{Pic}(X_2)$ with $c_1(\widetilde{M}) = {F^{-1}}^*H$. Then $c_1(f^*\widetilde{M}) = H$ and thus $L \otimes f^*\widetilde{M}^{-1} \in \mathrm{Pic}^0(X_1)$.

According to Proposition 1.4.3 the homomorphism $f^* : \mathrm{Pic}^0(X_2) \to \mathrm{Pic}^0(X_1)$ is surjective and so there is an $N \in \mathrm{Pic}^0(X_2)$ with $f^*N = L \otimes f^*\widetilde{M}^{-1}$. Now $M = \widetilde{M} \otimes N$ satisfies (i). The converse implication follows directly from Lemma 1.3.6. □

1.4.2 The Homomorphism $\phi_L : X \to \widehat{X}$

For any $L \in \mathrm{Pic}(X)$ and any point $x \in X$ the line bundle $t_x^*L \otimes L^{-1}$ has first Chern class 0. So identifying $\widehat{X} = \mathrm{Pic}^0(X)$ we get a map

$$\phi_L : X \to \widehat{X}, \qquad x \mapsto t_x^*L \otimes L^{-1},$$

which, according to the Theorem of the Square 1.3.5, is a homomorphism. In order to compute its analytic representation, suppose $X = V/\Lambda$ and $L = L(H, \chi)$.

Lemma 1.4.5 *The analytic representation of $\phi_L : X \to \widehat{X}$ is*

$$\phi_H : V \to \overline{\Omega}, \qquad v \mapsto H(v, \cdot).$$

Proof Applying Lemma 1.3.4, we get

$$t_{\overline{v}}^* L \otimes L^{-1} = L\left(0, \mathbf{e}(2\pi i \operatorname{Im} H(v, \cdot))\right) = L(0, \mathbf{e}(2\pi i \langle \phi_H(v), \cdot \rangle)).$$

Comparing this with the isomorphism $\widehat{X} \xrightarrow{\simeq} \operatorname{Hom}(\Lambda, \mathbb{C}_1)$ of Proposition 1.4.1 gives the assertion. □

Proposition 1.4.6

(a) *ϕ_L depends only on the first Chern class H of L, not on L itself.*
(b) *$\phi_{L \otimes M} = \phi_L + \phi_M$ for all $L, M \in \operatorname{Pic}(X)$.*
(c) *$\widehat{\phi_L} = \phi_L$ under the natural identification $\widehat{\widehat{X}} = X$.*
(d) *for any homomorphism $f : Y \to X$ of complex tori and any $L \in \operatorname{Pic}(X)$ the following diagram commutes*

$$\begin{array}{ccc} X & \xrightarrow{\phi_L} & \widehat{X} \\ {\scriptstyle f}\uparrow & & \downarrow{\scriptstyle \widehat{f}} \\ Y & \xrightarrow[f^*L]{} & \widehat{Y}. \end{array}$$

Proof (a), (b) and (d) follow immediately from the definitions. Finally, recall that ϕ_H^* is the analytic representation of $\widehat{\phi_L}$. So (c) follows from the fact that $\phi_H^* = \phi_H$ under the natural identification $\operatorname{Hom}_{\overline{\mathbb{C}}}(\overline{\Omega}, \mathbb{C}) = V$. □

The kernel of the map ϕ_L is denoted by $K(L)$. In order to describe $K(L)$, define

$$\Lambda(L) := \{v \in V \mid \operatorname{Im} H(v, \Lambda) \subseteq \mathbb{Z}\}.$$

Obviously $\Lambda(L) = \phi_H^{-1}(\widehat{\Lambda})$, which implies

$$K(L) = \Lambda(L)/\Lambda. \tag{1.14}$$

Since $K(L)$ and $\Lambda(L)$ depend only on the hermitian form H, we sometimes write $K(H)$ and $\Lambda(H)$ for these groups respectively. Compare Exercise 1.4.5 (5) for some elementary properties of $K(L)$.

According to Lemma 1.4.5 the homomorphism ϕ_L is an isogeny if and only if the hermitian form $H = c_1(L)$ is non-degenerate. This suggests the following definition: $L \in \operatorname{Pic}(X)$ is called a *non-degenerate line bundle* if the hermitian form $H = c_1(L)$ is non-degenerate, or equivalently if the alternating form $\operatorname{Im} H$ is non-degenerate (see Lemma 1.2.10). Equation 1.14 implies the first assertion of the following proposition.

Proposition 1.4.7 *A line bundle $L \in \mathrm{Pic}(X)$ is non-degenerate if and only if $K(L)$ is finite and we have for any L*

$$\deg \phi_L = \det(\mathrm{Im}\, H).$$

Proof (of the equation) We may assume that L is non-degenerate, since otherwise both sides are zero. Then $\deg \phi_L = (\phi_H^{-1}(\widehat{\Lambda}) : \Lambda) = (\Lambda(L) : \Lambda) = \det(\mathrm{Im}\, H)$, using an elementary result of Linear Algebra. □

1.4.3 The Seesaw Theorem

In this section we will prove the seesaw theorem, using some difficult results on complex spaces. Here it would be much easier to prove the theorem in the algebraic category, that is for abelian varieties, and in fact we will only use it in this setting. In this case one can use the analogous theorems in Hartshorne [61, Chapter III, Theorem 12.8 and Corollary 12.9].

Theorem 1.4.8 *Let X be a complex torus, Z a normal analytic space and $\mathcal{L}$ a line bundle on $X \times Z$.*

(a) *The set $Z_0 := \{z \in Z \mid \mathcal{L}|_{X\times\{z\}}$ is trivial$\}$ is Zariski closed in Z.*
(b) *If $q : X \times Z_0 \to Z_0$ denotes the projection map, then there is a holomorphic line bundle $\mathcal{M}$ on Z_0 such that*

$$\mathcal{L}_{X\times Z_0} \simeq q^*\mathcal{M}.$$

Actually the theorem and its proof are valid for any compact complex manifold X and any reduced analytic space, but we do not need it in this generality.

Proof (a): A holomorphic line bundle $\mathcal{N}$ is trivial if and only if $h^0(\mathcal{N}) > 0$ and $h^0(\mathcal{N}^{-1}) > 0$. Consequently

$$Z_0 = \left\{z \in Z \mid h^0(\mathcal{L}|_{X\times\{z\}}) > 0 \quad \text{and} \quad h^0(\mathcal{L}^{-1}|_{X\times\{z\}}) > 0\right\}$$

and the closedness of Z_0 follows from the semicontinuity Theorem (see Grauert–Remmert [52, Theorem 10.5.4]).

(b): According to the base change theorem (see Grauert [51, page 2 (2)]) the sheaf

$$\mathcal{M} := q_*(\mathcal{L}|_{X\times Z_0})$$

is invertible on Z_0 and the canonical base change homomorphism

$$\varphi(z) : \mathcal{M}(z) \to H^0(\mathcal{L}|_{X\times\{z\}}) \simeq \mathbb{C}$$

is an isomorphism for every $z \in Z_0$. It remains to show that the canonical map

$$q^*\mathcal{M} = q^*q_*(\mathcal{L}|_{X\times Z_0}) \to \mathcal{L}|_{X\times Z_0}$$

is an isomorphism. For this it suffices to show that the induced map

$$H^0(q^*\mathcal{M}|_{X\times\{z\}}) \to H^0(\mathcal{L}|_{X\times\{z\}})$$

is surjective for every $z \in Z_0$, since $\mathcal{L}|_{X\times\{z\}}$ is trivial. But this is a consequence of the surjectivity of $\varphi(z)$. □

The following corollary is called the *seesaw theorem* or sometimes the *seesaw principle* for obvious reasons. It is valid for any complex manifold X, where 0 has to be replaced by any point $x_0 \in X$.

Corollary 1.4.9 *Let X be a complex torus, Z a complex manifold and $\mathcal{L}$ a holomorphic line bundle on $X \times Z$. If $\mathcal{L}|_{X\times\{z\}}$ is trivial for all z in an open dense set of Z and $\mathcal{L}|_{\{0\}\times Z}$ is trivial, then $\mathcal{L}$ is trivial.*

Proof According to Theorem 1.4.8 (a), $Z_0 = Z$ and hence by (b) there is a line bundle $\mathcal{M}$ on Z such that $\mathcal{L} \simeq q^*\mathcal{M}$. Since $\mathcal{L}_{\{0\}\times Z}$ is trivial and $q : \{0\} \times Z \to Z$ is an isomorphism, the line bundle $\mathcal{M}$ is trivial, which implies that $\mathcal{L}$ is trivial. □

1.4.4 The Poincaré Bundle

Let $X = V/\Lambda$ be a complex torus. According to Proposition 1.4.1 the points of $\widehat{X}$ parametrize the line bundles of $\mathrm{Pic}^0(X)$. This suggests that there might be a line bundle on the product $X \times \widehat{X}$ which induces the line bundles of $\mathrm{Pic}^0(X)$. Such a bundle is called a Poincaré bundle. To be more precise, a holomophic line bundle $\mathcal{P}$ on $X \times \widehat{X}$ is called a *Poincaré bundle for* X if

(i) $\mathcal{P}|_{X\times\{L\}} \simeq L$ for every $L \in \mathrm{Pic}^0(X)$;
(ii) $\mathcal{P}|_{\{0\}\times\widehat{X}} \simeq \mathcal{O}_{\widehat{X}}$.

Condition (ii) serves for the sake of normalization.

Theorem 1.4.10 *There exists a unique Poincaré bundle $\mathcal{P}$ on $X \times \widehat{X}$.*

Proof Define a hermitian form H on the vector space $V \times \overline{\Omega}$ by

$$H((v_1, \ell_1), (v_2, \ell_2)) = \overline{\ell_2(v_1)} + \ell_1(v_2)$$

and define a semicharacter $\chi : \Lambda \times \widehat{\Lambda} \to \mathbb{C}_1$ for H by

$$\chi(\lambda, \ell_0) := \mathbf{e}\,(\pi i \operatorname{Im} \ell_0(\lambda))\,.$$

According to the Appell–Humbert Theorem 1.3.3 there is a unique line bundle $\mathcal{P}$ on $X \times \widehat{X}$ corresponding to the pair (H, χ). We have to check properties (i) and (ii). For this consider the canonical factor $a_{\mathcal{P}} : (\Lambda \times \widehat{\Lambda}) \times (V \times \overline{\Omega}) \to \mathbb{C}^*$ of $\mathcal{P}$

$$a_{\mathcal{P}}\,((\lambda, \ell_0), (v, \ell)) = \chi(\lambda, \ell_0)\,\mathbf{e}\left(\pi H\,((v, \ell), (\lambda, \ell_0)) + \frac{\pi}{2} H\,((\lambda, \ell_0), (\lambda, \ell_0))\right).$$

(i): suppose $L \in \widehat{X} = \mathrm{Pic}^0(X)$. There is an $\ell \in \overline{\Omega}$ such that $L = L(0, \mathbf{e}(2\pi i \operatorname{Im} \ell))$. The restriction $\mathcal{P}|_{X \times \{L\}}$ is given by the factor $a_{\mathcal{P}}|_{\Lambda \times \{0\} + V \times \{\ell\}}$. But

$$a_{\mathcal{P}}\left((\lambda, 0), (v, \ell)\right) = \chi(\lambda, 0)\, \mathbf{e}\left(\pi H\left((v, \ell), (\lambda, 0)\right) + \frac{\pi}{2} H\left((\lambda, 0), (\lambda, 0)\right)\right)$$
$$= \mathbf{e}\left(\pi \ell(\lambda)\right)$$

is equivalent to $a_{\mathcal{P}}\left((\lambda, 0), (v, \ell)\right) \mathbf{e}\left(\pi \overline{\ell(v+\lambda)}\right)^{-1} \mathbf{e}\left(\pi \overline{\ell(v)}\right) = \mathbf{e}\left(2\pi i \operatorname{Im} \ell(\lambda)\right)$, the canonical factor of L.

(ii): the restriction $\mathcal{P}|_{\{0\} \times \widehat{X}}$ is given by the factor $a_{\mathcal{P}}\left((0, \ell_0), (0, \ell)\right) = 1$ for all $\ell_0 \in \widehat{\Lambda}$ and $\ell \in \overline{\Omega}$, which is the canonical factor of the trivial line bundle on $\{0\} \times \widehat{X}$.

The uniqueness statement is a direct consequence of the seesaw theorem Corollary 1.4.9 above. □

The Poincaré bundle satisfies the following *universal property*.

Theorem 1.4.11 *For any normal analytic space T and any line bundle $\mathcal{L}$ on $X \times T$ satisfying the conditions*

(i) $\mathcal{L}|_{X \times \{t\}} \in \mathrm{Pic}^0(X)$ *for every* $t \in T$ *and*
(ii) $\mathcal{L}|_{\{0\} \times T}$ *is trivial,*

there is a unique holomorphic map $\psi : T \to \widehat{X}$ such that $\mathcal{L} \simeq (\mathbf{1}_X \times \psi)^* \mathcal{P}$.

Proof Define a map

$$\psi : T \to \widehat{X} \quad \text{by} \quad t \mapsto \mathcal{L}|_{X \times \{t\}}.$$

We first claim that ψ is holomorphic. For this consider the line bundle

$$\mathcal{N} := p_{12}^* \mathcal{L} \otimes p_{13}^* \mathcal{P}^{-1} \quad \text{on} \quad X \times T \times \widehat{X}.$$

Here p_{ij} denotes the projection onto the i-th and j-th factor. The set

$$\Gamma := \left\{(t, L) \in T \times \widehat{X} \mid \mathcal{N}|_{X \times \{(t, \mathcal{L})\}} \text{ is trivial}\right\}$$

is Zariski closed in $T \times \widehat{X}$ by Theorem 1.4.8 (a). But Γ is the graph of the map ψ, since $\mathcal{N}|_{X \times \{(t,L)\}} \simeq \mathcal{L}|_{X \times \{t\}} \otimes L^{-1}$. In particular the projection $p_1 : \Gamma \to T$ is a bijective holomorphic map. Since T is normal, we can apply Zariski's main theorem, which says in this case just that p_1 is biholomorphic. Therefore ψ is holomorphic. The fact that $\mathcal{L} \simeq (\mathbf{1}_X \times \psi)^* \mathcal{P}$ follows from the seesaw theorem Corollary 1.4.9.

Suppose $\psi' : T \to \widehat{X}$ is another holomorphic map with $\mathcal{L} \simeq (\mathbf{1}_X \times \psi')^* \mathcal{P}$. Then

$$\psi(t) = (\mathbf{1}_X \times \psi)^* \mathcal{P}|_{X \times \{t\}} = (\mathbf{1}_X \times \psi')^* \mathcal{P}|_{X \times \{t\}} = \psi'(t)$$

for all $t \in T$, which gives the uniqueness of ψ. □

Two line bundles L_1 and L_2 on X are called *analytically equivalent* if there is a connected complex manifold T, a line bundle $\mathcal{L}$ on $X \times T$ and points $t_1, t_2 \in T$ such that

$$\mathcal{L}|_{X\times\{t_i\}} \simeq L_i$$

for $i = 1$ and 2. We use the Poincaré bundle to give a criterion for this. Note that the notion is analogous the notion of *algebraic equivalence* (see Section 2.3.1 below).

Proposition 1.4.12 *For line bundles L_1, L_2 on X the following conditions are equivalent:*

(i) *L_1 and L_2 are analytically equivalent;*
(ii) $L_1 \otimes L_2^{-1} \in \mathrm{Pic}^0(X)$;
(iii) $\phi_{L_1} = \phi_{L_2}$;
(iv) $c_1(L_1) = c_1(L_2)$.

Proof It suffices to show (i) $\Leftrightarrow$ (ii), since the equivalence of (ii) with (iii) respectively (iv) follows immediately from Proposition 1.4.6 (b) and Exercise 1.4.5 (5) (b) respectively the Appell–Humbert Theorem 1.3.3.

(i) $\Rightarrow$ (ii): suppose L_1 and L_2 are analytically equivalent and let the notation be as in the definition. Then it is easy to see that the map $T \longrightarrow H^2(X, \mathbb{Z}),\ t \mapsto c_1(\mathcal{L}|_{X\times\{t\}})$ is continuous. It is even constant, since T is connected and $H^2(X, \mathbb{Z})$ is a discrete group. Thus $c_1(L_1) = c_1(L_2)$ and hence $L_1 \otimes L_2^{-1} \in \mathrm{Pic}^0(X)$.

(ii) $\Rightarrow$ (i): suppose $L_1 \otimes L_2^{-1} \in \mathrm{Pic}^0(X)$. Define $\mathcal{L} := \mathcal{P} \otimes p^* L_2$ where $p : X \times \widehat{X} \longrightarrow X$ denotes the projection. Then $\mathcal{L}|_{X\times\{L_1\otimes L_2^{-1}\}} \simeq L_1$ and $\mathcal{L}|_{X\times\{0\}} \simeq L_2$, so the line bundles L_1 and L_2 are analytically equivalent. □

It follows immediately from Proposition 1.4.12 that analytic equivalence is an equivalence relation and furthermore that the equivalence classes are just elements of the Néron–Severi group $\mathrm{NS}(X)$. In the special case that L is non-degenerate on X, we can say more.

Corollary 1.4.13 *Suppose $L_1, L_2 \in \mathrm{Pic}(X)$ with L_1 non-degenerate. Then L_1 and L_2 are analytically equivalent if and only if $L_2 \simeq t_x^* L_1$ for some $x \in X$.*

Proof By Proposition 1.4.12, $L_2 \otimes L_1^{-1} \in \mathrm{Pic}^0(X)$. Since L_1 is non-degenerate, the map $\phi_{L_1} : X \longrightarrow \mathrm{Pic}^0(X)$ is surjective. Hence there is an $x \in X$ such that $L_2 \otimes L_1^{-1} = \phi_{L_1}(x) = t_x^* L_1 \otimes L_1^{-1}$. The converse implication is obvious. □

Another application of the Poincaré bundle is a criterion for a homomorphism $f : X \longrightarrow \widehat{X}$ to be of the from ϕ_L for some line bundle L. For this we need the following lemma.

Lemma 1.4.14 *For any $M \in \mathrm{Pic}(X)$ and a positive integer n the following conditions are equivalent:*

(i) *there is an $L \in \mathrm{Pic}(X)$ such that $M = L^n$;*
(ii) $X_n \subset K(M)$.

Proof (i) $\Rightarrow$ (ii): let $M = L^n$. By definition $\Lambda(M) = \Lambda(L^n) = \frac{1}{n}\Lambda(L)$, which implies (ii).

(ii) $\Rightarrow$ (i): Suppose $X_n \subset K(M)$. If $M = L(H, \chi)$, this means

$$\frac{1}{n}\Lambda \subseteq \Lambda(M) = \{v \in V \mid \operatorname{Im} H(v, \Lambda) \subseteq \mathbb{Z}\}.$$

Hence $(nH)(\lambda_1, \lambda_2) \in H(\lambda_1, \Lambda) \subseteq \mathbb{Z}$ for all $\lambda_1, \lambda_2 \in \frac{1}{n}\Lambda$ and thus by Corollary 1.4.4 there is an $\widetilde{M} \in \operatorname{Pic}(X)$ with $M^n = n_X^*\widetilde{M}$. According to Proposition 1.3.7 the line bundles $M^n = n_X^*\widetilde{M}$ and $\widetilde{M}^{n^2}$ are analytically equivalent (for even n one can apply Corollary 1.3.8, for odd n this is slightly more complicated). Hence the same holds for M and $\widetilde{M}^n$. So $\widetilde{M}^n \otimes M^{-1} \in \operatorname{Pic}^0(X)$. Since $\operatorname{Pic}^0(X) = \widehat{X}$ is a divisible group, there is an $N \in \operatorname{Pic}^0(X)$ with $\widetilde{M}^n \otimes M^{-1} \simeq N^n$. Now $L = \widetilde{M} \otimes N^{-1}$ satisfies (i). □

Theorem 1.4.15 *Suppose $X = V/\Lambda$ and $f : X \to \widehat{X}$ is a homomorphism with analytic representation $F : V \to \overline{\Omega}$. F can be considered as a form $F : V \times V \to \mathbb{C}$ such that the following statements are equivalent:*

(i) *$f = \phi_L$ for some $L \in \operatorname{Pic}(X)$;*

(ii) *the form $F : V \times V \to \mathbb{C}$ is hermitian.*

Proof (ii) $\Rightarrow$ (i): Suppose the form $F : V \times V \to \mathbb{C}$ is hermitian. Let M denote the pullback of the Poincaré bundle $\mathcal{P}$ under the homomorphism $(\mathbf{1}_X, f) : X \to X \times \widehat{X}$.

We first claim that $2f = \phi_M$. If H denotes the hermitian form of $\mathcal{P}$ (see the proof of Theorem 1.4.10), then by assumption

$$\begin{aligned}(v, w) \mapsto (\mathbf{1}_V, F)^* H(v, w) &= H((v, F(v)) + (w, F(w)) \\ &= \overline{F(w)(v)} + F(v)(w) = 2F(v)(w)\end{aligned}$$

is the hermitian form of M. Since $(\mathbf{1}_V, F)^*\phi_H$ is the analytic representation of ϕ_M and $2F$ the analytic representation of $2f$, this implies $2f = \phi_M$.

According to Lemma 1.4.14 there is a line bundle L on X with $M = L^2$. By Proposition 1.4.6 (b) we have $2\phi_L = \phi_{L^2} = 2f$. Since $\operatorname{Hom}(X, \widehat{X})$ is torsion-free according to Proposition 1.1.8, this finally implies $f = \phi_L$.

(i) $\Rightarrow$ (ii): If $f = \phi_L$, the analytic representation F is just the hermitian form $c_1(L)$ by definition of ϕ_L. □

1.4.5 Exercises

(1) Let $f : X = V/\Lambda \to Y = V/\Gamma$ be an isogeny of complex tori with analytic representation F and $L \in \mathrm{Pic}(X)$ which descends to a line bundle M on Y.

(a) Any isomorphism $\varphi : \overline{\Omega} \xrightarrow{\sim} V$ induces an isomorphism $\phi_L(X) \simeq V/\Lambda(L)$.

(b) There is an isomorphism of groups $\mathrm{Ker}\,\widehat{f} \xrightarrow{\sim} \Lambda(L)/F^{-1}\Gamma(M)$. (Here $\Gamma(M)$ is defined analogously as $\Lambda(L)$ in Section 1.4.2.)

(2) For any complex torus X and any integer $n \neq 0$ the homomorphism $L \mapsto n_X^* L$ on $\mathrm{Pic}(X)$ induces the n-th power map on $\widehat{X}$ and the n^2-th power map on $\mathrm{NS}(X)$.

(3) (a) For any complex tori X_1 and X_2 there is a canonical isomorphism $(X_1 \times X_2)\widehat{\ } \simeq \widehat{X}_1 \times \widehat{X}_2$.

(b) Let $f_\nu : X_\nu \to Y_\nu$, $\nu = 1, 2$, be homomorphisms of complex tori. Show that $(f_1 \times f_2)\widehat{\ } = \widehat{f_1} \times \widehat{f_2}$ with respect to the canonical isomorphisms of (a).

(4) (a) For a complex torus X denote by $\Delta_X : X \to X \times X$ the diagonal map and by $\mu : X \times X \to X$ the addition map. Show that $\widehat{\mu} = \Delta_{\widehat{X}}$.

(b) Use (a) to show that $(f + g)\widehat{\ } = \widehat{f} + \widehat{g}$ for homomorphisms $f, g : X \to Y$.

(5) For any $L \in \mathrm{Pic}(X)$ let $K(L) = \mathrm{Ker}(\phi_L : X \to \widehat{X})$ as defined in equation (1.14). Show that

(a) $K(L \otimes P) = K(L)$ for any $P \in \mathrm{Pic}^0(X)$;
(b) $K(L) = X$ if and only if $L \in \mathrm{Pic}^0(X)$;
(c) $K(L^n) = n_X^{-1} K(L)$ for any $n \in \mathbb{Z}$;
(d) $K(L) = n_X K(L^n)$ for any $n \in \mathbb{Z}$.

(Hint: Use canonical factors.)

(6) Suppose $f : X \to Y$ is a homomorphism of complex tori of dimension g and g' respectively. As usual let $\rho_a : \mathrm{Hom}, (X, Y) \to \mathrm{M}(g' \times g, \mathbb{C})$ and $\rho_r : \mathrm{Hom}(X, Y) \to \mathrm{M}(2g' \times 2g, \mathbb{Z})$ denote the analytic and rational representations. Then

(a) $\rho_a(\widehat{f}) = {}^t\overline{\rho_a(f)}$,
(b) $\rho_r(\widehat{f}) = {}^t\rho_r(f)$.

(7) Let L_ν be a line bundle on the complex torus X_ν for $\nu = 1, 2$ and by $p_\nu : X_1 \times X_2 \to X_\nu$ the natural projection. Show that

$$\phi_{p_1^* L_1 \otimes p_2^* L_2} = \phi_{L_1} \times \phi_{L_2} : X_1 \times X_2 \to \widehat{X}_1 \times \widehat{X}_2.$$

(8) Let X be a complex torus. Show that the pair $(\widehat{X}, \mathcal{P})$ is uniquely determined (up to isomorphism) using only properties (i) and (ii) of the definition of the Poincaré bundle and the universal property Theorem 1.4.11.

(9) Let $\kappa : X \to \widehat{\widehat{X}}$ the canonical isomorphism for a complex torus X. Denote by $\mathcal{P}_X$ (respectively $\mathcal{P}_{\widehat{X}}$) the Poincaré bundle for X (respectively $\widehat{X}$) and by s the canonical isomorphism $\widehat{X} \times X \simeq X \times \widehat{X}$. Show that $(\mathbf{1}_{\widehat{\mathbf{X}}} \times \kappa)^* \mathcal{P}_{\widehat{\mathbf{X}}} \simeq \mathbf{s}^* \mathcal{P}_{\mathbf{X}}$ on $\widehat{X} \times X$.

(10) Let X be a complex torus and $\mathcal{P}$ the Poincaré bundle on $X \times \widehat{X}$. Denote by $p_1, p_2 : X \times X \to X$ the natural projections and by $\mu : X \times X \to X$ the addition map.

(a) Show that for any $L \in \operatorname{Pic}(X)$:

$$(\mathbf{1}_X \times \phi_L)^* \mathcal{P} \simeq \mu^* L \otimes p_1^* L^{-1} \otimes p_2^* L^{-1}.$$

(b) Conclude that $L \in \operatorname{Pic}^0(X)$ if and only if $\mu^* L \simeq p_1^* L \otimes p_2^* L$.

(11) Show that $c_1(\mathcal{P})$ can be considered as an isomorphism $H^1(X, \mathbb{Z})^* \to H^1(\widehat{X}, \mathbb{Z})$ and the following diagram is commutative:

$$\begin{array}{ccc} H^1(X,\mathbb{Z})^* & \xrightarrow{c_1(\mathcal{P})} & H^1(\widehat{X},\mathbb{Z}) \\ \downarrow \simeq & & \downarrow \simeq \\ \operatorname{Hom}(\Lambda,\mathbb{Z})^* & & \operatorname{Hom}(\widehat{\Lambda},\mathbb{Z}) \\ \downarrow = & & \downarrow = \\ \Lambda & \xrightarrow{\text{double duality}} & \widehat{\widehat{\Lambda}}. \end{array}$$

1.5 Theta Functions

1.5.1 Characteristics of Non-degenerate Line Bundles

The notion of characteristics played an important role in the classical theory of theta functions (see for example Krazer's book [77]). In this section we define characteristics of non-degenerate line bundles on any complex torus.

Let $X = V/\Lambda$ be a complex torus of dimension g and $L \in \operatorname{Pic}(X)$. Recall that the first Chern class $c_1(L)$ can be considered as a hermitian form H on V whose imaginary part $E = \operatorname{Im} H$ is an alternating form with integer values on Λ.

According to the elementary divisor theorem (see Frobenius [46] or Bourbaki [28, 5.1 Theorem 1]) there is a basis $\lambda_1, \ldots, \lambda_g, \mu_1, \ldots, \mu_g$ of Λ, with respect to which E is given by a matrix

$$\begin{pmatrix} 0 & D \\ -D & 0 \end{pmatrix}$$

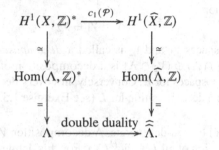

where $D = \text{diag}\,(d_1, \ldots, d_g)$ is a diagonal matrix with integers $d_i \geq 0$ satisfying $d_i | d_{i+1}$ for $i = 1, \ldots, g - 1$. The *elementary divisors* d_i are uniquely determined by E and Λ and thus by L. The vector $(d_1, \ldots, d_g)$ as well as the matrix D are called the *type* of the line bundle L. The basis $\lambda_1, \ldots, \lambda_g, \mu_1, \ldots, \mu_g$ of Λ is called a *symplectic* (or *canonical*) *basis* for L (or E).

Recall that $L \in \text{Pic}(X)$ is non-degenerate if the form H and thus E is non-degenerate. In terms of the type of L this means that $d_g > 0$ or equivalently $d_i > 0$ for all i.

A *decomposition for* L (or H or E respectively) is a direct sum decomposition

$$\Lambda = \Lambda_1 \oplus \Lambda_2 \tag{1.15}$$

if Λ_1 and Λ_2 are isotropic with respect to E, that is if $E|_{\Lambda_i} = 0$ for $i = 1, 2$. Such a decomposition always exists. In fact, if $\lambda_1, \ldots, \lambda_g, \mu_1, \ldots, \mu_g$ is a symplectic basis of Λ for L, then $\langle \lambda_1, \ldots, \lambda_g \rangle \oplus \langle \mu_1, \ldots, \mu_g \rangle$ is a decomposition for L. Conversely, it is easy to see that for every decomposition (1.15) there exists a symplectic basis such that $\Lambda_1 = \langle \lambda_1, \ldots, \lambda_g \rangle$ and $\Lambda_2 = \langle \mu_1, \ldots, \mu_g \rangle$.

A decomposition

$$V = V_1 \oplus V_2 \tag{1.16}$$

with real vector spaces V_1 and V_2 is called a *decomposition for* L (or H or E respectively) if $(V_1 \cap \Lambda) \oplus (V_2 \cap \Lambda)$ is a decomposition of Λ for L. Clearly V_1 and V_2 are isotropic subspaces for E. Conversely, not every decomposition into isotropic subspaces of V is a decomposition for L (see Exercise 1.5.5 (8)).

Let $H \in \text{NS}(X)$ be non-degenerate. A decomposition $V = V_1 \oplus V_2$ for H leads to an explicit description of all $L \in \text{Pic}^H(X)$. For this define a map χ_0 by

$$\chi_0 : V \to \mathbb{C}_1, \qquad v \mapsto \mathbf{e}\,(\pi i E(v_1, v_2)),$$

where $v = v_1 + v_2$ with $v_i \in V_i$. For the easy proof of the following lemma see Exercise 1.5.5 (1).

Lemma 1.5.1

(i) *For every* $v = v_1 + v_2,\ w = w_1 + w_2 \in V_1 \oplus V_2$ *we have*

$$\chi_0(v + w) = \chi_0(v)\chi_0(w)\,\mathbf{e}\left(\pi i E(v, w)\right)\mathbf{e}\left(-2\pi i E(v_2, w_1)\right).$$

In particular, $\chi_0|_\Lambda$ *is a semicharacter for* H.

(ii) $L_0 := L(H, \chi_0)$ *is the unique line bundle in* $\text{Pic}^H(X)$ *whose semicharacter is trivial on* $\Lambda_\nu = V_\nu \cap \Lambda$ *for* $\nu = 1, 2$.

Corollary 1.5.2 *For any $L = L(H, \chi)$ on X there is a point $c \in V$, uniquely determined up to translation by elements of $\Lambda(L)$, such that the following equivalent conditions are satisfied:*

(a) $L = t_{\overline{c}}^* L_0$;
(b) $\chi = \chi_0 \, \mathbf{e}\left(2\pi i(c, \cdot)\right)$.

Proof The existence of $c \in V$ is a direct consequence of Corollary 1.4.13. The uniqueness statement is a translation of the fact that $\operatorname{Ker} \phi_L = \Lambda(L)/\Lambda$. The equivalence of the two conditions follows from Lemma 1.3.4. □

The element $c \in V$ is called a *characteristic of the line bundle L with respect to the chosen decomposition* for H. When we speak of a characteristic c of L, we always mean that a decomposition for L is fixed and that c is its characteristic with respect to this decomposition. We will see in Lemma 3.3.5 below that this definition coincides with the classical notion.

Recall the canonical factor of automorphy $a_L : \Lambda \times V \to \mathbb{C}^*$ of Section 1.3.3. As an application of the notion of characteristics, we will extend it to a map $V \times V \to \mathbb{C}^*$, also denoted by a_L.

Suppose $L = L(H, \chi)$ is a non-degenerate line bundle on X and c a characteristic for L with respect to the decomposition $V = V_1 \oplus V_2$. Define $a_L : V \times V \to \mathbb{C}^*$ by

$$a_L(u, v) := \chi_0(u) \, \mathbf{e}\left(2\pi i E(c, u) + \pi H(v, u) + \frac{\pi}{2} H(u, u)\right).$$

According to Corollary 1.5.2 its restriction to $\Lambda \times V$ coincides with the canonical factor $a_L : \Lambda \times V \to \mathbb{C}^*$ as defined in Section 1.3.3. The following technical lemma gives some properties of a_L, for the easy proof of which we refer to Exercise 1.5.5 (2).

Lemma 1.5.3 *For all $u = u_1 + u_2$, $v = v_1 + v_2$ and $w \in V = V_1 \oplus V_2$:*

(a) $a_L(u, v + w) = a_L(u, v) \, \mathbf{e}\left(\pi H(w, u)\right)$;
(b) $a_L(u + v, w) = a_L(u, v + w) a_L(v, w) \, \mathbf{e}\left(2\pi i E(u_1, v_2)\right)$;
(c) $a_L(u, v)^{-1} = a_L(-u, v) \chi_0(u)^{-2} \, \mathbf{e}\left(-\pi H(u, u)\right)$;
(d) $a_{L'}(u, v) = a_L(u, v) \, \mathbf{e}\left(2\pi i E(w, u)\right)$ *for* $L' = t_{\overline{w}}^* L$.

1.5.2 Classical Theta Functions

Let $X = V/\Lambda$ be a complex torus with projection $\pi : V \to X$. According to Lemma 1.2.1 the pullback $\pi^* L$ of any line L on X is trivial. On the other hand, the lattice Λ acts naturally on $\pi^* L$. Hence $H^0(L)$ is isomorphic to the subspace $H^0(\mathcal{O}_V)^\Lambda$ of sections which are invariant under this action. Clearly the isomorphism depends on the choice of a factor of automorphy for L. To be more precise, let f be a factor of automorphy for L. Then $H^0(L)$ can be identified with the set of holomorphic functions $\theta : V \to \mathbb{C}$ satisfying

$$\theta(v + \lambda) = f(\lambda, v)\theta(v) \qquad \text{for all} \quad v \in V, \lambda \in \Lambda.$$

These functions are called *theta functions for the factor* f (see the beginning of Section 1.2.1).

In Section 1.3.3 we saw that for every line bundle there is a canonical factor a_L. Correspondingly the theta functions for a_L are called *canonical theta functions for* L.

In Section 1.5.3 we will determine the canonical theta function for L in order to compute the dimension $h^0(L)$ in the case when $H = c_1(L)$ is a positive hermitian form. For this it is convenient to introduce another factor for L, the *classical factor* of automorphy.

A line bundle $L = L(H, \chi)$ is called a *positive line bundle* if the hermitian form H is positive definite. Clearly every positive line bundle is non-degenerate. There exists a decomposition $V = V_1 \oplus V_2$ for L, which we fix.

Lemma 1.5.4 *The real vector space* V_2 *generates* V *as a* $\mathbb{C}$*-vector space.*

Proof The alternating form $E = \operatorname{Im} H$ vanishes identically on the complex vector space, $V_2 \cap iV_2$, since $E(iv, iw) = E(v, w)$ for all v, w. According to Lemma 1.2.10, the hermitian form H also vanishes identically on $V_2 \cap iV_2$. Since H is positive definite, this implies $V_2 \cap iV_2 = 0$, hence $V = V_2 + iV_2$ and thus the assertion. □

The hermitian form H is symmetric on V_2, since its imaginary part vanishes there. Define

$$B := \mathbb{C}\text{-bilinear extension of } H|_{V_2 \times V_2}.$$

By Lemma 1.5.4 the symmetric bilinear form B is defined on the whole of V.

Lemma 1.5.5

(a) $(H - B)(v, w) = \begin{cases} 0 & \textit{if } (v, w) \in V \times V_2, \\ 2iE(v, w) & \textit{if } (v, w) \in V_2 \times V. \end{cases}$

(b) $\operatorname{Re}(H - B)$ *is positive definite on* V_1.

Proof (a): $H - B = 0$ on $V \times V_2$, since H is $\mathbb{C}$-linear in its first component. Hence for $v \in V_2$ and $w \in V$,

$$(H - B)(v, w) = \overline{H(w, v)} - B(w, v) = (H - B)(w, v) - 2iE(w, v) = 2iE(v, w).$$

(b): Since $V = V_2 + iV_2$ and $V_2 \cap iV_2 = 0$, any $v_1 \in V_1$, $v_1 \neq 0$ can be uniquely written as $v_1 = v_2 + iv_2'$ with $v_2, v_2' \in V_2$ and $v_2' \neq 0$. Using (a), we have

$$\begin{aligned} \operatorname{Re}(H - B)(v_1, v_1) &= \operatorname{Re}\left(2iE(v_2, v_1) - 2E(v_2', v_1)\right) = 2iE(v_1, v_2') \\ &= 2E(iv_2', v_2') + 2iE(v_2', v_2') = 2H(v_2', v_2') > 0, \end{aligned}$$

since H is positive definite. □

The bilinear form B enables us to define the classical factor of automorphy for L in a coordinate-free way. Define the function $e_L : \Lambda \times V \to \mathbb{C}^*$ by

$$e_L(\lambda, v) := \chi(\lambda)\, \mathbf{e}\left(\pi(H-B)(v,\lambda) + \frac{\pi}{2}(H-B)(\lambda,\lambda)\right).$$

With an immediate computation one checks that for all $(\lambda, v) \in \Lambda \times V$,

$$e_L(\lambda, v) = a_L(\lambda, v)\, \mathbf{e}\left(\frac{\pi}{2}B(v,v)\right) \mathbf{e}\left(\frac{\pi}{2}B(v+\lambda, v+\lambda)\right)^{-1}. \tag{1.17}$$

This implies that e_L is a factor of automorphy for L equivalent to the canonical factor for L. It is called the *classical factor of automorphy for* L. Correspondingly the theta functions for the factor e_L are called *classical theta functions for* L. This terminology will be justified later (see Section 3.3.2). The classical theta functions have the advantage of being periodic with respect to the subgroup $\Lambda_2 = \Lambda \cap V_2$ of Λ.

1.5.3 Computation of $h^0(L)$ for a Positive Line Bundle L

Let $L \in \mathrm{Pic}(X)$ be positive of type $D = \mathrm{diag}(d_1, \ldots, d_g)$. So $\left(\begin{smallmatrix} 0 & D \\ -D & 0 \end{smallmatrix}\right)$ is the matrix of E with respect to a symplectic basis of the lattice Λ. Then

$$\mathrm{Pf}(E) := \det D$$

is called the *Pfaffian of* the alternating form E.

Lemma 1.5.6 $h^0(L) \leq \mathrm{Pf}(E)$.

Proof Suppose L is of characteristic c and $L_0 \in \mathrm{Pic}^H(X)$ the line bundle of characteristic 0 with respect to the decomposition $V = V_1 \oplus V_2$ for L.

We consider $H^0(L)$ and $H^0(L_0)$ as vector spaces of classical theta functions and claim that the map

$$h^0(L) \to h^0(L_0), \qquad \vartheta \mapsto \vartheta_0 := \mathbf{e}(\pi(H-B)(\cdot, c))\vartheta(\cdot - c) \tag{1.18}$$

is an isomorphism of vector spaces.

For this it suffices to show that ϑ_0 is a classical theta function, the isomorphism property of the map being clear: By Lemma 1.5.3 (d) the factors e_L and e_{L_0} are related by $e_L(\lambda, v) = e_{L_0}(\lambda, v)\, \mathbf{e}(2\pi i E(c, \lambda))$ for all $\lambda \in \Lambda, v \in V$. Using this and Lemma 1.5.3 (a) the assertion follows from an immediate computation.

Hence it suffices to show the inequality for L_0. Suppose $\vartheta \in H^0(L_0)$. According to Lemma 1.5.5 (a) and the definition of χ_0 we have $e_{L_0}(\lambda_2, v) = 1$ for all $\lambda_2 \in \Lambda_2, v \in V$; that is, ϑ is periodic with respect to Λ_2. Hence it admits a Fourier expansion. By the properties of $(H-B) : V \times V \to \mathbb{C}$ given in Lemma 1.5.5, the Fourier series of

ϑ can be written in the form

$$\vartheta(v) = \sum_{\lambda \in \Lambda(L_0)_1} \alpha_\lambda \mathbf{e}\left(\pi(H-B)(v,\lambda)\right)$$

for all $v \in V$ with uniquely determined coefficients $a_\lambda \in \mathbb{C}$. This also follows from Lemma 3.3.4 (b) below, which implies that this series of ϑ is the usual Fourier expansion.

The function ϑ satisfies the equation

$$\vartheta(v+\lambda_1) = e_{L_0}(\lambda_1, v)\vartheta(v) \text{ for } v \in V,\ \lambda_1 \in \Lambda_1.$$

The expansion of the left-hand side is

$$\vartheta(v+\lambda_1) = \sum_{\lambda \in \Lambda(L_0)_1} \alpha_\lambda \mathbf{e}\left(\pi(H-B)(\lambda_1,\lambda) + \pi(H-B)(v,\lambda)\right)$$

and the expansion of the right-hand side is

$$\begin{aligned} e_{L_0}(\lambda_1, v)\vartheta(v) &= \sum_{\lambda \in \Lambda(L_0)_1} \alpha_\lambda e_{L_0}(\lambda_1, 0)\, \mathbf{e}\left(\pi(H-B)(v,\lambda_1) + \pi(H-B)(v,\lambda)\right) \\ &= \sum_{\lambda \in \Lambda(L_0)_1} \alpha_{\lambda-\lambda_1} e_{L_0}(\lambda_1, 0)\, \mathbf{e}\left(\pi(H-B)(v,\lambda)\right). \end{aligned}$$

Comparing the coefficients gives

$$\alpha_{\lambda-\lambda_1} = \alpha_\lambda e_{L_0}(\lambda_1, 0)^{-1} \mathbf{e}\left(\pi(H-B)(\lambda_1,\lambda)\right) \qquad \text{for all} \quad \lambda \in \Lambda(L_0)_1, \lambda_1 \in \Lambda_1.$$

It follows that ϑ is determined by the coefficients α_λ, where $\lambda \in \Lambda(L_0)_1$ runs over a set of representatives of $K(L_0)_1 = \Lambda(L_0)_1/\Lambda_1$. So with Exercise 1.5.5 (5) this gives $h^0(L_0) \leq (\Lambda(L_0)_1 : \Lambda_1) = \mathrm{Pf}(E)$. □

In order to show that the above inequality is in fact an equality, we have to construct sufficiently many linearly independent theta functions of $H^0(L)$. According to the isomorphism of equation (1.18) it suffices to do this for L_0.

Recall that $L_0 = L(H, \chi_0)$ with $\chi_0(v) = \mathbf{e}\left(\pi i E(v_1, v_2)\right)$ for all $v = v_1 + v_2 \in V$ and that the canonical factor of L_0 is $a_{L_0}(\lambda, v) = \chi_0(\lambda)\, \mathbf{e}\left(\pi H(v,\lambda) + \frac{\pi}{2} H(\lambda,\lambda)\right)$. Define the function $\vartheta : V \to \mathbb{C}$ by

$$\vartheta(v) = \mathbf{e}\left(\frac{\pi}{2} B(v,v)\right) \sum_{\lambda \in \Lambda_1} \mathbf{e}\left(\pi(H-B)(v,\lambda) - \frac{\pi}{2}(H-B)(\lambda,\lambda)\right) \tag{1.19}$$

and with it for every $\overline{w} \in K(L_0) = \Lambda(L_0)/\Lambda$ with representative $w \in \Lambda(L_0)$ define a function

$$\vartheta_{\overline{w}}(v) = a_{L_0}(w, v)^{-1} \vartheta(v+w). \tag{1.20}$$

It is easy to see that $\vartheta_{\overline{w}}$ does not depend on the chosen representative w.

Lemma 1.5.7 *$\vartheta_{\overline{w}}$ is a canonical theta function for L_0 for every $\overline{w} \in K(L_0)$.*

Proof First we claim that the functions $\vartheta_{\overline{w}}$ are holomorphic. For this it suffices to show that the function

$$f(v) = \sum_{\lambda \in \Lambda_1} |\,\mathbf{e}\left(\pi(H-B)(v,\lambda) - \frac{\pi}{2}(H-B)(\lambda,\lambda)\right)|$$

converges uniformly on every compact subset of V.

Choose a norm map $||\cdot|| : V \to \mathbb{R}$ with $||\Lambda|| \subset \mathbb{Z}$. Since $\mathrm{Re}(H-B)$ is positive definite on V_1 by Lemma 1.5.5 (b), there is an $R > 0$ such that $|\,\mathbf{e}(\frac{\pi}{2}(H-B)(\lambda,\lambda))| \geq \mathbf{e}(R||\lambda||^2)$ for all $\lambda \in \Lambda_1$. Moreover, for every $r > 0$ there is an $R' > 0$ such that $|\,\mathbf{e}(\pi(H-B)(v,\lambda))| \leq \mathbf{e}(R'||\lambda||)$ for all $v \in V$ with $||v|| \leq r$. It follows that

$$f(v) \leq \sum_{\lambda \in \Lambda_1} \mathbf{e}(R'||\lambda||) - R||\lambda||^2) \leq k(\sum_{n \in \mathbb{Z}} \mathbf{e}(R'n - Rn^2))^{2g} < \infty$$

for all $v \in V$ with $||v|| \leq r$ and some constant $k > 0$. This implies the claim.

Then with an immediate computation using Lemma 1.5.1 (ii) and Lemma 1.5.5 (a) one checks that $\vartheta(v + \lambda_i) = a_{L_0}(\lambda_i, v)\vartheta(v)$ for $\lambda = \lambda_1 + \lambda_2$ and $i = 1, 2$. Then one checks $\vartheta(v+\lambda) = a_{L_0}(\lambda, v)\vartheta(v)$ using the cocycle relation.

Finally, using this and Lemma 1.5.3 and equation (1.20), we have for any $w \in \Lambda(L_0)$ and $\lambda \in \Lambda$,

$$\begin{aligned}
\vartheta_{\overline{w}}(v+\lambda) &= a_{L_0}(w, v+\lambda)^{-1}\vartheta(v+w+\lambda) \\
&= a_{L_0}(w,v)^{-1}\,\mathbf{e}\left(-\pi H(\lambda,w)\right) \cdot a_{L_0}(\lambda, v+w)\,\mathbf{e}\left(\pi H(w,\lambda)\right)\vartheta(v+w) \\
&= \mathbf{e}\left(-2\pi i E(\lambda,w)\right)a_{L_0}(\lambda,v)\vartheta_{\overline{w}}(v) \\
&= a_{L_0}(\lambda,v)\vartheta_{\overline{w}}(v),
\end{aligned}$$

where the last equation follows, since $w \in \Lambda(L_0)$. □

The proof of Lemma 1.5.7 actually shows more: It works even for $\lambda = \lambda_1 + \lambda_2 \in \Lambda_1 \oplus \Lambda(L_0)_2$. Hence, if M_0 denotes the line bundle of Exercise 1.5.5 (6) on $X_2 = V/(\Lambda_1 \oplus \Lambda(L_0)_2)$ such that

$$L_0 = p_2^* M_0,$$

we obtain the following corollary, first for L_0 and then for any translation $L = t_{\overline{c}}^* L_0$. We denote for any $w \in \Lambda(L)$ the translated canonical theta functions by

$$\theta^c_{\overline{w}} = a_L(w,\cdot)^{-1}\vartheta^c(\cdot + w) \tag{1.21}$$

and corresponding line bundle on X_2 by M_0^c.

Corollary 1.5.8 *For any $\overline{w} \in K(L)$ the function $\vartheta^c_{\overline{w}}$ is a canonical theta function for M_0^c.*

Using this we can show the main theorem of this section.

Theorem 1.5.9 *Suppose $L = L(H, \chi)$ is a positive line bundle on X. Then*

$$h^0(L) = \mathrm{Pf}(E).$$

Proof According to the isomorphism (1.18) it suffices to prove the theorem for $L_0 = L(H, \chi_0)$. It is easily checked using equation (1.17) that for any $w \in \Lambda(L_0)$ the function

$$\theta_w(v) := \mathbf{e}\left(-\frac{\pi}{2}B(v,v)\right)\vartheta_w(v)$$

is a classical theta function for L_0 and as such is periodic with respect to Λ_2.

Let $w_1, \dots, w_N \in \Lambda(L_0)_1$ denote a set of representatives of $K(L_0)_1 = \Lambda(L_0)_1/\Lambda_1$. In view of Lemma 1.5.6 it suffices to show that the functions θ_{w_ν}, $\nu = 1, \dots, N$, are linearly independent. We will do this by comparing the coefficients of their Fourier series. For all $v \in h^0(L_0)$ and $1 \le \nu \le N$, we have using equations (1.19) and (1.20)

$$\begin{aligned}
\theta_{w_\nu}(v) &= a_{L_0}(w_\nu, v)^{-1}\,\mathbf{e}\left(-\frac{\pi}{2}B(v,v) + \frac{\pi}{2}B(v+w_\nu, v+w_\nu)\right)\\
&\qquad \cdot \sum_{\lambda\in\Lambda_1}\mathbf{e}\left(\pi(H-B)(v+w_\nu,\lambda) - \frac{\pi}{2}(H-B)(\lambda,\lambda)\right)\\
&= \sum_{\lambda\in\Lambda_1}\mathbf{e}\Big(-\pi(H-B)(v,w_\nu) - \frac{\pi}{2}(H-B)(w_\nu,w_\nu) - \frac{\pi}{2}(H-B)(\lambda,\lambda)\\
&\qquad + \pi(H-B)(w_\nu,\lambda) + \pi(H-B)(v,\lambda)\Big)\\
&= \sum_{\lambda\in\Lambda_1}\mathbf{e}\Big(-\frac{\pi}{2}(H-B)(\lambda-w_\nu,\lambda-w_\nu) + \frac{\pi}{2}(H-B)(w_\nu,\lambda)\\
&\qquad - \frac{\pi}{2}(H-B)(\lambda,w_\nu) + \pi(H-B)(v,\lambda-w_\nu)\Big)\\
&= \sum_{\lambda\in\Lambda_1}\mathbf{e}\left(-\frac{\pi}{2}(H-B)(\lambda-w_\nu,\lambda-w_\nu) + \pi i E(w_\nu,\lambda)\right)\mathbf{e}\left(\pi(H-B)(v,\lambda-w_\nu)\right)\\
&= \sum_{\lambda\in\Lambda_1-w_\nu}\mathbf{e}\left(-\frac{\pi}{2}(H-B)(\lambda,\lambda)\right)\mathbf{e}\left(\pi(H-B)(v,\lambda)\right).
\end{aligned}$$

From this the assertion is obvious: a general Fourier series for the lattice Λ_2 is of the form $\sum_{\lambda\in\Lambda(L_0)_1}\alpha_\lambda\,\mathbf{e}(\pi(H-B)(v,\lambda))$ (see the proof of Lemma 1.5.6). But the sum for θ_{w_ν} runs only over the coset $\Lambda_1 - w_\nu$. Since these cosets are pairwise disjoint in $\Lambda(L_0)_1$, the functions θ_ν, $\nu = 1, \dots N$, are linearly independent. So with Exercise 1.5.5 (5) this gives $h^0(L_0) \ge (\Lambda(L_0)_1, \Lambda_1) = Pf(E)$. □

1.5.4 Computation of $h^0(L)$ for a Semi-positive L

A line bundle $L \in \mathrm{Pic}^H(X)$ on $X = V/\Lambda$ is called *semi-positive* if its first Chern class H is a positive semi-definite hermitian form; that is, if $H(v,v) \geq 0$ for all $v \in V$.

Let $L = L(H,\chi)$ be any line bundle on X. According to Section 1.4.2, L determines a homomorphism $\phi_L : X \to \widehat{X}$ with kernel $K(L) = \Lambda(L)/\Lambda$. Denote by $K(L)^0$ respectively $\Lambda(L)^0$ the connected component of $K(L)$ respectively $\Lambda(L)$ containing 0. Clearly

$$\Lambda(L)^0 = \{v \in V \mid H(v,V) = 0\} \tag{1.22}$$

is the radical of the hermitian form H. The group

$$K(L)^0 = (\mathrm{Ker}\,\phi_L)^0 = \Lambda(L)^0/(\Lambda(L)^0 \cap \Lambda)$$

is a complex subtorus of X. Let $\overline{X} = X/K(L)^0$ be the corresponding quotient complex torus and $p : X \to \overline{X}$ the natural map.

By definition of $K(L)^0$ there is a non-degenerate hermitian form $\overline{H}$ on $\overline{X}$ such that $p^*\overline{H} = H$. The following lemma gives a criterion for a line bundle $L \in \mathrm{Pic}(X)$ to be a pullback of a line bundle on $\overline{X}$.

Lemma 1.5.10 *Let $L \in \mathrm{Pic}(X)$ be any line bundle. There is a line bundle $\overline{L}$ on $\overline{X}$ with $L \simeq p^*\overline{L}$ if and only if $L|_{K(L)^0}$ is trivial.*

If $\overline{L}$ exists, it is non-degenerate with $h^0(L) = h^0(\overline{L})$.

Proof By definition $\overline{X} = \overline{V}/\overline{\Lambda}$ with $\overline{V} = V/\Lambda(L)^0$ and $\overline{\Lambda} = \Lambda/(\Lambda(L)^0 \cap \Lambda)$. Clearly the line bundle $L(H,\chi)$ descends to $\overline{X}$ if and only if H descends to $\overline{V}$ and χ descends to $\overline{\Lambda}$. This is the case if and only if $H|_{\Lambda(L)^0} = 0$ and $\chi|_{\Lambda(L)^0 \cap \Lambda}$ is trivial; that is, if $L|_{K(L)^0}$ is trivial.

Suppose now that $\overline{L}$ exists. By construction $\overline{L}$ is non-degenerate and $h^0(\overline{L}) \leq h^0(L)$. But here even equality holds, since otherwise L would admit a section which is non-trivial on $K(L)^0$. □

Let the alternating form E be of type $(d_1, \dots d_s, 0, \dots, 0)$ with $d_i > 0$ for $i = 1, \dots s$. Then

$$\mathrm{Pfr}(E) := \begin{cases} \prod_{\nu=1}^{s} d_i & \text{if } s > 0, \\ 1 & \text{if } s = 0 \end{cases}$$

is called the *reduced Pfaffian* of E.

Theorem 1.5.11 *Let $L = L(H,\chi) \in \mathrm{Pic}(X)$ be semi-positive and $E = \mathrm{Im}\,H$, then*

$$h^0(L) = \begin{cases} \mathrm{Pfr}(E) & \text{if } L|_{K(L)^0} \text{ is trivial,} \\ 0 & \text{if } L|_{K(L)^0} \text{ is non-trivial.} \end{cases}$$

Proof Suppose first that $L|_{K(L)^0}$ is trivial. By Lemma 1.5.10 L descends to a positive line bundle $\overline{L}$ on $\overline{X}$ with $h^0(L) = h^0(\overline{L})$. Denote by $\overline{E}$ the alternating form of $\overline{L}$. By construction $\mathrm{Pf}(\overline{E}) = \mathrm{Pfr}(E)$. Hence $h^0(L) = \mathrm{Pfr}(E)$ by Theorem 1.5.9.

Let now $L|_{K(L)^0}$ be non-trivial; that is, $\chi|_{\Lambda(L)^0 \cap \Lambda}$ is non-trivial. Suppose ϑ is a canonical theta function for L. Since for any $w \in V$ the canonical factor a_L of L restricted to $(\Lambda(L)^0 \cap \Lambda) \times V$ is $\chi|_{\Lambda(L)^0 \cap \Lambda}$, the function $t_w^*\vartheta$ satisfies

$$t_w^*\vartheta(v + \lambda) = \vartheta(v + w + \lambda) = a_L(\lambda, v + w)\vartheta(v + w) = \chi(\lambda)t_w^*\vartheta(v)$$

for all $\lambda \in \Lambda(L)^0 \cap \Lambda$ and $v \in \Lambda(L)^0$. Hence $t_w^*\vartheta$ is bounded and thus constant on $\Lambda(L)^0$. Since $\lambda_0 \neq 1$ for some $\lambda_0 \in \Lambda(L)^0 \cap \Lambda$, this implies $t_w^*\vartheta \equiv 0$ on $\Lambda(L)^0$. Since ϑ is a canonical theta function for L, this gives $\vartheta = 0$. □

1.5.5 Exercises

(1) Let $X = V/\Lambda$ be a complex torus and H a non-degenerate hermitian form on V with $E = \mathrm{Im}\, H_\Lambda$ integer-valued. Show that:

(i) $\chi_0 : \Lambda \longrightarrow \mathbb{C}_1 \quad \lambda \mapsto \mathbf{e}(\pi i E(\lambda_1, \lambda_2))$ is a semicharacter for H. Here $\lambda = \lambda_1 + \lambda_2$ is given by a decomposition of H;
(ii) $L_0 = L(H, \chi_0)$ is the unique line bundle in $\mathrm{Pic}^H(X)$ whose semicharacter is trivial on $\Lambda_\nu = V_\nu \cap \Lambda$ for $\nu = 1, 2$.

(2) With the assumptions of the previous exercise show that the extended canonical factor of automorphy $a_L : \Lambda \times V \longrightarrow \mathbb{C}^*$ satisfies for all $u = u_1 + u_2, v = v_1 + v_2$ and $w \in V$:

(a) $a_L(u, v + w) = a_L(u, v)\, \mathbf{e}\left(\pi H(w, u)\right)$;
(b) $a_L(u + v, w) = a_L(u, v + w) a_L(v, w)\, \mathbf{e}\left(2\pi i E(u_1, v_2)\right)$;
(c) $a_L(u, v)^{-1} = a_L(-u, v)\chi_0(u)^{-2}\, \mathbf{e}\left(-\pi H(u, u)\right)$;
(d) $a_{L'}(u, v) = a_L(u, v)\, \mathbf{e}\left(2\pi i E(w, u)\right)$ for $L' = t_{\overline{w}}^* L$.

(3) Let L and L_0 be line bundles in $\mathrm{Pic}^H(X)$ of characteristic c and 0 respectively with respect to some decomposition. Show that the corresponding canonical theta functions are related as follows

$$\vartheta^c = \mathbf{e}\big(-\pi H(\cdot, c) - \frac{\pi}{2} H(c, c)\big) t_c^* \vartheta^0.$$

(Hint: Use the previous exercise.)

(4) Let L be a positive line bundle on the abelian variety $X = V/\Lambda$ of characteristic c with respect to a decomposition of $\Lambda(L)$ for L. Deduce from the proof of Theorem 1.5.9 that the functions $\{\vartheta_{\overline{w}}^c \mid \overline{w} \in K(L)_1\}$ form a basis of $H^0(L)$.

(5) Let $L \in \operatorname{Pic}(X)$ be non-degenerate and $\Lambda = \Lambda_1 \oplus \Lambda_2$ a decomposition for L with induced decomposition $V = V_1 \oplus V_2$. Show that:

(a) $\Lambda(L) = \Lambda(L)_1 \oplus \Lambda(L)_2$ with $\Lambda(L)_i = V_i \cap \Lambda(L)$ for $i = 1, 2$;
(b) $K(L) = K_1 \oplus K_2$ with $K_i = \Lambda(L)_i/\Lambda_i$ for $i = 1, 2$;
(c) $K_\nu \simeq \mathbb{Z}^g/D\mathbb{Z}^g = \oplus_{\mu=1}^g \mathbb{Z}/d_\mu\mathbb{Z}$ for $\nu = 1, 2$, if the line bundle is of type $D = \operatorname{diag}(d_1, \dots, d_g)$.

(6) Let the notation be as in the previous exercise. It follows from it, that $\Lambda(L)_1 \oplus \Lambda_2$ and $\Lambda_1 \oplus \Lambda(L)_2$ are lattices in V. According to Lemma 1.5.3 (b) the map a_L restricted to $\Lambda(L)_1 \oplus \Lambda_2 \times V$ respectively $\Lambda_1 \oplus \Lambda(L)_2 \times V$ satisfies the cocycle relation.
This means: denote by

$$X_1 = V/(\Lambda(L)_1 \oplus \Lambda_2) = X/K_1 \quad \text{and} \quad X_2 = V/(\Lambda_1 \oplus \Lambda(L)_2) = X/K_2$$

the corresponding complex tori and $p_\nu : X \to X_\nu$ the induced isogenies. Then a_L determines line bundles M_ν on X_ν such that $L = p_\nu^* M_\nu$ for $\nu = 1, 2$. Varying the characteristic of L within $\Lambda(L)$, one obtains every descent M_ν of L to X_ν. (See Exercise 1.2.3 (9).)

(7) Show that for any $L \in \operatorname{Pic}(X)$ and positive integer n there is an $M \in \operatorname{Pic}(X)$ with $L = M^n$ if and only if L is of type $(n, d_2, \dots, d_g)$.
(Hint: Use Lemma 1.4.14.)

(8) Let $L = L(H, \chi)$ be a non-degenerate line bundle on a complex torus $X = V/\Lambda$. Give an example of a decomposition of V into maximal isotropic subvector spaces with respect to $\operatorname{Im} H$, which is not a decomposition for L.

(9) Let $L = L(H, \chi)$ be a positive line bundle on a complex torus $X = V/\Lambda$, of characteristic c with respect to a decomposition $\Lambda = \Lambda_1 \oplus \Lambda_2$.

(a) Show that for any $u,\ v = v_1 + v_2,\ w = w_1 + w_2 \in V$:

$$a_L(v, u)^{-1} a_L(w, u + v - w) = \mathbf{e}(2\pi i \operatorname{Im} H(w_1, w_2 - v_2)) a_L(v - w, u)^{-1}.$$

(b) Use (a) to generalize equation (1.21) to show that for any $v, w \in \Lambda(L)_1$:

$$\vartheta_{\overline{v}}^c = a_L(v - w, \cdot)^{-1} \vartheta_{\overline{w}}^c(\cdot + v - w).$$

(10) Let $L = L(H, \chi)$ be a positive line bundle on a complex torus $X = V/\Lambda$, of characteristic c with respect to a decomposition $\Lambda = \Lambda_1 \oplus \Lambda_2$. Show that $\vartheta^c : V \to \mathbb{C}$, defined by

$$\vartheta^c(v) = \mathbf{e}\big(-\pi H(v, c) - \frac{\pi}{2} H(c, c) + \frac{\pi}{2} B(v + c, v + c)\big) \cdot \\ \cdot \sum_{\lambda \in \Lambda_1} \mathbf{e}\left(\pi (H - B)(v + c, \lambda) - \frac{\pi}{2}(H - B)(\lambda, \lambda)\right),$$

is the canonical theta function for L with respect to the decomposition.
(Hint: Use canonical factors and Corollary 1.5.2.)

(11) Suppose $L = L(H,\chi)$ and $L' = L(H,\chi')$ are positive line bundles with characteristics c and c' with respect to a decomposition. Let $\tau : V \to \mathbb{C}^*$ be the holomorphic function

$$\tau(v) = \mathbf{e}\left(\pi i \operatorname{Im} H(c',c) - \pi H(v,c'-c) - \frac{\pi}{2} H(c'-c,c'-c)\right).$$

Show that $\vartheta \mapsto \tau \cdot t^*_{c'-c}\vartheta$ defines an isomorphism of vector spaces $H^0(L) \to H^0(L')$.
(Hint: Using the definition of $\vartheta^c_{\overline{w}}$ and Lemma 1.5.3 check first that $\vartheta^c_{\overline{w}}$ and $\vartheta^0_{\overline{w}}$ are related as follows: $\vartheta^c_{\overline{w}}(v) = \mathbf{e}\left(-\pi H(v,c) - \frac{\pi}{2} H(c,c)\right) \vartheta^0_{\overline{w}}(v+c)$ for all $v \in V$.)

(12) Use the notation of Section 1.5.3 to show the following generalization of Lemma 1.5.7: for any $\overline{w} \in K(L)_1$ the function $\vartheta_{\overline{w}}$ is a canonical theta function for $t^*_{\overline{w}} M_2$, where M_2 is a descent of L to $X_2 = X/K(L)_2$.
(Use Exercise (5) above.)

1.6 Cohomology of Line Bundles

Let again L be an arbitrary line bundle on the complex torus $X = V/\Lambda$. The aim of this section is to compute $h^q(L)$ for all q. For this we prove the vanishing theorem of Mumford–Kempf (see Kempf [75]). Its proof uses the theory of harmonic forms with values in L.

1.6.1 Harmonic Forms with Values in L

In this section we define the vector space of harmonic forms $\mathcal{H}^q(L)$ with values in the line bundle L and explain (without complete proof) its relation with the cohomology group $H^q(L)$. Moreover we prove some properties of these forms needed in the next section.

The *Dolbeault resolution* of $\mathcal{O}_X$ is the resolution

$$0 \to \mathcal{O}_X \to \mathcal{A}^{0,0}_X \xrightarrow{\bar\partial} \mathcal{A}^{0,1}_X \xrightarrow{\bar\partial} \mathcal{A}^{0,2}_X \xrightarrow{\bar\partial} \cdots$$

where $\mathcal{A}^{0,q}_X$ denotes the sheaf of complex-valued differential C^∞-forms of type $(0,q)$

and $\overline{\partial}$ the operator defined by differentiation with respect to $\overline{z}$ (see Griffiths–Harris [55, p. 45]). Tensoring with L we obtain the following resolution of L,

$$0 \to L \to \mathcal{A}_X^{0,0}(L) \xrightarrow{\overline{\partial}} \mathcal{A}_X^{0,1}(L) \xrightarrow{\overline{\partial}} \mathcal{A}_X^{0,2}(L) \xrightarrow{\overline{\partial}} \cdots$$

where $\mathcal{A}_X^{0,q}(L) = \mathcal{A}_X^{0,q} \otimes_{\mathcal{O}_X} L$ and for simplicity we denote by $\overline{\partial}$ also the induced operator $\mathcal{A}_X^{0,q}(L) \xrightarrow{\overline{\partial}} \mathcal{A}_X^{0,q+1}(L)$. Taking global sections we get the following complex

$$0 \to H^0(L) \to A_X^{0,0}(L) \xrightarrow{\overline{\partial}} A_X^{0,1}(L) \xrightarrow{\overline{\partial}} A_X^{0,2}(L) \xrightarrow{\overline{\partial}} \cdots \tag{1.23}$$

of L where $A_X^{0,q}(L) = H^0(\mathcal{A}_X^{0,q}(L))$. Its cohomology groups $H^{0,q}(L)$ are called the *Dolbeault cohomology groups of L*. According to Griffiths–Harris [55, p. 150] they are isomorphic to the usual cohomology group $H^q(L)$.

In order to describe its elements, we introduce a *hermitian metric on L*. This is a hermitian inner product on the fibres $L(x)$ depending differentiably on $x \in X$. We will see that such a metric together with a suitable Kähler metric on X induces a global inner product (,) on the vector space $A_X^{0,q}(L)$ in a natural way. Let $\overline{\delta} : A_X^{0,q+1}(L) \to A_X^{0,q}(L)$ denote the adjoint operator of $\overline{\partial}$ with respect to (,) and

$$\Delta = \overline{\partial}\overline{\delta} + \overline{\delta}\overline{\partial} : A_X^{0,q}(L) \to A_X^{0,q}(L)$$

the corresponding Laplacian. The elements of

$$\mathcal{H}^q(L) := \operatorname{Ker} \Delta$$

are called *harmonic q-forms with values in L* In Griffiths–Harris [55, p. 152] it is shown that the metrics induce an isomorphism $H^{0,q}(L) \simeq \mathcal{H}^q(L)$. Combining this with the above isomorphism, we get the following theorem.

Theorem 1.6.1 *A suitable hermitian metric on L and a compatible Kähler metric on X induce an isomorphism*

$$H^q(L) \xrightarrow{\simeq} \mathcal{H}^q(L)$$

for all q.

In order to make the description of harmonic forms more precise, we start by defining a hermitian form on L. Suppose $L = L(H, \chi)$. Consider the elements of $A^{0,0}(L)$ as C^∞-functions $f : V \to \mathbb{C}$ satisfying

$$f(v + \lambda) = a_L(v, \lambda) f(v) \qquad \text{for} \quad (\lambda, v) \in \Lambda \times V.$$

They are called *differentiable theta functions for L*. Define for $f, g \in A^{0,0}(L)$

$$\langle f, g \rangle(v) := f(v)\overline{g(v)}\,\mathbf{e}\left(-\pi H(v, v)\right). \tag{1.24}$$

Obviously $\langle f, g\rangle$ is a C^∞-function on V, periodic with respect to the lattice Λ. Hence

$$\langle\ ,\ \rangle : A^{0,0}(L) \times A^{0,0}(L) \to A^{0,0}(\mathcal{O}_X)$$

defines a hermitian metric on L.

Next we define a Kähler metric on X: Fix a basis $e_1, \ldots, e_g$ of V with respect to which the matrix of the hermitian form $H = c_1(L)$ is diagonal and let $v_1, \ldots, v_g$ denote the corresponding coordinate functions on V, so there are $h_\nu \in \mathbb{R}$ such that

$$H(v, v) = \sum_{\nu=1}^{g} h_\nu v_\nu \bar{v}_\nu.$$

For any positive real numbers $k_1, \ldots, k_g$ we define a translation-invariant hermitian metric on V by

$$\mathrm{d}s^2 = \sum_{\nu=1}^{g} k_\nu \mathrm{d}v_\nu \otimes \mathrm{d}\bar{v}_\nu.$$

It defines a translation-invariant hermitian metric on X denoted by the same symbol. Its associate $(1, 1)$-form is $\omega = \frac{i}{2} \sum_{\nu=1}^{g} \mathrm{d}v_\nu \wedge \mathrm{d}\bar{v}_\nu$ which is d-closed, meaning that $\mathrm{d}s^2$ is a *Kähler metric on* X. In the next section we will choose the coefficients k_ν suitably, in order to ensure that metric $\mathrm{d}s^2$ is compatible with the hermitian form H. The volume element corresponding to $\mathrm{d}s^2$ is

$$\mathrm{d}v := \left(\frac{i}{2}\right)^g \prod_{\nu=1}^{g} k_\nu \mathrm{d}v_1 \wedge \mathrm{d}\bar{v}_1 \wedge \cdots \wedge \mathrm{d}v_g \wedge \mathrm{d}\bar{v}_g$$

and the global inner product $(\ ,\ ) : A^{0,0}(L) \times A^{0,0}(L) \longrightarrow \mathbb{C}$ associated to the metric $\langle\ ,\ \rangle$ is given by

$$(f, g) = \int_X \langle f, g\rangle \mathrm{d}v. \tag{1.25}$$

In analogy to the elements of $A^{0,0}(L)$ any $\omega \in A^{0,q}(L)$ may be considered as a C^∞-form of type $(0, q)$ on the vector space V satisfying

$$t_\lambda^* \omega = a_L(\lambda, \cdot)\omega \qquad \text{for all} \quad \lambda \in \Lambda.$$

Clearly ω can be uniquely written in the form

$$\omega = \sum_I \varphi_I \mathrm{d}\bar{v}_I,$$

where the sum is taken over all multi-indices $I = (i_1 < \cdots < i_q)$ and each $\varphi_I \in A^{0,0}(L)$.

Let $\sum_I \psi_I d\overline{v}_I$ be another $(0,q)$-form in $A^{0,q}(L)$. Define a global inner product on $A^{0,q}(L)$ by

$$\left(\sum_I \varphi_I d\overline{v}_I, \sum_I \psi_I d\overline{v}_I\right) = \sum_I k^I(\varphi_I, \psi_I), \qquad \text{where} \quad k^I = \prod_{\nu \in I} k_\nu^{-1}.$$

We use the abbreviations $\partial_\nu = \frac{\partial}{\partial v_\nu}$ and $\overline{\partial}_\nu = \frac{\partial}{\partial \overline{v}_\nu}$. Since $a_L(\lambda, v)$ is holomorphic in v, the partial derivative $\overline{\partial}_\nu$ is a linear operator of $A^{0,0}(L)$ into itself. This shows that the differential operator $\overline{\partial} : A^{0,q}(L) \longrightarrow A^{0,q+1}(L)$ in the complex (1.23) is given by

$$\overline{\partial}(\varphi d\overline{v}_I) = \sum_{\nu=1}^{g} \overline{\partial}_\nu \varphi \, d\overline{v}_\nu \wedge d\overline{v}_I.$$

Let

$$\overline{\delta}_\nu : A^{0,0}(L) \to A^{0,0}(L) \qquad \text{and} \qquad \overline{\delta} : A^{0,q+1}(L) \to A^{0,q}(L)$$

denote the adjoint operators of $\overline{\partial}_\nu$ and $\overline{\partial}$ with respect to the chosen inner products.

Lemma 1.6.2 *Let $\varphi \in A^{0,0}(L)$. Then*

(a) $\overline{\delta}_\nu \varphi = -\partial_\nu \varphi + \pi h_\nu \overline{v}_\nu \varphi$;
(b) $\overline{\delta}(\varphi d\overline{v}_I) = \sum_{\nu=1}^{q+1} (-1)^{\nu-1} k_{j_\nu}^{-1} \overline{\delta}_{j_\nu} \varphi \, d\overline{v}_{J-j_\nu}$ *for* $J = (j_1 < \cdots < j_{q+1})$.

Proof Denote by δ'_ν the right-hand side of (a). We have to show $(\overline{\partial}_\nu \varphi, \psi) = (\varphi, \delta'_\nu \psi)$ for $\varphi, \psi \in A^{0,0}(L)$. But one checks that $\langle \overline{\partial}_\nu \varphi, \psi \rangle - \langle \varphi, \delta'_\nu \psi \rangle = \overline{\partial}_\nu \langle \varphi, \psi \rangle$, which gives

$$\int_X \overline{\partial}_\nu \langle \varphi, \psi \rangle dv = -\left(\frac{i}{2}\right)^g \prod_{\nu=1}^{g} k_\nu \int_X d\left(\langle \varphi, \psi \rangle dv_1 \wedge d\overline{v}_1 \wedge \cdots \wedge \overset{\vee}{d\overline{v}_\nu} \wedge \cdots, \wedge d\overline{v}_g\right),$$

which is equal to 0 by Stokes' theorem, since X is compact. This proves (a). For (b) it suffices to check that

$$\left(\overline{\partial}(\psi d\overline{v}_{J-j_\mu}), \varphi d\overline{v}_J\right) = \left(\psi d\overline{v}_{J-j_\mu}, \sum_{\nu=1}^{q+1} (-1)^{\nu-1} k_{j_\nu}^{-1} \overline{\delta}_{j_\nu} \varphi d\overline{v}_{J-j_\nu}\right),$$

and this is a consequence of (a). □

Using Lemma 1.6.2 we are in a position to compute the Laplacian $\Delta = \overline{\delta}\overline{\partial} + \overline{\partial}\overline{\delta} : A^{0,q}(L) \to A^{0,q}(L)$ and hence the harmonic q-forms with values in L.

Proposition 1.6.3 *For all $\varphi d\overline{v}_I \in A^{0,q}(L)$ and $I = (1 \leq i_1 < \cdots < i_q \leq g)$*

$$\Delta(\varphi d\overline{v}_I) = \sum_{\nu=1}^{g} k_\nu^{-1} \overline{\delta}_\nu \overline{\partial}_\nu \varphi d\overline{v}_I + \pi \sum_{\nu=1}^{q} k_{i_\nu}^{-1} h_{i_\nu} \varphi d\overline{v}_I$$

Proof First we compute $\overline{\delta}(\varphi d\overline{v}_\mu \wedge d\overline{v}_I)$. If $\mu \in I$, this is zero. So suppose $i_{\lambda-1} < \mu < i_\lambda$ and let $J = (j_1 < \cdots < j_{q+1}) = (i_1 < \cdots < i_{\lambda-1} < \mu < i_\lambda < \cdots < i_q)$. Applying Lemma 1.6.2 gives

$$\begin{aligned}\overline{\delta}(\varphi d\overline{v}_\mu \wedge d\overline{v}_I) = \overline{\delta}\big((-1)^{\lambda-1}\varphi d\overline{v}_J\big) &= \sum_{\nu=1}^{q+1}(-1)^{\lambda+\nu}k_{j_\nu}^{-1}\overline{\delta}_{j_\nu}\varphi d\overline{v}_{J-j_\nu} \\ &= \sum_{\nu=1}^{\lambda-1}(-1)^{\lambda+\nu}k_{i_\nu}^{-1}\overline{\delta}_{i_\nu}\varphi d\overline{v}_{J-i_\nu} + k_\mu^{-1}\overline{\delta}_\mu\varphi d\overline{v}_{J-\mu} \\ &\quad + \sum_{\nu=\lambda}^{q}(-1)^{\lambda+\nu-1}k_{i_\nu}^{-1}\overline{\delta}_{i_\nu}\varphi d\overline{v}_{J-i_\nu} \\ &= k_\mu^{-1}\overline{\delta}_\mu\varphi d\overline{v}_I + \sum_{\nu=1}^{q}(-1)^{\nu}k_{i_\nu}^{-1}\overline{\delta}_{i_\nu}\varphi d\overline{v}_\mu \wedge \overline{v}_{I-i_\nu}.\end{aligned}$$

An immediate computation using Lemma 1.6.2 (a) shows that $\overline{\delta}_{i_\nu}\overline{\partial}_\mu - \overline{\partial}_\mu\overline{\delta}_{i_\nu} = 0$ for $i_\nu \neq \mu$ and $\overline{\partial}_{i_\nu}\overline{\delta}_{i_\nu}\varphi = \overline{\delta}_{i_\nu}\overline{\partial}_{i_\nu}\varphi + \pi h_{i_\nu}\varphi$. So we obtain

$$\begin{aligned}\Delta(\varphi d\overline{v}_I) &= \overline{\delta}\Big(\sum_{\substack{\mu=1\\ \mu\notin I}}^{g}\overline{\partial}_\mu\varphi d\overline{v}_\mu \wedge d\overline{v}_I\Big) + \overline{\partial}\Big(\sum_{\nu=1}^{q}(-1)^{\nu-1}k_{i_\nu}^{-1}\overline{\delta}_{i_\nu}\varphi d\overline{v}_{I-i_\nu}\Big) \\ &= \sum_{\substack{\mu=1\\ \mu\notin I}}^{g}k_\mu^{-1}\overline{\delta}_\mu\overline{\partial}_\mu\varphi d\overline{v}_I + \sum_{\nu=1}^{q}\sum_{\substack{\mu=1\\ \mu\notin I}}^{g}(-1)^{\nu}k_{i_\nu}^{-1}\overline{\delta}_{i_\nu}\overline{\partial}_\mu\varphi d\overline{v}_\mu \wedge d\overline{v}_{I-i_\nu} \\ &\quad + \sum_{\nu=1}^{q}\sum_{\mu=1}^{g}(-1)^{\nu-1}k_{i_\nu}^{-1}\overline{\partial}_\mu\overline{\delta}_{i_\nu}\varphi d\overline{v}_\mu \wedge d\overline{v}_{I-i_\nu} \\ &= \sum_{\substack{\mu=1\\ \mu\notin I}}^{g}k_\mu^{-1}\overline{\delta}_\mu\overline{\partial}_\mu\varphi d\overline{v}_I + \sum_{\nu=1}^{q}k_{i_\nu}^{-1}\big(\overline{\delta}_{i_\nu}\overline{\partial}_{i_\nu}\varphi + \pi h_{i_\nu}\varphi\big)d\overline{v}_I.\end{aligned}$$

This implies the assertion. □

1.6.2 The Vanishing Theorem

Our approach to the following theorem, called the *vanishing theorem*, was first given independently by Deligne [36] and with a slightly different method by Umemura [134].

Theorem 1.6.4 (Vanishing Theorem) *Let X be a complex torus of dimension g and $L = L(H, \chi) \in \mathrm{Pic}(X)$ such that the hermitian form H has r positive and s negative eigenvalues. Then*

$$H^q(L) = 0 \quad for \quad q > g - r \quad and \quad q < s.$$

The *index* of a non-degenerate line bundle is by definition the number of negative eigenvalues of its first Chern class considered as a hermitian form. Using this definition we get as a direct consequence:

Corollary 1.6.5 (Mumford's Index Theorem) *Let L be a non-degenerate line bundle of index s. Then $H^i(L) = 0$ for all $i \neq q$.*

For the proof of the Vanishing Theorem we use the notations and results of the last section, but need some more. Recall that the basis of V was given in such a way that H is in diagonal form. Denoting by r and s the number of positive and negative eigenvalues of H respectively, we may in addition assume that $h_\nu = 1$ for $\nu \leq r$ and $h_\nu = -1$ for $r + 1 \leq \nu \leq r + s$. So

$$H(v, w) = \sum_{\nu=1}^{r} v_\nu \overline{w}_\nu - \sum_{\nu=r+1}^{r+s} v_\nu \overline{w}_\nu.$$

Recall that the numbers k_ν defining ds^2 were arbitrary positive. Now we choose them as follows:

$$k_\nu = \begin{cases} \frac{1}{s+1} & \text{if } \nu \leq r, \\ 1 & \text{if } \nu > r. \end{cases}$$

Moreover, define for any multi-index I,

$$R_I = \#I \cap \{1, \ldots, r\} \quad \text{and} \quad S_I = \#I \cap \{r + 1, \ldots, r + s\}.$$

With these notations we have the following for the Laplacian Δ.

Lemma 1.6.6 *For every $\varphi d\overline{v}_I \in A^{0,q}(L)$*

$$\big(\Delta(\varphi d\overline{v}_I), \varphi d\overline{v}_I\big) \geq \pi\big((s + 1)R_I - S_I\big)(\varphi d\overline{v}_I, \varphi d\overline{v}_I).$$

Proof By Proposition 1.6.3 we have

$$\big(\Delta(\varphi d\overline{v}_I), \varphi d\overline{v}_I\big) = \sum_{\nu=1}^{g} k_\nu^{-1} \big(\overline{\delta}_\nu \overline{\partial}_\nu \varphi d\overline{v}_I, \varphi d\overline{v}_I\big) + \pi\big((s + 1)R_I - S_I\big)(\varphi d\overline{v}_I, \varphi d\overline{v}_I),$$

which implies the assertion, since $(\overline{\delta}_\nu \overline{\partial}_\nu \varphi d\overline{v}_I, \varphi d\overline{v}_I) = (\overline{\partial}_\nu \varphi d\overline{v}_I, \overline{\partial}_\nu \varphi d\overline{v}_I) \geq 0$ for all ν. □

Proposition 1.6.3 shows in particular that Δ acts on the subvector spaces of $A^{0,q}(L)$ of monomial forms $\varphi d\bar{v}_I$ for every fixed I. So if $\mathcal{H}^q_I(L)$ denotes the corresponding subvector space of $\mathcal{H}^q(L)$, we get a direct sum decomposition

$$\mathcal{H}^q(L) = \bigoplus_{\#I=q} \mathcal{H}^q_I(L). \tag{1.26}$$

Corollary 1.6.7 $\mathcal{H}^q_I(L) = 0$ *if* $R_I > 0$.

Proof Suppose $\varphi d\bar{v}_I \in \mathcal{H}^q_I(L)$. Applying Lemma 1.6.6 we get

$$0 = (0, \varphi d\bar{v}_I) = (\Delta(\varphi d\bar{v}_I), \varphi d\bar{v}_I) \geq \pi\big((s+1)R_I - S_I\big)(\varphi d\bar{v}_I, \varphi d\bar{v}_I).$$

Now $\big((s+1)R_I - S_I\big) \geq 1$ and $(\varphi d\bar{v}_I, \varphi d\bar{v}_I) \geq 0$ imply $\varphi d\bar{v}_I = 0$. □

Proof (of the Vanishing Theorem) By Theorem 1.6.1 it suffices to show the assertion for $\mathcal{H}^q(L)$. Suppose first $q > g - r$. Then every multi-index I of length q intersects $\{1, \dots, r\}$, that is $R_I > 0$. So Corollary 1.6.7 implies $\mathcal{H}^q(L) = 0$ for all $q > g - r$.

For $q < s$ we apply Kodaira–Serre duality (see Griffiths–Harris [55, p. 153]). It gives an isomorphism $H^q(L) \simeq H^{g-q}(K_X \otimes L^{-1})$, where K_X denotes the *canonical line bundle of* X, which according to Griffiths–Harris [55, p. 146] is defined as the g-fold wedge product of the dual of the tangent bundle of X. But the tangent bundle of X is trivial, which is seen using the translations t_x. Hence we get

$$K_X = \mathcal{O}_X. \tag{1.27}$$

Hence the duality gives

$$H^q(L) \simeq H^{g-q}(L^{-1}).$$

But since $c_1(L^{-1}) = -c_1(L)$, the hermitian form of L^{-1} has s positive eigenvalues. So the first part of the proof gives $H^{g-q}(L^{-1}) = 0$ for $g - q > g - s$, which implies $H^q(L) = 0$ for $q < s$. □

1.6.3 Computation of the Cohomology

Let $L = L(H, \chi) \in \operatorname{Pic}(X)$ such that H admits exactly r positive and s negative eigenvalues and let $E = \operatorname{Im} H$. Under these hypotheses we prove the following theorem.

Theorem 1.6.8

$$h^q(L) = \begin{cases} \binom{g-r-s}{q-s} \operatorname{Pfr}(E) & \textit{if } s \leq q \leq g - r \textit{ and } L|_{K(L)^0} \textit{ is trivial;} \\ 0 & \textit{otherwise.} \end{cases}$$

We prove this in several steps. We use the notations of the beginning of Section 1.6.1. By the Vanishing Theorem it remains to compute $h^q(L)$ for $s \leq q \leq g-r$. The following lemma shows that it suffices to consider $H^s(L)$.

Lemma 1.6.9 $h^q(L) = \binom{g-r-s}{q-s} h^s(L)$ *for* $s \leq q \leq g-r$.

Proof First we claim that $\mathcal{H}^q(L) = \bigoplus \mathcal{H}_I^q(L)$, where the direct sum runs over all multi-indices $I = (i_1 < \cdots < i_q)$ with $R_I = 0$ and $S_I = s$.

According to equation (1.26) and Corollary 1.6.7 it remains to verify that $\mathcal{H}_I^q(L) = 0$ for $R_I = 0$ and $S_I < s$.

Suppose first $R_I = 0$ and $S_I \leq s$. By Proposition 1.6.3 and our choice of the h_i and k_ν we have

$$\Delta(\varphi d\overline{v}_I) = \psi d\overline{v}_I \qquad \text{with} \quad \psi = \sum_{\nu=1}^{g} k_\nu^{-1} \overline{\delta}_\nu \overline{\partial}_\nu \varphi - \pi S_I \varphi.$$

If $J = I \cap \{r+1, \ldots, r+s\}$, then $R_J = R_I = 0$ and $S_J = S_I$, such that

$$\Delta(\varphi d\overline{v}_J) = \psi d\overline{v}_J.$$

Hence the map

$$\mathcal{H}_I^q(L) \to \mathcal{H}_J^{S_I}(L), \qquad \varphi d\overline{v}_I \mapsto \varphi d\overline{v}_J \tag{1.28}$$

is an isomorphism. Moreover, if we assume $S_I < s$, the Vanishing Theorem 1.6.4 implies that $\mathcal{H}_J^{S_I}(L) = 0$. This proves the above assertion.

Using this assertion twice and the isomorphism (1.28) we have with $J = (r+1 < \cdots < r+s)$,

$$\begin{aligned} H^q(L) \simeq \mathcal{H}^q(L) = \bigoplus_{\substack{\#I=q \\ R_I=0, S_I=s}} \mathcal{H}_I^q(L) &\simeq \bigoplus_{\substack{\#I=q \\ R_I=0, S_I=s}} \mathcal{H}_J^s(L) \\ &\simeq \mathcal{H}_J^s(L) \otimes \mathbb{C}^{\binom{g-r-s}{q-s}} \simeq H^s(L) \otimes \mathbb{C}^{\binom{g-r-s}{q-s}}, \end{aligned}$$

which was to be shown. □

It remains to compute $h^s(L)$. The idea for this is to associate to L on X a positive semi-definite line bundle M on another complex torus Y, for which there exists an isomorphism

$$H^s(X, L) \simeq H^0(Y, M),$$

and then apply Theorem 1.5.11.

Recall the radical $\Lambda(L)^0$ of H defined in (1.22). By the choice of the basis $e_1, \ldots, e_g$, the vector space V decomposes as

$$V = V_+ \oplus V_- \oplus \Lambda(L)^0$$

with $V_+ = \langle e_1, \ldots, e_r \rangle$, $V_- = \langle e_{r+1}, \ldots, e_{r+s} \rangle$ and $\Lambda(L)^0 = \langle e_{r+s+1}, \ldots, e_g \rangle$ and where H has positive respectively negative eigenvalues on V_+ respectively V_-.

Let W denote the underlying real vector space of V and j the complex structure on W defining V. So j is the $\mathbb{R}$-linear automorphism of W given by multiplication by $i = \sqrt{-1}$ in V. Clearly the decomposition of V induces a direct sum decomposition

$$W = W_+ \oplus W_- \oplus W^0 \tag{1.29}$$

over $\mathbb{R}$. Now define a new complex structure $k : W \to W$ by

$$k(w) = \begin{cases} j(w) & \text{if } w \in W_+ \oplus W^0, \\ -j(w) & \text{if } w \in W_-. \end{cases}$$

Let $U = (W, k)$ denote the complex vector space defined by k. Then $\{e_1, \dots, e_g\}$ is also a basis of U and the corresponding complex coordinate functions $u_1, \dots, u_g$ satisfy

$$u_\nu = \begin{cases} v_\nu & \text{if } \nu \in \{1, \dots, r, r+s+1, \dots, g\}; \\ \overline{v}_\nu & \text{if } \nu \in \{r+1, \dots, r+s\}. \end{cases} \tag{1.30}$$

The lattice Λ in V does not depend on the complex structure, so Λ is also a lattice in U and

$$Y = U/\Lambda$$

is a complex torus of dimension g. Recall that $L = L(H, \chi)$. According to Lemma 1.2.10

$$\widetilde{H}(v, w) = \operatorname{Im} H(k(v).w) + i \operatorname{Im} H(v, w)$$

is a hermitian form on U. By construction it is positive semidefinite with $\operatorname{Im} \widetilde{H}(\Lambda, \Lambda) \subseteq \mathbb{Z}$. Moreover χ is a semicharacter for $\widetilde{H}$, since $\operatorname{Im} \widetilde{H} = \operatorname{Im} H$. Let

$$M = L(\widetilde{H}, \chi)$$

denote the corresponding line bundle on Y. Note that M is a semi-positive line bundle.

Proposition 1.6.10 *There is an isomorphism of $\mathbb{C}$-vector spaces*

$$H^s(L) \to H^0(M).$$

Proof **Step 1**: As we saw in the proof of Lemma 1.6.9, we have $H^s(L) = \mathcal{H}^s_J(L)$ with $J = (r+1 < \cdots < r+s)$. Define

$$A^{0,s}_J(L) = \{\varphi d\overline{v} \mid \varphi \in A^{0,0}(L)\}.$$

Consider the map

$$f : W \to \mathbb{C}^*, \qquad f(w) = \mathbf{e}\left(\pi \widetilde{H}(w_-, w_-)\right),$$

where $w = w_+ + w_- + w_0$ is the decomposition given in (1.29).

We claim: the map

$$A_J^{0,s}(L) \to A^{0,0}(M), \qquad \varphi d\overline{v}_J \mapsto \varphi f \tag{1.31}$$

is an isomorphism of $\mathbb{C}$-vector spaces.

Note that $\varphi d\overline{v}$ is a form on V, whereas φf is considered as a $\mathbb{C}^\infty$-function on U. However, the claim makes sense, since the underlying real vector space is W in both cases.

For the proof of the assertion we have to show that $\varphi(w+\lambda) = a_L(\lambda, w)\varphi(w)$ if and only if $(\varphi f)(w+\lambda) = a_M(\lambda, w)(\varphi f)(w)$ for all $w \in W, \lambda \in \Lambda$. Using the definition of $\widetilde{H}$ and the fact that the decomposition (1.29) is orthogonal for $\widetilde{H}$ and H, we have

$$\begin{aligned}
&a_M(\lambda, w) f(w+\lambda)^{-1} f(w) \\
&= \chi(\lambda)\, \mathbf{e}\left(\pi\widetilde{H}(w,\lambda) + \frac{\pi}{2}\widetilde{H}(\lambda,\lambda) - \pi\widetilde{H}(w_- + \lambda_-, w_- + \lambda_-) + \pi\widetilde{H}(w_-.w_-)\right) \\
&= \chi(\lambda)\, \mathbf{e}\left(\pi\widetilde{H}(w_+,\lambda_+) - \pi\widetilde{H}(\lambda_-, w_-) + \frac{\pi}{2}\widetilde{H}(\lambda_+,\lambda_+) - \frac{\pi}{2}\widetilde{H}(\lambda_-,\lambda_-)\right) \\
&= \chi(\lambda)\, \mathbf{e}\left(\pi H(w_+,\lambda_+) + \pi H(\lambda_-, w_-) + \frac{\pi}{2} H(\lambda_+,\lambda_+) + \frac{\pi}{2} H(\lambda_-,\lambda_-)\right) \\
&= a_L(\lambda, w).
\end{aligned}$$

This implies that the map (1.31) is an isomorphism.

Step 2: It remains to show that the isomorphism (1.31) restricts to an isomorphism $\mathcal{H}_J^s(L) \to \mathcal{H}^0(M)$. In other words, we have to show that $\varphi d\overline{v}_J$ is a harmonic form with values in L if and only if φf is a harmonic differentiable function with values in M.

The function $\varphi d\overline{v}$ is harmonic if and only if $\frac{\partial}{\partial \overline{u}_\nu}(\varphi d\overline{v}) = 0$ for $\nu = 1, \ldots, g$. Since $f(u) = \mathbf{e}\left(\pi \sum_{\nu=r+1}^{r+s} u_\nu \overline{u}_\nu\right)$, we obtain applying (1.30),

$$\frac{\partial}{\partial \overline{u}_\nu}(\varphi f) = \frac{\partial \varphi}{\partial \overline{u}_\nu} f + \frac{\partial f}{\partial \overline{u}_\nu}\varphi = \begin{cases} \left(\frac{\partial \varphi}{\partial \overline{v}_\nu} + \pi \overline{v}_\nu \varphi\right) f & \text{if } \nu \in J = \{r+1, \ldots, r+s\}, \\ \frac{\partial \varphi}{\partial \overline{v}_\nu} f & \text{if } \nu \notin J = \{r+1, \ldots, r+s\}. \end{cases}$$

Hence it suffices to show that $\varphi d\overline{v}_J$ is harmonic if and only if $\frac{\partial \varphi}{\partial \overline{v}_\nu} + \pi \overline{v}_\nu \varphi = 0$ for $\nu \in J$ and $\frac{\partial \varphi}{\partial \overline{v}_\nu} = 0$ otherwise. But this is a consequence of Lemma 1.6.2, since a form ω is harmonic if and only if $\overline{\delta}\omega = \overline{\partial}\omega = 0$. □

Proposition 1.6.11 $h^s(L) = \begin{cases} \mathrm{Pf}r(L) & \textit{if } L|_{K(L)^0} \textit{ is trivial,} \\ 0 & \textit{if } L|_{K(L)_0} \textit{ is non-trivial.} \end{cases}$

Proof This is a consequence of Proposition 1.6.10 and Theorem 1.5.11. One only has to check that $L|_{K(L)^0} = 0$ is trivial if and only if $M|_{K(M)^0} = 0$ is trivial. But this is obvious, since the semi-character is the same in both cases. □

Finally, combining Lemma 1.6.9 and Proposition 1.6.11 gives the proof of Theorem 1.6.8.

1.6.4 Exercises

(1) For any complex torus X compute the cohomology groups of the trivial line bundle $\mathcal{O}_X$.

(2) Let $L = L(H, \chi) \in \operatorname{Pic}(X)$, $K(L)^0$ be the connected component of $K(L)$ containing 0 and s be the number of negative eigenvalues of H. Show that for all $q \geq s$ there is an isomorphism $H^q(L) \simeq H^s(L) \otimes H^{q-s}(\mathcal{O}_{K(L)^0})$.

(3) Let $L = L(H, \chi)$ be a non-degenerate line bundle of index s on $X = V/\Lambda$. Let V_+ (respectively V_-) denote the sum of the eigenspaces of H with positive (respectively negative) eigenvalues. Show that $H^s(L)$ can be identified with the space of C^∞-theta functions with respect to the canonical factor of L, holomorphic on V_+ and anti-holomorphic on V_-.
(Hint: Use the methods of Section 1.6.3.)

(4) *(Cohomology of the Poincaré bundle)* Let X be a complex torus of dimension g. Show that for the Poincaré bundle $\mathcal{P}$ on $X \times \widehat{X}$:

$$h^q(\mathcal{P}) = \begin{cases} \mathbb{C} \text{ if } q = g, \\ 0 \text{ if } q \neq g. \end{cases}$$

(Hint: Show first that $\mathcal{P}$ is non-degenerate of index g.)

(5) Show that, identifying $(X \times \widehat{X})\widehat{\ } = \widehat{X} \times X$, we have for the dual of the Poincaré bundle,

$$\widehat{\mathcal{P}}_X = \mathcal{P}_{\widehat{X}}^{-1}.$$

(6) *(Poincaré's Reducibility Theorem for Complex Tori)*

(a) Let X be a complex torus admitting a non-degenerate line bundle L. For any complex subtorus Y of X such that $L|_Y$ is non-degenerate, there exists a complex subtorus Z of X such that $Y \cap Z$ is finite and $Y + Z = X$. In other words, the addition map $\mu : Y \times Z \to X$ is an isogeny.

(b) Consider the complex torus $X = \mathbb{C}^2/\Pi\mathbb{Z}^4$ with $\Pi = \begin{pmatrix} i & \sqrt{2} & 1 & 0 \\ 0 & i & 0 & 1 \end{pmatrix}$ and $Y = \mathbb{C}/(i, 1)\mathbb{Z}^2$, embedded as a complex subtorus of X via $z \mapsto (z, 0)$. Show that X admits no complex subtorus Z such that $Y \times Z$ is isogenous to X.

1.7 The Riemann–Roch Theorem

For a line bundle L on a g-dimensional complex torus $X = V/\Lambda$ the alternating sum

$$\chi(L) = \sum_{\nu=0}^{g} (-1)^\nu h^\nu(L)$$

is called the *Euler–Poincaré characteristic of L*. The Riemann–Roch Theorem gives a formula for it. There are two (equivalent) versions of the Riemann–Roch Theorem.

1.7.1 The Analytic Riemann–Roch Theorem

Theorem 1.7.1 (Analytic Riemann–Roch) *Let L be a line bundle whose first Chern class has s negative eigenvalues. Then*

$$\chi(L) = (-1)^s \operatorname{Pf}(E).$$

In other words, if L is of type $D = \operatorname{diag}(d_1, \dots, d_g)$ and s is the number of negative eigenvalues of $H = c_1(L)$, then

$$\chi(L) = (-1)^s d_1 \cdots d_g.$$

In particular, for degenerate L we have $\chi(L) = 0$.

Proof If $L|_{\Lambda(L)^0}$ is non-trivial, all cohomology groups of L vanish according to Theorem 1.6.8 and $\operatorname{Pf}(E) = 0$, since L is necessarily degenerate in this case. So assume that $L|_{\Lambda(L)^0}$ is trivial. Let r be the number of positive eigenvalues of H. Then Theorem 1.6.8 gives

$$\chi(L) = \sum_{q=s}^{g-r} (-1)^q \binom{g-r-s}{q-s} \operatorname{Pfr}(E)$$

$$= \begin{cases} (-1)^s \operatorname{Pf}(E) & \text{if } g = r+s, \\ 0 & \text{if } g > r+s, \end{cases}$$

where we used the binomial formula $\sum_{i=0}^{n} (-1)^i \binom{n}{i} = 0$ and that $\binom{0}{0} = 1$. This implies the assertion. □

Since $\deg \phi_L = \det E = \operatorname{Pf}(E)^2$, we obtain the following as an immediate consequence.

Corollary 1.7.2 *For every* $L \in \operatorname{Pic}(X)$,

$$\deg \phi_L = \chi(L)^2.$$

1.7.2 The Geometric Riemann–Roch Theorem

The geometric version of Riemann–Roch expresses the Euler–Poincaré characteristic of a line bundle in terms of its self-intersection number. For this recall that the *self-intersection number* (L^g) of a line bundle L on X is defined as

$$(L^g) := \int_X \wedge^g c_1(L).$$

Here the first Chern class $c_1(L)$ is considered as a 2-form on X via the de Rham isomorphism $H^2(X, \mathbb{C}) \simeq H^2_{\mathrm{dR}}(X)$ (see (1.5) and also Exercise 1.3.4 (8)).

Theorem 1.7.3 (Geometric Riemann–Roch) *For any line bundle L on X,*

$$\chi(L) = \frac{1}{g!}(L^g).$$

For the proof we need two lemmas. The first lemma computes $c_1(L)$ as an element of $H^2_{\mathrm{dR}}(X)$. Suppose $L = L(H, \chi)$. Choose a symplectic basis of Λ for $E = \operatorname{Im} H$ and denote by $x_1, \dots, x_g, y_1, \dots, y_g$ the corresponding real coordinate functions of V.

Lemma 1.7.4 *If L is of type $D = \operatorname{diag}(d_1, \dots, d_g)$, then*

$$c_1(L) = -\sum_{\nu=1}^{g} d_\nu \mathrm{d}x_\nu \wedge y_\nu.$$

Proof By definition the canonical isomorphism

$$\gamma_2 : \mathrm{Alt}^2_{\mathrm{dR}}(V, \mathbb{C}) \to H^2_{\mathrm{dR}}(X)$$

sends the alternating form E to the 2-form $\gamma_2(E) = -\sum_{\nu=1}^{g} d_\nu \mathrm{d}x_\nu \wedge \mathrm{d}y_\nu$ (see Exercise 1.2.3 (7)). This implies the assertion, since $\gamma_2(E) = c_1(E)$ by Lemma 1.2.6 and Proposition 1.2.8.
(For a different proof, see Exercise 1.3.4 (8)). □

The minus sign in Lemma 1.7.4 arises from the particular choice of the coordinate functions. The corresponding orientation is not always positive, as the following lemma shows.

Lemma 1.7.5 *If L is non-degenerate of index s and $x_1, \dots, x_g, y_1, \dots, y_g$ are the real coordinate functions of V corresponding to a symplectic basis of Λ for L, then*

$$\int_X \mathrm{d}x_1 \wedge \mathrm{d}y_1 \wedge \cdots \wedge \mathrm{d}x_g \wedge \mathrm{d}y_g = (-1)^{g+s}.$$

Proof We have to show that the volume form $(-1)^{g+s}\mathrm{d}x_1 \wedge \mathrm{d}y_1 \wedge \cdots \wedge \mathrm{d}x_g \wedge \mathrm{d}y_g$ represents the natural positive orientation of the complex vector space V. For this let $\lambda_1, \dots, \lambda_g, \mu_1, \dots, \mu_g$ denote the chosen symplectic basis. Recall from Lemma 1.5.4

that $\mu_1, \dots, \mu_g$ is a basis of the complex vector space V and denote by $v_1, \dots, v_g$ the corresponding complex coordinate functions. By definition the natural positive orientation of V is given by the volume form

$$(\frac{i}{2})^g dv_1 \wedge d\overline{v}_1 \wedge \cdots \wedge dv_g \wedge d\overline{v}_g.$$

In order to compare both volume forms, let $(Z_g, \mathbf{1}_g)$ denote the period matrix with respect to the chosen bases. Since

$$(\frac{i}{2})^g dv_1 \wedge d\overline{v}_1 \wedge \cdots \wedge dv_g \wedge d\overline{v}_g = (-1)^g \det(\operatorname{Im} Z) dx_1 \wedge dy_1 \wedge \cdots \wedge dx_g \wedge dy_g,$$

it remains to show that $(-1)^s \det(\operatorname{Im} Z) > 0$.

To see this, suppose $Y \in M_g(\mathbb{C})$ is the matrix of the hermitian form H of L with respect to the basis $\mu_1, \dots, \mu_g$. Then ${}^t(Z, \mathbf{1}_g)Y(\overline{Z}, \mathbf{1}_g)$ is the matrix of H with respect to the $\mathbb{R}$-basis $\lambda_1, \dots, \mu_g$ and

$$\begin{pmatrix} 0 & D \\ -D & 0 \end{pmatrix} = \operatorname{Im} \begin{pmatrix} {}^tZY\overline{Z} & {}^tZY \\ Y\overline{Z} & Y \end{pmatrix},$$

since $\lambda_1, \dots, \mu_g$ is a symplectic basis of $E = \operatorname{Im} H$. It follows that Y is real and hence $D = (\operatorname{Im} {}^tZ)Y$. Since by assumption Y has s negative eigenvalues, this completes the proof. □

Proof (of Theorem 1.7.3) The theorem is certainly true for any degenerate line bundle, since in this case $\wedge^g c_1(L) = 0$ and thus $(L^g) = 0$. So suppose L is non-degenerate. Then the theorem follows from Theorem 1.7.1 using Lemmas 1.7.4 and 1.7.5, since

$$\wedge^g c_1(L) = (-1)^g g! d_1 \cdots d_g dx_1 \wedge dy_1 \wedge \cdots \wedge dx_g \wedge dy_g$$

with the notation as above. □

As a consequence we get a formula for the Euler–Poincaré characteristic of the pullback of a line bundle under a homomorphism.

Corollary 1.7.6 *Let $f : Y \to X$ be a surjective homomorphism of complex tori and $L \in \operatorname{Pic}(X)$. Then*

$$\chi(f^*L) = \deg f \cdot \chi(L).$$

Proof If f is not an isogeny, then f^*L is degenerate and thus both sides of the equation are zero. So assume f is an isogeny. If $X = V/\Lambda$ and $Y = V'/\Lambda'$, we have according to Proposition 1.1.13, $\deg f = (\Lambda : \rho_r(f)(\Lambda'))$. So we get using the transformation formula and the geometric Riemann–Roch Theorem

$$g!\chi(f^*L) = (f^*L^g) = \int_Y \wedge^g c_1(f^*L)$$
$$= (\Lambda : \rho_r(f)(\Lambda')) \int_X \wedge^g c_1(L) = \deg f \cdot (L^g) = \deg f \cdot g!\chi(L),$$

which gives the assertion. □

1.7.3 Exercises

(1) *(Mumford's Index Theorem)* Let L_0 be a positive definite line bundle and L a non-degenerate line bundle on a complex torus X. Consider $\chi(L_0^n \otimes L)$ as a polynomial in n.

(a) This polynomial has only real roots.

(b) The index of L is the number of positive roots of this polynomial, counted with multiplicity.

(Hint: use the analytic Riemann–Roch Theorem.)

(2) *(Riemann–Roch for vector bundles on an elliptic curve)* Define the *degree* of a vector bundle E of rank r on an elliptic curve X by $\deg E := \deg \wedge^r E$. Then

$$\chi(E) = h^0(E) - h^1(E) = \deg E.$$

(Hint: Use a filtration by subbundles and induction.)

(3) *(Riemann–Roch for vector bundles on a complex torus)* Let X be a complex torus of dimension g and E a holomorphic vector bundle of rank r on X with Chern polynomial $c_t(E) = \sum_{i=0}^r c_i(E)t^i \in H^\bullet_{\mathrm{dR}}(X)[t]$. Writing $c_t(E) = \prod_{i=1}^r (1+a_i t)$, where the a_i are just formal symbols, the *Chern character* of E is defined as

$$\mathrm{ch}(E) = \sum_{i=1}^r \sum_{k=0}^\infty \frac{a_i^k}{k!}.$$

Denote by $\mathrm{ch}(E)_g$ the image in $\mathbb{C}$ of the component of degree g of $\mathrm{ch}(E)$ under the natural isomorphism $H^{2g}_{\mathrm{dR}}(X) \to \mathbb{C}$, $\omega \mapsto \int_X \omega$.

(a) Deduce from the general Hirzebruch–Riemann–Roch formula (see Hirzebruch [62])

$$\chi(E) = \sum_{i=1}^g (-1)^i h^i(E) = \mathrm{ch}(E)_g.$$

(Hint: Use that the tangent bundle of X is trivial.)

(b) Write $c_i = c_i(E)$ with $c_i(E) = 0$ for $i > r$. Then

$$\text{for } g = 1: \ \chi(E) = c_1 = \deg(E);$$
$$\text{for } g = 2: \ \chi(E) = \frac{1}{2}(c_1^2 - 2c_2);$$
$$\text{for } g = 3: \ \chi(E) = \frac{1}{3!}(c_1^3 - 3c_1c_2 + 3c_3);$$
$$\text{for } g = 4: \ \chi(E) = \frac{1}{4!}(c_1^4 - 4c_1^2c_2 + 4c_1c_3 + 2c_2^2 - 4c_4).$$

(c) Deduce the geometric Riemann–Roch Theorem: $\chi(L) = \frac{(L^g)}{g!}$ for all $L \in \mathrm{Pic}(X)$.

Chapter 2
Abelian Varieties

An *abelian variety* is by definition a complex torus admitting a positive line bundle or equivalently a projective embedding. The Riemann Relations are necessary and sufficient conditions for a complex torus to be an abelian variety. They were introduced by Riemann in the special case of a Jacobian variety of a curve. A general statement was given by Poincaré and Picard in [104] and Frobenius in [46], although it was apparently known to Riemann and Weierstraß.

This chapter contains the main results on abelian varieties. In Section 2.1 we define the polarization of an abelian variety as the first Chern class of a positive line bundle L. By a slight abuse of notation we denote a polarized abelian variety as a pair (X, L). We prove the Theorem of Lefschetz which says that if L is a positive line bundle, the n-th power of L is very ample for any $n \geq 3$. Finally, we give a proof of the Riemann Relations.

The second section deals with the Decomposition Theorem, which says that a polarized abelian variety (X, L) splits off all polarized abelian subvarieties associated to irreducible fixed components of the linear system $|L|$. Furthermore we show that for a positive line bundle L without a fixed component, already the second power is very ample. For this a version of Bertini's Theorem is proved as well as some properties of the Gauss map.

Concerning the behaviour of the linear system of the second power of a line bundle L, it remains to consider the case of an irreducible principal polarization L. This is done in the third section. For this we use a symmetric line bundle L, which by definition is a line bundle invariant under the action of $-\mathbf{1}_X$. Such a line bundle is contained in any algebraic equivalence class. The *Kummer variety* K_X of X is defined as the quotient of X by the action of $-\mathbf{1}_X$. It is easy to see that the projective map associated to L^2 factors via K_X. The main theorem is that the induced map on K_X is a projective embedding.

Section 2.4 contains a proof of Poincaré's Reducibility Theorem, which says that any abelian variety is isogenous to a product of *simple* abelian varieties, which by definition do not contain any non-trivial abelian subvarieties. For the proof we use a polarization L on X. It induces an anti-involution $'$ on the endomorphism algebra, called the *Rosati (anti-) involution*. Moreover, using $'$, we associate to any

H. Lange, *Abelian Varieties over the Complex Numbers*, Grundlehren Text Editions,
https://doi.org/10.1007/978-3-031-25570-0_2

abelian subvariety Y of X its norm-endomorphism N_Y (which is a generalization of the norm map in number theory) and consequently a symmetric idempotent of the endomorphism algebra of X. The main result is that any polarization induces an isomorphism of $\mathrm{NS}_{\mathbb{Q}}(X) = \mathrm{NS}(X) \otimes \mathbb{Q}$ with the space of symmetric elements of $\mathrm{End}_{\mathbb{Q}}(X)$.

The last three sections contain some more special results on abelian varieties. In Section 2.5 we introduce the dual of a polarization, give a result on maps of smooth varieties into an abelian variety, which implies that an abelian variety does not contain any rational curves, and finally study the Pontryagin product on the homology group $H_\bullet(X, \mathbb{Z})$. Section 2.6 contains a proof of the classification of endomorphism algebras of a simple abelian variety, due to Albert [3]. In the last section we introduce the theta group of a line bundle L and show that its commutator map is a generalization of the Weil pairing on $K(L)$.

2.1 Algebraicity of Abelian Varieties

In this section we define abelian varieties as complex tori admitting a positive line bundle and show that these varieties are exactly the algebraic complex tori; that is, which admit an embedding into a projective space.

2.1.1 Polarized Abelian Varieties

Let $X = V/\Lambda$ be a complex torus. Recall that a positive line bundle on X is by definition a line bundle on X whose first Chern class is a positive definite hermitian form on V. A *polarization* on X is by definition the first Chern class $H = c_1(L)$ of a positive line bundle L on X. By abuse of notation we sometimes consider the line bundle L itself as a polarization. The type of L (see Section 1.5.1) is called the *type of the polarization*. A polarization is called *principal* if it is of type $(1, \dots, 1)$. An *abelian variety* is by definition a complex torus X admitting a polarization $H = c_1(L)$. The pair (X, H) is called a *polarized abelian variety*. Again we often write (X, L) instead of (X, H).

A polarization L on X defines an isogeny $\phi_L \colon X \to \widehat{X}$ (see Section 1.4.2). Conversely, according to Exercise 2.1.6 (2) a homomorphism $\varphi \colon X \to \widehat{X}$ is of the form ϕ_L for some polarization L on X if and only if its analytic representation $V \to \overline{\Omega} = \mathrm{Hom}_{\overline{\mathbb{C}}}(V, \mathbb{C})$ is given by a positive definite hermitian form. Hence one could define equivalently a polarization to be an isogeny $X \to \widehat{X}$ with this property.

A *homomorphism of polarized abelian varieties* $f \colon (Y, M) \to (X, L)$ is a homomorphism of complex tori $f \colon Y \to X$ such that $f^* c_1(L) = c_1(M)$. According to Proposition 1.4.12 this means that $f^* L$ and M are analytically equivalent. Note that f necessarily has a finite kernel, since otherwise $f^* L$ would be degenerate.

Conversely, if $f: Y \to X$ is a homomorphism of complex tori with finite kernel and L a polarization on X, then f^*L defines a polarization on Y, which is called the *induced polarization*. This proves

Proposition 2.1.1

(a) *A complex subtorus of an abelian variety is an abelian variety.*
(b) *A complex torus isogenous to an abelian variety is an abelian variety.*

As a special case of (b) we note that the dual complex torus $\widehat{X}$ of an abelian variety X is also an abelian variety. Therefore it makes sense to speak of the *dual abelian variety*.

The question, whether a polarization on X descends via an isogeny, is answered by Corollary 1.4.4. Moreover, we have:

Proposition 2.1.2 *Every polarization is induced by a principal polarization via an isogeny.*

Proof Let (X, L) be a polarized abelian variety of type $(d_1, \dots, d_g)$ and let $p_1: X \to X_1$ be the isogeny of Exercise 1.5.5 (6). As we saw, there is an $M_1 \in \operatorname{Pic}(X_1)$ with $L = p_1^* M_1$. According to Exercise 1.5.5 (5) the isogeny p_1 is of degree $\prod d_\nu$. From the Riemann–Roch Theorem and Corollary 1.7.6 we obtain $\prod d_\nu = \chi(L) = \prod d_\nu \cdot \chi(M_1)$, that is, $\chi(M_1) = 1$. Applying Riemann–Roch again this implies the assertion, since M_1 is positive definite. □

As a first example we will show that every elliptic curve is an abelian variety.

Example 2.1.3 Suppose $X = \mathbb{C}/\Lambda$ is an elliptic curve (see Section 1.1.1). Without loss of generality we may assume that $\{z, 1\}$, with a complex number z, $\operatorname{Im} z > 0$, is a basis for Λ. Define

$$H: \mathbb{C} \times \mathbb{C} \to \mathbb{C}, \quad (v, w) \mapsto \frac{v \cdot \overline{w}}{\operatorname{Im} z}.$$

It is easy to check that H is a hermitian form with $\operatorname{Im} H(\Lambda, \Lambda) \subseteq \mathbb{Z}$, so $H \in \operatorname{NS}(X)$. Since H is positive definite, X is an abelian variety. So every elliptic curve is an abelian variety. However, not every complex torus of dimension $g \geq 2$ is an abelian variety. For examples, see Exercise 5.1.5 (3) (below).

Let $X = V/\Lambda$ be an abelian variety of dimension g and $L \in \operatorname{Pic}(X)$ a polarization. According to Theorem 1.5.9, $h^0(L) > 0$. Hence the line bundle L induces in the usual way a meromorphic map

$$\varphi : X \dashrightarrow \mathbb{P}^n$$

for some $n \geq 0$, defined as follows: if $\sigma_0, \dots, \sigma_n$ is a basis of $H^0(L)$, then

$$\varphi_L(x) = (\sigma_0(x) : \cdots : \sigma_n(x)),$$

whenever $\sigma_\nu(x) \neq 0$ for some ν. If $\sigma_\nu(x) = 0$ for all ν, then φ_L is not defined at x.

Choosing a factor of automorphy f for L we may consider $H^0(L)$ as the vector space of theta functions on V with respect to f. Let $\vartheta_0, \dots, \vartheta_n$ denote the basis of theta functions for the factor f. Then the map φ_L is given by

$$\varphi_L(\overline{v}) = (\vartheta_0(v) : \cdots : \vartheta_n(v)),$$

whenever defined. Note that φ_L does not depend on the choice of the factor f, however it depends on the choice of the basis of $H^0(L)$. Changing this basis means modifying φ_L by a projective transformation of $\mathbb{P}^n$.

We want to study the map φ_L. In particular, we want to give sufficient conditions for φ_L to be an embedding. For this it turns out to be convenient to use the language of linear systems of divisors (see Hartshorne [61]).

Let $|L|$ denote the complete linear system associated to the line bundle L. A divisor on X is a linear combination of codimension-one subvarieties with integer coefficients. A *divisor associated* to L is the zero-set of a section $\sigma \in H^0(L)$, considered as a codimension-one subvariety (with multiplicities) of X. Every meromorphic function f on X induces a divisor on X, denoted by (f), namely the difference of the codimension-one subspaces of the zero-set and the set where f is not defined, both with multiplicity. Two divisors D_1, D_2 are *linear equivalent* if they differ by the divisor of a meromorphic function, in symbols

$$D_1 \sim D_2 \quad \Leftrightarrow \quad \exists \text{ a meromorphic function } f \text{ on } X \text{ such that } D_1 = (f) + D_2.$$

Clearly linear equivalence defines an equivalence relation on the set of all divisors associated to L.

The following lemma is a generalization of the Theorem of the Square 1.3.5. It turns out to be an important tool for studying the map φ_L.

Lemma 2.1.4 *For $\overline{v}_1, \dots, \overline{v}_m \in X$ with $\sum_{\nu=1}^m \overline{v}_\nu = 0$ and $D \in |L|$,*

$$\sum_{\nu=1}^{m} t_{\overline{v}_\nu}^* D \sim mD \quad \textit{or equivalently} \quad \bigotimes_{\nu=1}^{m} t_{\overline{v}_\nu}^* L \simeq L^m.$$

Proof Suppose $L = L(H, \chi)$. Using Lemma 1.3.4 and Exercise 1.3.4 (3) we have

$$\bigotimes_{\nu=1}^{m} t_{\overline{v}_\nu}^* L \simeq L\left(mH, \prod_{\nu=1}^{m} \chi\, \mathbf{e}(2\pi i \operatorname{Im} H(v_\nu, \cdot))\right) \simeq L(mH, \chi^m) \simeq L^m.$$

Note that the assumption on the $\overline{v}_\nu$ implies $\prod_{\nu=1}^m \mathbf{e}(2\pi i \operatorname{Im} H(v_\nu, \lambda)) = 1$ for all $\lambda \in \Lambda$. □

Another easy observation, which will be applied many times, is the following: by definition

$$t_x^* D = D - x,$$

so

$$y \in t_x^* D \quad \Leftrightarrow \quad x \in t_y^* D \tag{2.1}$$

for every $x, y \in X$. As a first application we get

Proposition 2.1.5 *If L is a positive line bundle on X of type $(d_1, \ldots, d_g)$ and $d_1 \geq 2$, then φ_L is a holomorphic map.*

According to Lemma 1.4.14 and Exercise 1.5.5 (7) the assumption $d_1 \geq 2$ means that L is the d_1-th power of some positive line bundle on X.

Proof Suppose $x \in X$. We have to show that there exists a divisor in the linear system $|L|$ not containing x.

Choose $M \in \mathrm{Pic}(X)$ with $L \simeq M^{d_1}$. Since M is a positive line bundle, $|M|$ contains a divisor D. By continuity of the addition map, there exist points $x_1, \ldots, x_{d_1-1} \notin t_x^* D$ such that $x_{d_1} := -\sum_{\nu=1}^{d_1-1} x_\nu$ is not contained in $t_x^* D$. Lemma 2.1.4 implies $\sum_{\nu=1}^{d_1} t_{x_\nu}^* D \sim d_1 D \in |L|$. By choice of $x_1, \ldots, x_{d_1}$ and (2.1) we obtain $x \notin \sum_{\nu=1}^{d_1} t_{x_\nu}^* D$. □

A divisor is called *reduced* if it is the sum of distinct subvarieties. Another consequence of Lemma 2.1.4 is:

Proposition 2.1.6 *For any positive line bundle L on X a general member of $|L|$ is reduced.*

Proof Suppose $D = nE + F \in |L|$ with $n \geq 2$ and $E > 0$, $F \geq 0$. According to Lemma 2.1.4 we have $nE \sim \sum_{\nu=1}^{n} t_{x_\nu}^* E$ for all $x_1, \ldots, x_n \in X$ satisfying $\sum_{\nu=1}^{n} x_\nu = 0$. This implies the assertion, since we can choose $n-1$ of these points arbitrarily in X and so not every divisor of $|L|$ can be of this form, and hence a general divisor cannot be of this form. □

It is easy to see that the linear system $|L|$ of any positive line bundle has the structure of a projective space, namely $|L| \simeq \mathbb{P}^n$ with $n = h^0(L) - 1$. Hence it makes sense to speak of an open set of $|L|$.

Proposition 2.1.7 *Let $L \in \mathrm{Pic}(X)$ be positive. There is an open dense set $U \subset |L|$ such that for every $D \in U$ the identity $t_x^* D = D$ only holds for $x = 0$.*

Proof Suppose $0 \neq x \in X$ and $t_x^* D = D$ for an open dense set of divisors D in $|L|$ and hence for all D in $|L|$. In particular $t_x^* L \simeq L$ and x is contained in the finite group $K(L)$. The point x generates a finite subgroup S of X, say of order $n \geq 2$.

Let $Y = X/S$ and $p \colon X \to Y$ be the canonical projection map. The assumption implies that every $D \in |L|$ descends to an effective divisor E on Y. In particular $L \simeq p^* \mathcal{O}_Y(E)$. On the other hand, according to Proposition 1.4.3 there are only finitely many line bundles on Y, say $M_1, \ldots, M_n$, such that $L \simeq p^* M_\nu$. Since every divisor of $|L|$ descends to Y, this gives $H^0(L) = \bigcup_{\nu=1}^{n} p^* H^0(M_\nu)$.

But this implies $H^0(L) = p^* H^0(M_\nu)$ for some ν and we get using Riemann–Roch and Corollary 1.7.6 the contradiction $h^0(M_\nu) = h^0(L) = n h^0(M_\nu) > h^0(M_\nu)$. □

For later use we mention the following special case:

Corollary 2.1.8 *Let (X, L) be a principally polarized abelian variety. Then $|L|$ consists only of one divisor, say D, and $t_x^*D \neq D$ for all $x \in X$, $x \neq 0$.*

2.1.2 The Gauss Map

Let $X = V/\Lambda$ be an abelian variety of dimension g and L a positive line bundle on X. Suppose $D \in |L|$ is a reduced divisor (which exists according to Proposition 2.1.6). By D_s we denote the smooth part of D. Then for every $\overline{w} \in D_s$ the tangent space $T_{D,\overline{w}}$ is a $(g-1)$-dimensional vector space and its translation to zero is a well-defined $(g-1)$-dimensional subvector space of $T_{X,0} = V$. Consider $H^0(L)$ as a vector space of theta functions on V with respect to some factor for L. Then there is a theta function $\vartheta \in H^0(L)$, uniquely determined up to a constant, such that $\pi^*D = (\vartheta)$.

Let $v_1, \ldots, v_g$ denote the coordinate functions with respect to some complex basis of V. The equation of the tangent space $T_{D,\overline{w}}$ at a point $\overline{w} \in D$ is

$$\sum_{\nu=1}^{g} \frac{\partial \vartheta}{\partial v_\nu}(w)(v_\nu - w_\nu) = 0.$$

So the 1-dimensional subspace of the dual vector space V^* determined by $T_{D,\overline{w}}$ is generated by the vector $\left(\frac{\partial \vartheta}{\partial v_1}(w), \ldots, \frac{\partial \vartheta}{\partial v_g}(w)\right)$ (in coordinates with respect to the dual basis). The *Gauss map* of D is defined by

$$G \colon D_s \to \mathbb{P}^*_{g-1} = \mathbb{P}(V^*), \qquad G(\overline{w}) = \left(\frac{\partial \vartheta}{\partial v_1}(w) : \cdots : \frac{\partial \vartheta}{\partial v_g}(w)\right).$$

Obviously G is a holomorphic map (on D_s), neither depending on the choice of ϑ nor on the choice of the factor for L. In coordinate-free terms, the Gauss map $G : D_s \to \mathbb{P}^*_{g-1}$ is defined by associating to every point $\overline{w} \in D_s$ the translate to the origin of the projectivized tangent hyperplane $P(T_{\overline{w}}D)$ (see also Lemma 2.2.7 below).

In the next section we need the following property of the Gauss map G.

Proposition 2.1.9 *For any reduced divisor $D \in |L|$, with L a positive line bundle on X, the image of the Gauss map is not contained in a hyperplane.*

Proof Assume the contrary. This means there is a nonzero tangent vector $t \in V$ contained in $T_{D,\overline{v}}$ for every $\overline{v} \in D$. We may choose the basis of V in such a way that $t = (1, 0, \ldots, 0)$. Moreover we may assume that the function ϑ corresponding to D is a theta function with respect to the canonical factor $a_L = a_{L(H,\chi)}$ for L. Then the assumption means that $\frac{\partial \vartheta}{\partial v_1}(v) = 0$ for all $v \in V$ with $\vartheta(v) = 0$. Since D is reduced, the function

$$f = \frac{1}{\vartheta} \frac{\partial \vartheta}{\partial v_1}$$

is holomorphic on V, and the functional equation of ϑ translated to f is

$$f(v+\lambda) = f(v) + \pi H(t,\lambda)$$

for all $v \in V, \lambda \in \Lambda$. This implies that $\mathrm{d}f$ is the pullback of a holomorphic differential on X which according to Proposition 1.1.20 is an invariant one-form. Hence it is of the form

$$\mathrm{d}f = \sum_{\nu=1}^{g} \alpha_\nu \mathrm{d}v_\nu$$

for some $\alpha_\nu \in \mathbb{C}$. Integrating we obtain $f = \sum_{\nu=1}^{g} \alpha_\nu v_\nu + c$, where c is a constant. Inserting this into the functional equation of f, we get $\sum_{\nu=1}^{g} \alpha_\nu \lambda_\nu = \pi H(t,\lambda)$, where $\lambda_1, \ldots, \lambda_g$ denote the coordinates of λ with respect to the given basis. Hence

$$f(v) = \pi H(t, v) + c.$$

Since f is holomorphic and H is non-degenerate and $\mathbb{C}$-antilinear in the second variable, this implies $t = 0$, a contradiction. □

2.1.3 Theorem of Lefschetz

Let (X, L) be a polarized abelian variety of type $(d_1, \ldots, d_g)$. In this section we study the map φ_L. In Proposition 2.1.5 we saw that φ_L is a holomorphic map, if $d_1 \geq 2$. Here we show that φ_L is an embedding if $d_1 \geq 3$.

Theorem 2.1.10 (Theorem of Lefschetz) *If L is a positive line bundle on X of type $(d_1, \ldots, d_g)$ with $d_1 \geq 3$, then $\varphi_L \colon X \longrightarrow \mathbb{P}^n$ is an embedding.*

Proof We have to show that (i): φ_L is injective and (ii): the differential $\mathrm{d}\varphi_{L,x}$ is injective for every $x \in X$.

(i): assume $y_1, y_2 \in X$ with $\varphi_L(y_1) = \varphi_L(y_2)$. So $y_1 \in D$ if and only if $y_2 \in D$ for any $D \in |L|$. According to Exercise 1.5.5 (7) there is a positive definite $M \in \operatorname{Pic}(X)$ with $L \simeq M^{d_1}$. By Propositions 2.1.6 and 2.1.7 there is a reduced divisor $D_M \in |M|$ such that $t_x^* D_M = D_M$ only for $x = 0$.

Suppose $x_1 \in t_{y_1}^* D_M$. By continuity of the addition map and since $d_1 \geq 3$, there are $x_2, \ldots, x_{d_1} \in X$ with $x_1 + \cdots + x_{d_1} = 0$ such that $y_2 \notin t_{x_\nu}^* D_M$ for $\nu = 2, \ldots, d_1$. Since $\sum_{\nu=1}^{d_1} t_{x_\nu}^* D_M$ is a divisor in $|L|$ (see Lemma 2.1.4) containing y_1, we have by assumption $y_2 \in \sum_{\nu=1}^{d_1} t_{x_\nu}^* D_M$ and hence $y_2 \in t_{x_1}^* D_M$ by construction. So $x_1 \in t_{y_2}^* D_M$ and this holds for an arbitrary $x_1 \in t_{y_1}^* D_M$. Since D_M is reduced, we obtain $t_{y_1}^* D_M \subset t_{y_2}^* D_M$ and thus $t_{y_1}^* D_M = t_{y_2}^* D_M$, the situation being symmetric. Applying Proposition 2.1.7 we conclude $y_1 = y_2$.

(ii): suppose $t \neq 0$ is a tangent vector at $x \in X$. It suffices to show that there is a divisor $D \in |L|$ passing through x such that t is not tangent at D in x.

Assume the contrary, that is, t is tangent at D in x for all $D \in |L|$ containing x. Fix a reduced divisor $D_M \in |M|$. For $x_1 \in t_x^* D_M$ we can choose as above $x_2, \dots, x_{d_1} \in X$ with $x_1 + \dots + x_{d_1} = 0$ such that $x \notin t_{x_\nu}^* D_M$ for $\nu = 2, \dots, d_1$. Since $x \in \sum_{\nu=1}^{d_1} t_{x_\nu}^* D_M \in |L|$, we have by assumption that t is tangent at the divisor $\sum_{\nu=1}^{d_1} t_{x_\nu}^* D_M$ in x and hence t is tangent at $t_{x_1}^* D_M$ in x. This holds for all $x_1 \in t_x^* D_M$. Hence t is tangent to D_M at all points of D_M. But this implies that the image of the Gauss map for D_M is contained in a hyperplane, contradicting Proposition 2.1.9. □

Before we study the case $d_1 = 2$ (in Section 2.2.4), we deduce some consequences. Recall that by definition a line bundle L is *ample* if L^n is very ample for some $n \geq 1$; that is, if the map φ_{L^n} is an embedding. We have the following characterizations for a line bundle to be ample.

Proposition 2.1.11 *For a line bundle L on X the following statements are equivalent:*

(i) *L is ample;*
(ii) *L is positive;*
(iii) *$H^0(L) \neq 0$ and $K(L)$ is finite;*
(iv) *$H^0(L) \neq 0$ and $(L^g) > 0$.*

Proof (i) ⇒ (ii): Suppose φ_{L^n} is an embedding for some $n \geq 1$. It follows that $h^0(L^n) \neq 0$ and L^n is positive semidefinite according to the Vanishing Theorem 1.6.4. But L^n and as such L itself is even positive, since otherwise φ_{L^n} would not be injective by Theorem 1.5.11.

(ii) ⇒ (iii): is an immediate consequence of Theorem 1.5.9 and the definition of $K(L)$.

(iii) ⇒ (iv): L is non-degenerate, since $K(L)$ is finite. According to the Vanishing Theorem 1.6.4 we have $h^q(L) \neq 0$ only for $q = 0$. Hence the assertion is a consequence of the Geometric Riemann–Roch Theorem 1.7.3.

(iv) ⇒ (i): From the Riemann–Roch Theorems together with Theorem 1.6.8 we get $h^0(L) = \chi(L) = d_1 \cdot \ldots \cdot d_g > 0$, where $(d_1, \dots, d_g)$ is the type of L. Consequently L is positive and thus ample by the Theorem of Lefschetz 2.1.10. □

As a consequence we obtain the familiar version of the theorem of Lefschetz:

Corollary 2.1.12 *If $L \in \operatorname{Pic}(X)$ is ample, L^n is very ample for $n \geq 3$.*

Proposition 2.1.11 leads to the following criterion for a complex torus X to be an abelian variety. For this recall that the transcendence degree of the field of meromorphic functions on X is called the *algebraic dimension of* X, denoted by $a(X)$. Necessarily $a(X) \leq \dim X$, as for any connected compact complex manifold.

Theorem 2.1.13 *For a complex torus X the following conditions are equivalent:*

(i) *X is an abelian variety;*
(ii) *X admits the structure of a projective variety;*
(iii) *$a(X) = \dim X$.*

Proof (i) $\Rightarrow$ (ii): Recall that by definition an abelian variety is a complex torus admitting a positive line bundle. According to Proposition 2.1.11 this means that X is an abelian variety if and only if it can be analytically embedded into projective space. But by the Theorem of Chow 2.1.16 any closed analytic subvariety of $\mathbb{P}^n$ is algebraic.

It remains to show (iii) $\Rightarrow$ (i), the implication (ii) $\Rightarrow$ (iii) being trivial. Suppose $f_1, \dots, f_g$, with $g = \dim X$, are algebraically independent meromorphic functions on X. Denote by D_i the polar divisor of f_i and $L = \mathcal{O}_X(D_1 + \dots + D_g)$. Let σ be a section in $H^0(L)$ corresponding to the divisor $D_1 + \dots + D_g$. For every i there exists a uniquely determined section $\sigma_i \in H^0(L)$ such that $f_i = \frac{\sigma_i}{\sigma}$. The line bundle L is positive semidefinite, since $h^0(L) > 0$. Let $p\colon X \to \overline{X} = X/K(L)^0$ be the natural map of Section 1.5.4. According to Lemma 1.5.10 there is a positive definite line bundle $\overline{L}$ on $\overline{X}$ such that $L = p^*\overline{L}$, again since $h^0(L) > 0$. Moreover, $p^*\colon H^0(\overline{L}) \to H^0(L)$ is an isomorphism. This implies that $f_i = p^*h_i$ for some meromorphic function h_i on $\overline{X}$. Certainly $h_1, \dots, h_g$ are algebraically independent and $g \le a(\overline{X}) \le \dim \overline{X} \le \dim X = g$. So p is an isomorphism; that is, L is positive definite. □

2.1.4 Algebraic Varieties and Complex Analytic Spaces

In the sequel we will deal exclusively with abelian varieties. Hence we can work either in the analytic or the algebraic category. In Chapter 1 we had to work of course in the analytic category. It seems more natural to consider abelian varieties as algebraic varieties rather than analytic varieties and we will do this without further notice.

In this section, for the convenience of the reader we compile without proof some results about the interrelations between algebraic varieties over $\mathbb{C}$ and complex analytic spaces, mainly due to Serre [122].

To any algebraic variety X over the complex numbers one can associate a complex analytic space X_{hol} in a natural way. This construction is functorial: for any morphism $f: Z \to X$ of algebraic varieties over $\mathbb{C}$ there is a natural morphism of complex analytic spaces $f_{\text{hol}}: Z_{\text{hol}} \to X_{\text{hol}}$. Moreover, many of the important notions in both categories are preserved under the functor $X \to X_{\text{hol}}$. For example: an algebraic variety X is complete if and only if X_{hol} is compact, and X is smooth if and only if X_{hol} is a complex manifold. Conversely, a complex analytic space $\mathcal{X}$ is called *algebraic* if there is an algebraic variety X over $\mathbb{C}$ such that $\mathcal{X} \simeq X_{\text{hol}}$.

In both the algebraic and the complex analytic category one has the notion of a vector bundle and its associated locally free sheaf. For our purposes it is convenient not to distinguish between these objects: we always speak of vector bundles, even if we mean the associated locally free sheaf. In particular, for a line bundle L on a complex analytic space (respectively algebraic variety) X we denote by $H^i(X, L)$, or $H^i(L)$ if there is no ambiguity about the base space X, the i-th cohomology group of X with values in the locally free sheaf associated to L.

With a similar construction as above one can associate to an algebraic vector bundle (or more generally an algebraic coherent sheaf) F on the algebraic variety X over $\mathbb{C}$ a holomorphic vector bundle (respectively a holomorphic coherent sheaf) F_{hol} over the complex analytic space X_{hol}. Again this process is functorial.

Not every holomorphic vector bundle on an arbitrary algebraic complex analytic space comes from an algebraic one. However in the case of complete algebraic varieties we have the following comparison theorems due to Serre [122].

Theorem 2.1.14 *Let X be a complete algebraic variety over $\mathbb{C}$. For any holomorphic vector bundle (respectively holomorphic coherent sheaf) $\mathcal{F}$ on X_{hol} there exists a unique algebraic vector bundle (respectively algebraic coherent sheaf) F such that $F_{\mathrm{hol}} \simeq \mathcal{F}$. An analogous statement is true for any homomorphism $f : \mathcal{F} \to \mathcal{G}$ of holomorphic coherent sheaves over X_{hol}.*

Theorem 2.1.15 *Let X be a complete algebraic variety over $\mathbb{C}$. For any coherent sheaf F on X the natural maps $H^i(X, F) \to H^i(X_{\mathrm{hol}}, F_{\mathrm{hol}}), i \in \mathbb{Z}$, are isomorphisms of $\mathbb{C}$-vector spaces.*

Thus in the case of a complete algebraic variety X we need not distinguish between algebraic vector bundles on X and holomorphic vector bundles on X_{hol}.

Finally, recall the following theorem (see Griffiths–Harris [55]).

Theorem 2.1.16 (Theorem of Chow) *Suppose X is a complete algebraic variety over $\mathbb{C}$ and $\mathcal{Z}$ a closed analytic subset of X_{hol}, then there is an algebraic subvariety Z of X such that $Z_{\mathrm{hol}} \simeq \mathcal{Z}$.*

We obtain from this,

Corollary 2.1.17 *Suppose X and Z are complete algebraic varieties over $\mathbb{C}$ and $\mathsf{f} : \mathsf{Z}_{\mathrm{hol}} \to \mathsf{X}_{\mathrm{hol}}$ is a holomorphic map. Then there is an algebraic map $f : Z \to X$ with $f_{\mathrm{hol}} = \mathsf{f}$.*

As a consequence of Theorem 2.1.13, Theorems 2.1.14 to 2.1.16 and Corollary 2.1.17 apply to abelian varieties. In the proof of Theorem 2.1.13 we used already the theorem of Chow, but note that we used it only for the projective space $\mathbb{P}^n$.

2.1.5 The Riemann Relations

Before we proceed, we work out in terms of period matrices what it means that a complex torus is an abelian variety.

Let $X = V/\Lambda$ be a complex torus of dimension g. Choose bases $e_1, \ldots, e_g$ of V and $\lambda_1, \ldots, \lambda_{2g}$ of Λ and let Π be the corresponding period matrix. With respect to these bases we have

$$X = \mathbb{C}^g / \Pi\mathbb{Z}^{2g}.$$

The aim of this section is to prove the following

Theorem 2.1.18 *X is an abelian variety if and only if there is a non-degenerate alternating matrix $A \in \mathrm{M}_{2g}(\mathbb{Z})$ such that*

(i) $\Pi A^{-1}\,{}^t\Pi = 0$,
(ii) $i\Pi A^{-1}\,{}^t\overline{\Pi} > 0$.

The conditions (i) and (ii) are called *Riemann Relations*. By definition the complex torus X is an abelian variety if and only if X admits a polarization. It turns out that A is the matrix of the alternating form defining the polarization.

The most important versions of the Riemann relations are given by the following two special cases:

Corollary 2.1.19 *Let $\Pi = (Z_1, Z_2)$ be the period matrix of X with respect to a symplectic basis $\lambda_1, \dots, \lambda_g, \mu_1, \dots, \mu_g$ of Λ and a basis $e_1, \dots, e_g$ of V. According to the definition (see Section 1.5) the matrix A is of the form $A = \begin{pmatrix} 0 & D \\ -D & 0 \end{pmatrix}$ with $D = \mathrm{diag}(d_1, \dots, d_g)$ and $d_i | d_{i+1}$ for $i \leq g-1$. Then the Riemann Relations are:*

(i) $\Pi_2 D^{-1}\,{}^t\Pi_1 = \Pi_1 D^{-1}\,{}^t\Pi_2$;
(ii) $i\big(\Pi_2 D^{-1}\,{}^t\overline{\Pi}_1 - \Pi_1 D^{-1}\,{}^t\overline{\Pi}_2\big) > 0$.

Corollary 2.1.20 *Choose the basis of Λ as in the previous corollary. If in addition we choose the basis $\{e_i = d_i^{-1}\mu_i;\ i = 1, \dots, g\}$ of V, the period matrix of X is of the form (Z, D) and the Riemann Relations with respect to these bases are:*

(i) ${}^tZ = Z$;
(ii) $\mathrm{Im}\, Z$ *positive definite.*

The proof of both corollaries are immediate. Only note that $d_1^{-1}\mu_1, \dots d_g^{-1}\mu_g$ is a basis of V according to Lemma 1.5.4.

For the proof of Theorem 2.1.18 we start with an arbitrary non-degenerate alternating form E on Λ extended to $\Lambda \otimes \mathbb{R} = \mathbb{C}^g$. Denote by A the matrix of the alternating form E with respect to the basis $\lambda_1, \dots, \lambda_{2g}$ of Λ. Define the corresponding hermitian form $H \colon \mathbb{C}^g \times \mathbb{C}^g \to \mathbb{C}$ by

$$H(u, v) = E(iu, v) + iE(u, v).$$

The theorem is a direct consequence of the following two lemmas, which work out conditions for H to be a positive definite hermitian form.

Lemma 2.1.21 *H is a hermitian form on $\mathbb{C}^g$ if and only if $\Pi A^{-1}\,{}^t\Pi = 0$.*

Proof According to Lemma 1.2.10 the form H is hermitian if and only if $E(iu, iv) = E(u, v)$ for all $u, v \in \mathbb{C}^g$. In order to analyse this condition in terms of matrices define

$$I = \begin{pmatrix} \Pi \\ \overline{\Pi} \end{pmatrix}^{-1} \begin{pmatrix} i\mathbf{1_g} & \\ & -i\mathbf{1_g} \end{pmatrix} \begin{pmatrix} \Pi \\ \overline{\Pi} \end{pmatrix}.$$

Clearly $I^4 = \mathbf{1_{2g}}$ and one checks that I satisfies $i\Pi = \Pi I$. In fact,

$$\begin{pmatrix} \Pi \\ \overline{\Pi} \end{pmatrix} I = \begin{pmatrix} i\mathbf{1_g} & \\ & -i\mathbf{1_g} \end{pmatrix} \begin{pmatrix} \Pi \\ \overline{\Pi} \end{pmatrix} = \begin{pmatrix} i\Pi \\ -i\overline{\Pi} \end{pmatrix}.$$

Since

$$E(\Pi x, \Pi y) = {}^t x A y \qquad \text{for all} \quad x, y \in \mathbb{R}^{2g},$$

the form H is hermitian if and only if ${}^t I A I = A$ or equivalently

$$\begin{pmatrix} i\mathbf{1_g} & \\ & -i\mathbf{1_g} \end{pmatrix} \left(\begin{pmatrix} \Pi \\ \overline{\Pi} \end{pmatrix} A^{-1} ({}^t\Pi \; {}^t\overline{\Pi}) \right)^{-1} \begin{pmatrix} i\mathbf{1_g} & \\ & -i\mathbf{1_g} \end{pmatrix} = \left(\begin{pmatrix} \Pi \\ \overline{\Pi} \end{pmatrix} A^{-1} ({}^t\Pi \; {}^t\overline{\Pi}) \right)^{-1}.$$

Comparing the $g \times g$-blocks of both sides and using that A is a real matrix, one sees that this is the case if and only if $\Pi A^{-1}\, {}^t\Pi = 0$. □

In order to complete the proof of Theorem 2.1.18 we compute the matrix of H with respect to the basis $e_1, \ldots, e_g$ under the assumption that H is hermitian.

Lemma 2.1.22 *Suppose the form H is hermitian. Then $2i(\overline{\Pi} A^{-1}\, {}^t\Pi)^{-1}$ is the matrix of H with respect to the given basis. In particular H is positive definite if and only if $i\Pi A^{-1}\, {}^t\overline{\Pi} > 0$.*

Proof Write $u = \Pi x$ and $v = \Pi y$ with $x, y \in \mathbb{R}^{2g}$. With the notation as in the proof of Lemma 2.1.21 and using $x = \binom{\Pi}{\overline{\Pi}}^{-1} \binom{u}{\overline{u}}$ similarly for y as well as $\Pi A^{-1}\, {}^t\Pi = 0$ we get

$$\begin{aligned} E(iu, v) &= {}^t x\, {}^t I A y = {}^t \begin{pmatrix} u \\ \overline{u} \end{pmatrix} \begin{pmatrix} i\mathbf{1_g} & \\ & -i\mathbf{1_g} \end{pmatrix} \left(\begin{pmatrix} \Pi \\ \overline{\Pi} \end{pmatrix} A^{-1} ({}^t\Pi \; {}^t\overline{\Pi}) \right)^{-1} \begin{pmatrix} v \\ \overline{v} \end{pmatrix} \\ &= {}^t \begin{pmatrix} u \\ \overline{u} \end{pmatrix} \begin{pmatrix} 0 & i(\overline{\Pi} A^{-1}\, {}^t\Pi)^{-1} \\ -i(\Pi A^{-1}\, {}^t\overline{\Pi})^{-1} & 0 \end{pmatrix} \begin{pmatrix} v \\ \overline{v} \end{pmatrix} \\ &= i\, {}^t u (\overline{\Pi} A^{-1}\, {}^t\Pi)^{-1} \overline{v} - i\, {}^t\overline{u} (\Pi A^{-1}\, {}^t\overline{\Pi})^{-1} v. \end{aligned}$$

Similarly one computes

$$E(u, v) = {}^t u\, (\overline{\Pi} A^{-1}\, {}^t\Pi)^{-1}\, \overline{v} + {}^t\overline{u}\, (\Pi A^{-1}\, {}^t\overline{\Pi})^{-1}\, v.$$

So

$$H(u, v) = E(iu, v) + iE(u, v) = 2i\, {}^t u\, (\overline{\Pi} A^{-1}\, {}^t\Pi)^{-1}\, \overline{v}.$$

The last assertion follows from the fact that a matrix is positive definite if and only if its inverse is positive definite. □

2.1.6 Exercises

(1) Let $X = V/\Lambda$ be an abelian variety and D a reduced effective divisor on X with Gauss map $G : D_s \to \mathbb{P}^{g-1}$ and an associated theta function ϑ on V. Let $v_1, \dots, v_g$ denote coordinate functions on $V = T_{X,0}$. For any $w \in T_{X,0}$ let $\partial_w = \sum_{\nu=1}^g w_\nu \frac{\partial}{\partial v_\nu}$. Then $\partial_w \vartheta$ is the derivative of ϑ in the direction of w and $\partial_w \vartheta|_{\pi^* D}$ can be considered as a section of $\mathcal{O}_X(D)|_D$. Show that the Gauss map is given by the linear system given by the subvector space

$$\{\partial_w \vartheta|_D \mid w \in V = T_{X,0}\} \subset H^0(D, \mathcal{O}_X(D)|_D.$$

(2) Let $X = V/\Lambda$ be a complex torus and $f : X \to \widehat{X}$ a homomorphism. Show that $f = \phi_L$ for some polarization L on X if and only if the analytic representation of f is given by a positive definite hermitian form.
(Hint: Use Theorem 1.4.15.)

(3) Suppose X is an abelian variety with period matrix $\Pi \in \mathrm{M}(g \times 2g, \mathbb{C})$ and $A \in \mathrm{M}_{2g}(\mathbb{Z})$ the alternating matrix defining a polarization as in Theorem 2.1.18. Show that there is a matrix $\Lambda \in \mathrm{M}(g \times 2g, \mathbb{C})$ such that

$$A = {}^t\Pi\Lambda - {}^t\Lambda\Pi.$$

(4) Suppose X and X' are abelian varieties with period matrices $\Pi \in \mathrm{M}(g \times 2g, \mathbb{C})$ and $\Pi' \in \mathrm{M}(g' \times 2g', \mathbb{C})$ respectively. There is a non-trivial homomorphism $X \to X'$ if and only if there is a matrix $Q \neq 0$ in $\mathrm{M}(2g' \times 2g, \mathbb{Q})$ with $\Pi' Q\, {}^t\Pi = 0$.
(Hint: Use the previous exercise).

(5) (*Real Riemann Matrices according to H. Weyl [146]*) Let $X = V/\Lambda$ be an abelian variety of dimension g and $\Pi \in \mathrm{M}(g \times 2g, \mathbb{C})$ a period matrix for X. Suppose $A \in \mathrm{M}_{2g}(\mathbb{Z})$ is an alternating matrix satisfying the Riemann Relations (i) and (ii) of Theorem 2.1.18. Show that the matrix

$$R = R(\Pi) = \begin{pmatrix} \Pi \\ \overline{\Pi} \end{pmatrix}^{-1} \begin{pmatrix} -i\mathbf{1_g} & \mathbf{0} \\ \mathbf{0} & i\mathbf{1_g} \end{pmatrix} \begin{pmatrix} \Pi \\ \overline{\Pi} \end{pmatrix}$$

satisfies:

(a) R is independent of the chosen basis of V.
(b) The matrix R satisfies the following properties:
  (1) R is real;
  (2) $R^2 = -\mathbf{1_{2g}}$;
  (3) AR is positive definite and symmetric.
(c) Conversely, for any $R \in \mathrm{M}_{2g}(\mathbb{R})$ satisfying (1), (2) and (3) there is a period matrix Π satisfying (i) and (ii) of Theorem 2.1.18 with $R = R(\Pi)$. The matrix Π is uniquely determined by R up to multiplication by a nonsingular matrix from the left.

A matrix $R \in M_{2g}(\mathbb{R})$ satisfying (1), (2) and (3) is called a *real Riemann matrix for X*. The main advantage of a real Riemann matrix is that an endomorphism of X may be described in a simpler way:

(d) A matrix $M \in M_{2g}(\mathbb{Z})$ is the rational representation of an endomorphism of X if and only if $RM = MR$ for some real Riemann matrix R for X.

(6) Let $X = V/\Lambda$ be an abelian variety of dimension g and D a reduced effective divisor on X. Show that the Gauss map $G\colon D_s \to \mathbb{P}^{g-1}$ is given as follows: If $z_1, \dots, z_g$ denote complex coordinate functions on V and $\omega_\nu = dz_1 \wedge \dots \wedge dz_{\nu-1} \wedge dz_{\nu+1} \wedge \dots \wedge dz_g$, $\nu = 1, \cdots, g$, then

$$G(\overline{v}) = \big(\omega_1(v) : \cdots : \omega_g(v)\big)$$

for every $\overline{v} \in D_s$ with representative $v \in V$.

(7) Let $X = E_1 \times \dots \times E_g$ be a product of elliptic curves. Consider the divisor $D = \sum_{\nu=1}^g E_1 \times \dots \times E_{\nu-1} \times \{0\} \times E_{\nu+1} \times \dots \times E_g$ on X. Show that the image of the corresponding Gauss map $G\colon D_s \to \mathbb{P}^{g-1}$ consists of g points spanning $\mathbb{P}^{g-1}$.

(8) (*The maximal quotient abelian variety X_a of a complex torus X*) Let X be a complex torus. Recall that for a line bundle L on X the connected component of $K(L) = \ker \phi_L$ containing 0 is denoted by $K(L)^0$.

(a) Show that $K(L_1 \otimes L_2)^0 = K(L_1)^0 \cap K(L_2)^0$ for any positive semidefinite $L_1, L_2 \in \mathrm{Pic}(X)$.
(b) Conclude that there is a positive semidefinite line bundle L_a on X such that $K(L_a)^0 \subseteq K(L)^0$ for all positive semidefinite $L \in \mathrm{Pic}(X)$.
(c) Show that $X_a := X/K(L_a)^0$ is the maximal abelian quotient variety of X.
(d) (*Universal Property of X_a*) Denote by $p\colon X \to X_a$ the natural projection. For any homomorphism $f\colon X \to Y$ into an abelian variety Y there exists a unique homomorphism $g\colon X_a \to Y$ such that $f = gp$.
(e) The homomorphism $p\colon X \to X_a$ induces an isomorphism between the divisor groups of X and X_a.
(f) The homomorphism $p\colon X \to X_a$ induces an isomorphism between the fields of meromorphic functions on X and X_a.
(g) Give an example of a complex torus $X \neq 0$ with $X_a = 0$.

(9) (a) Show that if $(Z, \mathbf{1}_g)$ is a period matrix of a complex torus X, then $({}^tZ, \mathbf{1}_g)$ is a period matrix of $\widehat{X}$.
(b) Conclude that there exists a complex torus X not isogenous to its dual $\widehat{X}$.

(10) Let $X = V/\Lambda$ be a complex torus of algebraic dimension $a(X)$. Consider $\mathrm{NS}(X)$ as the group of hermitian forms on V, whose imaginary part is integer-valued on Λ. Show

(a) $a(X) = \max\{\operatorname{rk} H \mid H \in \mathrm{NS}(X), H \geq 0\}$.
(Hint: use Section 1.5.4 and Theorem 2.1.13.)

(b) $\rho(X) := \operatorname{rk} \mathrm{NS}(X) = 0$ implies $a(X) = 0$.

2.2 Decomposition of Abelian Varieties and Consequences

2.2.1 The Decomposition Theorem

Let (X, L), $L \in \mathrm{Pic}(X)$, be a polarized abelian variety. In this section we decompose (X, L) as a polarized abelian variety. This will turn out to be convenient for studying the associated map $\varphi_L : X \to \mathbb{P}^N$.

The linear system $|L|$ has a unique decomposition

$$|L| = |M| + F_1 + \cdots + F_r, \tag{2.2}$$

where $|M|$ is the moving part of $|L|$ and $F_1 + \cdots + F_r$ is the decomposition of the fixed part of $|L|$ into irreducible components. Note that we do not assume that there is a moving part, that is, there might be no line bundle M. Note moreover that $F_\nu \neq F_\mu$ for $\nu \neq \mu$. Define

$$N_\nu = \mathcal{O}_X(F_\nu)$$

for $\nu = 1, \ldots, r$. The line bundles M and $N_1, \ldots, N_r$ are semi-positive with $h^0(M) > 1$ and $h^0(N_\nu) = 1$ for $\nu = 1, \ldots, r$. So according to Theorem 1.5.11 the restrictions of M and N_ν to the subtori $K(M)^0$ and $K(N_\nu)^0$ of X respectively are trivial. Denote by $p_M : X \to X_M := X/K(M)^0$ and $p_{N_\nu} : X \to X_{N_\nu} := X/K(N_\nu)^0$ the canonical projections. Then Lemma 1.5.10 provides positive line bundles $\overline{M}$ on X_M and $\overline{N}_\nu$ on X_{N_ν} with

$$M = p_M^* \overline{M} \quad \text{and} \quad N_\nu = p_{N_\nu}^* \overline{N}_\nu \quad \text{for } \nu = 1, \ldots, r.$$

The pairs $(X_M, \overline{M})$ and $(X_{N_\nu}, \overline{N}_\nu)$ are polarized abelian varieties. In particular, the $\overline{N}_\nu$'s define principal polarizations on the abelian varieties X_{N_ν}, since $h^0(\overline{N}_\nu) = h^0(N_\nu) = 1$. Consider the product $X_M \times X_{N_1} \times \cdots \times X_{N_r}$ and denote by q_M and q_{N_ν} the projections of $X_M \times X_{N_1} \times \cdots \times X_{N_r}$ onto its factors. Moreover denote by

$$p := (p_M, p_{N_1}, \ldots, p_{N_r}) : X \to X_M \times X_{N_1} \times \cdots \times X_{N_r}.$$

With this notation we can state

Theorem 2.2.1 (Decomposition Theorem) *The homomorphism p is an isomorphism of polarized abelian varieties:*

$$p : (X, L) \longrightarrow (X_M \times X_{N_1} \times \cdots \times X_{N_r}, q_M^* \overline{M} \otimes q_{N_1}^* \overline{N}_1 \otimes \cdots \otimes q_{N_r}^* \overline{N}_r).$$

For the proof we need some preliminaries: generalizing the definition of the self-intersection number in Section 1.7.2 we define the *intersection number* $(L_1 \cdot \ldots \cdot L_g)$ of the line bundles $L_1, \ldots, L_g$ on X by

$$(L_1 \cdot \ldots \cdot L_g) := \int_X c_1(L_1) \wedge \cdots \wedge c_1(L_g).$$

If $L_1 \simeq \cdots \simeq L_\nu$ and $L_{\nu+1} \simeq \cdots \simeq L_g$, we write $(L_1^\nu \cdot L_g^{g-\nu})$ instead of $(L_1 \cdot \ldots \cdot L_1 \cdot L_g \cdot \ldots \cdot L_g)$. Moreover, since the intersection number depends only on the first Chern class, it makes sense to define

$$(H_1 \cdot \ldots \cdot H_g) := (L_1 \cdot \ldots \cdot L_g)$$

for hermitian forms $H_\nu \in \mathrm{NS}(X)$ and line bundles $L_\nu \in \mathrm{Pic}^{H_\nu}(X)$. In the sequel we freely apply some elementary properties of intersection numbers, for which we refer to Griffiths–Harris [55]. Furthermore we need

Lemma 2.2.2 *Let L_1 and L_2 be line bundles on $X = V/\Lambda$ and $H_i = c_1(L_i)$ the associated hermitian forms on V for $i = 1, 2$.*

(a) *Suppose L_1 and L_2 are semi-positive.*

  (i) *If H_1 and H_2 can be diagonalized simultaneously, then $(L_1^\nu \cdot L_2^{g-\nu}) \geq 0$ for $\nu = 0, \ldots, g$. The assumption is fulfilled for example if one of the line bundles is positive.*

  (ii) *If L_1 and L_2 are positive, then $(L_1^\nu \cdot L_2^{g-\nu}) > 0$ for $\nu = 0, \ldots, g$.*

(b) *If L_1 is positive and $(L_1^\nu \cdot L_2^{g-\nu}) > 0$ for $\nu = 0, \ldots, g$, then L_2 is also positive.*

More generally one can show that (see Exercise 2.2.5 (2)) $(L_1 \cdot \ldots \cdot L_g) \geq 0$ for any semi-positive line bundles $L_1, \ldots, L_g$ on X, but we do not need this fact.

Proof (a): According to Exercise 2.2.5 (1) any two hermitian forms on a finite-dimensional complex vector space, one of which is positive definite, can be diagonalized simultaneously. So for the whole proof we can choose a basis of V with respect to which $H_1 = \mathrm{diag}(h_1, \ldots, h_g)$ and $H_2 = \mathrm{diag}(k_1, \ldots, k_g)$ with nonnegative real numbers h_i and real numbers k_i. Denoting by $v_1, \ldots, v_g$ the complex coordinate functions with respect to the chosen basis we have

$$c_1(L_1) = \frac{i}{2} \sum_{\mu=1}^{g} h_\mu dv_\mu \wedge d\bar{v}_\mu \quad \text{and} \quad c_1(L_2) = \frac{i}{2} \sum_{\mu=1}^{g} k_\mu dv_\mu \wedge d\bar{v}_\mu$$

(see Exercise 1.3.4 (8)). Then

$$\bigwedge^{\nu} c_1(L_1) = \left(\frac{i}{2}\right)^{\nu} \sum_{i_1,\ldots,i_\nu=1}^{g} h_{i_1} \cdot \ldots \cdot h_{i_\nu} dv_{i_1} \wedge d\bar{v}_{i_1} \wedge \cdots \wedge dv_{i_\nu} \wedge d\bar{v}_{i_\nu}.$$

A similar formula holds for $\bigwedge^{g-\nu} c_1(L_2)$. Using this we get by the definition of the intersection numbers

$$\begin{aligned}(L_1^\nu \cdot L_2^{g-\nu}) \\ &= \int_X \left(\frac{i}{2}\right)^g \sum_{\sigma \in S_g} h_{\sigma(1)} \cdot \ldots \cdot h_{\sigma(\nu)} k_{\sigma(\nu+1)} \cdot \ldots \cdot k_{\sigma(g)} \mathrm{d}v_1 \wedge \mathrm{d}\overline{v}_1 \wedge \cdots \wedge \mathrm{d}v_g \wedge \mathrm{d}\overline{v}_g \\ &= c \cdot \sum_{\sigma \in S_g} h_{\sigma(1)} \cdot \ldots \cdot h_{\sigma(\nu)} k_{\sigma(\nu+1)} \cdot \ldots \cdot k_{\sigma(g)}\end{aligned}$$

with $c = \int_X \left(\frac{i}{2}\right)^g \mathrm{d}v_1 \wedge \mathrm{d}\overline{v}_1 \wedge \cdots \wedge \mathrm{d}v_g \wedge \mathrm{d}\overline{v}_g$. The constant c is positive, since the volume element $(\frac{i}{2})^g \mathrm{d}v_1 \wedge \mathrm{d}\overline{v}_1 \wedge \cdots \wedge \mathrm{d}v_g \wedge \mathrm{d}\overline{v}_g$ represents the natural positive orientation of V.

In case (i) the entries h_i and k_i, $i = 1, \ldots, g$, are all nonnegative, so $(L_1^\nu \cdot L_2^{g-\nu}) \geq 0$, for $\nu = 0, \ldots, g$. Similarly assertion (ii) holds, since then the entries are all positive. This shows (a).

(b): We may assume that $h_1 = \cdots = h_g = 1$. Let s_ν denote the νth elementary symmetric polynomial of degree g. We compute as above

$$\frac{1}{\nu!(g-\nu)!}(L_1^\nu \cdot L_2^{g-\nu}) = c \sum_{1 \leq i_1 < \cdots < i_\nu \leq g} k_{i_1} \cdot \ldots \cdot k_{i_\nu} = c\, s_\nu(k_1, \ldots, k_g).$$

So $s_\nu(k_1, \ldots, k_g) > 0$ for all ν by assumption. The polynomial

$$\prod_{\nu=1}^{g}(t - k_\nu) = \sum_{\nu=0}^{g}(-1)^\nu s_\nu(k_1, \ldots, k_g) t^{g-\nu}$$

has real roots and alternating coefficients. This implies that the roots $k_1, \ldots, k_g$ are positive and that L_2 is positive. □

As an immediate consequence we obtain an improvement of the *Nakai–Moishezon Criterion* (see Hartshorne [61, Appendix A, Theorem 5.1]) in the special case of an abelian variety.

Corollary 2.2.3 *Suppose L_0 is a line bundle defining a polarization on X. A line bundle L on X defines a polarization if and only if $(L^\nu \cdot L_0^{g-\nu}) > 0$ for $\nu = 1, \ldots, g$.*

For the proof of the Decomposition Theorem we need another lemma. Let M_ν denote a line bundle on X with $h^0(M_\nu) \geq 1$ for $\nu = 1, 2$. In particular M_ν is semi-positive. According to Theorem 1.5.11 and Lemma 1.5.10 the line bundle M_ν descends to a positive line bundle $\overline{M}_\nu$ on $X_{M_\nu} := X/K(M_\nu)^0$ via the natural projections $p_{M_\nu} \colon X \to X_{M_\nu}$ for $\nu = 1, 2$.

Lemma 2.2.4 *Suppose $M_1 \otimes M_2$ is positive and the homomorphism $(p_{M_1}, p_{M_2}) \colon X \to X_{M_1} \times X_{M_2}$ is not surjective and has finite kernel. Then*

$$h^0(M_1 \otimes M_2) \geq h^0(M_1) + h^0(M_2).$$

Proof Writing $g_\nu = \dim X_{M_\nu}$ for $\nu = 1, 2$, we have by assumption $g < g_1 + g_2$. Then

$$\begin{aligned} h^0(M_1 \otimes M_2) &= \frac{1}{g!}\left((M_1 \otimes M_2)^g\right) = \frac{1}{g!}\left((p_{M_1}^* \overline{M}_1 \otimes p_{M_2}^* \overline{M}_2)^g\right) \\ &= \sum_{\nu=g-g_2}^{g_1} \left(\frac{(p_{M_1}^* \overline{M}_1)^\nu}{\nu!} \cdot \frac{(p_{M_2}^* \overline{M}_2)^{g-\nu}}{(g-\nu)!} \right), \end{aligned}$$

since the intersection products

$$\left((p_{M_1}^* \overline{M}_1)^\nu\right) = p_{M_1}^*(\overline{M}_1^\nu) \quad \text{and} \quad \left((p_{M_2}^* \overline{M}_2)^{g-\nu}\right) = p_{M_2}^*(\overline{M}_2^{g-\nu})$$

vanish for $\nu > g_1$ respectively for $g - \nu > g_2$. For the summand with index $\nu = g_1$ we have

$$\begin{aligned} &\left(\frac{(p_{M_1}^* \overline{M}_1)^{g_1}}{g_1!} \cdot \frac{(p_{M_2}^* \overline{M}_2)^{g-g_1}}{(g-g_1)!}\right) \\ &= \left(\frac{(p_{M_1}^* \overline{M}_1)^{g_1}}{g_1!(g-g_1)!} \cdot \left\{(p_{M_2}^* \overline{M}_2)^{g-g_1} + \sum_{\mu=1}^{g-g_1} \binom{g-g_1}{\mu} (p_{M_1}^* \overline{M}_1)^\mu \cdot (p_{M_2}^* \overline{M}_2)^{g-g_1-\mu}\right\}\right) \\ &\qquad \text{(since the intersection product } \left((p_{M_1}^* \overline{M}_1)^{g_1+\mu}\right) \text{ vanishes for } \mu > 0) \\ &= \left(\frac{(p_{M_1}^* \overline{M}_1)^{g_1}}{g_1!} \cdot \frac{(M_1 \otimes M_2)^{g-g_1}}{(g-g_1)!}\right) \\ &= h^0(\overline{M}_1) \frac{\left(p_{M_1}^*(\text{point}) \cdot (M_1 \otimes M_2)^{g-g_1}\right)}{(g-g_1)!} \\ &\qquad \text{(by Riemann–Roch, since } \overline{M}_1 \text{ is positive)} \\ &= h^0(\overline{M}_1) \frac{\left((M_1 \otimes M_2|_{K(M_1)^0})^{g-g_1}\right)}{(g-g_1)!} \\ &= h^0(\overline{M}_1) \cdot h^0(M_1 \otimes M_2|_{K(M_1)^0}) \\ &\qquad \text{(by Riemann–Roch, since } M_1 \otimes M_2 \text{ is positive)} \\ &\geq h^0(\overline{M}_1) = h^0(M_1). \end{aligned}$$

Similarly we have for the summand with index $\nu = g - g_2$

$$\left(\frac{(p_{M_1}^* \overline{M}_1)^{g-g_2}}{(g-g_2)!} \cdot \frac{(p_{M_2}^* \overline{M}_2)^{g_2}}{g_2!}\right) \geq h^0(M_2).$$

Now by assumption $g_1 \neq g - g_2$. By Lemma 2.2.2 (a) all other summands are nonnegative. This implies the assertion. □

Proof (of Theorem 2.2.1) Define $L_r = M \otimes N_1 \otimes \cdots \otimes N_{r-1}$. According to the decomposition (2.2) we have

$$L \simeq L_r \otimes N_r \quad \text{with} \quad h^0(L) = h^0(L_r) \geq 1.$$

Hence L_r descends via the natural projection $p_{L_r}\colon X \to X_{L_r}$ to a positive line bundle $\overline{L}_r$ on $X_{L_r} := X/K(L_r)^0$ such that $h^0(L_r) = h^0(\overline{L}_r)$. Denote by q_{L_r} and q_{N_r} the projections of $X_{L_r} \times X_{N_r}$ onto its factors.

We claim that

$$(p_{L_r}, p_{N_r})\colon (X, L) \to (X_{L_r} \times X_{N_r}, q_{L_r}^* \overline{L}_r \otimes q_{N_r}^* \overline{N}_r)$$

is an isomorphism of polarized abelian varieties.

Suppose we have proven this. Applying this argument to (X_{L_r}, L_r) instead of (X, L) we can split off the principal polarized abelian variety $(X_{N_{r-1}}, \overline{N}_{r-1})$ in the same way. Repeating this process we finally obtain the asserted decomposition.

For the proof of the claim it suffices to show that (p_{L_r}, p_{N_r}) is an isomorphism. The kernel $K(L_r)^0 \cap K(N_r)^0$ of (p_{L_r}, p_{N_r}) is finite, since it is contained in the finite group $K(L)$. If (p_{L_r}, p_{N_r}) were not surjective, we could apply Lemma 2.2.4 to get

$$h^0(L) \geq h^0(L_r) + h^0(N_r) = h^0(L) + 1,$$

a contradiction. Hence (p_{L_r}, p_{N_r}) is an isogeny. By definition we have

$$L = (p_{L_r}, p_{N_r})^*(q_{L_r}^* \overline{L}_r \otimes q_{N_r}^* \overline{N}_r).$$

Applying Riemann–Roch, Corollary 1.7.6 and the Künneth formula we get

$$\begin{aligned}\frac{1}{\deg(p_{L_r}, p_{N_r})}(L^g) &= ((q_{L_r}^* \overline{L}_r \otimes q_{N_r}^* \overline{N}_r)^g) = g!\, h^0(q_{L_r}^* \overline{L}_r \otimes q_{N_r}^* \overline{N}_r) \\ &= g!\, h^0(\overline{L}_r)\, h^0(\overline{N}_r) = g! h^0(L) = (L^g),\end{aligned}$$

which implies the assertion. □

2.2.2 Bertini's Theorem for Abelian Varieties

The general *Bertini's theorem* (see Hartshorne [61, Theorem II, 8.18]) says that if $L \in \operatorname{Pic}(X)$ is very ample on a smooth projective variety X, then a general member of the linear system $|L|$ is smooth. If in addition $\dim X \geq 2$, then a general member of $|L|$ is irreducible.

We show in this section that in the case of an abelian variety X the irreducibility statement is true more generally, in fact for any positive line bundle L on X of dimension ≥ 2 without fixed components, even if $\dim \varphi_L(X) = 1$. We need the following theorem for the investigation of φ_{L^2} in Section 2.2.4 below.

Theorem 2.2.5 *Let L be a positive line bundle on an abelian variety X of dimension ≥ 2 without any fixed components. Then a general member of $|L|$ is irreducible.*

Proof Suppose this is not the case, that is

$$L = M \otimes N$$

with $h^0(M) \geq 2, h^0(N) \geq 2$ and the map

$$|M| \times |N| \longrightarrow |L|, \quad (D_1, D_2) \mapsto D_1 + D_2$$

is surjective. Note that, if $h^0(M) = 1$ or $h^0(N) = 1$, the linear system $|L|$ would have a fixed component under this assumption.

As in Section 2.2.1 there are ample line bundles $\overline{M}$ on $X_M = X/K(M)^0$ and $\overline{N}$ on $X_N = X/K(N)^0$ such that $M = p^*\overline{M}$ and $N = q^*\overline{N}$ where p and q are the corresponding surjections. The map

$$(p, q) : X \to X_M \times X_N$$

has finite kernel, since $K(M)^0 \cap K(N)^0 \subseteq \operatorname{Ker} \phi_L$ and L is positive.

If (p, q) is not surjective, we get from Lemma 2.2.4 and $\dim |M| + \dim |N| \geq \dim |L|$ that $h^0(L) \geq h^0(M) + h^0(N) \geq h^0(L) + 1$, a contradiction.

Hence (p, q) is an isogeny. Denoting by p_1 and p_2 the projections of $X_M \times X_N$ and applying Riemann–Roch, Corollary 1.7.6 and the Künneth formula, we get

$$\begin{aligned} h^0(M) + h^0(N) - 1 \geq h^0(L) &= \frac{1}{g!}(L^g) \\ &= \frac{1}{g!} \deg(p, q) \cdot (p_1^*\overline{M} \otimes p_2^*\overline{N})^g \\ &= \deg(p, q) \cdot h^0(p_1^*\overline{M} \otimes p_2^*\overline{N}) \\ &= \deg(p, q) \cdot h^0(M) \cdot h^0(N) \\ &\geq h^0(M) \cdot h^0(N) \end{aligned}$$

and hence

$$1 \leq (h^0(M) - 1)(h^0(N) - 1) = h^0(M) \cdot h^0(N) - h^0(N) - h^0(N) + 1 \leq 0,$$

a contradiction. □

2.2.3 Some Properties of the Gauss Map

In Section 2.1.2 we defined the Gauss map $G : D_s \to \mathbb{P}^{g-1} = \mathbb{P}(V^*)$ on the smooth part D_s of any reduced divisor D in the linear system $|L|$ of a positive line bundle on an abelian variety $X = V/\Lambda$ of dimension g. In Exercise 2.1.6 (7) we saw that the image of the Gauss map is not always irreducible. The following proposition shows that, in the case of an irreducible divisor, we can say more.

Proposition 2.2.6 *Let L be a positive line bundle on X. For any irreducible reduced divisor $D \in |L|$ the Gauss map $G : D_s \to \mathbb{P}^{g-1}$ is dominant.*

Proof For any $x \in D$ denote by B_x the maximal complex subtorus of X with $x + B_x \subset D$. Since X admits only countably many complex subtori (see Exercise 1.1.6 (2)), and D is irreducible, there is a complex subtorus B of X such that $B_x = B$ for almost all $x \in D$ and thus $D + B = D$. In particular this implies $B \subset K(L) = K(\mathcal{O}_X(D))$. But by assumption $K(L)$ is finite, so $B = 0$.

Suppose now that G is not dominant. Then the image of G is contained in a subvariety Y of codimension 1 in $\mathbb{P}^{g-1}$ and all fibres of G are of dimension ≥ 1. Let $\overline{w}$ denote a general point of D_s, where general means that $\overline{w}$ is smooth in the fibre $G^{-1}(G(\overline{w}))$ and $G(\overline{w})$ is smooth in Y. By what we said above, it suffices to show that $B_{\overline{w}} \neq 0$.

We may choose the basis of V in such a way that $G(\overline{w}) = (1 : 0 : \cdots : 0)$. As above we denote by $v_1, \dots, v_g$ the corresponding coordinate functions and by ϑ a theta function corresponding to D. Since $\frac{\partial \vartheta}{\partial v_1}(w) \neq 0$, we may apply the implicit function theorem to get that $\pi^* D$ is given locally around w by an equation

$$v_1 + f(v_2, \dots, v_g) = 0.$$

It follows that for every vector v near w in the inverse image under $\pi : V \to X$ of the fibre $G^{-1}(G(\overline{w}))$,

$$G(\overline{v}) = \left(1 : \frac{\partial f}{\partial v_2}(v) : \cdots : \frac{\partial f}{\partial v_g}(v)\right) = (1 : 0 : \cdots : 0). \tag{2.3}$$

This implies that

$$\frac{\partial f}{\partial v_\nu} = 0, \quad \nu = 2, \dots, g \tag{2.4}$$

are equations for $\pi^{-1} G^{-1}(G(\overline{w}))$ locally around w. If $x_1, \dots, x_g$ denote the given homogenous coordinates in $\mathbb{P}^{g-1}$, then

$$z_\nu = \frac{x_\nu}{x_1}, \quad \nu = 2, \dots, g$$

is a set of affine coordinates of $\mathbb{P}^{g-1} - \{x_1 = 0\} \simeq \mathbb{C}^{g-1}$, such that $G(\overline{w}) = (0, \dots, 0)$. By assumption the variety Y is smooth near $G(\overline{w})$. So we may assume that (after a suitable linear transformation of $z_2, \dots, z_g$) it is given locally around $G(\overline{w})$ by an equation

$$z_g = h(z_2, \ldots, z_g) \tag{2.5}$$

with a power series h vanishing of order ≥ 2 in $G(\overline{w}) = (0, \ldots 0)$. Note that applying the same linear transformation to $v_2, \ldots, v_g$, the equations (2.3) and (2.4) remain valid. So by definition of the Gauss map and (2.5) we get

$$\frac{\partial f}{\partial v_g} = \frac{\partial f}{\partial v_1} \cdot h\left(\frac{\partial f}{\partial v_2} / \frac{\partial f}{\partial v_1}, \ldots, \frac{\partial f}{\partial v_g} / \frac{\partial f}{\partial v_1}\right)$$

near w. Hence for $\nu = 2, \ldots, g$,

$$\begin{aligned}
\frac{\partial}{\partial v_g} \frac{\partial f}{\partial v_\nu} &= \frac{\partial}{\partial v_\nu} \frac{\partial f}{\partial v_g} \\
&= \frac{\partial^2 f}{\partial v_\nu \partial v_1} \cdot h\left(\frac{\partial f}{\partial v_2} / \frac{\partial f}{\partial v_1}, \ldots, \frac{\partial f}{\partial v_g} / \frac{\partial f}{\partial v_1}\right) + \\
&\qquad + \frac{\partial f}{\partial v_1} \cdot \frac{\partial}{\partial v_\nu} h\left(\frac{\partial f}{\partial v_2} / \frac{\partial f}{\partial v_1}, \ldots, \frac{\partial f}{\partial v_g} / \frac{\partial f}{\partial v_1}\right) = 0
\end{aligned}$$

on $\pi^{-1}G^{-1}G(\overline{w})$ near w by equation (2.4), since h vanishes of order ≥ 2 in $(0, \ldots, 0)$. This means that $\pi^{-1}G^{-1}(G(\overline{w}))$ is invariant under translations in the direction of v_g. Denoting by A the complex subtorus of X generated by $\{\overline{v} \mid v_1 = \cdots = v_{g-1} = 0\}$ we get

$$\overline{w} + A \subset G^{-1}(G(\overline{w})) + A \subset G^{-1}(G(\overline{w})) \subset D.$$

This completes the proof, since $0 \neq A \subset B_{\overline{w}}$. □

In order to understand the linear system defining the Gauss map, consider the derivative of a theta function in the direction of a tangent vector. As above let $v_1, \ldots, v_g$ be coordinate functions on $V = T_{X,0}$. For a tangent vector $w = (w_1, \ldots, w_g) \in T_{X,0}$ denote by

$$\partial_w := \sum_{\nu=1}^{g} w_\nu \frac{\partial}{\partial v_\nu}$$

the corresponding derivation. Then $\partial_w \vartheta$ is the derivative of the theta function ϑ in the direction of w. Note that if $\pi^* D = (\vartheta)$, then $\partial_w \vartheta|_{\pi^* D}$ can be considered as a section of $L|_D$ which we denote by $\partial_w \vartheta|_D$. In fact, the equation $\vartheta(\lambda + v) = a_L(\lambda, v)\vartheta(v)$ implies

$$\partial_w(\vartheta(\lambda + v)) = (\partial_w a_L(\lambda, v))\vartheta(v) + a_L(\lambda, v)\partial_w \vartheta(v) = a_L(\lambda, v)\partial_w \vartheta(v)$$

for any $v \in \pi^* D$ and $\lambda \in \Lambda$.

Lemma 2.2.7 *The linear system defining the Gauss map $G : D_s \to \mathbb{P}^{g-1}$ is given by the linear system corresponding to the subvector space*

$$\left\{\partial_w \vartheta|_D \;\middle|\; w \in V = T_{X,0}\right\} \subset H^0(D, L|_D).$$

Proof Every hyperplane of $\mathbb{P}^{g-1} = \mathbb{P}(V^*)$ is of the form

$$H_w := \left\{(x_1 : \ldots : x_g) \in \mathbb{P}^{g-1} \middle| \sum_{i=1}^{g} x_i w_i = 0\right\} \quad \text{for} \quad w \in V.$$

Hence for $\overline{v} \in D_s$ we have $G(\overline{v}) \in H_w$ if and only if $\sum_{i=1}^{g} \frac{\partial \vartheta}{\partial v_i}(v) w_i = 0$; that is, $\partial_w \vartheta(v) = 0$. □

2.2.4 Projective Embeddings with L^2

Let X be an abelian variety and M be an ample line bundle on X of type $(2, d_2, \ldots, d_g)$. According to Exercise 1.5.5 (7) $M = L^2$ for an ample line bundle L on X. So let L be an arbitrary ample line bundle on X. According to the Decomposition Theorem 2.2.1, for the investigation of the map $\varphi_{L^2} : X \to \mathbb{P}^n$ it suffices to consider the following two cases: Either L is a polarization such that $|L|$ does not admit any fixed component or L is an irreducible principal polarization.

In this section we will study the first case. The principally polarized case will be studied in Section 2.3.6. The following theorem was proved by Ramanan [107] for a generic abelian variety and by Obuchi [103] in general. Here we follow the proof given in Lange–Narasimhan [82].

Theorem 2.2.8 *If L is an ample line bundle without fixed components, then L^2 is very ample.*

Proof As in the proof of the Theorem of Lefschetz 2.1.10 we have to show (i) that $\varphi_{L^2} : X \to \mathbb{P}^N$ is injective and (ii) that the differential $d\varphi_{L^2,x}$ is injective for every $x \in X$.

(i): Assume $y_1, y_2 \in X$ with $\varphi_{L^2}(y_1) = \varphi_{L^2}(y_2)$. So $y_1 \in D$ if and only if $y_2 \in D$ for all $D \in |L^2|$. According to Propositions 2.1.6 and 2.1.7 and Theorem 2.2.5 there is an irreducible reduced $D_1 \in |L|$ such that $t_x^* D_1 = D_1$ only for $x = 0$. Again, since $|L|$ has no fixed component, there is an irreducible $D_2 \in |L|$ with $D_2 \neq (-1)^* t_{y_1+y_2}^* D_1$.

For $x \in t_{y_1}^* D_1$ we have

$$y_1 \in t_x^* D_1 \subset t_x^* D_1 + t_{-x}^* D_2 \in |L^2|.$$

By assumption this implies $y_2 \in t_x^* D_1 + t_{-x}^* D_2$, which in turn is equivalent to

$$x \in t_{y_2}^* D_1 + (-1)^* t_{y_2}^* D_2.$$

This holds for every $x \in t_{y_1}^* D_1$. Since D_1 is reduced, it follows that $t_{y_1}^* D_1 \subset t_{y_2}^* D_1 + (-1)^* t_{y_2}^* D_2$ or equivalently

$$D_1 = t_{-y_1}^* t_{y_1}^* D_1 \subset t_{-y_1+y_2}^* D_1 + (-1)^* t_{y_1+y_2}^* D_2.$$

Since $D_1 \neq (-1)^* t_{y_1+y_2}^* D_2$ by construction, and since the divisors D_1 and D_2 are irreducible, it follows that $D_1 = t_{-y_1+y_2}^* D_1$. By assumption on D_1 this implies $y_1 = y_2$.

(ii): Suppose $t \neq 0$ is a tangent vector at $x \in X$. It suffices to show that there is a divisor $D \in |L^2|$ containing x such that t is not tangent at D in x.

Assume the contrary: the vector t is tangent at D in x for all $D \in |L^2|$ containing x. According to Proposition 2.1.6, Theorem 2.2.5 and the assumption that $|L|$ has no fixed component, there are irreducible reduced divisors D_1 and D_2 in $|L|$ with $t_x^* D_1 \neq (-1)^* t_x^* D_2$. Let $D_{1,s}$ denote the smooth part of D_1. For any $y \in t_x^* D_{1,s} \setminus (-1)^* t_x^* D_2$ we have

$$x \in t_y^* D_{1,s} \subset t_y^* D_1 + t_{-y}^* D_2 \in |L^2|.$$

By assumption this implies that t is tangent to $t_y^* D_1 + t_{-y}^* D_2$ in x. But x is not contained in $t_{-y}^* D_2$, since otherwise $y \in (-1)^* t_x^* D_2$. Hence t is tangent to $t_y^* D_{1,s}$ in x. This is true for every y in an open dense subset of $t_x^* D_{1,s}$. So it implies that t is tangent to D_1 in every point of D_1. But this means that the image of the Gauss map for D_1 is contained in a hyperplane, contradicting Proposition 2.1.9. □

2.2.5 Exercises

(1) Let V be a finite-dimensional complex vector space and H_1 and H_2 hermitian forms on V with H_1 positive definite. Show that there exists a basis of V with respect to which H_1 and H_2 are in diagonal form.
(Hint: The proof is only a slight improvement of the usual diagonalization method.)

(2) Show that for any semi-positive line bundles $L_1, \ldots, L_g$ on an abelian variety X we have $(L_1 \cdots L_g) \geq 0$.

(3) Let $f : X \to Y$ be an isogeny of abelian varieties of dimension ≥ 2.

(a) If D is a positive irreducible and smooth divisor on Y, show that $f^* D$ is also smooth and irreducible.
(Hint: Use Corollary 2.2.3 and that f is étale.)

(b) Give an example of an irreducible D, such that $f^* D$ is not irreducible.

(4) Let X be an abelian variety of dimension $g \geq 2$ and L be a positive line bundle without fixed components on X. If $h^0(L) \geq 3$, then $\dim \varphi_L(X) \geq 2$.
(Hint: Use Bertini's Theorem 2.2.5.)

(5) Let L be an ample line bundle on the abelian variety X. Show that for any smooth divisor $D \in |L|$ the Gauss map is a finite morphism.
(Hint: Use Lemma 2.2.7 and the fact that any hyperplane in $\mathbb{P}^n$, $n \geq 2$, intersects a non-degenerate curve in more that one point.)

(6) (*Generalized Gauss Map*) Let $X = V/\Lambda$ be an abelian variety of dimension g and Y a subvariety of dimension n. For any smooth point y of Y the translation to the origin of the tangent space at Y in y is an n-dimensional subvector space of $T_{X,0} = V$. This defines a holomorphic map G of the smooth part Y_s of Y into the Grassmannian $\mathrm{Gr}(n, V)$ of n-dimensional subvector spaces of V. If the canonical sheaf of Y is an ample line bundle, G is generically one to one (see Ran [110]).

2.3 Symmetric Line Bundles and Kummer Varieties

As we saw in the last section, for the investigation of the map φ_{L^2} it suffices to consider the cases where L has no fixed components and where L is an irreducible principal polarization. We treated the first case in Theorem 2.2.8. For the latter case it turns out to be convenient to take a particular line bundle in its algebraic equivalence class. We need some preliminaries. First recall the definition of algebraic equivalence.

2.3.1 Algebraic Equivalence of Line Bundles

Two line bundles L_1 and L_2 on an abelian variety X are called *algebraically equivalent* if there is a connected algebraic variety T, a line bundle $\mathcal{L}$ on $X \times T$ and points $t_1, t_2 \in T$ such that

$$\mathcal{L}|_{X\times\{t_i\}} \simeq L_i \quad \text{for} \quad i = 1 \text{ and } 2.$$

Note that the definition is analogous to the definition of analytic equivalence of line bundles (see Section 1.4.4), only the parameter space T has to be algebraic. In particular algebraically equivalent line bundles are analytically equivalent. So Corollary 1.4.13 immediately gives the following lemma.

Lemma 2.3.1 *Let* $L_1, L_2 \in \mathrm{Pic}(X)$ *be on an abelian variety* X *with* L_1 *non-degenerate. Then* L_1 *and* L_2 *are algebraically equivalent if and only if there is a point* $x \in X$ *such that* $L_2 \simeq t_x^* L_1$.

The most important case is when L_1 is ample, since then $h^0(L_1) > 0$ so that the (meromorphic) map $\varphi_{L_1} : X \to \mathbb{P}^n$ given by $H^0(L_1)$ (see Section 2.1.1) exists. Then we have to compare φ_{L_1} and $\varphi_{t_x^* L_1}$. In fact, the next lemma follows immediately from the definitions.

Lemma 2.3.2 *For an ample line bundle L on X and a point $x \in X$ the following diagram commutes up to an automorphism of $\mathbb{P}^n$*

$$\begin{array}{ccc} X & \xrightarrow{t_x} & X \\ & \searrow^{\varphi_{t_x^* L}} \quad \swarrow^{\varphi_L} & \\ & \mathbb{P}^n . & \end{array}$$

Notice that the maps φ_L and $\varphi_{t_x^* L}$ depend on the choice of bases of $H^0(L)$ and $H^0(t_x^* L)$. If one chooses the bases in a compatible way, the diagram actually commutes: let $\vartheta_0, \dots, \vartheta_n$ be a basis of $H^0(L)$, then $t_x^* \vartheta_0, \dots, t_x^* \vartheta_n$ is a basis of $H^0(t_x^* L)$ according to Exercise 1.5.5 (11) and we have $\varphi_{t_x^* L} = \varphi_L \circ t_x$.

2.3.2 Symmetric Line Bundles

Let L be an ample line bundle on an abelian variety X. Lemma 2.3.2 implies that, up to an automorphism of $\mathbb{P}^n$, the image $\overline{X}$ of φ_L in $\mathbb{P}^n$ does not depend on L itself, but only on the algebraic equivalence class of L. This reflects the fact that the map φ_L and the chosen point 0 of the group X are independent of each other. So, in order to investigate the projective variety $\overline{X}$ in $\mathbb{P}^n$, we may choose the line bundle L suitably within its algebraic equivalence class. Good candidates for this are the symmetric line bundles on X, introduced in Section 1.3.3.

Recall that a line bundle L on X is called *symmetric* if $(-1)^* L \simeq L$.

Lemma 2.3.3 *The line bundle $L = L(H, \chi)$ on $X = V/\Lambda$ is symmetric if and only if $\chi(\Lambda) \subset \{\pm 1\}$.*

Proof This is a direct consequence of $(-1)^* L(H, \chi) = L(H, \chi^{-1})$. □

If L is non-degenerate, a decomposition $\Lambda = \Lambda_1 \oplus \Lambda_2$ for L distinguishes a line bundle in the algebraic equivalence class $\operatorname{Pic}^H(X)$, namely the bundle $L_0 = L(H, \chi_0)$ of characteristic 0 (see Section 1.5.1).

Lemma 2.3.4 *For any non-degenerate $H \in \operatorname{NS}(X)$ the line bundle $L_0 = L(H, \chi_0)$ is symmetric.*

Proof The semicharacter χ_0 was defined by

$$\chi_0 : \lambda \to \mathbb{C}_1, \qquad \lambda \mapsto \lambda_1 + \lambda_2 \mapsto e(\pi i \operatorname{Im} H(\lambda_1, \lambda_2)),$$

where $\lambda_\nu \in \Lambda_\nu$, $\nu = 1, 2$. This shows that L_0 is symmetric. □

For any $H \in \operatorname{NS}(X)$ denote by $\operatorname{Pic}_s^H(X)$ the set of symmetric line bundles in the algebraic class $\operatorname{Pic}^H(X)$. The sets $\operatorname{Pic}_s^H(X)$ have the following structure.

Lemma 2.3.5

(a) $\mathrm{Pic}_s^0(X)$ *is a vector space of dimension* $2g$ *over the field* $\mathbb{Z}/2\mathbb{Z}$.
(b) *For any nonzero* $H \in \mathrm{NS}(X)$ *the set* $\mathrm{Pic}_s^H(X)$ *is a principal homogeneous space over* $\mathrm{Pic}_s^0(X)$.

Proof (a) follows from the fact that $\mathrm{Pic}_s^0(X)$ is just the set of 2-division points in the dual abelian variety.

(b): We first claim that $\mathrm{Pic}_s^H(X) \neq \emptyset$. By what we have said above, $\mathrm{Pic}_s^H(X)$ is non-empty for any non-degenerate H. So it remains to consider the case when H is degenerate. As was shown in Section 1.5.4, there is a surjective homomorphism of abelian varieties $p\colon X \to \overline{X}$ and a non-degenerate $\overline{H} \in \mathrm{NS}(\overline{X})$ with $H = p^*\overline{H}$. With $\overline{L}_0 \in \mathrm{Pic}_s^{\overline{H}}(\overline{X})$ also the pullback $p^*\overline{L}_0$ is symmetric. So $\mathrm{Pic}_s^H(X) \neq \emptyset$.

Now suppose $L \in \mathrm{Pic}_s^H(X)$. Obviously the bijective map $\mathrm{Pic}^0(X) \to \mathrm{Pic}^H(X)$, $P \mapsto L \otimes P$ induces a bijection $\mathrm{Pic}_s^0(X) \to \mathrm{Pic}_s^H(X)$, defining the structure of a principal homogeneous space. □

If H is non-degenerate, we can interpret the action of $\mathrm{Pic}_s^0(X)$ on $\mathrm{Pic}_s^H(X)$ in terms of the characteristics with respect to the chosen decomposition of Λ: let $L_0 \in \mathrm{Pic}_s^H(X)$ be the line bundle with characteristic 0 as above. One easily sees that the line bundle with characteristic c

$$t_{\overline{c}}^* L_0 = L\Big(H, \chi_0\, \mathbf{e}(2\pi i \operatorname{Im} H(c,\cdot))\Big) \tag{2.6}$$

is symmetric if and only if the character $\mathbf{e}(2\pi i \operatorname{Im} H(c,\cdot))$ on Λ takes only values in $\{\pm 1\}$. This is the case if and only if $c \in \frac{1}{2}\Lambda(H)$. Hence the 2^{2g} line bundles $t_{\overline{c}}^* L_0$ with characteristic c in $\frac{1}{2}\Lambda(H)$ (modulo $\Lambda(H)$) build up the principal homogeneous space $\mathrm{Pic}_s^H(X)$ and the action of $\mathrm{Pic}_s^0(X)$ on $\mathrm{Pic}_s^H(X)$ is induced by the map

$$\frac{1}{2}\Lambda(H) \to \mathrm{Pic}_s^H(X), \qquad c \mapsto t_{\overline{c}}^* L_0.$$

Let L be any symmetric line bundle on X. A biholomorphic map $\varphi : L \to L$ is called an *isomorphism of* L *over* $(-1)_X$ if the diagram

$$\begin{array}{ccc} L & \xrightarrow{\varphi} & L \\ \downarrow & & \downarrow \\ X & \xrightarrow{(-1)_X} & X \end{array}$$

commutes and for every $x \in X$ the induced map $\varphi(x)\colon L(x) \to L(-x)$ is $\mathbb{C}$-linear. Here $L(x)$ denotes the fibre of L over the point x and $\varphi(x)$ is the restriction of L to $L(x)$. The isomorphism φ is called *normalized* if the induced map $\varphi(0) : L(0) \to L(0)$ is the identity.

Lemma 2.3.6 *Any symmetric line bundle $L \in \mathrm{Pic}(X)$ admits a unique normalized isomorphism $(-1)_L : L \to L$ over $(-1)_X$.*

Proof The biholomorphic map $(-1)\times 1 \colon V\times\mathbb{C} \to V\times\mathbb{C}$ is an isomorphism of the trivial line bundle on V over the multiplication by (-1) on V. Since L is symmetric, its canonical factor a_L satisfies $a_L(-\lambda, -v) = a_L(\lambda, v)$ for all $(\lambda, v) \in \Lambda\times V$ according to Lemma 1.3.6. This implies that the action of Λ on $V\times\mathbb{C}$ via a_L defining L (see Section 1.3.1) is compatible with $(-1)\times 1$. Hence $(-1)\times 1$ descends to an isomorphism $(-1)_L : L \to L$ over $(-1)_X$. Certainly $(-1)_L$ is normalized, because $(-1)\times 1$ induces the identity on the fibre $\{0\}\times\mathbb{C}$. The uniqueness of $(-1)_L$ follows from the fact that any two automorphisms of L over X differ by a nonzero constant. □

Suppose now $L = L(H,\chi)$ is an ample symmetric line bundle on X. The normalized isomorphism $(-1)_L$ induces an involution on the vector space of canonical theta functions for L

$$(-1)_L^* \colon H^0(L) \to H^0(L)\,, \quad \vartheta \mapsto (-1)_V^*\vartheta.$$

Denote by $H^0(L)_+$ its $(+1)$-eigenspace in $H^0(L)$ and by $H^0(L)_-$ its (-1)-eigenspace in $H^0(L)$. For the computation of the dimensions $h^0(L)_+$ and $h^0(L)_-$ we need to work out how $(-1)_L$ acts on $H^0(L)$. For this choose a decomposition $\Lambda = \Lambda_1 \oplus \Lambda_2$ for L.

Proposition 2.3.7 (Inverse Formula) *Let $\{\vartheta^c_{\overline{w}} \mid \overline{w} \in K(L)_1\}$ denote the basis of $H^0(L)$ given in Exercise 1.5.5 (4) and $c = c_1 + c_2$ the decomposition of the characteristic $c \in \frac{1}{2}\Lambda(L)$ of L. Then*

$$(-1)_V^*\vartheta^c_{\overline{w}} = \mathbf{e}(4\pi i \,\mathrm{Im}\, H(w+c_1, c_2))\vartheta^c_{-\overline{w}-2\overline{c}_1}.$$

In particular, if L is of characteristic 0, then

$$(-1)_L^*\vartheta^0_{\overline{w}} = \vartheta^0_{-\overline{w}} \quad \text{for all} \quad \overline{w} \in K(L)_1.$$

Proof Using the definition of $\vartheta^c_{\overline{w}}$ and Lemma 1.5.3 one checks that $\vartheta^c_{\overline{w}}$ and $\vartheta^0_{\overline{w}}$ are related as follows (see also Exercise 1.5.5 (3))

$$\vartheta^c_{\overline{w}}(v) = \mathbf{e}\left(-\pi H(v,c) - \frac{\pi}{2}H(c,c)\right)\vartheta^0_{\overline{w}}(v+c) \qquad \text{for all} \quad v \in V$$

and similarly

$$\vartheta^0_{\overline{w}} = a_{L_0}(w,\cdot)\vartheta^0(\cdot + w).$$

So for all $v \in V$ we get

$$
\begin{aligned}
(-1)^*_V \vartheta^c_{\overline{w}}(v) &= \vartheta^c_{\overline{w}}(-v) = \mathbf{e}\left(\pi H(v,c) - \frac{\pi}{2} H(c,c)\right) \vartheta^0_{\overline{w}}(-v+c) \\
&= \mathbf{e}\left(\pi H(v,c) - \frac{\pi}{2} H(c,c)\right) a_{L_0}(w, -v+c)^{-1} \vartheta^0_0(-v+c+w) \\
&= \mathbf{e}\left(\pi H(v,c) - \frac{\pi}{2} H(c,c)\right) a_{L_0}(w, -v+c)^{-1} \vartheta^0_0(v-c-w) \quad (\text{since } \vartheta^0_0 \text{ is even}) \\
&= \mathbf{e}\left(\pi H(v,c) - \frac{\pi}{2} H(c,c)\right) a_{L_0}(w, -v+c)^{-1} a_{L_0}(-2c_2, v-c_1+c_2-w) \\
&\qquad \cdot \vartheta^0_0(v-c_1+c_2-w) \quad (\text{by Corollary 1.5.8}) \\
&= \mathbf{e}\left(\pi H(v,c) - \frac{\pi}{2} H(c,c)\right) a_{L_0}(w, -v+c)^{-1} a_{L_0}(-2c_2, v-c_1+c_2-w) \\
&\qquad \cdot a_{L_0}(-w-2c_1, v+c) \vartheta^0_{-\overline{w}-2\overline{c}_1}(v+c) \\
&= \mathbf{e}(2\pi H(v,c)) a_{L_0}(w, -v+c)^{-1} a_{L_0}(-2c_2, v-c_1+c_2-w) \\
&\qquad \cdot a_{L_0}(-w-2c_1, v+c) \vartheta^c_{-\overline{w}-2\overline{c}_1}(v),
\end{aligned}
$$

where for the last two equations one uses again the equations of the beginning of the proof. Then using the definition of a_{L_0} and the fact that $E(w, c_1) = 0$ (c_1 and w both are contained in the subspace V_1, which is isotropic for E), one easily deduces the assertion. □

Using the Inverse Formula, we can compute the dimensions of the eigenspaces of the action of $(-1)^*$ on $H^0(L)$.

Proposition 2.3.8 *Let $L \in \operatorname{Pic}^H(X)$ be an ample symmetric line bundle on X of characteristic c with respect to a decomposition of $\Lambda(L)$ for L. Write $c = c_1 + c_2$ and define*

$$S = \{\overline{w} \in K(L)_1 \mid 2\overline{w} = -2\overline{c}_1\} \text{ and } S^{\pm} = \{\overline{w} \in S \mid \mathbf{e}(4\pi i \operatorname{Im} H(w + c_1, c_2)) = \pm 1\}.$$

Then

$$h^0(L)_{\pm} = \frac{1}{2}\left(h^0(L) - \#S\right) + \#S^{\pm}.$$

Proof According to Exercise 1.5.5 (4) the functions $\{\vartheta^c_{\overline{w}} \mid \overline{w} \in K(L)_1\}$ are a basis of $H^0(L)$. Define for any $\overline{w} \in K(L)_1$

$$\theta^{\pm}_{\overline{w}} = \mathbf{e}(-4\pi i \operatorname{Im} H(w + c_1, c_2)) \vartheta^c_{\overline{w}} \pm \vartheta^c_{-\overline{w}-2\overline{c}_1}.$$

It follows immediately from the Inverse Formula that $\theta^+_{\overline{w}}$ is an even function and $\theta^-_{\overline{w}}$ is odd. Since $\{\theta^+_{\overline{w}}, \theta^-_{\overline{w}} \mid \overline{w} \in K(L)_1\}$ spans the vector space $H^0(L)$, the theta functions $\theta^+_{\overline{w}}, \overline{w} \in K(L)_1$ span $H^0(L)_+$. By definition

$$\theta^+_{-\overline{w}-2\overline{c}_1} = \mathbf{e}(4\pi i \operatorname{Im} H(w + c_1, c_2)) \theta^+_{\overline{w}}.$$

So for $\overline{w} \in K(L)_1 - S$ the functions $\theta^+_{\overline{w}}$ and $\theta^+_{-\overline{w}-2\overline{c}_1}$ are linearly dependent over $\mathbb{C}$.

Moreover, for $\overline{w} \in S$ we have

$$\theta^+_{\overline{w}} = \Big(\mathbf{e}(-4\pi i \operatorname{Im} H(w + c_1, c_2)) + 1\Big)\vartheta^c_{\overline{w}} = \begin{cases} 2\vartheta^c_{\overline{w}} & \text{if } \ \overline{w} \in S^+, \\ 0 & \text{if } \ \overline{w} \in S^-. \end{cases}$$

To see this, note that S is the disjoint union of the sets S^+ and S^-, since $\overline{w} \in S$ implies $2w+2c_1 \in \Lambda$ such that $\mathbf{e}(-4\pi i \operatorname{Im} H(w+c_1, c_2)) = \mathbf{e}(-\pi i \operatorname{Im} H(2w+2c_1, 2c_2)) = \pm 1$.

Choosing for every $\overline{w} \in K(L)_1 - S$ one function out of $\{\theta^+_{\overline{w}}, \theta^+_{-\overline{w}-2\overline{c}_1}\}$, these functions together with the functions $\vartheta^c_{\overline{w}}$ for $\overline{w} \in S^+$ obviously form a basis for $H^0(L)_+$. Noting that $\# K_1 = h^0(L)$ this implies the assertion. □

One can determine the sets S and S^+ in terms of the characteristic and the type of the line bundle L. In this way we get explicit formulas for $h^0(L)_+$ and $h^0(L)_-$. For the general case compare Exercise 2.3.7 (3). Here we only outline the most important case, the line bundle of characteristic 0.

Corollary 2.3.9 *Let L_0 denote the ample line bundle of type $(d_1, \dots, d_g)$, with characteristic* 0 *with respect to some decomposition for L_0. Suppose $d_1, \dots, d_s$ are odd and $d_{s+1}, \dots, d_g$ are even for $0 \le s \le g$, then*

$$h^0(L_0)_\pm = \frac{1}{2} h^0(L_0) \pm 2^{g-s-1}.$$

Proof For the proof note that $S = S^+ = K(L_0)_1 \cap X_2$ and $\#(K(L_0)_1 \cap X_2) = 2^{g-s}$, since $K(L_0)_1 \simeq \oplus_{r=1}^g \mathbb{Z}/d_r\mathbb{Z}$. □

It is easy to derive analogous formulas for any symmetric line bundle L with $h^0(L) > 0$ using the reduction to the ample case of Section 1.5.4.

2.3.3 The Weil Form on X_2

Let $X = V/\Lambda$ be an abelian variety of dimension g. Our next aim is to compute the number of 2-division points $x \in X_2$ with even (respectively odd) multiplicity on certain divisors on X. For this we need a quadratic form on the $\mathbb{Z}/2\mathbb{Z}$-vector space X_2, sometimes called the *Weil form*, which we introduce in this section.

For any $H \in \operatorname{NS}(X)$ define a map

$$\mathrm{e}^H \colon X_2 \times X_2 \to \{\pm 1\} \quad \text{by} \quad \mathrm{e}^H(\overline{v}, \overline{w}) = \mathbf{e}\left(\pi i \operatorname{Im} H(2v, 2w)\right).$$

This definition does not depend on the choice of the representatives v of $\overline{v}$ and w of $\overline{w}$ in V and e^H takes values in $\{\pm 1\}$, since $\operatorname{Im} H(\Lambda \times \Lambda) \subseteq \mathbb{Z}$. So e^H is a symmetric bilinear form on the $2g$-dimensional $\mathbb{Z}/2\mathbb{Z}$-vector space X_2. It is called the *Weil pairing on* X_2 associated to H.

By definition a *quadratic form associated to* e^H is a map

$$q\colon X_2 \longrightarrow \{\pm 1\} \quad \text{satisfying} \quad q(x)q(y)q(x+y) = \mathrm{e}^H(x,y) \tag{2.7}$$

for all $x, y \in X_2$.

Lemma 2.3.10 *For every symmetric line bundle* $L = L(H,\chi)$ *on* X *the map*

$$q_L : X_2 \to \{\pm 1\}, \qquad q_L(\overline{v}) = \chi(2v)$$

is a quadratic form on X_2 *associated to* e^H.

Each q_L is called a *Weil form for* e^H. According to Lemma 2.3.5 we get in this way 2^{2g} quadratic forms for e^H.

Proof The map $q_L\colon X_2 \to \{\pm 1\}$ is well defined, since χ takes only values in $\{\pm 1\}$, L being symmetric. Moreover, the defining equation for a semicharacter, namely

$$\chi(\lambda)\chi(\mu) = \chi(\lambda+\mu)\,\mathbf{e}(\pi i \operatorname{Im} H(\lambda,\mu))$$

for every $\lambda, \mu \in \Lambda$, translates just to equation (2.7). Thus q_L is a quadratic form associated to e^H. □

Remark 2.3.11 It is easy to see that the quadratic form q_L coincides with the form e_*^L defined in Mumford [96] (see Exercise 2.7.4 (4)(c) below). For another definition of q_L see Exercise 2.3.7 (1).

We need the following elementary lemma on quadratic forms in characteristic 2.

Lemma 2.3.12 *Let* U *be a* $\mathbb{Z}/2\mathbb{Z}$*-vector space of dimension* $2g$ *and suppose that* $e\colon U \times U \longrightarrow \{\pm 1\}$ *is a symmetric bilinear form of rank* $2s$ *with radical* $K = \{u \in U \mid e(u,\cdot) \equiv 1\}$. *Suppose* $q\colon U \longrightarrow \{\pm 1\}$ *is a quadratic form associated to* e.

(a) *If* $q|_K$ *is trivial, then either*

(i) $\#q^{-1}(1) = 2^{2g-s-1}(2^s+1)$ *and* $\#q^{-1}(-1) = 2^{2g-s-1}(2^s-1)$ *or*
(ii) $\#q^{-1}(1) = 2^{2g-s-1}(2^s-1)$ *and* $\#q^{-1}(-1) = 2^{2g-s-1}(2^s+1)$.

(b) *If* $q|_K$ *is non-trivial, then* $\#q^{-1}(1) = \#q^{-1}(-1) = 2^{2g-1}$.

Proof **Step I**: Suppose e is non-degenerate; that is, $s = g$ and $K = \{0\}$: According to the elementary divisor theorem (see Bourbaki [28, Alg. IX.5.1 Th. 1]) there is a basis $u_1, \dots, u_g, u_1', \dots, u_g'$ of U such that $e(u_i,u_j) = e(u_i',u_j') = 1$ and $e(u_i,u_j') = (-1)^{\delta_{ij}}$ for $1 \le i,j \le g$.

Suppose first that $g = 1$: Then

$$q(u_1)q(u_1')q(u_1+u_1') = e(u_1,u_1') = -1,$$

Hence $\#q^{-1}(1) = 3$ or 1 and $\#q^{-1}(-1) = 1$ or 3, since $q(0) = 1$ in any case.

Now suppose $g > 1$ and that the assertion holds for all $g' < g$. Define subvector spaces $U_{g-1} = \langle u_i, u_i' | i = 2, \dots, g\rangle$ and $U_1 = \langle u_1, u_1'\rangle$. The restrictions $q_{g-1} = q|_{U_{g-1}}$ and $q_1 = q|_{U_1}$ are quadratic forms with non-degenerate associated bilinear forms $e|_{U_{g-1}\times U_{g-1}}$ and $e|_{U_1\times U_1}$. Any $v \in U$ decomposes uniquely as $v = v_{g-1} + v_1$ with $v_{g-1} \in U_{g-1}$ and $v_1 \in U_1$. Since $e(v_{g-1}, v_1) = 1$, we get $q(v) = q_{g-1}(v_{g-1})q_1(v_1)$. It follows that

$$\#q^{-1}(1) = \#q_{g-1}^{-1}(1)\#q_1^{-1}(1) + \#q_{g-1}^{-1}(-1)\#q_1^{-1}(-1).$$

If we are in case (i) for q_{g-1} and q_1, then

$$\#q^{-1}(1) = 2^{g-2}(2^{g-1}+1)\cdot 3 + 2^{g-2}(2^{g-1}-1)\cdot 1 = 2^{g-1}(2^g+1).$$

Hence $\#q^{-1}(-1) = 2^{2g} - \#q^{-1}(1) = 2^{g-1}(2^g - 1)$ and we are again in case (i). Similarly one checks the other possibilities to see that one ends up in case (i) or in case (ii).

Step II: Suppose e is trivial; that is, $s = 0$: If q is trivial, then $\#q^{-1}(1) = 2^{2g}$ and we are in case (i) of (a). Otherwise q is a surjective homomorphism $U \to \{\pm 1\}$, which implies the assertion.

Step III: Suppose $0 < s < g$: Let W denote an orthogonal complement of K in U with respect to the bilinear form e. The restriction $q_W = q|_W$ is a quadratic form as in Step I and $q_K = q|_K$ is a quadratic form as in Step II. Every $u \in U$ admits a unique decomposition $u = w + k$ with $w \in W$ and $k \in K$ such that $q(u) = q_W(w)\cdot q_K(k)$. It follows that

$$\#q^{-1}(1) = \#q_W^{-1}(1)\#q_K^{-1}(1) + \#q_W^{-1}(-1)\#q_K^{-1}(-1).$$

Inserting the results of Step I and Step II for q_W and q_K, we obtain the assertion. □

2.3.4 Symmetric Divisors

Let $X = V/\Lambda$ be an abelian variety of dimension g. A divisor D on X is called *symmetric* if $(-1)_X^* D = D$. The main result of this section is Proposition 2.3.15, where we compute the number of 2-division points, at which a symmetric divisor has even or odd multiplicity.

Let D be a symmetric divisor on X. Certainly the line bundle $L = \mathcal{O}_X(D)$ is also symmetric. Suppose L is ample, so that the linear system $|L|$ is non-empty. Clearly the divisors D in $|L|$ correspond one to one to canonical theta functions ϑ for L modulo $\mathbb{C}^*$ via $\pi^* D = (\vartheta)$. In order to determine the theta functions corresponding to symmetric divisors, we observe that, if ϑ is an even or odd theta function, the corresponding divisor D is symmetric. The following lemma shows that the converse is also true.

Lemma 2.3.13 *For $D \in |L|$ and $\vartheta \in H^0(L)$ with $\pi^* D = (\vartheta)$ the following conditions are equivalent:*

(i) *D is symmetric;*
(ii) *$\vartheta \in H^0(L)_+$ or $\vartheta \in H^0(L)_-$.*

Proof If one considers $H^0(L)$ as the space of sections of the line bundle L, the statement is obvious by the construction of the normalized isomorphism $(-1)_L$ (see Section 2.3.2). Let us also include a proof in terms of canonical theta functions.

It suffices to show (i) $\Rightarrow$ (ii): Since D is symmetric, there is a nowhere vanishing holomorphic function ϵ_D on V such that $\vartheta(-v) = \epsilon_D(v)\vartheta(v)$ for all $v \in V$. On the other hand $a_L(\lambda, v) = a_L(-\lambda, -v)$ for all $v \in V$ and $\lambda \in \Lambda$, since the line bundle L is symmetric. Hence we have

$$\begin{aligned}\epsilon_D(v+\lambda)\vartheta(v) &= \epsilon_D(v+\lambda)\vartheta(v+\lambda)a_L(\lambda, v)^{-1} \\ &= \vartheta(-v-\lambda)a_L(-\lambda,-v)^{-1} \\ &= \vartheta(-v) = \epsilon_D(v)\vartheta(v)\end{aligned}$$

for all $v \in V$ and $\lambda \in \Lambda$. This means ϵ_D is $2g$-fold periodic on V. So ϵ_D is constant by Liouville's theorem. Since $(-1)_V$ is an involution, $\epsilon_D = +1$ or $\epsilon_D = -1$. □

The lemma shows in particular that for any symmetric $L \in \text{Pic}(X)$ with $h^0(L) > 0$ there is an effective symmetric divisor D with $L = \mathcal{O}_X(D)$.

For an arbitrary, not necessarily effective divisor D on X, denote by $\text{mult}_x(D)$ the multiplicity of D at a point $x \in X$. A symmetric divisor D is called *even* (respectively *odd*) if $\text{mult}_0(D)$ is even (respectively odd). If D is moreover effective and ϑ a corresponding theta function, $\text{mult}_{\overline{v}}(D)$ is just the subdegree of the Taylor expansion of ϑ in $v \in V$, and D is even (respectively odd) if and only if ϑ is even (respectively odd).

Proposition 2.3.14 *Let $L = L(H, \chi)$ be a symmetric line bundle. For any symmetric divisor D on X with $L = \mathcal{O}_X(D)$ we have*

$$(-1)^{\text{mult}_x(D)} = \chi(\lambda)(-1)^{\text{mult}_0(D)}$$

for every 2-division point $x = \pi(\frac{1}{2}\lambda) \in X_2$.

Proof Without loss of generality we may assume that D is effective. Let $\vartheta \in H^0(L)$ be a corresponding canonical theta function. Then we have

$$\vartheta(-v) = (-1)^{\text{mult}_0(D)}\vartheta(v)$$

for all $v \in V$. On the other hand, $\text{mult}_x(D) = \text{mult}_0(t_x^* D)$ and according to Exercise 1.5.5 (11), $\widetilde{\vartheta} := \mathbf{e}(-\frac{\pi}{2}H(\cdot, \lambda))\vartheta(\cdot + \frac{1}{2}\lambda))$ is a canonical theta function for $t_x^* L$ corresponding to $t_x^* D$. For every $v \in V$,

$$\begin{aligned}(-1)^{\mathrm{mult}_x(D)}\widetilde{\vartheta}(v) &= \widetilde{\vartheta}(-v)\\ &= \mathbf{e}\left(\frac{\pi}{2}H(v,\lambda)\right)\vartheta(-v+\frac{1}{2}\lambda)\\ &= \mathbf{e}\left(\frac{\pi}{2}H(v,\lambda)\right)\vartheta(v+\frac{1}{2}\lambda-\lambda)\\ &= (-1)^{\mathrm{mult}_0(D)}\,\mathbf{e}\left(\frac{\pi}{2}H(v,\lambda)\right)\\ &\quad\cdot\chi(-\lambda)\,\mathbf{e}\left(\pi H(v+\frac{1}{2}\lambda,-\lambda)+\frac{\pi}{2}H(\lambda,\lambda)\right)\vartheta(v+\frac{1}{2}\lambda)\\ &= (-1)^{\mathrm{mult}_0(D)}\chi(\lambda)\widetilde{\vartheta}(v),\end{aligned}$$

since $\chi(\lambda) = \chi(-\lambda)$. This implies the assertion. □

For any not necessarily effective symmetric divisor $D \neq 0$ on X, define

$$X_2^+(D) = \{x \in X_2 \mid \mathrm{mult}_x(D) \equiv 0 \pmod 2\} \quad \text{and}$$
$$X_2^-(D) = \{x \in X_2 \mid \mathrm{mult}_x(D) \equiv 1 \pmod 2\}.$$

Obviously X_2 is the disjoint union of $X_2^+(D)$ and $X_2^-(D)$. The following proposition gives the cardinalities of these sets.

Proposition 2.3.15 *Let D be a non-trivial symmetric divisor on X and $L = \mathcal{O}_X(D)$. Suppose $L = L(H,\chi)$ is of type $(d_1,\ldots,d_g)$ with $d_1,\ldots,d_s$ odd and $d_{s+1},\ldots,d_g$ even.*

(a) *If $\chi|_{2\Lambda(L)\cap\Lambda}$ is trivial, then either*

(i) $\#X_2^+(D) = 2^{2g-s-1}(2^s+1)$ *and* $\#X_2^-(D) = 2^{2g-s-1}(2^s-1)$, *or*
(ii) $\#X_2^+(D) = 2^{2g-s-1}(2^s-1)$ *and* $\#X_2^-(D) = 2^{2g-s-1}(2^s+1)$.

(b) *If $\chi|_{2\Lambda(L)\cap\Lambda}$ is non-trivial, then* $\#X_2^+(D) = \#X_2^-(D) = 2^{2g-1}$.

Proof According to Proposition 2.3.14 we have for the quadratic form $q_L : X_2 \to \{\pm 1\}$ of Lemma 2.3.10,

$$q_L(x) = (-1)^{\mathrm{mult}_x(D)-\mathrm{mult}_0(D)}$$

for all $x \in X_2$ and hence $q_L^{-1}(1) = X_2^+(D)$, if D is even, and $q_L^{-1}(1) = X_2^-(D)$, if D is odd. It is easy to see that the rank of the bilinear form e^H is s. So in order to apply Lemma 2.3.12, it remains to show that q_L is trivial on the radical K of e^H if and only if $\chi|_{2\Lambda(L)\cap\Lambda}$ is trivial. Let $\pi\colon V \to X$ denote the canonical projection, then by definition of e^H

$$\pi^{-1}(K) = \left\{v \in \frac{1}{2}\Lambda \mid \mathrm{Im}\,H(v,\Lambda) \subseteq \mathbb{Z}\right\} = \Lambda(L) \cap \frac{1}{2}\Lambda.$$

Since $q_L(\overline{v}) = \chi(2v)$, the form q_L is trivial on K if and only if χ is trivial on $2(\Lambda(L) \cap \frac{1}{2}\Lambda) = 2\Lambda(L) \cap \Lambda$. □

In case (a) of the proposition there is still an ambiguity. In order to decide which of the possibilities (i) or (ii) holds, one has to consider the characteristic of the line bundle and the parity of the multiplicity of the divisor D in 0. A general formula would be messy. We give here a precise statement only in the case which we need later. Let L be a non-degenerate symmetric line bundle of type $(d_1, \dots, d_g)$. As always let $L_0 = L(H, \chi_0) \in \mathrm{Pic}^H(X)$ denote the line bundle of characteristic 0 with respect to some decomposition for H.

Corollary 2.3.16 *Suppose d_1 is even and D is a symmetric divisor on X with $\mathcal{O}_X(D) = L$.*

(a) *If $L = L_0$, then* $\#X_2^+(D) = \begin{cases} 2^{2g} & \text{if } D \text{ is even,} \\ 0 & \text{if } D \text{ is odd.} \end{cases}$

(b) *If $L \neq L_0$, then $\#X_2^+(D) = \#X_2^-(D) = 2^{2g-1}$.*

In particular, this includes the case of a Kummer polarization $(2, \dots, 2)$, which will be studied in Section 2.3.6.

Proof (a): With the notation of Proposition 2.3.15 we have $s = 0$ and $2\Lambda(L) \cap \Lambda = \Lambda$. By assumption the alternating form $\mathrm{Im}\, H$ on Λ is divisible by 2, such that $\chi_0(\lambda) = \mathbf{e}(\pi i \,\mathrm{Im}\, H(\lambda_1, \lambda_2)) = 1$ for all $\lambda = \lambda_1 + \lambda_2 \in \Lambda$. Hence in case $L = L_0$, Proposition 2.3.15 (a) gives $\#X_2^+(D) = 2^{2g}$ or 0. Since by definition $0 \in X_2^+(D)$ for an even divisor D, this proves (a).

(b): Suppose $L = t_c^* L_0$ with $c \in \frac{1}{2}\Lambda(L) \setminus \Lambda(L)$. One immediately sees that the semicharacter $\chi_0\, \mathbf{e}(\pi i \,\mathrm{Im}\, H(2c, \cdot))$ of L is non-trivial on Λ. So Proposition 2.3.15 (b) gives the assertion. □

A formula for the cardinality of $X_2^+(D)$ in the case when d_g is odd is given in Exercise 2.3.7 (4).

2.3.5 Quotients of Algebraic Varieties

In this section let us recall a result about complex algebraic varieties, which we need later in full generality. In the next section we need only the special case of an action of a finite group. The result is in fact valid even as a result about quotients of complex analytic spaces (see Cartan [30]).

Let G be a group acting as a group of isomorphisms on a complex algebraic variety X (with Euclidean topology). The quotient X/G, endowed with the quotient topology, admits the structure of a ringed space in a natural way: denote by $\pi : X \to X/G$ the canonical projection. Then by definition $\mathcal{O}_{X/G}(U)$, for $U \subseteq X/G$ open, is the set of functions $f : U \to \mathbb{C}$, for which $f\pi$ is an element of $\mathcal{O}_X(\pi^{-1}U)$.

Moreover, recall that the group G *acts properly and discontinuously* on X if for any pair of compact subsets K_1, K_2 of X the set $\{g \in G \mid gK_1 \cap K_2 \neq \emptyset\}$ is finite. Then

Theorem 2.3.17 *Suppose X is a complex quasiprojective algebraic variety and G is a group, acting properly and discontinuously on X. The quotient X/G is also an algebraic variety. Moreover, if X is normal, so is X/G.*

In the special case when X is smooth quasiprojective there is a criterion for the quotient X/G also to be smooth. Note that the action of G on X is said to be *free* if $gx = x$ for some $x \in X$ and $g \in G$ implies $g = \mathbf{1}_G$, in other words, if all stabilizers of the action are trivial.

Corollary 2.3.18 *Let X be smooth and suppose G is a group acting freely and properly discontinuously on X. Then X/G is also smooth.*

2.3.6 Kummer Varieties

As we saw at the beginning of Section 2.2.4, for the investigation of the map φ_{L^2} it suffices to consider the cases when L has no fixed components, which we treated in Theorem 2.2.8, and when L is an irreducible principal polarization. In this section we study the latter case.

Let $X = V/\Lambda$ be an abelian variety of dimension g. According to the last section, the quotient

$$K_X := X/(\pm\mathbf{1}_X)$$

is an algebraic variety. It is called the *Kummer variety* associated to X. Looking at the action of $(-\mathbf{1}_X)$ locally around a 2-division point, one immediately checks the following lemma.

Lemma 2.3.19 *The Kummer variety K_X is an algebraic variety of dimension g, smooth for $g = 1$ and smooth apart from 2^{2g} singular points, the images of the 2-division points of X under the natural map $p : X \to K_X$ for $g \geq 2$.*

Let $L = L(H, \chi)$ be an ample symmetric line bundle on X defining an irreducible principal polarization. Since $\chi(\Lambda) \subseteq \{\pm 1\}$ by Lemma 2.3.3, the semicharacter χ^2 of L^2 is identically equal to 1 on Λ. This implies that L^2 is of characteristic 0 with respect to any decomposition. According to Corollary 2.3.9 all theta functions in $H^0(L^2)$ are even. Hence there is a map $\psi = \psi_{L^2} \colon K_X \to \mathbb{P}^{2^g-1}$ such that the following diagram commutes

$$\begin{array}{ccc} X & \xrightarrow{\varphi=\varphi_{L^2}} & \mathbb{P}^{2^g-1} \\ & {\scriptstyle p}\searrow \quad \nearrow{\scriptstyle \psi} & \\ & K_X. & \end{array}$$

Theorem 2.3.20 *If $L \in \operatorname{Pic}(X)$ is symmetric and defines an irreducible principal polarization on X, then $\psi \colon K_X \to \mathbb{P}^{2^g-1}$ is an embedding.*

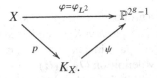

Proof Denote by Θ the unique (necessarily symmetric) divisor in the linear system $|L|$.

Step I: *ψ is injective*: Suppose $x, y \in X$ with $x \neq \pm y$. We have to show that there is a divisor $D \in |L^2|$ with $x \in D$ and $y \notin D$.

Since L is a principal polarization, $\phi_L \colon X \to \widehat{X}$ is an isomorphism and consequently $\Theta \neq t^*_{y-x}\Theta$ and $\Theta \neq t^*_{-(x+y)}\Theta$. Hence there is an element $z \in \Theta$ with $z \notin t^*_{y-x}\Theta \cup t^*_{-(x+y)}\Theta$. Consider the divisor

$$D = t^*_{z-x}\Theta + t^*_{x-z}\Theta \in |L^2|.$$

It satisfies $x = z + (x - z) \in t^*_{z-x}\Theta \subset D$. Moreover $y \notin t^*_{z-x}\Theta$, since otherwise $z \in t^*_{y-x}\Theta$, and $y \notin t^*_{x-z}\Theta$, since otherwise $z \in t^*_{-(x+y)}(-1)^*\Theta = t^*_{-(x+y)}\Theta$. Hence $y \notin D$.

Step II: *The differential $d\psi_q$ is injective for all smooth points $q \in K_X$.* Suppose $x \in X$, $2x \neq 0$, and $t \neq 0$ is a tangent vector to X at x. We have to show that there is a divisor $D \in |L^2|$ containing x such that t is not tangent to D at x.

According to Proposition 2.1.9 the image of the Gauß map for Θ is not contained in a hyperplane. This implies that there is a point $y \in \Theta$ with $y \notin t^*_{-2x}\Theta$ such that t is not tangent to Θ at y. The divisor

$$D = t^*_{y-x}\Theta + t^*_{x-y}\Theta \in |L^2|$$

contains x, since $x = y + (x - y) \in t^*_{y-x}\Theta \subset D$. By choice of y the vector t is not tangent to $t^*_{y-x}\Theta$ at x. Moreover $x \notin t^*_{x-y}\Theta$, since otherwise $y \in t^*_{-2x}(-1)^*\Theta = t^*_{-2x}\Theta$. Thus t is not tangent to D at x.

Step III: *The differential $d\psi_q$ is injective for any singular point $q \in K_X$.* Without loss of generality we may assume that q is the image of $0 \in X$; that is, $q = p(0)$. Since we identified $T_{X,0}$ with V, the tangent space of K_X at q can be identified with the symmetric product S^2V. If $v_1, \dots, v_g$ denote coordinate functions of V, $\{\frac{\partial}{\partial v_1}, \dots, \frac{\partial}{\partial v_g}\}$ is a basis of $T_{X,0} = V$ and $\{\frac{\partial^2}{\partial v_\nu \partial v_\mu} \mid 1 \leq \nu \leq \mu \leq g\}$ is a basis of $T_{K_X,q} = S^2V$.

According to Grothendieck [56, n° 221, Corollaire 5.3] the tangent space of $\mathbb{P}^{2^g-1} = P(H^0(L^2)^*)$ at a point P (considered as a hyperplane in $H^0(L^2)$) is

$$T_{\mathbb{P}_{2^g-1},P} = \mathrm{Hom}(P, H^0(L^2)/P).$$

We have to show that the natural map

$$d\psi_q \colon S^2V \to \mathrm{Hom}\left(\varphi(0), H^0(L^2)/\varphi(0)\right)$$

is injective. It is defined as follows: choose an isomorphism $H^0(L^2)/\varphi(0) \simeq \mathbb{C}$ such that $\mathrm{Hom}(\varphi(0), H^0(L^2)/\varphi(0)) = \mathrm{Hom}(\varphi(0), \mathbb{C})$. By definition $\varphi(0) = \{\vartheta \in H^0(L^2) \mid \vartheta(0) = 0\}$ and we have

$$d\psi_q\left(\sum_{\nu\le\mu}\alpha_{\nu\mu}\frac{\partial^2}{\partial v_\nu\partial v_\mu}\right) = \begin{cases} \varphi(0) & \longrightarrow & \mathbb{C} \\ \vartheta & \mapsto & \sum_{\nu\le\mu}\alpha_{\nu\mu}\frac{\partial^2\vartheta}{\partial v_\nu\partial v_\mu}(0). \end{cases}$$

Given $0 \neq \sum_{\nu\le\mu}\alpha_{\nu\mu}\frac{\partial^2}{\partial v_\nu\partial v_\mu} \in S^2V$, we have to show that there is a theta function $\vartheta \in \varphi(0)$ with $\sum \alpha_{\nu\mu}\frac{\partial^2\vartheta}{\partial v_\nu\partial v_\mu}(0) \neq 0$.

For this denote by Q the quadric in $P(V^*)$ defined by $\sum \alpha_{\nu\mu}\frac{\partial}{\partial v_\nu}\frac{\partial}{\partial v_\mu} = 0$. Since the Gauss map $G: \Theta_s \to P(V^*)$ (Θ_s = smooth part of Θ) is dominant by Proposition 2.2.6, there is an element $\overline{y} \in \Theta_s$ such that $G(\overline{y}) \notin Q$. In other words

$$\sum_{\nu\le\mu}\alpha_{\nu\mu}\frac{\partial\theta}{\partial v_\nu}(y)\frac{\partial\theta}{\partial v_\mu}(y) \neq 0\,,$$

where θ denotes a theta function associated to Θ. Now consider the theta function

$$\vartheta = \theta(\cdot + y)\theta(\cdot - y) \in H^0(L^2).$$

It is an element of $\varphi(0)$, since $\vartheta(0) = \theta(y)\theta(-y) = 0$. Moreover, we have

$$\frac{\partial^2\vartheta}{\partial v_\nu\partial v_\mu}(0) = \frac{\partial\theta}{\partial v_\nu}(y)\frac{\partial\theta}{\partial v_\mu}(-y) + \frac{\partial\theta}{\partial v_\nu}(-y)\frac{\partial\theta}{\partial v_\mu}(y)$$

$$= \begin{cases} -2\frac{\partial\theta}{\partial v_\nu}(y)\frac{\partial\theta}{\partial v_\mu}(y) & \text{if } \theta \text{ is even,} \\ +2\frac{\partial\theta}{\partial v_\nu}(y)\frac{\partial\theta}{\partial v_\mu}(y) & \text{if } \theta \text{ is odd.} \end{cases}$$

This implies that $\sum_{\nu\le\mu}\alpha_{\nu\mu}\frac{\partial^2\theta}{\partial v_\nu\partial v_\mu}(0) = \pm 2\sum_{\nu\le\mu}\alpha_{\nu\mu}\frac{\partial\theta}{\partial v_\nu}(y)\frac{\partial\theta}{\partial v_\mu}(y) \neq 0$. This completes the proof of the theorem. □

Let L be again an arbitrary ample symmetric line bundle on X. Combining the Decomposition Theorem 2.2.1, Theorem 2.2.8 and the previous theorem as well as Poincaré's Complete Reducibility Theorem 2.4.25 below, we obtain the following result on the map φ_{L^2}: Let

$$(X, L) = (X_1, L_1) \times \cdots \times (X_r, L_r)$$

denote the decomposition of the polarized abelian variety (X, L) into a product of irreducible polarized abelian varieties. To be precise, we assume that L is isomorphic to the exterior tensor product of the line bundles L_ν, $\nu = 1, \ldots, r$. In particular all L_ν's are symmetric. Suppose that (X_ν, L_ν) is principally polarized for $\nu = 1, \ldots, s$ and is not principally polarized for $\nu = s+1, \ldots, r$. For $\nu = 1, \ldots, s$ let $p_\nu : X_\nu \to K_{X_\nu}$ be the natural projection onto the Kummer variety K_{X_ν} associated to X_ν. Define

$$K = K_{X_1} \times \cdots \times K_{X_s} \times X_{s+1} \times \cdots \times X_r$$

and $p = p_1 \times \cdots \times p_s \times \mathbf{1}_{X_{s+1}} \times \cdots \times \mathbf{1}_{X_r} : X \to K$. Then $\varphi = \varphi_{L^2}$ factors as

$$\begin{array}{ccc} X & \xrightarrow{\varphi} & \mathbb{P}^n \\ & {\scriptstyle p}\searrow \quad \nearrow{\scriptstyle \psi} & \\ & K & \end{array}$$

with a holomorphic mapping ψ. Thus we obtain as a consequence of Theorems 2.2.8 and 2.3.20:

Theorem 2.3.21 *ψ is an embedding.*

We observe that φ is of degree 2^s onto its image. In particular, if none of the components (X_ν, L_ν) are principally polarized, φ is an embedding. Using Lemma 2.3.2, Theorem 2.3.21 generalizes easily to an arbitrary (not necessarily symmetric) ample line bundle L. One has only to modify the projection map $p\colon X \to K$ slightly.

2.3.7 Exercises

(1) Let L be a symmetric line bundle on X. For any $x \in X_2$ the normalized isomorphism $(-1)_L$ induces an automorphism $(-1)_L(x)$ of the fibre $L(x)$, which is multiplication by a constant denoted by $e_*^L(x) \in \mathbb{C}$.

(a) e_*^L is a map on X_2 with values in $\{\pm 1\}$.
(b) e_*^L coincides with q_L, the quadratic form defined in Section 2.3.3.

(2) Let L be a symmetric line bundle on X of type $(d_1, \dots, d_g)$ with $d_1, \dots, d_s$ odd and $d_{s+1}, \dots, d_g$ even. Denote by X_2^+ (respectively X_2^-) the set of 2-division points $x \in X_2$ such that the normalized isomorphism $(-1)_L$ acts on the fibre $L(x)$ by multiplication with $+1$ (respectively -1). Show that

(a) if $q_L|K(L) \cap X_2$ is trivial, then either
 (i) $\#X_2^+ = 2^{2g-s-1}(2^s + 1)$ and $\#X_2^- = 2^{2g-s-1}(2^s - 1)$, or
 (ii) $\#X_2^+ = 2^{2g-s-1}(2^s - 1)$ and $\#X_2^- = 2^{2g-s-1}(2^s + 1)$;
(b) if $q_L|K(L) \cap X_2$ is non-trivial, then $\#X_2^+ = \#X_2^- = 2^{2g-1}$.

(Hint: Use the previous exercise and Lemma 2.3.12.)

(3) Let H be a polarization of type $D = \operatorname{diag}(d_1, \dots, d_g)$ on $X = V/\Lambda$ with $d_1, \dots, d_s$ odd and $d_{s+1}, \dots, d_g$ even, and $L \in \operatorname{Pic}^H(X)$ symmetric. Suppose L is of characteristic $c = c_1 + c_2 \in \frac{1}{2}\Lambda(L)$ with respect to the decomposition defined by a symplectic basis of Λ. The symplectic basis induces a homomorphism

$$\psi : K(L) \to (\mathbb{Z}^g/D\mathbb{Z}^g)^2/2(\mathbb{Z}^g/D\mathbb{Z}^g)^2 = (\mathbb{Z}/2\mathbb{Z})^{2(g-s)}.$$

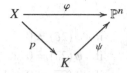

Show that

$$h^0(L)_\pm = \begin{cases} \frac{1}{2}h^0(L) & \text{if} \quad \psi(2\overline{c}) \neq 0, \\ \frac{1}{2}h^0(L) \pm 2^{g-s-1} & \text{if} \quad \psi(2\overline{c}) = 0 \text{ and } \mathbf{e}(4\pi i \operatorname{Im} H(c_1, c_2)) = 1, \\ \frac{1}{2}h^0(L) \mp 2^{g-s-1} & \text{if} \quad \psi(2\overline{c}) = 0 \text{ and } \mathbf{e}(4\pi i \operatorname{Im} H(c_1, c_2)) = -1. \end{cases}$$

(Hint: Use Proposition 2.3.8.)

(4) Let H be a polarization on an abelian variety $X = V/\Lambda$ of type $(d_1, \dots, d_g)$ with d_g odd.

(a) $2\Lambda(H) \cap \Lambda = 2\Lambda$.

(b) There are $2^{g-1}(2^g \pm 1)$ symmetric line bundles $L \in \operatorname{Pic}_s^H(X)$ such that

$$\# X_2^\pm(D) = 2^{g-1}(2^g \pm 1)$$

for all even symmetric divisors D on X with $\mathcal{O}(D) = L$.

(c) If L is of characteristic zero, and D a symmetric divisor with $L = \mathcal{O}(D)$,

$$\# X_2^\pm(D) = \begin{cases} 2^{g-1}(2^g \pm 1) & \text{if } D \text{ is even}, \\ 2^{g-1}(2^g \mp 1) & \text{if } D \text{ is odd}. \end{cases}$$

(5) Let (X, H) be a principally polarized abelian variety of dimension g. A subset of 2-division points $A \subset X_2$ is called *azygetic* if $e^H(x + y, x + z) = -1$ for all pairwise different points $x, y, z \in A$. For the definition of $e^H(\cdot, \cdot)$ see Section 2.3.3.

(a) Show that $\# A \leq 2g + 2$.

An azygetic subset $A \subset X_2$ with $2g + 2$ elements is called a *fundamental system* (see Krazer [77, p. 283]).

(b) There exist exactly $\frac{2^{2g+g^2}}{(2g+2)!}(2^{2g} - 1)(2^{2g-2} - 1) \cdots (2^2 - 1)$ fundamental systems.

(c) Suppose X is an abelian surface, $L \in \operatorname{Pic}^H(X)$ is of characteristic zero and D is the unique divisor in the linear system $|L|$. The set $X_2^-(D)$ (see Section 2.3.3) is a fundamental system. Moreover $\sum_{x \in X_2^-(D)} x = 0$.

(6) Let X be a simple abelian variety; that is, X admits no non-trivial abelian subvariety. Show that any algebraic subvarieties V and W of X with $\dim V + \dim W \geq \dim X$ have a non-empty intersection.

2.4 Poincaré's Complete Reducibility Theorem

This section contains a proof of Poincaré's Complete Reducibility Theorem. For this we need some preparations, namely the Rosati involution and the description of abelian subvarieties in terms their norm-endomorphisms, both depending on a polarization. In the corresponding three sections we prove a bit more than is actually needed for the proof of Poincaré's theorem, which is proved in Section 2.4.4.

2.4.1 The Rosati Involution

Let $X = V/\Lambda$ be an abelian variety of dimension g. In Section 1.1.2 we introduced the endomorphism algebra $\mathrm{End}_{\mathbb{Q}}(X)$ and its analytic and rational representations ρ_a and ρ_r. In this section we show that every polarization on X induces an anti-involution, called the Rosati involution, and a positive definite bilinear form on $\mathrm{End}_{\mathbb{Q}}(X)$.

Fix a polarization L on X. It induces an isogeny $\phi_L \colon X \to \widehat{X}$ depending only on the class of L in $\mathrm{NS}(X)$. The exponent $e(L)$ of the finite group $K(L) = \mathrm{Ker}\,\phi_L$ is called the *exponent of the polarization L*. According to Proposition 1.1.15 there exists a unique isogeny $\psi_L \colon \widehat{X} \to X$ such that

$$\psi_L \phi_L = e(L)_X \qquad \text{and} \qquad \phi_L \psi_L = e(L)_{\widehat{X}},$$

the multiplications by the integer $e(L)$ on X and $\widehat{X}$ respectively. Thus ϕ_L has an inverse in $\mathrm{Hom}_{\mathbb{Q}}(\widehat{X}, X)$, namely

$$\phi_L^{-1} = \frac{1}{e(L)} \psi_L.$$

Every $f \in \mathrm{End}_{\mathbb{Q}}(X)$ can be written in the form rh with $h \in \mathrm{End}(X)$ and $r \in \mathbb{Q}$. Then the dual of $f = rh$ is defined as

$$\widehat{f} := r\widehat{h} \in \mathrm{End}_{\mathbb{Q}}(\widehat{X}).$$

Clearly this definition does not depend on the choice of r and h. Consider the map

$$' \colon \mathrm{End}_{\mathbb{Q}}(X) \to \mathrm{End}_{\mathbb{Q}}(X), \quad f' = \phi_L^{-1} \widehat{f} \phi_L.$$

For the proof of the following lemma we refer to Exercise 2.4.5 (1).

Lemma 2.4.1 *Given a polarization on X, the map ′ satisfies*

$$(rf + sg)' = rf' + sg',$$
$$(fg)' = g'f' \quad \textit{and} \quad f'' = f$$

for all $f, g \in \mathrm{End}_{\mathbb{Q}}(X)$ *and* $r, s \in \mathbb{Q}$.

So $'$ is an anti-involution on $\mathrm{End}_{\mathbb{Q}}(X)$, called the *Rosati involution* with respect to the polarization L, although it is an anti-involution.

Suppose $L = L(H, \chi)$ and $E = \operatorname{Im} H$. The following proposition shows that the Rosati involution is the adjoint operator with respect to the hermitian form H as well as with respect to the alternating form E.

Proposition 2.4.2 *Suppose* $f \in \mathrm{End}_{\mathbb{Q}}(X)$.

(a) $E(\rho_r(f)(\lambda), \mu) = E(\lambda, \rho_r(f')(\mu))$ *for all* $\lambda, \mu \in \Lambda$.
(b) $H(\rho_a(f)(v), w) = H(v, \rho_a(f')(w))$ *for all* $v, w \in V$.

Proof At the beginning of Section 1.4.1 we saw that the canonical bilinear form

$$\langle\,,\,\rangle : \overline{\Omega} \times V \to \mathbb{R}, \quad \langle l, v\rangle = \operatorname{Im} l(v)$$

is non-degenerate. According to Lemma 1.4.5, the map $\phi_H : V \to \overline{\Omega}$, $v \to H(v, \cdot)$ is the analytic representation of ϕ_L, and finally, by the definition of the dual $\widehat{f}$ (see diagram (1.13)) we have that $\rho_a(\widehat{f}) = \rho_a(f)^*$.

This implies $\rho_a(f') = \phi_H^{-1}\rho_a(f)^*\phi_H$. Hence for all $v, w \in V$

$$\begin{aligned} E\big(\rho_a(f')(v), w\big) &= \langle \phi_H\big(\rho_a(f')(v)\big), w\rangle = \langle \rho_a(f)^*\phi_H(v), w\rangle \\ &= \langle \phi_H(v), \rho_a(f)(w)\rangle = E\big(v, \rho_a(f)(w)\big). \end{aligned}$$

Since $\rho_r(f) = \rho_a(f)|_\Lambda$ and $f'' = f$, this implies (a). For the proof of (b), see Exercise 2.4.5 (2). □

For any $f \in \mathrm{End}_{\mathbb{Q}}(X)$ the characteristic polynomial P^r_f of the rational representation $\rho_r(f)$ is

$$P^r_f(t) = \det\big(t\mathbf{1}_\Lambda - \rho_r(f)\big).$$

It is a monic polynomial in t of degree $2g$ with rational coefficients. Similarly the characteristic polynomial P^a_f of the analytic representation $\rho_a(f)$,

$$P^a_f(t) = \det\big(t\mathbf{1}_V - \rho_a(f)\big),$$

is a monic polynomial in t of degree g with complex coefficients. The polynomials P^a_f and P^r_f are related as follows:

Proposition 2.4.3 *For any* $f \in \mathrm{End}_{\mathbb{Q}}(X)$,

(a) $P^r_f = P^a_f \cdot \overline{P^a_f}$;
(b) $P^r_f(n) = \deg(n_X - f)$ *for all* $n \in \mathbb{Z}$.

Proof (a) is a consequence of Proposition 1.1.9 which states that $\rho_r \simeq \rho_a \oplus \bar{\rho}_a$.

(b): Recall from Proposition 1.1.13 that the degree of an endomorphism is equal to the determinant of its rational representation. Hence

$$\deg(n_X - f) = \det \rho_r(n_X - f) = \det\big(n\mathbf{1}_\Lambda - \rho_r(f)\big) = P^r_f(n). \qquad \square$$

Suppose $f \in \mathrm{End}_{\mathbb{Q}}(X)$ and

$$P^r_f(t) = \sum_{\nu=0}^{2g}(-1)^\nu r_\nu t^{2g-\nu} \quad \text{and} \quad P^a_f(t) = \sum_{\nu=0}^{g}(-1)^\nu a_\nu t^{g-\nu}$$

with coefficients $r_\nu \in \mathbb{Q}$, $r_0 = 1$ and $a_\nu \in \mathbb{C}$, $a_0 = 1$. The *rational* and the *analytic trace* of f are defined by

$$\mathrm{Tr}_r(f) = r_1 \quad \text{and} \quad \mathrm{Tr}_a(f) = a_1.$$

Similarly the *rational* and the *analytic norm* of f are defined by

$$N_r(f) = r_{2g} \quad \text{and} \quad N_a(f) = a_g.$$

As an immediate consequence of Proposition 2.4.3 we get

Corollary 2.4.4 *For any $f \in \mathrm{End}_{\mathbb{Q}}(X)$,*

(a) $N_r(f) = |N_a(f)|^2 = \deg(f)$,
(b) $\mathrm{Tr}_r(f) = 2\,\mathrm{Re}\,\mathrm{Tr}_a(f)$.

The analytic trace and norm of f and its Rosati involution f' are related as follows:

Lemma 2.4.5 *For any $f \in \mathrm{End}_{\mathbb{Q}}(X)$ we have $P^a_{f'}(t) = \overline{P^a_f(t)}$. In particular*

$$\mathrm{Tr}_a(f') = \overline{\mathrm{Tr}_a(f)} \quad \text{and} \quad N_a(f') = \overline{N_a(f)}.$$

Proof It suffices to prove the first assertion.

$$\begin{aligned} P^a_{f'}(t) &= \det\big(t\,\mathbf{1}_V - \rho_a(\phi_L^{-1})\rho_a(\hat{f})\rho_a(\phi_L)\big) \\ &= \det\big(\rho_a(\phi_L)^{-1}(t\,\mathbf{1}_{\overline{\Omega}} - \rho_a(\hat{f}))\rho_a(\phi_L)\big) \\ &= \det\big(t\,\mathbf{1}_{\overline{\Omega}} - \rho_a(\hat{f})\big) \\ &= \det\big(t\,\mathbf{1}_{\mathbb{C}^g} - {}^t\overline{\rho_a(f)}\big) = \overline{P^a_f(t)}, \end{aligned}$$

where for the last equation we used Exercise 1.4.5 (6). □

Before we proceed, we compute the rational trace of an endomorphism f of X in terms of intersection numbers. For this we need some further notation: For any line bundle M on X define $D_M(f)$ to be the line bundle

$$D_M(f) := (f + \mathbf{1}_X)^* M \otimes f^* M^{-1} \otimes M^{-1}.$$

Proposition 2.4.6 *Let (M^g) and $(D_M(f) \cdot M^{g-1})$ denote the intersection numbers of the corresponding line bundles. Then*

$$(M^g)\mathrm{Tr}_r(f) = g \cdot (D_M(f) \cdot M^{g-1}).$$

Proof Comparing first Chern classes and semicharacters, one easily checks that for all integers n we have $(n_X - f)^*M \equiv D_M(f)^{-n} \otimes f^*M \otimes M^{n^2}$. So we get for the self-intersection number

$$\left(((n_X - f)^*M)^g\right) = ((D_M(f)^{-n} \otimes f^*M \otimes M^{n^2})^g)$$
$$= (M^g)n^{2g} - g\big(D_M(f) \cdot M^{g-1}\big)n^{2g-1} + \cdots .$$

On the other hand, according to Corollary 1.7.6 we have $\chi((n_X - f)^*M) = \deg(n_X - f) \cdot \chi(M)$. Hence Riemann–Roch (applied twice), Corollary 1.7.6 and Proposition 2.4.3 give

$$\left(((n_X - f)^*M)^g\right) = g!\chi((n_X - f)^*M)$$
$$= g!\deg(n_X - f) \cdot \chi(M) = g!P^r_f(n) \cdot \chi(M)$$
$$= P^r_f(n) \cdot (M^g).$$

Comparing coefficients gives the assertion. □

As above let L be a polarization on X with Rosati involution $'$. According to Lemma 2.4.5 and Corollary 2.4.4 (b)

$$(f, g) \mapsto \mathrm{Tr}_r(f'g) = \mathrm{Tr}_a(f'g) + \mathrm{Tr}_a(g'f)$$

defines a symmetric bilinear form on $\mathrm{End}_{\mathbb{Q}}(X)$ with values in $\mathbb{Q}$. We claim that the associated quadratic form $f \mapsto \mathrm{Tr}_r(f'f)$ is positive definite. To see this, we give, more generally, a geometric interpretation of the coefficients of the polynomial $P^r_{f'f}$.

Lemma 2.4.7 *For all $f \in \mathrm{End}(X)$ and $n \in \mathbb{Z}$*

$$\chi(L)\, P^a_{f'f}(n) = \chi(f^*L^{-1} \otimes L^n).$$

Proof According to Proposition 1.4.6 we have $\phi_{f^*L} = \widehat{f}\phi_L f$. Since $(f'f)' = f'f$, Proposition 2.4.3 (a) and Lemma 2.4.5 yield $P^r_{f'f} = (P^a_{f'f})^2$. So applying the Riemann–Roch Theorem, Proposition 1.4.6 and Proposition 2.4.3 (b) we get

$$\chi(f^*L^{-1} \otimes L^n)^2 = \deg \phi_{f^*L^{-1} \otimes L^n}$$
$$= \deg(\phi_{f^*L^{-1}} + \phi_{L^n}) = \deg(n\phi_L - \widehat{f}\phi_L f)$$
$$= \deg(n\phi_L - \phi_L f'f) = \deg\phi_L \deg(n_X - f'f)$$
$$= \chi(L)^2 P^r_{f'f}(n) = \big(\chi(L)\, P^a_{f'f}(n)\big)^2.$$

Hence $\chi(f^*L^{-1} \otimes L^n) = \pm\chi(L)\, P^a_{f'f}(n)$ as polynomials in n. But for large n both sides are positive, since L is ample. □

We obtain the following geometric interpretation of the coefficients of the polynomial $P^a_{f'f}$.

Corollary 2.4.8 *Suppose* $f \in \mathrm{End}(X)$ *and* $P^a_{f'f}(t) = \sum_{v=0}^{g}(-1)^v a_v t^{g-v}$. *For* $v = 0, \dots, g,$

$$a_v = \binom{g}{v}\frac{(f^*L^v \cdot L^{g-v})}{(L^g)} \geq 0.$$

Proof Applying Riemann–Roch we conclude from the previous lemma

$$P^a_{f'f}(n) = \frac{((f^*L^{-1} \otimes L^n)^g)}{(L^g)} = \sum_{v=0}^{g}(-1)^v\binom{g}{v}\frac{(f^*L^v \cdot L^{g-v})}{(L^g)}n^{g-v},$$

and the equality of the coefficients holds. All intersection numbers are nonnegative by Lemma 2.2.2. □

For any nonzero endomorphism f of X and ample $L \in \mathrm{Pic}(X)$, the restriction $L|_{\mathrm{Im}\,f}$ is ample and $f^*: H^0(L|_{\mathrm{Im}\,f}) \to H^0(f^*L)$ is injective. Hence there is a non-trivial effective divisor D on X with $f^*L = O_X(D)$ and we get

$$\mathrm{Tr}_r(f'f) = 2a_1 = 2g\frac{(f^*L \cdot L^{g-1})}{(L^g)} = 2g\frac{((L|_D)^{g-1})}{(L^g)} > 0,$$

since $L|_D$ is ample on D. Thus we get as a consequence:

Theorem 2.4.9 $(f,g) \mapsto \mathrm{Tr}_r(f'g)$ *is a positive definite symmetric bilinear form on the* $\mathbb{Q}$-*vector space* $\mathrm{End}_{\mathbb{Q}}(X)$.

By what we have said above, this implies that also the form $(f,g) \mapsto \mathrm{Tr}_a(f'g)$ is positive definite. Finally we give some applications of Theorem 2.4.9.

The group of automorphisms of an abelian variety need not be finite. For an example, see Exercise 1.1.6 (14).

Recall that an *automophism of a polarized abelian variety* (X, L) is an automorphism f of X which respects the polarization, meaning that $f^*L \sim L$ or equivalently $c_1(f^*L) = c_1(L)$. Clearly these automorphisms form a group.

Corollary 2.4.10 *The group of automorphisms of any polarized abelian variety* (X, L) *is finite.*

Proof Suppose f is an automorphism of (X, L). Then $f^*L \otimes L^{-1} \in \mathrm{Pic}^0(X)$ so that $\phi_L = \phi_{f^*L} = \widehat{f}\phi_L f$. This implies $f'f = \phi_L^{-1}\widehat{f}\phi_L f = \phi_L^{-1}\phi_L = 1_X$. Consequently

$$f \in \mathrm{End}(X) \cap \{\varphi \in \mathrm{End}(X) \otimes_{\mathbb{Z}} \mathbb{R} \mid \mathrm{Tr}_a(\varphi'\varphi) = g\}.$$

Since the group $\mathrm{End}(X)$ is discrete in $\mathrm{End}(X) \otimes \mathbb{R}$ (see Proposition 1.1.8) and since moreover the set

$$\{\varphi \in \mathrm{End}(X) \otimes_{\mathbb{Z}} \mathbb{R} \mid \mathrm{Tr}_a(\varphi'\varphi) = g\}$$

is compact according to Theorem 2.4.9, this intersection is finite. □

Corollary 2.4.11 *Let* f *be an automorphism of a polarized abelian variety* (X, L) *and* $n \geq 3$ *an integer. If* $f|_{X_n} = 1_{X_n}$, *then* $f = 1_X$.

Proof Assume the contrary; that is, $f \neq \mathbf{1}_X$. According to Corollary 2.4.10 the automorphism f has finite order. By eventually passing to a power of f we may assume that f is of order p for some prime p. Since the only unipotent automorphism of (X, L) is the identity, there is an eigenvalue ξ of f which is a primitive p-th root of unity. By assumption $X_n \subset \ker(\mathbf{1}_X - f)$. Hence there is a $g \in \mathrm{End}(X)$ such that $ng = \mathbf{1}_X - f$. This implies that there is an algebraic integer η, namely an eigenvalue of g, such that

$$n\eta = 1 - \xi.$$

Applying the norm of the field extension $\mathbb{Q}(\xi)|\mathbb{Q}$ we get

$$n^{p-1} N_{\mathbb{Q}(\xi)|\mathbb{Q}}(\eta) = N_{\mathbb{Q}(\xi)|\mathbb{Q}}(1-\xi) = (1-\xi) \cdot \ldots \cdot (1-\xi^{p-1}) = p.$$

This is impossible, since p is a prime and $n \geq 3$. □

According to Corollary 2.4.11 the restriction to X_n induces an embedding

$$\mathrm{Aut}(X, L) \hookrightarrow \mathrm{Aut}_{\mathbb{Z}/n\mathbb{Z}}(X_n) = GL_{2g}(\mathbb{Z}/n\mathbb{Z})$$

for any $n \geq 3$. This gives an easy bound for the order of the group of automorphisms of a polarized abelian variety.

2.4.2 Polarizations

Recall that by definition a polarization on an abelian variety $X = V/\Lambda$ is the class of an ample line bundle L in $\mathrm{NS}(X)$. By abuse of notation we often write L instead of its class in $\mathrm{NS}(X)$. If L is of type $(d_1, \ldots, d_g)$, we define the *degree of the polarization* L to be the product $d = d_1 \cdot \ldots \cdot d_g$. In this section we study the subset of $\mathrm{NS}(X)$ of polarizations of a given degree. The aim is to give a formula for the number of isomorphism classes of such polarizations.

Fix a polarization L_0 on X. The inverse $\phi_{L_0}^{-1}$ of $\phi_{L_0} : X \to \widehat{X}$ exists in $\mathrm{Hom}_{\mathbb{Q}}(\widehat{X}, X)$. Hence for every line bundle L on X the product $\phi_{L_0}^{-1}\, \phi_L$ is an element of $\mathrm{End}_{\mathbb{Q}}(X)$ depending only on the class of L in $\mathrm{NS}(X)$. Denoting $\mathrm{NS}_{\mathbb{Q}}(X) = \mathrm{NS}(X) \otimes_{\mathbb{Z}} \mathbb{Q}$, the polarization L_0 induces in this way a homomorphism of abelian groups

$$\mathrm{NS}_{\mathbb{Q}}(X) \longrightarrow \mathrm{End}_{\mathbb{Q}}(X), \quad L \mapsto \phi_{L_0}^{-1}\phi_L.$$

Consider the Rosati involution $f \mapsto f'$ on $\mathrm{End}_{\mathbb{Q}}(X)$ with respect to the polarization L_0. An element $f \in \mathrm{End}_{\mathbb{Q}}(X)$ is called *symmetric* (with respect to L_0), if $f' = f$. Let $\mathrm{End}^s_{\mathbb{Q}}(X)$ (respectively $\mathrm{End}^s(X)$) denote the subset of $\mathrm{End}_{\mathbb{Q}}(X)$ (respectively $\mathrm{End}(X)$) of symmetric elements. $\mathrm{End}^s(X)$ is an additive group and $\mathrm{End}^s_{\mathbb{Q}}(X)$ is a $\mathbb{Q}$-vector space and we have

$$\mathrm{End}^s_{\mathbb{Q}}(X) \simeq \mathrm{End}^s(X) \otimes_{\mathbb{Z}} \mathbb{Q}.$$

Proposition 2.4.12 *Let (X, L_0) be a polarized abelian variety.*

(a) *The map*

$$\varphi\colon \mathrm{NS}_{\mathbb{Q}}(X) \to \mathrm{End}^s_{\mathbb{Q}}(X), \quad L \mapsto \phi_{L_0}^{-1}\phi_L$$

is an isomorphism of $\mathbb{Q}$-vector spaces.

(b) *If L_0 is a principal polarization, φ restricts to an isomorphism of (additive) groups*

$$\varphi\colon \mathrm{NS}(X) \to \mathrm{End}^s(X).$$

Proof According to Exercise 1.4.5 (5) (b) the map φ is injective. Hence it suffices to show that $f \in \mathrm{End}_{\mathbb{Q}}(X)$ is in the image of φ if and only if f is symmetric with respect to L_0. But $f \in \mathrm{Im}\,\varphi$ means that $\phi_{L_0} f = \phi_L$ for some $L \in \mathrm{Pic}(X)$. According to Theorem 1.4.15 this is the case if and only if the bilinear form $(v, w) \mapsto \rho_a(\phi_{L_0} f)(v, w) = H_0(\rho_a(f)(v), w)$ is hermitian, where $H_0 = c_1(L_0)$. By Proposition 2.4.2 (b) the form $H_0(\rho_a(f)(\cdot), \cdot)$ is hermitian if and only if

$$H_0\big(\rho_a(f)(v), w\big) = \overline{H_0\big(\rho_a(f)(w), v\big)} = \overline{H_0\big(w, \rho_a(f')(v)\big)} = H_0\big(\rho_a(f')(v), w\big).$$

Since H_0 is non-degenerate, this is fulfilled if and only if $f' = f$. This completes the proof of (a).

For (b) we only note that for a principal polarization the map ϕ_{L_0} is an isomorphism, that is $\phi_{L_0}^{-1} \phi_L \in \mathrm{End}(X)$ for all L. □

Suppose $f = \varphi(L)$ is a symmetric endomorphism. The following proposition gives a geometric interpretation for the coefficients of the analytic characteristic polynomial P^a_f in terms of L.

Proposition 2.4.13 *Let $f = \phi_{L_0}^{-1} \phi_L \in \mathrm{End}^s_{\mathbb{Q}}(X)$ with characteristic polynomial $P^a_f(t) = \sum_{\nu=0}^g (-1)^\nu a_\nu t^{g-\nu}$. Then*

$$d_0\, a_\nu = \frac{(L_0^{g-\nu} \cdot L^\nu)}{(g-\nu)!\nu!} \quad \textit{for} \quad \nu = 0, \dots, g\,,$$

where d_0 denotes the degree of the polarization L_0.

Proof Applying Riemann–Roch and Proposition 2.4.3 we get

$$\begin{aligned}\chi(L_0^n \otimes L^{-1})^2 &= \deg \phi_{L_0^n \otimes L^{-1}} = \deg(n\phi_{L_0} - \phi_L)\\ &= \deg \phi_{L_0} \deg(n_X - \phi_{L_0}^{-1} \phi_L) = d_0^2 \deg(n_X - f)\\ &= d_0^2\, P^r_f(n) = d_0^2\, \big(P^a_f(n)\big)^2.\end{aligned}$$

The last equation follows from Lemma 2.4.5, since f is symmetric. Now the Euler–Poincaré characteristic $\chi(L_0^n \otimes L^{-1})$ is positive for large n as L_0 is ample. So

$$\chi(L_0^n \otimes L^{-1}) = d_0\, P^a_f(n).$$

On the other hand we get by Riemann–Roch

$$\chi(L_0^n \otimes L^{-1}) = \frac{1}{g!}((L_0^n \otimes L^{-1})^g) = \sum_{\nu=0}^{g}(-1)^\nu \frac{(L_0^{g-\nu} \cdot L^\nu)}{(g-\nu)!\nu!} n^{g-\nu}.$$

Comparing coefficients gives the assertion. □

One can use this proposition to determine the subset of $\mathrm{NS}(X)$ of polarizations of a given degree in terms of the endomorphism algebra. An endomorphism in $\mathrm{End}(X)$ is called *totally positive* if the zeros of its analytic characteristic polynomial P_f^a are all positive.

Theorem 2.4.14 *For a principal polarization L_0 on X the isomorphism $\varphi\colon \mathrm{NS}(X) \to \mathrm{End}^s(X)$ induces a bijection between the sets of*

(a) *polarizations of degree d on X, and*
(b) *totally positive symmetric endomorphisms of X with analytic norm d.*

Proof Identify $X = \widehat{X}$ via the isomorphism ϕ_{L_0}. Then $f = \phi_L$ and according to Lemma 1.4.5 its analytic characteristic polynomial P_f^a coincides with the characteristic polynomial of the hermitian form $H = c_1(L)$. In particular, the zeros of P_f^a are the eigenvalues of H. So L is a polarization; that is, positive definite, if and only if f is totally positive in $\mathrm{End}(X)$. Moreover, by Riemann–Roch and Proposition 2.4.13

$$\deg L = \frac{(L^g)}{g!} = a_g = N_a(f).$$

This completes the proof. □

Pulling back a line bundle by an endomorphism of X defines an action of $\mathrm{End}(X)$ on $\mathrm{NS}(X)$. Given a principal polarization L_0 this induces an action of $\mathrm{End}(X)$ on $\mathrm{End}^s(X)$ via the diagram

$$\begin{array}{ccc} \mathrm{End}(X) \times \mathrm{NS}(X) & \longrightarrow & \mathrm{NS}(X) \\ {\scriptstyle (\mathrm{id},\varphi)}\downarrow & & \downarrow{\scriptstyle \varphi} \\ \mathrm{End}(X) \times \mathrm{End}^s(X) & \xrightarrow{\ \tau\ } & \mathrm{End}^s(X). \end{array}$$

Lemma 2.4.15 *If L_0 defines a principal polarization, then*

$$\tau(\alpha, f) = \alpha' f \alpha \quad \textit{for all} \quad \alpha \in \mathrm{End}(X) \textit{ and } f \in \mathrm{End}^s(X).$$

Proof By Proposition 2.4.12 there is an $L \in \mathrm{NS}(X)$ with $f = \phi_{L_0}^{-1}\phi_L$. Using Proposition 1.4.6 (d) we get

$$\begin{aligned}\varphi(\alpha^* L) &= \phi_{L_0}^{-1}\phi_{\alpha^* L} = \phi_{L_0}^{-1}\widehat{\alpha}\phi_L\alpha \\ &= (\phi_{L_0}^{-1}\widehat{\alpha}\phi_{L_0})(\phi_{L_0}^{-1}\phi_L)\alpha = \alpha' f \alpha.\end{aligned}$$

□

Two polarizations L and L' on X are called *isomorphic* if there is an automorphism α of X such that $L' = \alpha^* L$ in $\mathrm{NS}(X)$. This defines an equivalence relation on the set of polarizations of given degree on X. Using Theorem 2.4.14 and Lemma 2.4.15 one can translate this equivalence relation into terms of $\mathrm{End}^s(X)$.

Corollary 2.4.16 *For a principal polarization L_0 on X the isomorphism $\varphi\colon \mathrm{NS}(X) \to \mathrm{End}^s(X)$ induces a bijection between the sets of*

(a) *isomorphism classes of polarizations of degree d on X, and*
(b) *equivalence classes of totally positive symmetric endomorphisms with analytic norm d with respect to the equivalence relation:*

$$f_1 \sim f_2 \Leftrightarrow f_1 = \alpha' f_2 \alpha \quad \textit{for some} \quad \alpha \in \mathrm{End}(X).$$

Remark 2.4.17 One can use Corollary 2.4.16 in order to determine the set of isomorphism classes of polarizations of degree d explicitly in special cases (see Exercises 2.4.5 (15) and (16)). By a theorem of Narasimhan–Nori this set is always finite (see Exercise 2.4.5. (14)).

2.4.3 Abelian Subvarieties and Symmetric Idempotents

In this section we describe the set of abelian subvarieties of an abelian variety X in terms of the endomorphism algebra $\mathrm{End}_{\mathbb{Q}}(X)$. Given a polarization L on X we associate to every abelian subvariety Y of X an endomorphism N_Y, the norm-endomorphism, and a symmetric idempotent ε_Y. We will see that the symmetric idempotents are in one to one correspondence to the abelian subvarieties of X. This leads to a criterion for an endomorphism to be a norm-endomorphism. One of the various consequences is that $\mathrm{End}_{\mathbb{Q}}(X)$ is a semisimple $\mathbb{Q}$-algebra.

Let (X, L) be a polarized abelian variety and Y an abelian subvariety of X with canonical embedding $\iota\colon Y \hookrightarrow X$. Define the *exponent* of the abelian subvariety Y to be the exponent $e(\iota^* L)$ of the induced polarization on Y and write $e(Y) = e(\iota^* L)$. We have (as in Section 2.4.1) the isogeny

$$\psi_{\iota^* L} = e(Y)\phi_{\iota^* L}^{-1} : \widehat{Y} \to Y.$$

With this notation define the *norm-endomorphism of X associated to Y (with respect to L)* by

$$N_Y = \iota \psi_{\iota^* L} \hat{\iota} \phi_L ,$$

that is, as the composition

$$X \overset{\phi_L}{\to} \widehat{X} \overset{\hat{\iota}}{\to} \widehat{Y} \overset{\psi_{\iota^* L}}{\to} Y \overset{\iota}{\to} X.$$

The name norm-endomorphism comes from the theory of Jacobian varieties. In fact, it is a generalization of the usual notion of a norm map associated to a covering of algebraic curves (see Section 4.5.2).

Lemma 2.4.18 *For any abelian subvariety Y of X*

$$N_Y' = N_Y \quad \textit{and} \quad N_Y^2 = e(Y)N_Y\,,$$

where ′ denotes the Rosati involution with respect to the polarization L.

Proof $N_Y' = \phi_L^{-1}(\widehat{\phi_L}\,\iota\,\widehat{\psi_{\iota^*L}}\,\hat{\iota})\phi_L = N_Y$, since $\widehat{\phi_L} = \phi_L$ and $\widehat{\psi_{\iota^*L}} = \psi_{\iota^*L}$ by Proposition 1.4.6. The second assertion follows by a similar computation using $\hat{\iota}\phi_L\,\iota = \phi_{\iota^*L}$. □

We will show that these conditions characterize norm-endomorphisms. For this note that for the norm-endomorphism N_Y the element

$$\varepsilon_Y := \frac{1}{e(Y)} N_Y = \iota\phi_{\iota^*L}^{-1}\hat{\iota}\phi_L$$

of $\mathrm{End}^s_{\mathbb{Q}}(X)$ satisfies

$$\varepsilon_Y' = \varepsilon_Y \quad \text{and} \quad \varepsilon_Y^2 = \varepsilon_Y\,.$$

In other words, given a polarization L on X, we associate to every abelian subvariety Y of X a symmetric idempotent ε_Y in $\mathrm{End}_{\mathbb{Q}}(X)$. Conversely, if ε is a symmetric idempotent in $\mathrm{End}_{\mathbb{Q}}(X)$, there is an integer $n > 0$ such that $n\varepsilon \in \mathrm{End}(X)$. Define

$$X^{\varepsilon} = \mathrm{Im}(n\varepsilon).$$

Certainly this definition does not depend on the choice of n. Thus to every symmetric idempotent ε we associate an abelian subvariety X^{ε} of X.

Theorem 2.4.19 *The assignments $\varphi\colon Y \mapsto \varepsilon_Y$ and $\psi\colon \varepsilon \mapsto X^{\varepsilon}$ are inverse to each other and give a bijection between the sets of*

(a) *abelian subvarieties of X, and*
(b) *symmetric idempotents in* $\mathrm{End}_{\mathbb{Q}}(X)$.

Proof By definition we have $\psi\varphi(Y) = Y$ for any abelian subvariety Y of X. It remains to show that ψ is injective. Suppose that ε_1 and ε_2 are symmetric idempotents in $\mathrm{End}_{\mathbb{Q}}(X)$ with $X^{\varepsilon_1} = X^{\varepsilon_2}$. We have to show that $\varepsilon_1 = \varepsilon_2$.

Choose a positive integer n such that $f_i = n\varepsilon_i \in \mathrm{End}^s(X)$ for $i = 1$ and 2. Then $f_1^2 = nf_1$ and $f_2^2 = nf_2$. This means that f_ν is multiplication by n on $X^{\varepsilon_\nu} = \mathrm{Im}\, f_\nu$ implying $f_2 f_1 = nf_1$ and $f_1 f_2 = nf_2$. So

$$(f_1 - f_2)^2 = nf_1 - nf_2 - nf_1 + nf_2 = 0$$

and hence

$$\mathrm{Tr}_r\big((f_1 - f_2)'(f_1 - f_2)\big) = \mathrm{Tr}_r\big((f_1 - f_2)^2\big) = \mathrm{Tr}_r(0) = 0.$$

According to Theorem 2.4.9 this implies $f_1 = f_2$ and thus $\varepsilon_1 = \varepsilon_2$. □

As a direct consequence we obtain the following criterion for an endomorphism to be a norm-endomorphism with respect to a polarization L.

Corollary 2.4.20 *For $f \in \operatorname{End}(X)$ and $Y = \operatorname{Im} f$ the following statements are equivalent*

(i) $f = N_Y$,
(ii) $f' = f$ *and* $f^2 = e(Y)f$.

In general it is not easy to compute the exponent $e(Y)$. So Corollary 2.4.20 is not very useful in practice. In case of a principal polarization L we have a better criterion. Recall that an endomorphism $f \neq 0$ is called *primitive* if $f = ng$ for some $g \in \operatorname{End}(X)$ holds only for $n = \pm 1$. Equivalently, f is primitive if and only if its kernel does not contain a subgroup X_n of n-division points of X for some $n \geq 2$. (See Exercise 2.4.5 (20).)

Theorem 2.4.21 (Norm-endomorphism Criterion) *Let L be a principal polarization on X. For $f \in \operatorname{End}(X)$ the following statements are equivalent:*

(i) $f = N_Y$ *for some abelian subvariety Y of X.*
(ii) *The following three conditions hold:*

   (a) f *is either primitive or* $f = 0$;
   (b) $f = f'$;
   (c) $f^2 = ef$ *for some positive integer e.*

Proof It suffices to show that the norm-endomorphism of a non-trivial abelian subvariety Y of the principally polarized abelian variety (X, L) is primitive. Since ϕ_L is an isomorphism and $\iota: Y \hookrightarrow X$ is an embedding, it suffices to show that the kernel of $\psi_{\iota^*L}\,\widehat{\iota}$ does not contain $\widehat{X}_n$ for any $n \geq 2$. But

$$\widehat{(\psi_{\iota^*L}\widehat{\iota})} = \iota\,\widehat{\psi_{\iota^*L}} = \iota\psi_{\iota^*L}$$

does not contain $\widehat{Y}_n$ for any $n \geq 2$ by definition of ψ_{ι^*L} and since ι is an embedding. This implies the assertion. □

2.4.4 Poincaré's Theorem

Theorem 2.4.19 has some important applications. Note that the set of symmetric idempotents in $\operatorname{End}_{\mathbb{Q}}(X)$ admits a canonical involution, namely

$$\varepsilon \mapsto 1 - \varepsilon.$$

So by Theorem 2.4.19 the polarization L of X induces a canonical involution on the set of abelian subvarieties of X:

$$Y \mapsto Z := X^{1-\varepsilon_Y}.$$

We call Z the *complementary abelian subvariety of* Y *in* X (with respect to the polarization L). Of course Y is also the complementary abelian subvariety of Z in X. Hence it makes sense to call (Y, Z) a *pair of complementary abelian subvarieties of* X (with respect to the polarization L). In general the exponents $e(Y)$ of Y and $e(Z)$ of Z are different (for an example see Exercise 5.3.4 (1)). However, if L is a principal polarization, then $e(Y) = e(Z)$ (see Corollary 5.3.2 below).

The following lemma is an immediately consequence of the definitions and Lemma 2.4.18.

Lemma 2.4.22 *Let* (X, Y) *be a pair of complementary abelian subvarieties of* X *with respect to a polarization* L*. The norm-endomorphisms of* Y *and* Z *have the following properties:*

(1) $N_Y|_Y = e(Y)\mathbf{1}_Y$;
(2) $N_Y|_Z = 0$;
(3) $N_Y N_Z = 0$;
(4) $e(Y)N_Z + e(Z)N_Y = e(Z)e(Y)\mathbf{1}_X$.

For a proof see Exercise 2.4.5 (4). This leads to:

Theorem 2.4.23 (Poincaré's Reducibility Theorem) *Let* (X, L) *be a polarized abelian variety and* (Y, Z) *a pair of complementary abelian subvarieties of* X*. Then the map*

$$(N_Y, N_Z)\colon X \to Y \times Z$$

is an isogeny.

Proof The map (N_Y, N_Z) has finite kernel, since by lemma 2.4.22 (4) the kernel of (N_Y, N_Z) consists of $e(Y)e(Z)$-division points. In order to show that it is surjective, suppose $(y, z) \in Y \times Z$. There are $y_1, z_1 \in X$ such that

$$y = N_Y\Big(e(Y)e(Z)y_1\Big) \quad \text{and} \quad z = N_Z\Big(e(Y)e(Z)z_1\Big).$$

Using Lemma 2.4.22 this gives

$$(N_Y, N_Z)\Big(e(Z)N_Y(y_1) + e(Y)N_Z(z_1)\Big) = (y, z).$$ □

Poincaré's Reducibility Theorem has several important consequences.

Corollary 2.4.24 *For any pair* (Y, Z) *of complementary abelian subvarieties of* X *the addition map*

$$\mu\colon (Y, L|_Y) \times (Z, L|_Z) \to (X, L)$$

is an isogeny of polarized abelian varieties.

Proof Lemma 2.4.22 (4) means

$$\mu\big(e(Z)\mathbf{1}_Y \times e(Y)\mathbf{1}_Z\big)(N_Y, N_Z) = e(Y)e(Z)\mathbf{1}_X.$$

So with (N_Y, N_Z), $\big((e(Z)\mathbf{1}_Y \times e(Y)\mathbf{1}_Z\big)$, and $e(Y)e(Z)\mathbf{1}_X$ also μ is an isogeny of abelian varieties.

It remains to show that the induced polarization μ^*L on $Y \times Z$ splits. For this consider the isogeny

$$\phi_{\mu^*L} = \phi_{(\iota_Y+\iota_Z)^*L} = (\widehat{\iota_Y} + \widehat{\iota_Z})\,\phi_L\,(\iota_Y + \iota_Z).$$

As a map $Y \times Z \to \widehat{Y} \times \widehat{Z}$ it is given by the matrix

$$\begin{pmatrix} \alpha & \beta \\ \gamma & \delta \end{pmatrix} := \begin{pmatrix} \widehat{\iota_Y}\,\phi_L\,\iota_Y & \widehat{\iota_Y}\,\phi_L\,\iota_Z \\ \widehat{\iota_Z}\,\phi_L\,\iota_Y & \widehat{\iota_Z}\,\phi_L\,\iota_Z \end{pmatrix}.$$

First we claim that $\beta : Z \longrightarrow \widehat{Y}$ is the zero map: Using Lemma 2.4.22 (3) we have

$$0 = N_Y N_Z = \iota_Y\,\psi_{\iota_Y^* L}\,\widehat{\iota_Y}\,\phi_L\,\iota_Z\,\psi_{\iota_Z^* L}\,\widehat{\iota_Z}\,\phi_L = \iota_Y\,\psi_{\iota_Y^* L}\,\beta\,\psi_{\iota_Z^* L}\,\widehat{\iota_Z}\,\phi_L.$$

But $Z = \operatorname{Im}\psi_{\iota_Z^* L}\,\widehat{\iota_Z}$, the homomorphism ι_Y is a closed immersion, and $\psi_{\iota_Y^* L}$ and ϕ_L are isogenies. So $\beta = \widehat{\iota_Y}\,\phi_L\,\iota_Z = 0$. In the same way we obtain $\gamma = \widehat{\iota_Z}\,\phi_L\,\iota_Y = 0$. Finally we have $\alpha = \widehat{\iota_Y}\,\phi_L\,\iota_Y = \phi_{\iota_Y * L}$ and similarly $\delta = \phi_{\iota_Z * L}$. This this implies the assertion. □

An abelian variety X is called *simple* if it does not contain any abelian subvariety apart from X and 0. By induction one immediately obtains:

Theorem 2.4.25 (Poincaré's Complete Reducibility Theorem) *Given an abelian variety X there is an isogeny*

$$X \to X_1^{n_1} \times \cdots \times X_r^{n_r}$$

with simple abelian varieties X_ν not isogenous to each other. Moreover the abelian varieties X_ν and the integers n_ν are uniquely determined up to isogenies and permutations.

Corollary 2.4.26 $\mathrm{End}_{\mathbb{Q}}(X)$ *is a semisimple $\mathbb{Q}$-algebra. To be more precise: if $X \to X_1^{n_1} \times \cdots \times X_r^{n_r}$ is an isogeny as in the previous theorem, then*

$$\mathrm{End}_{\mathbb{Q}}(X) \simeq \mathrm{M}_{n_1}(F_1) \oplus \cdots \oplus \mathrm{M}_{n_r}(F_r)\,,$$

where $F_\nu = \mathrm{End}_{\mathbb{Q}}(X_\nu)$ are skew fields of finite dimension over $\mathbb{Q}$.

Proof Without loss of generality we may assume $X = X_1^{n_1} \times \cdots \times X_r^{n_r}$.

Since $\mathrm{Hom}(X_\nu^{n_\nu}, X_\mu^{n_\mu}) = 0$ for $\nu \neq \mu$, we obtain

$$\mathrm{End}_{\mathbb{Q}}(X) = \oplus_{\nu=1}^{r}\,\mathrm{End}_{\mathbb{Q}}(X_\nu^{n_\nu}).$$

Certainly $\mathrm{End}_{\mathbb{Q}}(X_\nu^{n_\nu})$ equals the ring of $(n_\nu \times n_\nu)$-matrices with entries in $\mathrm{End}_{\mathbb{Q}}(X_\nu)$. For the simple abelian variety X_ν every nonzero endomorphism is an isogeny and hence invertible in $\mathrm{End}_{\mathbb{Q}}(X)$. This proves that $\mathrm{End}_{\mathbb{Q}}(X)$ is a skew field over $\mathbb{Q}$. It is of finite dimension by Proposition 1.1.8. □

Corollary 2.4.27 *For any abelian variety X the Néron–Severi group* $\mathrm{NS}(X)$ *is a free abelian group of finite rank.*

This is a special case of Exercise 1.3.4 (10). Another proof: this is a consequence of Corollary 2.4.26, Proposition 2.4.12 and the fact that $\mathrm{NS}(X)$ is torsion-free. It also follows from the injectivity of the map $\mathrm{NS}(X) \longrightarrow \mathrm{Hom}(X, \widehat{X})$, $L \mapsto \phi_L$ (see Proposition 1.4.12), and the property of $\mathrm{Hom}(X, \widehat{X})$ to be a free $\mathbb{Z}$-module of finite rank.

For any symmetric idempotent ε in $\mathrm{End}_{\mathbb{Q}}(X)$ one can compute the dimension of the corresponding abelian subvariety X^ε:

Corollary 2.4.28 $\dim X^\varepsilon = \mathrm{Tr}_a(\varepsilon)$.

Proof Let $Y = X^\varepsilon$ and Z be the complementary abelian subvariety of X. Using Lemma 2.4.22 (1) and (2) we see that the following diagram is commutative

$$\begin{array}{ccc} X & \xrightarrow{(N_Y, N_Z)} & Y \times Z \\ {\scriptstyle N_Y}\downarrow & & \downarrow {\scriptstyle \begin{pmatrix} e(Y)\mathbf{1}_Y & 0 \\ 0 & 0 \end{pmatrix}} \\ X & \xrightarrow{(N_Y, N_Z)} & Y \times Z. \end{array}$$

By Poincaré's Reducibility Theorem (N_Y, N_Z) is an isogeny, so we have in $\mathrm{End}_{\mathbb{Q}}(X)$:

$$N_Y = (N_Y, N_Z)^{-1} \begin{pmatrix} e(Y)\mathbf{1}_Y & 0 \\ 0 & 0 \end{pmatrix} (N_Y, N_Z).$$

This gives

$$\mathrm{Tr}_a(\varepsilon) = \frac{1}{e(Y)} \mathrm{Tr}_a(N_Y) = \frac{1}{e(Y)} \mathrm{Tr}_a\left(\begin{pmatrix} e(Y)\mathbf{1}_Y & 0 \\ 0 & 0 \end{pmatrix} \right) = \dim Y.$$ □

Finally, we give an estimate for the degree of the isogeny $\mu : Y \times Z \longrightarrow X$ for a pair of complementary abelian subvarieties (Y, Z) of a polarized abelian variety (X, L).

Proposition 2.4.29 $\ker \mu \subset Y_{e(Y)} \cap Z_{e(Z)}$.

Proof By equations Lemma 2.4.22 (2),(3), and (4) we have $Z = \left(\ker N_Y\right)^0$. So we get using Lemma 2.4.22 (1),

$$\ker \mu = Y \cap Z \subset Y \cap \ker N_Y = \ker N_Y \iota_Y = \ker e(Y)\mathbf{1}_Y = Y_{e(Y)}.$$

Similarly $\ker \mu \subset Z_{e(Z)}$. This implies the assertion. □

Corollary 2.4.30 *If* $\gcd(e(Y), e(Z)) = 1$, *then* μ *is an isomorphism.*

Corollary 2.4.31 *If* Y *is an abelian subvariety of* (X, L) *with* $\deg L|_Y = 1$, *then there is an abelian subvariety* $Z \subset X$ *and an isomorphism of polarized abelian varieties*

$$(X, L) \simeq (Y, L|_Y) \times (Z, L|_Z).$$

Proof By Riemann–Roch $h^0(L|_Y) = \deg L|_Y) = 1$ if and only if $e(Y) = 1$. Hence Corollary 2.4.30 implies the assertion. □

2.4.5 Exercises

(1) Let (X, L) be a polarized abelian variety. Show that the Rosati involution $'$ satisfies

$$(rf + sg)' = rf' + sg',$$
$$(fg)' = g'f' \quad \text{and} \quad f'' = f$$

for all $f, g \in \mathrm{End}_{\mathbb{Q}}(X)$ and $r, s \in \mathbb{Q}$.
(Hint: Use Exercise 1.4.5 (4) (b).)

(2) Let H denote the first Chern class of a polarization L on X with Rosati involution $'$. Show that

$$H(\rho_a(f)(v), w) = H(v, \rho_a(f')(w))$$

for all $f \in \mathrm{End}_{\mathbb{Q}}(X)$ and $v, w \in V$.

(3) Let X be an abelian variety of positive dimension. Show that the abelian variety $X \times X$ admits infinitely many automorphisms.

(4) Let (X, Y) be a pair of complementary abelian subvarieties of X with respect to a polarization L of X. Show that the norm-endomorphisms of Y and Z have the properties:

(1) $N_Y|_Y = e(Y)\mathbf{1}_Y$;
(2) $N_Y|_Z = 0$;
(3) $N_Y N_Z = 0$;
(4) $e(Y)N_Z + e(Z)N_Y = e(Z)e(Y)\mathbf{1}_X$.

(5) Let $\varphi : \mathrm{NS}_{\mathbb{Q}}(X) \to \mathrm{End}^s_{\mathbb{Q}}(X)$ denote the isomorphism of Proposition 2.4.12. Show that the inverse map $\varphi^{-1} : \mathrm{End}^s_{\mathbb{Q}}(X) \to \mathrm{NS}_{\mathbb{Q}}(X)$ is given as follows: If $f \in \mathrm{End}^s_{\mathbb{Q}}(X)$, the element $\varphi^{-1}(f) \in \mathrm{NS}_{\mathbb{Q}}(X)$ is uniquely determined by $\phi_{L_0} f \in \mathrm{Hom}_{\mathbb{Q}}(X, \widehat{X})$.
(Hint: Use Theorem 1.4.15 which gives the element of $\mathrm{NS}_{\mathbb{Q}}(X)$ corresponding to $\phi_{L_0} f$.)

(6) Suppose S is a commutative subring of $\mathrm{End}^s_{\mathbb{Q}}(X)$. The multiplication on S induces a multiplication "$\circ$" on its preimage $\widetilde{S}$ under the isomorphism $\varphi : \mathrm{NS}_{\mathbb{Q}}(X) \longrightarrow \mathrm{End}^s_{\mathbb{Q}}(X)$, considered as a subspace of $\mathrm{Hom}_{\mathbb{Q}}(X, \widetilde{X})$. For $\phi_1, \phi_2 \in \widetilde{S}$ we have $\phi_1 \circ \phi_2 = \phi_2 \phi_{L_0}^{-1} \phi_1$.

(7) Suppose L_0 is an arbitrary not necessarily principal polarization. Generalize Theorem 2.4.14 to the following statement: Call an element $l \in \mathrm{NS}_{\mathbb{Q}}(X)$ a *polarization* if ml is represented by an ample line bundle on X for a suitable integer $m > 0$. Then the map φ: $\mathrm{NS}_{\mathbb{Q}}(X) \longrightarrow \mathrm{End}^s_{\mathbb{Q}}(X)$ induces a bijection between the sets of

(a) polarizations in $\mathrm{NS}_{\mathbb{Q}}(X)$, and
(b) totally positive symmetric elements in $\mathrm{End}^s_{\mathbb{Q}}(X)$.

(8) Let X_ν be an abelian variety with polarization L_ν of degree d_ν for $\nu = 1, 2$. Show that $p_1^* L_1 \otimes p_2^* L_2$ is a polarization on $X_1 \times X_2$ of degree $d_1 d_2$.

(9) (*Zarhin's trick*) Let L be a polarization of exponent e on an abelian variety $X = V/\Lambda$.

(a) Suppose there is an $f \in \mathrm{End}(X)$ such that:
 (i) $f(K(L)) \subseteq K(L)$;
 (ii) $\rho_a(f'f)|_{\frac{1}{e}\Lambda} \equiv -\mathbf{1}_{\frac{1}{e}\Lambda} \pmod{\Lambda}$, where $'$ denotes the Rosati involution with respect to L.
 Then $X \times \widehat{X}$ is principally polarized.
(b) Conclude that for any abelian variety X the abelian variety $(X \times \widehat{X})^4$ is principally polarized.

(10) Let f be an automorphism of order n of an abelian variety of dimension g. Show that $\varphi(n) \leq 2g$, where φ is the Euler function of elementary number theory.
In particular, $n \leq 6$ for $g = 1$, $n \leq 12$ for $g = 2$ and $n \leq 18$ for $g = 3$.

(11) Let X be an abelian variety. There exists an integer $n = n(X)$ such that for any abelian subvariety Y of X there exists an abelian subvariety Z of X with $Y + Z = X$ and $\# Y \cap Z \leq n$ (see Bertrand [20]).

(12) Let X be an abelian variety and L a line bundle on X. For any $f_1, f_2 \in \mathrm{End}(X)$ define a line bundle on X by

$$D_L(f_1, f_2) = (f_1 + f_2)^* L \otimes f_1^* L^{-1} \otimes f_2^* L^{-1}.$$

Note that $D_L(f, \mathbf{1}_X) = D_L(f)$ as defined in Proposition 2.4.6.

(a) D_L: $\mathrm{End}(X) \times \mathrm{End}(X) \longrightarrow \mathrm{Pic}(X)$ is symmetric and bilinear.
(b) The map D_L depends only on the class of L in $\mathrm{NS}(X)$: $D_L = D_{L \otimes P}$ for all $P \in \mathrm{Pic}^0(X)$.
(c) $\phi_{D_L(f_1, f_2)} = \widehat{f_1} \phi_L f_2 + \widehat{f_2} \phi_L f_1$.

(d) Suppose $M \in \operatorname{Pic}(X)$ defines a principal polarization on X, then $D_M(\phi_M^{-1}\widehat{f_1}\phi_L f_2) = D_L(f_1, f_2)$.

(13) Let X be an abelian variety of dimension g and $L \in \operatorname{Pic}(X)$ a polarization. Then

$$\operatorname{Tr}_r(f_1' f_2) = \frac{g}{(L^g)}(D_L(f_1, f_2) \cdot L^{g-1}) \quad \text{for all} \quad f_1, f_2 \in \operatorname{End}(X).$$

(Hint: generalize the proof of Proposition 2.4.6)

(14) (*The Narasimhan–Nori Theorem*) Let X be an abelian variety and d a positive integer. The number of isomorphism classes of polarizations of degree d is finite.
(Hint: show that $\operatorname{Aut}(X)$ is an arithmetic group and apply a result of Borel (see Borel [27, Théorème 9.11], Narasimhan–Nori [102]).)

(15) (*The number $\pi(X)$ of isomorphism classes of principal polarizations of X*) Let X be an abelian variety of dimension g with $\operatorname{End}(A) = \mathcal{O}$ the maximal order of a totally real number field K of degree g over $\mathbb{Q}$. Let U denote the group of units in $\mathcal{O}$ and U^+ the subgroup of totally positive units. If h denotes the class number and h^+ the narrow class number of K (that is, the order of the factor group of the group of ideals modulo the subgroup of totally positive principal ideals), then we have for the number of isomorphism classes $\pi(X)$ of principal polarizations of X:

$$\pi(X) = 0 \quad \text{or} \quad \pi(X) = \#U^+/U^2 = \frac{h^+}{h}.$$

(See Lange [80].)

(16) Let the notation be as in the previous exercise and assume that X admits a principal polarization.

(a) If $\sigma_0, \ldots, \sigma_{g-1}$ denote the real embeddings $K \hookrightarrow \mathbb{R}$, and if $\eta_1, \ldots, \eta_{g-1}$ is a system of fundamental units of $\mathcal{O}$ and $\eta_0 = -1$, then

$$\pi(X) = 2^{g-\operatorname{rk}M},$$

where M denotes the matrix $(\operatorname{sign}\sigma_i(\eta_j))$.

(b) Conclude the following result of Humbert [66]: If $g = 2$, then

$$\pi(X) = \begin{cases} 1 & \text{if } \operatorname{Aut}(X) \text{ contains an element of negative norm,} \\ 2 & \text{if } \operatorname{Aut}(X) \text{ contains no element of negative norm.} \end{cases}$$

(c) If $g = 3$, we have $\pi(X) = 1, 2$ or 4. Suppose $K = \mathbb{Q}(\omega)$. Show that $\pi(X) = 2$ if ω is the root of $x^3 + 12x^2 + 32x - 1 = 0$ with $0 < \omega < \frac{1}{2}$. Show that $\pi(X) = 4$ if ω is the root of $x^3 - 12x^2 + 26x - 1 = 0$.

(d) Let $g = 4$ and $K = \mathbb{Q}(\sqrt{6}, \sqrt{7})$. Show that $\pi(X) = 2$.
(Hint: use the set of fundamental units $\eta_1 = 8 + 3\sqrt{7}, \eta_2 = \sqrt{6} + \sqrt{7}$ and $\eta_3 = \frac{1}{2}(6 + 6\sqrt{6} + 2\sqrt{7} + \sqrt{42})$.)

(17) For any integer $d \geq 3$ give an example of an abelian surface X admitting two principal polarizations L_0 and L_1 such that $(L_0 \cdot L_1) = d$.
(Hint: Use the previous exercise and Proposition 2.4.13.)

(18) Let (X, L) be a polarized abelian variety. Consider the isomorphism $\varphi\colon \mathrm{NS}_{\mathbb{Q}}(X) \to \mathrm{End}^s_{\mathbb{Q}}(X)$ of Proposition 2.4.12. Show that for any abelian subvariety Y of X with norm-endomorphism N_Y with respect to L:

$$\varphi^{-1}(N_Y) = e(Y)^{-1} N_Y^* L.$$

(19) (*A Second Proof of Corollary 2.4.20*) Let L be a polarization on the abelian variety X and $f \in \mathrm{End}(X)$, $Y = \mathrm{im} f$, with $f' = f$ and $f^2 = e(i_Y^* L) f$. Let $Z = \phi_L(Y)$ and $\iota_Z\colon Z \to \widehat{X}$ be the canonical embedding.

(a) Show that there is a $\varphi \in \mathrm{Hom}(X, Z)$ such that

$$\phi_L f = \iota_Z \varphi.$$

(b) Show that $e(L)\phi_L^{-1}\widehat{\varphi}(\widehat{Z}) = Y$.
(c) There is a $\psi \in \mathrm{Hom}(\widehat{Z}, Y)$ such that

$$e(L)\phi_L^{-1}\widehat{\varphi} = \iota_Y \psi.$$

Show that

$$\phi_L^{-1}\iota_Z\widehat{\psi} = e(L)e(\iota_Y^* L)\iota_Y \phi^{-1}_{\iota_Y^* L}.$$

(d) Use (a), (b) and (c) to conclude that $f = N_Y$.

(See Birkenhake–Lange [21].)

(20) Show that an endomorphism f of an abelian variety X is primitive if and only if its kernel does not contain a subgroup X_n of n-division points of X for some $n \geq 2$.

(21) Let E be an elliptic curve with $\mathrm{End}(E) = \mathbb{Z}$ and $X = E \times E$. The line bundle $L = p_1^* \mathcal{O}_E(0) \otimes p_2^* \mathcal{O}_E(0)$ defines a principal polarization on X. Show

(a) There is a canonical isomorphism $\varphi\colon \mathrm{End}(X) \xrightarrow{\simeq} \mathrm{M}_2(\mathbb{Z})$ such that for the Rosati involution $'$ (with respect to L) is

$$\varphi(f') = {}^t\varphi(f).$$

(b) The preimages under φ of

$$\frac{1}{1+r^2}\begin{pmatrix} 1 & r \\ r & r^2 \end{pmatrix}, \quad \frac{1}{1+r^2}\begin{pmatrix} r^2 & -r \\ -r & 1 \end{pmatrix},$$

where $r \in \mathbb{Q}$ arbitrary, are all the symmetric idempotents $\neq 0, 1$ in $\mathrm{End}_{\mathbb{Q}}(X)$.

(22) Let (X, L) be a polarized abelian variety and $\varepsilon_1, \ldots, \varepsilon_{r+s}$ be symmetric idempotents of $\mathrm{End}_{\mathbb{Q}}(X)$ with $\sum_{\nu=1}^{r} \varepsilon_\nu = \sum_{\nu=1}^{s} \varepsilon_{r+\nu}$. Show that the abelian varieties $\times_{\nu=1}^{r} X^{\varepsilon_\nu}$ and $\times_{\nu=1}^{s} X^{\varepsilon_{r+\nu}}$ are isogenous.

2.5 Some Special Results

In the first section we introduce the dual of a polarization of an abelian variety. The second section contains a result on rational maps from a smooth projective variety to an abelian variety, namely that it is defined everywhere. Finally, the Pontryagin product, which was already introduced in Exercise 1.1.6 (11), is defined in a slightly different way and investigated in more detail.

2.5.1 The Dual Polarization

Let (X, L) be a polarized abelian variety of dimension g. In Sections 1.4.1 and 1.4.4 the dual abelian variety $\widehat{X}$ and the Poincaré bundle $\mathcal{P}$ on $X \times \widehat{X}$ were introduced. In this section we show that the polarization L induces a polarization L_δ on the dual abelian variety $\widehat{X}$, which can be considered as the dual polarization of L, in the sense that the double dual polarization coincides with the polarization L. The reason why we do not denote it by $\widehat{L}$ is that $\widehat{L}$ already denotes the Fourier-Mukai transform of L. In Section 6.1.2 below we will define $\widehat{L}$ and see in Exercise 6.1.4 (7) how it is related to L_δ.

Proposition 2.5.1 *Suppose the polarization L is of type $(d_1, \ldots, d_g)$. There is a unique polarization L_δ on $\widehat{X}$ characterized by the following equivalent properties*

$$(i) \quad \phi_L^* L_\delta \equiv L^{d_1 d_g}, \qquad (ii) \quad \phi_{L_\delta} \phi_L = d_1 d_g \mathbf{1}_X.$$

The polarisation L_δ is of type $(d_1, \frac{d_1 d_g}{d_{g-1}}, \ldots, \frac{d_1 d_g}{d_2}, d_g)$.

We call L_δ the *dual polarization* and the pair $(\widehat{X}, L_\delta)$ the *dual polarized abelian variety*.

Proof The isogeny $\phi_L : X \longrightarrow \widehat{X}$ is of exponent d_g. So

$$\psi_L := d_1 d_g \phi_L^{-1} : \widehat{X} \longrightarrow X$$

is also an isogeny. By Lemma 1.4.5 its analytic representation is given by the hermitian form $d_1 d_g c_1(L)^{-1}$. According to Theorem 1.4.15 there is a line bundle L_δ on $\widehat{X}$ such that $\psi_L = \phi_{L_\delta}$. In particular $c_1(L_\delta) = d_1 d_g c_1(L)^{-1}$, which is positive definite. So L_δ defines a polarization on $\widehat{X}$. By definition L_δ satisfies *ii*). The equivalence (i) $\Leftrightarrow$ (ii) follows from

$$\phi_{\phi_L^* L_\delta} = \widehat{\phi}_L \phi_{L_\delta} \phi_L = d_1 d_g \phi_L = \phi_{L^{d_1 d_g}}$$

using Proposition 1.4.12. The polarization L_δ is uniquely determined, since the Néron–Severi group is torsion-free. The type of the line bundle L_δ is given by the elementary divisors of the alternating form $\operatorname{Im} c_1(L_\delta) = d_1 d_g \operatorname{Im} c_1(L)^{-1}$, so L_δ is of type $(\frac{d_1 d_g}{d_g}, \frac{d_1 d_g}{d_{g-1}}, \dots, \frac{d_1 d_g}{d_1})$. □

An immediate consequence is

Corollary 2.5.2 $(L_\delta)_\delta \equiv L$.

Clearly the proof of Proposition 2.5.1 also works for $g \leq 2$, but perhaps it should be noted that for $g = 1$, we have $d_g = d_1$. So we get:

Corollary 2.5.3 *If* $\dim X = g \leq 2$, *then* $\deg L_\delta = \deg L$.

In general this need not be the case. The notion of a dual polarization behaves almost functorially with respect to isogenies: Suppose

$$f : (Y, M) \longrightarrow (X, L)$$

is an isogeny of polarized abelian varieties. In Section 1.4.1 the dual homomorphism $\widehat{f} : \widehat{Y} \to \widehat{X}$ is defined. If L is of type $(d_1, \dots, d_g)$ and M of type $(d_1', \dots, d_g')$, it is obvious that $d_1 | d_1'$ and $d_g | d_g'$. With this notation we have:

Proposition 2.5.4 $\widehat{f}^* M_\delta \equiv L_\delta^d$ *with* $d := \frac{d_1' d_g'}{d_1 d_g}$.

For the easy proof, see Exercise 2.5.4 (1).

Remark 2.5.5 Since the proof of Proposition 2.5.1 uses only results of Chapter 1, it also works for non-degenerate line bundles, say of index k. So the line bundle L_δ exists. For the proof that it is also non-degenerate of index k, see Exercise 2.5.4 (3) (Birkenhake–Lange [22]).

2.5.2 Morphisms into Abelian Varieties

In this section we compile some properties of morphisms of algebraic varieties into abelian varieties. For this we need the following general lemma.

Lemma 2.5.6 (Rigidity Lemma) *Let $f\colon Y \times Z \to X$ be a morphism of algebraic varieties. Suppose Y is complete. If*

$$f(\{y_0\} \times Z) = x_0 = f(Y \times \{z_0\})$$

for some $y_0 \in Y$, $z_0 \in Z$ and $x_0 \in X$, then f is constant.

Proof Let $U \subset X$ be an open affine neighbourhood of x_0 and $q\colon Y \times Z \to Z$ the natural projection. The variety Y being complete implies that the set

$$A := qf^{-1}(X \setminus U)$$

is closed in Z. Note that a point z is contained in $Z \setminus A$ if and only if $f(Y \times \{z\}) \subset U$. In particular $z_0 \in Z \setminus A$. Hence $Z \setminus A$ is open and dense in Z. Using the fact that any morphism of a complete variety into an affine variety is constant, we conclude that the image of $Y \times \{z\}$ under f is a point whenever $z \in Z \setminus A$. So

$$f(Y \times \{z\}) = f(\{y_0\} \times \{z\}) = x_0 \quad \text{for all} \quad z \in Z \setminus A.$$

This implies the assertion, since $Y \times (Z \setminus A)$ is open and dense in $Y \times Z$. □

Corollary 2.5.7 *Let Y and Z be algebraic varieties, one of them complete, and let X be an abelian variety.*

If $f\colon Y \times Z \to X$ is a morphism with $f(Y \times \{z_0\}) = 0$ for some $z_0 \in Z$, then there is a uniquely determined morphism $g\colon Z \to X$ such that $f(y, z) = g(z)$ for all $(y, z) \in Y \times Z$.

Proof Choose $y_0 \in Y$ and define

$$f'\colon Y \times Z \to X, \qquad (y, z) \mapsto f(y, z) - f(y_0, z).$$

Then $f'(\{y_0\} \times Z) = 0 = f'(Y \times \{z_0\})$ and by the Rigidity Lemma $f' \equiv 0$. Thus $g(z) := f(y_0, z)$ satisfies the assertion. □

Corollary 2.5.8 *Let Y and Z be algebraic varieties, one of them complete, and let X be an abelian variety. Suppose $f\colon Y \times Z \to X$ is a morphism with $f(y_0, z_0) = 0$ for some $(y_0, z_0) \in Y \times Z$.*

Then there are uniquely determined morphisms $g\colon Y \to X$ and $h\colon Z \to X$ with $g(y_0) = 0 = h(z_0)$ such that for all $(y, z) \in Y \times Z$,

$$f(y, z) = g(y) + h(z).$$

Proof Define $f' \colon Y \times Z \to X$ by $f'(y,z) = f(y,z) - f(y,z_0) - f(y_0,z)$. Since $f'(\{y_0\} \times Z) = 0 = f'(Y \times \{z_0\})$, the Rigidity Lemma gives $f' \equiv 0$. So $g(y) := f(y,z_0)$ and $h(z) := f(y_0,z)$ satisfy the assertion. The uniqueness of g and h follows by fixing $y = y_0$ and $z = z_0$ respectively in the equation $f(y,z) = g(y) + h(z)$. □

Recall that a *rational map* $f : Y \dashrightarrow X$ of smooth complex algebraic varieties is an equivalence class of pairs (U, f_U) with an open dense subset U of Y and a morphism $f_U \colon U \to X$. Such pairs (U, f_U) and (V, f_V) are equivalent if $f_U|_{U\cap V} = f_V|_{U\cap V}$. The map f is said to be *defined at a point* $y \in Y$ if there is a pair (U, f_U) as above with $y \in U$.

Let $\mathbb{C}(Y)$ denote as usual the field of rational functions on Y. A rational map $f : Y \dashrightarrow X$ induces a homomorphism of local rings $\mathcal{O}_{X,x} \to \mathbb{C}(Y)$ for every $x \in X$. It is easy to see that f is defined at a point $y \in Y$ if and only if there is an $x \in X$ such that the induced map $\mathcal{O}_{X,x} \to \mathbb{C}(Y)$ factorizes via $\mathcal{O}_{Y,y}$. Note that then necessarily $x = f(y)$.

Theorem 2.5.9 *Any rational map $f : Y \dashrightarrow X$ from a smooth variety Y to an abelian variety X is defined on the whole of Y.*

Proof First recall that f is not defined at most in a subvariety of codimension ≥ 2, since Y is smooth and X is complete (see Grothendieck–Dieudonné [58, Corollaire 8.2.12]). Assume f is not defined at one point. It suffices to show that there is a subvariety D of codimension 1 in Y such that f is not defined on the whole of D.

Consider the rational map $F : Y \times Y \dashrightarrow X$ given by

$$F(y_1, y_2) = f(y_1) - f(y_2)$$

whenever f is defined at y_1 and y_2.

We claim that F is defined at a point (y, y) if and only if f is defined at y. For the proof suppose F is defined at $(y_0, y_0) \in Y \times Y$ and let (U, F_U) be a representative of F with $(y_0, y_0) \in U$. There is a point $y_1 \in Y$, at which f is defined, such that y_0 is contained in the open set $V = \{y \in Y \mid (y_1, y) \in U\}$. Then $f(y) = f(y_1) - F(y_1, y)$ for all $y \in V$; in particular f is defined at y_0. The converse implication is obvious.

Suppose now f is not defined at a point $y \in Y$. Then F is not defined at (y, y). By what we have said above, this means that the homomorphism $\phi : \mathcal{O}_{X,0} \to \mathbb{C}(Y \times Y)$ induced by F satisfies

$$\operatorname{Im} \phi \not\subset \mathcal{O}_{Y\times Y,(y,y)}. \tag{2.8}$$

Since Y is smooth, $\mathcal{O}_{Y\times Y,(y,y)}$ is the subring of all functions φ in $\mathbb{C}(Y \times Y)$ which are defined at (y, y). So by equation 2.8 there is a $\varphi \in \operatorname{Im}\phi \subset \mathbb{C}(Y \times Y)$ such that (y, y) is contained in the polar divisor $(\varphi)_\infty$. Consider the intersection D of $(\varphi)_\infty$ with the diagonal Δ in $Y \times Y$. It is of codimension ≤ 1 in $\Delta \simeq Y$, since Δ is a complete intersection. Clearly F is not defined at every $(y', y') \in D$. Hence by what we have said above, f is not defined at all y' contained in a subvariety of codimension ≤ 1 in Y. This completes the proof. □

Proposition 2.5.10 *Every rational map $f : \mathbb{P}^n \dashrightarrow X$ from projective space to an abelian variety X is constant.*

In particular an abelian variety does not contain any rational curves.

Proof According to Theorem 2.5.9 the map f is defined everywhere. Since any two points in $\mathbb{P}^n$ are joined by a line $\mathbb{P}^1$, it suffices to show that any morphism $f: \mathbb{P}^1 \longrightarrow X$ is constant.

Let $X = V/\Lambda$ and $v_1, \ldots, v_g$ be a basis of V. Then $\mathrm{d}v_1, \ldots, \mathrm{d}v_g$ is a basis of holomorphic differentials of X. Since $\mathbb{P}^1$ does not admit any non-zero global holomorphic differential, we have $f^*\mathrm{d}v_\nu = 0$ for $\nu = 1, \ldots, g$. Thus f is constant. □

2.5.3 The Pontryagin Product

Let $X = V/\Lambda$ be an abelian variety of dimension g. The addition on X induces a multiplication on the homology ring $H_\bullet(X, \mathbb{Z})$, the Pontryagin product. In this section we show that the Pontryagin product is dual to the cup product in $H^\bullet(X, \mathbb{Z})$. As a first application we derive a formula which expresses the homology class of an ample divisor by a symplectic basis of $\Lambda = H_1(X, \mathbb{Z})$ in terms of the Pontryagin product. The results of this section generalize immediately to any non-degenerate line bundle on a complex torus.

Let $\times\colon H_p(X, \mathbb{Z}) \times H_q(X, \mathbb{Z}) \to H_{p+q}(X \times X, \mathbb{Z})$ denote the exterior homology product. The addition map $\mu\colon X \times X \longrightarrow X$ induces a homomorphism $\mu_*\colon H_{p+q}(X \times X, \mathbb{Z}) \longrightarrow H_{p+q}(X, \mathbb{Z})$. The *Pontryagin product* on X is defined to be the composition

$$\star : H_p(X, \mathbb{Z}) \times H_q(X, \mathbb{Z}) \xrightarrow{\times} H_{p+q}(X \times X, \mathbb{Z}) \xrightarrow{\mu_*} H_{p+q}(X, \mathbb{Z}).$$

For the definition of the Pontryagin product in terms of cycles see Exercise 1.1.6 (11).

Lemma 2.5.11 *The Pontryagin product is anti-commutative; that is,*

$$\sigma \star \tau = (-1)^{p+q} \tau \star \sigma$$

for all $\sigma \in H_p(X, \mathbb{Z})$ and $\tau \in H_q(X, \mathbb{Z})$.

Proof This follows from the fact that the exterior homology product is anti-commutative (see Greenberg–Harper [53, Corollary 29.29]). □

Let L be an ample line bundle on X. Fix a symplectic basis $\lambda_1, \ldots, \lambda_{2g}$ of $\Lambda = H_1(X, \mathbb{Z})$ for L and denote by $x_1, \ldots, x_{2g}$ the corresponding real coordinate functions on V. As we saw in Section 1.1.4, the differentials $\mathrm{d}x_1, \ldots, \mathrm{d}x_{2g}$ define a basis of $H^1(X, \mathbb{Z})$ dual to $\lambda_1, \ldots, \lambda_{2g}$, that is $\int_{\lambda_i} \mathrm{d}x_j = \delta_{ij}$. In this and the following section a multi-index is always an ordered multi-index in $\{1, \ldots, 2g\}$. For $I = (i_1 < \cdots < i_p)$

we write

$$\lambda_I = \lambda_{i_1} \star \cdots \star \lambda_{i_p} \quad \text{and} \quad dx_I = dx_{i_1} \wedge \cdots \wedge dx_{i_p}.$$

According to Proposition 1.1.20 the set $\{dx_I \mid \#I = p\}$ is a basis of $H^p(X, \mathbb{Z})$. This will be used to prove the following lemma.

Lemma 2.5.12 *The set $\{\lambda_I \mid \#I = p\}$ is a basis of $H_p(X, \mathbb{Z})$, dual to the basis $\{dx_I \mid \#I = p\}$ of $H^p(X, \mathbb{Z})$ with respect to the natural isomorphism $H_p(X, \mathbb{Z}) \to H^p(X, \mathbb{Z})^*$, $\sigma \mapsto \int_\sigma$.*

Proof We have to show that $\int_{\lambda_I} dx_J = \delta_{IJ}$ for all multi-indices I and J of length p. We will do this by induction on p. For $p = 1$ this is clear. Assume the assertion is proved for all multi-indices of length p and let $I = (i_1 < \cdots < i_p < i_{p+1})$ and $J = (j_1 < \cdots < j_{p+1})$. Denote by I_p the multi-index $(i_1 < \cdots < i_p)$. If $p_i \colon X \times X \to X$ is the natural projection onto the i-th factor, then $p_1 + p_2 = \mu$ is the addition map on X and

$$\begin{aligned}
\int_{\lambda_I} dx_J &= \int_{\mu(\lambda_{I_p} \times \lambda_{i_{p+1}})} dx_J = \int_{\lambda_{I_p} \times \lambda_{i_{p+1}}} \mu^* dx_J \\
&= \int_{\lambda_{I_p} \times \lambda_{i_{p+1}}} \wedge_{\nu=1}^{p+1} (p_1^* dx_{j_\nu} + p_2^* dx_{j_\nu}) \\
&= \int_{\lambda_{I_p} \times \lambda_{i_{p+1}}} \sum_{\nu=1}^{p+1} (-1)^{p+1-\nu} p_1^* dx_{J-j_\nu} \wedge p_2^* dx_{j_\nu} \\
&= \sum_{\nu=1}^{p+1} (-1)^{p+1-\nu} \int_{\lambda_{I_p}} dx_{J-j_\nu} \cdot \int_{\lambda_{i_{p+1}}} dx_{j_\nu} = \delta_{IJ}.
\end{aligned}$$

In the fourth equation the other summands vanish by Fubini's Theorem, since $\lambda_{i_{p+1}}$ is a 1-cycle on X. For the last equation we used the fact that $\delta_{I_p, J-j_\nu} \cdot \delta_{i_{p+1}, j_\nu} = 0$ unless $\nu = p + 1$, since the multi-indices are ordered. □

According to Lemma 2.5.12, the map $D \colon H_p(X, \mathbb{Z}) \to H^p(X, \mathbb{Z})$, defined by $\lambda_I \mapsto dx_I$, is an isomorphism for every p. Note that this isomorphism depends not only on the line bundle L, but also on the choice of the symplectic basis. However, it shows the following proposition.

Proposition 2.5.13 *The Pontryagin product in homology is dual to the cup product in cohomology.*

To be more precise, one immediately checks from the definition of the maps (see Exercise 2.5.4 (3)) that the following diagram is commutative.

$$\begin{array}{ccc}
H_p(X, \mathbb{Z}) \times H_q(X, \mathbb{Z}) & \xrightarrow{\star} & H_{p+q}(X, \mathbb{Z}) \\
\downarrow{\scriptstyle D \times D} & & \downarrow{\scriptstyle D} \\
H^p(X, \mathbb{Z}) \times H^q(X, \mathbb{Z}) & \xrightarrow{\wedge} & H^{p+q}(X, \mathbb{Z}).
\end{array} \tag{2.9}$$

Let $P\colon H_p(X,\mathbb{Z}) \to H^{2g-p}(X,\mathbb{Z})$ denote the isomorphism induced by Poincaré Duality. We want to compute P explicitly in terms of the bases $\{\lambda_I\}$ and $\{\mathrm{d}x_I\}$. If $I = (i_1 < \cdots < i_p)$ is a multi-index in $\{1,\ldots,2g\}$, denote by $I^o := (i_1^o < \cdots < i_{2g-p}^o)$ the ordered multi-index, for which as a set $I \cup I^o = \{1,\ldots,2g\}$, and define the sign $\varepsilon(I)$ of I by

$$\varepsilon(I)\,\mathrm{d}x_I \wedge \mathrm{d}x_{I^o} = \mathrm{d}x_1 \wedge \mathrm{d}x_{g+1} \wedge \cdots \wedge \mathrm{d}x_g \wedge \mathrm{d}x_{2g}.$$

Lemma 2.5.14 *For every multi-index I of length p we have*

$$P(\lambda_I) = (-1)^{g+p}\varepsilon(I)\,\mathrm{d}x_{I^o}.$$

Proof By definition of Poincaré Duality we have $\int_X P(\lambda_I) \wedge \varphi = \int_{\lambda_I} \varphi$ for any p-form φ on X. Since $\int_{\lambda_I} \mathrm{d}x_J = \delta_{IJ}$ and the differentials $\mathrm{d}x_J$ with $\#J = p$ form a basis of $H^p(X,\mathbb{Z})$, the Poincaré dual of λ_I is necessarily of the form $P(\lambda_I) = c\mathrm{d}x_{I^o}$ for some $c \in \mathbb{C}^*$.

It remains to compute the constant c. According to Lemma 1.7.5 we have $\int_X \mathrm{d}x_1 \wedge \mathrm{d}x_{g+1} \wedge \cdots \wedge \mathrm{d}x_g \wedge \mathrm{d}x_{2g} = (-1)^g$, since L is assumed to be ample. So

$$\begin{aligned} 1 &= \int_X P(\lambda_I) \wedge \mathrm{d}x_I = c \int_X \mathrm{d}x_{I^o} \wedge \mathrm{d}x_I \\ &= c\,(-1)^p \varepsilon(I) \int_X \mathrm{d}x_1 \wedge \mathrm{d}x_{g+1} \wedge \cdots \wedge \mathrm{d}x_g \wedge \mathrm{d}x_{2g} = c\,(-1)^{g+p}\varepsilon(I), \end{aligned}$$

which implies the assertion. □

Recall that $H_\bullet(X,\mathbb{Z})$ is a ring with respect to the intersection product (see Griffiths–Harris [55]). For $\sigma \in H_p(X,\mathbb{Z})$ and $\tau \in H_q(X,\mathbb{Z})$ we denote by $\sigma \cdot \tau$ their intersection product in $H_{p+q-2g}(X,\mathbb{Z})$. If in particular $\tau \in H_{2g-p}(X,\mathbb{Z})$, then $\sigma \cdot \tau$ is an element of $H_0(X,\mathbb{Z})$, which is canonically isomorphic to $\mathbb{Z}$. The *intersection number* $(\sigma \cdot \tau)$ is by definition the image of $\sigma \cdot \tau$ in $\mathbb{Z}$. If V is a p-cycle on X, we denote its homology class by

$$\{V\} \in H_p(X,\mathbb{Z}).$$

Note that we consider the cycles λ_i already as homology classes via the natural identification $\Lambda = H_1(X,\mathbb{Z})$.

Suppose D is a divisor in the linear system $|L|$. If L is of type $(d_1,\ldots,d_g)$, then we have

Lemma 2.5.15 $(\lambda_i \star \lambda_j \cdot \{D\}) = -d_i\delta_{g+i,j}$ *for all $i \le j$.*

Proof We may assume $i < j$, since $\lambda_i \star \lambda_i = 0$. According to Griffiths–Harris [55, p. 141] the Poincaré dual of $\{D\}$ is the first Chern class $c_1(L)$ of L. Recall from Lemma 1.7.4 that $c_1(L) = -\sum_{\nu=1}^{g} d_\nu \mathrm{d}x_\nu \wedge \mathrm{d}x_{g+\nu}$. Since the intersection product in homology is Poincaré dual to the cup product in cohomology, this implies using Lemma 2.5.12,

$$
\begin{aligned}
(\lambda_i \star \lambda_j \cdot \{D\}) &= \int_X P(\lambda_i \star \lambda_j) \wedge c_1(L) = \int_{\lambda_i \star \lambda_j} c_1(L) \\
&= -\sum_{\nu=1}^{g} d_\nu \int_{\lambda_i \star \lambda_j} \mathrm{d}x_\nu \wedge \mathrm{d}x_{g+\nu} \\
&= -\sum_{\nu=1}^{g} d_\nu \, \delta_{\{i,j\},\{\nu,g+\nu\}} = -d_i \, \delta_{i,g+i}.
\end{aligned}
$$
□

Preserving the notation of the above, we can state the main result of this section.

Theorem 2.5.16 *Let D be a divisor in $|L|$ with L of type $(d_1, \ldots, d_g)$. Then for all $0 \le p \le g$,*

$$
\{D\}^{g-p} = (-1)^p (g-p)! \sum_S \Big(\prod_{\nu \notin S} d_\nu\Big) \lambda_{s_1} \star \lambda_{g+s_1} \star \cdots \star \lambda_{s_p} \star \lambda_{g+s_p}.
$$

Here the sum is to be taken over all ordered subsets $S = \{s_1, \ldots, s_p\}$ of $\{1, \ldots, g\}$.

Proof Since the intersection product in homology is Poincaré dual to the cup product in cohomology, we have by Lemma 1.7.4,

$$
\{D\}^{g-p} = P^{-1}\Big(\wedge^{g-p} c_1(L)\Big) = (-1)^{g-p} P^{-1}\Big(\wedge^{g-p} \sum_{\nu=1}^{g} d_\nu \mathrm{d}x_\nu \wedge \mathrm{d}x_{g+\nu}\Big)
$$

for all $0 \le p \le g$. Now the assertion follows by an easy computation using Lemma 2.5.14. □

The following corollary lists the most important cases.

Corollary 2.5.17

(a) $\{D\}^0 = \{X\} = (-1)^g \lambda_1 \star \lambda_{g+1} \star \cdots \star \lambda_g \star \lambda_{2g}$.

(b) $\{D\} = (-1)^{g-1} \sum_{\nu=1}^{g} d_\nu \lambda_1 \star \lambda_{g+1} \star \cdots \star \check{\lambda}_\nu \star \check{\lambda}_{g+\nu} \star \cdots \star \lambda_g \star \lambda_{2g}$.

(c) $\{D\}^{g-1} = -(g-1)! \sum_{\nu=1}^{g} d_1 \cdot \ldots \cdot \check{d}_\nu \cdot \ldots \cdot d_g \, \lambda_\nu \star \lambda_{g+\nu}$.

(d) $(\{D\}^g) = d_1 \cdot \ldots \cdot d_g \, g! = (L^g)$.

The last equation uses the Geometric Riemann–Roch Theorem 1.7.3.

2.5.4 Exercises and a few Words on Applications

(1) Let $f : (Y, M) \to (X, L)$ be an isogeny of polarized abelian varieties with type $L = (d_d, \ldots, d_g)$ and type $M = (d'_1, \ldots, d'_g)$. If $d = \frac{d'_1 d'_g}{d_1 d_g}$, then

$$\widehat{f}^* M_\delta \equiv L_\delta^d.$$

(Hint: Use Propositions 1.4.6 and 1.4.12.)

(2) Give an example of a polarization L such that type $L \neq$ type L_δ.

(3) Give a proof of Remark 2.5.5.

(4) Identifying $(X \times \widehat{X})\widehat{} = \widehat{X} \times X$, the dual of the Poincaré bundle (in the sense of Remark 2.5.5) is

$$(\mathcal{P}_X)_\delta = \mathcal{P}_{\widehat{X}}.$$

(Hint: Use Proposition 2.5.1, Exercise 1.6.4 (5) and Lemma 6.1.8 below.)

(5) Show that diagram (2.9) commutes.

(6) Let (X, L) be a polarized abelian variety of dimension g and degree d and $(\widehat{X}, L_\delta)$ its dual.

(a) Show that $c_1(L_\delta) = \frac{1}{(g-1)! d_2 \cdots d_{g-1}} \alpha(c_1(L))$, where α denotes the composed map

$$\alpha : H^2(X, \mathbb{Z}) \xrightarrow{L^{g-2}} H^{2g-2}(X, \mathbb{Z}) \xrightarrow{P^{-1}} H_2(X, \mathbb{Z}) \xrightarrow{\epsilon} H^2(\widehat{X}, \mathbb{Z}),$$

with Lefschetz operator L and Poincaré duality P as in Section 5.4.1 (below) and Section 2.5.3 respectively and canonical duality ϵ induced by the canonical pairing $\langle\, , \rangle : \overline{\Omega} \times V \to \mathbb{R}, \langle l, v\rangle := \operatorname{Im} l(v)$ of Section 1.4.1.

(b) Show that $c_1(L_\delta) = \frac{d_1 d_g}{d^2} c_1(\mathcal{O}_{\widehat{X}}(\phi_{L*}(D)))$, where D is a divisor in the linear system $|L|$.

(See Birkenhake–Lange [23, Theorem], but note that there the dual polarization is defined slightly differently.)

Finally, let me say a few words on the role of abelian varieties in other subjects. Abelian varieties and their theta functions occur in many other parts of mathematics and theoretical physics. Let me mention only mirror symmetry in string theory. An explanation of it in this introductory book would be too long. Too many things would have to be explained.

But let me mention how abelian varieties arise in the theory of integrable systems. Many integrable systems of classical mechanics admit a complexification of the phase space and the time variable. The precise mathematical notion is that of an *algebraic complete integrable system* introduced by Adler and van Moerbeke in [1]. This gives many integrable systems which can be linearized on open sets of abelian varieties.

To give a complete definition here together with motivation and some results would lead us too far away. So we refer the interested reader to the book [2] by Adler, van Moerbeke and Vanhaecke as well as the article [40] by Donagi and Markman.

2.6 The Endomorphism Algebra of a Simple Abelian Variety

Let X be a simple abelian variety of dimension g and L a polarization on X. According to Corollary 2.4.26 the algebra

$$F = \mathrm{End}_{\mathbb{Q}}(X)$$

is a skew field of finite dimension over $\mathbb{Q}$. The Rosati involution $f \mapsto f'$ with respect to the polarization L is an anti-involution on F such that the map

$$\mathrm{End}_{\mathbb{Q}}(X) \to \mathbb{Q}, \qquad f \mapsto \mathrm{Tr}_r(f'f) = 2\,\mathrm{Tr}_a(f'f)$$

is a positive definite quadratic form on F (see Theorem 2.4.9). In this section we give the classification of simple polarized abelian varieties with positive anti-involution (for the definition see the beginning of Section 2.6.2). Although we do not need this very much, we include it because of its importance in Number Theory.

2.6.1 The Classification Theorem

Here we give a proof of the classification theorem using the results of the next section.

First we express the quadratic form $\mathrm{Tr}_r(f'f)$ in terms of $(F, \ ')$. Let K denote the centre of the skew field F. The degree $[F : K]$ of F over K is a square, say d^2. The characteristic polynomial of any $f \in F$ over K is a d'th power of a polynomial

$$t^d - a_1 t^{d-1} + \cdots + (-1)^d a_0 \in K[t]$$

of degree d, called the *reduced characteristic polynomial of f over K*. In these terms the *reduced trace of f over K* is defined as

$$\mathrm{tr}_{F|K}(f) = a_1.$$

For any subfield $k \subseteq K$ we define the *reduced trace of f over k* by

$$\mathrm{tr}_{F|k}(f) = \mathrm{tr}_{K|k}\big(\mathrm{tr}_{F|K}(f)\big),$$

where $\mathrm{tr}_{K|k}$ denotes the usual trace for the field extension $K|k$.

Lemma 2.6.1 *The quadratic form*

$$F \to \mathbb{Q}, \qquad f \mapsto \mathrm{tr}_{F|\mathbb{Q}}(f'f)$$

is positive definite.

Proof According to Corollary 2.4.8 and Theorem 2.4.9 the alternating coefficients a_ν of the analytic characteristic polynomial of $f'f$ ($\neq 0$) are all nonnegative rational numbers with $a_1 > 0$. It follows that the zeros of the minimal polynomial of $f'f$ over $\mathbb{Q}$ are all non-negative with at least one positive. This immediately implies the assertion. □

Corollary 2.4.26 and Theorem 2.4.9 reduce the classification of simple polarized abelian varieties with positive anti-involution to the classification of pairs $(F, ')$ with F a skew-field of finite dimension over $\mathbb{Q}$ with positive anti-involution $'$. This classification is due to Albert [4]. The proof is part of Algebraic Number Theory and since probably not every reader wants to study it, it is given in an extra section, the next one. In this section we give its consequences for the endomorphism algebras of abelian varieties, using the results of the next section.

Let the notation be as at the beginning of this section. So (X, L) is a simple polarized abelian variety of dimension g. Then $F = \mathrm{End}_{\mathbb{Q}}(X)$ is a skew field of finite dimension over $\mathbb{Q}$ with positive anti-involution $x \mapsto x'$, the Rosati involution with respect to L.

Let K denote the centre of F and K_0 the fixed field of the anti-involution restricted to K. Let

$$[F : K] = d^2, \quad [K : \mathbb{Q}] = e, \quad [K_0 : \mathbb{Q}] = e_0 \quad \text{and} \quad \mathrm{rk}\ \mathrm{NS}(X) = \rho.$$

Then we have the following theorem giving more details on F as well as restrictions for these values. For this recall that the skew field F is called a *quaternion algebra over K* if it is of dimension 4 over its centre K. A quaternion algebra F over a totally real number field $K = K_0$ is called a *totally indefinite quaternion algebra*, respectively a *totally definite quaternion algebra*, if for every embedding $\sigma : K \hookrightarrow \mathbb{R}$ we have

$$F \otimes_\sigma \mathbb{R} \simeq \mathrm{M}_2(\mathbb{R}) \qquad \text{respectively} \qquad F \otimes_\sigma \mathbb{R} \simeq \mathbb{H},$$

where $\mathbb{H}$ denotes the usual Hamiltonian quaternions. Finally, the pair $(F, ')$ is called of *the second kind* if the anti-involution does not restrict to the identity of K.

For these values we have the following restrictions:

Theorem 2.6.2

$F = \mathrm{End}_{\mathbb{Q}}(X)$	d	e_0	ρ	*restriction*
totally real number field	1	e	e	$e \mid g$
totally indefinite quaternion algebra	2	e	$3e$	$2e \mid g$
totally definite quaternion algebra	2	e	e	$2e \mid g$
$(F, {}')$ *of the second kind*	d	$\frac{1}{2}e$	$e_0 d^2$	$e_0 d^2 \mid g$

Proof The results of the first and second column of the table follow from Theorems 2.6.5 and 2.6.8.

In order to compute the Picard number ρ recall from Proposition 2.4.12 that

$$\rho = \dim_{\mathbb{R}} \mathrm{End}^s_{\mathbb{Q}}(X) \otimes_{\mathbb{Q}} \mathbb{R}.$$

Obviously $\rho = e$ in the totally real number field case. In the case that $\mathrm{End}_{\mathbb{Q}}(X)$ is a totally indefinite quaternion algebra we have

$$\mathrm{End}_{\mathbb{Q}}(X) \otimes_{\mathbb{Q}} \mathbb{R} \simeq \prod_{\nu=1}^{e} \mathrm{M}_2(\mathbb{R})$$

by Theorem 2.6.5 and Lemma 2.6.4 such that the anti-involution translates to transposition on the factors. So $\rho = 3e$. Similarly in the totally definite quaternion algebra case

$$\mathrm{End}_{\mathbb{Q}}(X) \otimes_{\mathbb{Q}} \mathbb{R} \simeq \prod_{\nu=1}^{e} \mathbb{H}$$

such that the anti-involution translates to quaternion conjugation on the factors and thus $\rho = e$. Finally in the last case, by Theorem 2.6.8 we have an isomorphism $\mathrm{End}_{\mathbb{Q}}(X) \otimes_{\mathbb{Q}} \mathbb{R} \simeq \prod_{\nu=1}^{e_0} \mathrm{M}_d(\mathbb{C})$ carrying the anti-involution to $X \mapsto {}^t\overline{X}$ on every factor and thus $\rho = e_0 d^2$.

As for the restrictions, note first that $\dim_{\mathbb{Q}} F = ed^2$ divides $2g$, since $\mathrm{End}_{\mathbb{Q}}(X)$ admits a faithful representation in the vector space $\Lambda \otimes \mathbb{Q}$ and thus $\Lambda \otimes \mathbb{Q}$ is a vector space over the skew field F. This gives the restrictions for the last three lines.

It remains to show that $e \mid g$, if $(\mathrm{End}_{\mathbb{Q}}(X), {}') = (K, \mathbf{1}_K)$. For this consider the isomorphism

$$\varphi\colon \mathrm{NS}_{\mathbb{Q}}(X) \to \mathrm{End}^s_{\mathbb{Q}}(X) = \mathrm{End}_{\mathbb{Q}}(X) = K$$

from Proposition 2.4.12. Define a map

$$p\colon \mathrm{End}(X) \to \mathbb{Q} \qquad f \mapsto \frac{\chi(\varphi^{-1}(f))}{\chi(L)}.$$

According to the Geometric Riemann–Roch Theorem 1.7.3, p is a homogeneous polynomial function of degree g on the $\mathbb{Z}$-module $\mathrm{End}(X)$. Hence, setting $p(\frac{f}{n}) = n^{-g}p(f)$, we can extend it to a homogeneous polynomial function of degree g on the whole $\mathbb{Q}$-vector space K, which we also denote by p.

We claim that $p\colon K \to \mathbb{Q}$ is multiplicative. For the proof let $f_1, f_2 \in \mathrm{End}(X)$. Applying Exercise 2.4.5 (6) and Corollary 1.7.2 we have

$$\begin{aligned} p(f_1f_2)^2 &= \frac{\deg(\varphi^{-1}(f_1f_2))}{\deg \phi_L} \\ &= \frac{\deg(\varphi^{-1}(f_1)\phi_L^{-1}\varphi^{-1}(f_2))}{\deg \phi_L} \\ &= \frac{\deg \varphi^{-1}(f_1)}{\deg \phi_L} \cdot \frac{\deg \varphi^{-1}(f_2)}{\deg \phi_L} = p(f_1)^2 p(f_2)^2. \end{aligned}$$

Since p is a polynomial function and $p(1 \cdot f) = +p(1)p(f)$ for all $f \in \mathrm{End}^s_{\mathbb{Q}}(X)$, this proves the claim.

Finally it is well known that the degree of any multiplicative homogeneous polynomial function on the $\mathbb{Q}$-vector space K is divisible by the dimension of K over $\mathbb{Q}$ (see Mumford [97, Lemma p. 179]). Hence $e = [K : \mathbb{Q}]$ divides g. □

2.6.2 Skew Fields with an Anti-involution

In this section we give a proof of the results on simple algebras with an anti-involution, which were applied in the previous section.

We need some preliminaries. We call an anti-involution $x \mapsto x'$ on a semisimple algebra over $\mathbb{Q}$ or $\mathbb{R}$ *positive* if the quadratic form $\mathrm{tr}(x'x)$ is positive definite. Here tr denotes the reduced trace over $\mathbb{Q}$ or $\mathbb{R}$. It is well known that any finite-dimensional simple $\mathbb{R}$-algebra is isomorphic to either $\mathrm{M}_r(\mathbb{R})$ or $\mathrm{M}_r(\mathbb{C})$ or $\mathrm{M}_r(\mathbb{H})$ for some r, where $\mathbb{H}$ denotes the skew field of Hamiltonian quaternions. For any of these algebras there is a natural anti-involution, namely

$$x^* = \begin{cases} {}^t x & \text{for } \mathrm{M}_r(\mathbb{R}) \\ {}^t\overline{x} & \text{for } \mathrm{M}_r(\mathbb{C}) \text{ and } \mathrm{M}_r(\mathbb{H}). \end{cases} \tag{2.10}$$

Here $x \mapsto \overline{x}$ means complex (respectively quaternion) conjugation. The following lemma shows that up to isomorphism * is the unique positive anti-involution.

Lemma 2.6.3 *For any simple $\mathbb{R}$-algebra A of finite dimension with positive anti-involution $x \mapsto x'$ there is an isomorphism φ of A onto one of the matrix algebras as above such that for every $x \in A$*

$$\varphi(x') = \varphi(x)^*.$$

Proof We may assume that A is either $\mathrm{M}_r(\mathbb{R})$ or $\mathrm{M}_r(\mathbb{C})$ or $\mathrm{M}_r(\mathbb{H})$ for some r. The two anti-involutions $x \mapsto x'$ and $x \mapsto x^*$ on A differ by an automorphism of A. By the Skolem–Noether Theorem (see Jacobson [73, II, p. 222]) this implies that there is an $a \in A$ such that

$$x' = a^{-1}x^*a.$$

It suffices to show that $a = \pm b^*b$ for some $b \in A$, since then the isomorphism $\varphi\colon x \mapsto bxb^{-1}$ satisfies $\varphi(x') = \varphi(x)^*$ for every $x \in A$.

Since $x = x'' = a^{-1}a^*x(a^{-1}a^*)^{-1}$ for any $x \in A$, the element $\lambda := a^{-1}a^*$ is in the centre of A, that is, in $\mathbb{R}$ or $\mathbb{C}$. Moreover we have $|\lambda| = 1$, since $a = a^{**} = \bar{\lambda}\lambda a$.

We claim that we may assume that $\lambda = 1$ and thus $a^* = a$. For the proof suppose first that $A = \mathrm{M}_r(\mathbb{C})$ or $\mathrm{M}_r(\mathbb{H})$ and $\lambda \neq 1$. There is a $\mu \in \mathbb{C}$ with $\lambda = \mu^2$. Replacing a by μa gives the assertion in this case.

If $A = \mathrm{M}_r(\mathbb{R})$ we proceed as follows. Suppose $\lambda = -1$. Then a is an alternating matrix and thus

$$a = c^*\begin{pmatrix} I & 0 \\ 0 & * \end{pmatrix}c$$

with $I := \left(\begin{smallmatrix} 0 & 1 \\ -1 & 0 \end{smallmatrix}\right)$ and some $c \in \mathrm{M}_r(\mathbb{R})$. Denoting $c^{-1}\operatorname{diag}(1, -1, 0, \dots, 0)c$ by x_0 we get

$$\begin{aligned} 0 < \operatorname{tr}(x_0'x_0) &= \operatorname{tr}\left(c^{-1}\begin{pmatrix} I & 0 \\ 0 & * \end{pmatrix}^{-1} c^{*-1}x_0^*c^*\begin{pmatrix} I & 0 \\ 0 & * \end{pmatrix}cx_0\right) \\ &= \operatorname{tr}\left(\begin{pmatrix} I & 0 \\ 0 & * \end{pmatrix}^{-1} (cx_0c^{-1})^*\begin{pmatrix} I & 0 \\ 0 & * \end{pmatrix}(cx_0c^{-1})\right) \\ &= \operatorname{tr}\left(\begin{pmatrix} 0 & -1 \\ 1 & 0 \end{pmatrix}\begin{pmatrix} 1 & 0 \\ 0 & -1 \end{pmatrix}\begin{pmatrix} 0 & 1 \\ -1 & 0 \end{pmatrix}\begin{pmatrix} 1 & 0 \\ 0 & -1 \end{pmatrix}\right) = -2, \end{aligned}$$

a contradiction. Hence $a^* = a$, which completes the proof of the claim.

From Linear Algebra we know that there is a matrix $b \in A$ such that

$$a = b^* \operatorname{diag}(\varepsilon_1, \dots, \varepsilon_r)b$$

with $\varepsilon_\nu \in \{\pm 1\}$ for $1 \le \nu \le r$. It remains to show that $\varepsilon_\mu = \varepsilon_\nu$ for $1 \le \mu, \nu \le r$.

Suppose $\varepsilon_\mu \neq \varepsilon_\nu$. Denoting by $e(\mu, \nu)$ the matrix whose (μ, ν)-th entry is 1 and the others are 0, we have

$$\begin{aligned} 0 &< \operatorname{tr}\Big(\big((b^{-1}e(\mu,\nu)b\big)'\big((b^{-1}e(\mu,\nu)b\big)\Big) \\ &= \operatorname{tr}\big(\operatorname{diag}(\varepsilon_1, \dots, \varepsilon_r)\ e(\nu,\mu)\ \operatorname{diag}(\varepsilon_1, \dots, \varepsilon_r)\ e(\mu,\nu)\big) = \varepsilon_\mu\varepsilon_\nu = -1\,, \end{aligned}$$

a contradiction. So $\varphi(x) = bxb^{-1}$ satisfies the assertion of the lemma. □

Recall that $(F, {}')$ denotes a skew field of finite dimension over $\mathbb{Q}$ with positive anti-involution $x \mapsto x'$. The anti-involution $x \mapsto x'$ restricts to an involution on the centre K of F, whose fixed field we denote by K_0.

Lemma 2.6.4 *K_0 is a totally real number field; that is, every embedding $K_0 \hookrightarrow \mathbb{C}$ factorizes via $\mathbb{R}$.*

Proof Let $\sigma_1, \dots, \sigma_r$ be the real embeddings and $\sigma_{r+1}, \bar{\sigma}_{r+1}, \dots, \sigma_{r+s}, \bar{\sigma}_{r+s}$ the nonreal complex embeddings of K_0 and assume $s \geq 1$.

The Approximation Theorem (see for example van der Waerden [136, II, p. 234]) implies that for any $\varepsilon > 0$ there is an $x \in K_0$ such that $|\sigma_\nu(x)| < \varepsilon$ for $1 \leq \nu < r+s$ and $|\sigma_{r+s}(x) - i| < \varepsilon$. For small ε the term $2\,\mathrm{Re}\,\sigma_{r+s}(x^2) \approx -2$ is dominant in

$$\mathrm{tr}_{K_0|\mathbb{Q}}(x'x) = \mathrm{tr}_{K_0|\mathbb{Q}}(x^2) = \sum_{\nu=1}^{r} \sigma_\nu(x^2) + 2\sum_{\nu=1}^{s} \mathrm{Re}\,\sigma_{r+\nu}(x^2).$$

Hence $0 < \mathrm{tr}_{F|\mathbb{Q}}(x'x) = \mathrm{tr}_{K_0|\mathbb{Q}}(\mathrm{tr}_{F|K_0}(x'x)) = [F : K_0]\ \mathrm{tr}_{K_0|\mathbb{Q}}(x^2) < 0$, a contradiction. □

The pair $(F, {}')$ (respectively the anti-involution $'$) is said to be *of the first kind* if the anti-involution is trivial on the centre K of F; that is, if $K = K_0$, and *of the second kind* otherwise. We first consider the case when $(F, {}')$ is of the first kind.

A quaternion algebra F over K admits a canonical anti-involution, namely

$$x \mapsto \bar{x} = \mathrm{tr}_{F|K}(x) - x.$$

Theorem 2.6.5 *Let F be a skew field of characteristic 0 and $x \mapsto x'$ an anti-involution of F with centre K.*

$(F, {}')$ is a skew field of finite dimension over $\mathbb{Q}$ with positive anti-involution of the first kind if and only if K is a totally real number field and one of the following cases holds:

(a) *$F = K$ and $x' = x$ for all $x \in F$.*
(b) *F is a quaternion algebra over K and for every embedding $\sigma : K \hookrightarrow \mathbb{R}$*

$$F \otimes_\sigma \mathbb{R} \simeq \mathrm{M}_2(\mathbb{R}).$$

Moreover there is an element $a \in F$ with $a^2 \in K$ totally negative such that the anti-involution $x \mapsto x'$ is given by $x' = a^{-1}\bar{x}a$.

c) *F is a quaternion algebra over K and for every embedding $\sigma : K \hookrightarrow \mathbb{R}$*

$$F \otimes_\sigma \mathbb{R} \simeq \mathbb{H}.$$

Moreover the anti-involution $x \mapsto x'$ is given by $x' = \bar{x}$.

Proof In Steps I—IV we show that a skew field $(F, {}')$ of finite dimension over $\mathbb{Q}$ with positive anti-involution of the first kind is of type (a), (b) or (c). In Step V we will prove the converse.

Step I: The anti-involution $x \mapsto x'$ on F may be considered as an isomorphism between F and its opposite algebra F^{op}, defined by the product $x \circ y = yx$ on the K-vector space F. Since the elements in the Brauer group $\mathrm{Br}(K)$ corresponding to

F and F^{op} are inverse to each other, F has order 1 or 2 considered as an element in $\mathrm{Br}(K)$. Using a result on the Brauer group (see Jacobson [73, II, Theorem 9.23]) this implies that the rank of F over K is either $1^2 = 1$ or $2^2 = 4$. It follows that F is either K, and we are in case (a), or a quaternion algebra over K.

Step II: Suppose F is a quaternion algebra over K. As in the proof of Lemma 2.6.3 there is an $a \in F$ such that

$$x' = a^{-1}\overline{x}a \quad \text{with} \quad \overline{a} = \lambda a \text{ and } |\lambda| = 1$$

for all $x \in F$. Since K is totally real, $\lambda = \pm 1$. In particular $a^2 \in K$. Denote by $\sigma_1, \dots, \sigma_e$ the different embeddings of K and write $\mathbb{R}_{\sigma_\nu} = \mathbb{R}$ when $\mathbb{R}$ is considered as a K-algebra via σ_ν. Then

$$F \otimes_{\mathbb{Q}} \mathbb{R} \simeq F \otimes_K (K \otimes_{\mathbb{Q}} \mathbb{R}) \simeq F \otimes_K (\mathbb{R}_{\sigma_1} \times \dots \times \mathbb{R}_{\sigma_e}) = \times_{\nu=1}^{e} F \otimes_{\sigma_\nu} \mathbb{R},$$

with

$$F \otimes_{\sigma_\nu} \mathbb{R} = \mathrm{M}_2(\mathbb{R}) \quad \text{or} \quad \mathbb{H}$$

for $1 \le \nu \le e$. Denote by x_ν the image of $x \in F$ in $F \otimes_{\sigma_\nu} \mathbb{R}$. The anti-involutions $x \mapsto x'$ and $x \mapsto \overline{x}$ extend in a natural way to anti-involutions $x_\nu \mapsto x'_\nu$ and $x_\nu \mapsto \overline{x}_\nu$ on $F \otimes_{\sigma_\nu} \mathbb{R}$ for $1 \le \nu \le e$. Note that in the case $F \otimes_{\sigma_\nu} \mathbb{R} = \mathbb{H}$ the anti-involution $x_\nu \mapsto \overline{x}_\nu$ coincides with the usual Hamiltonian conjugation, whereas in the case $F \otimes_{\sigma_\nu} \mathbb{R} = \mathrm{M}_2(\mathbb{R})$ the matrix $\overline{x}_\nu = \mathrm{tr}_{F \otimes_{\sigma_\nu} \mathbb{R}/\mathbb{R}}(x_\nu) - x_\nu$ is the adjoint of the matrix x_ν.

Since F is dense in $F \otimes_{\sigma_\nu} \mathbb{R}$, we have $\mathrm{tr}(x'_\nu x_\nu) \ge 0$ for every $x_\nu \in F \otimes_{\sigma_\nu} \mathbb{R}$ by continuity. But the nullspace of this quadratic form must be a rational subspace, since it is the orthogonal complement of the whole space. Hence it is 0 and $\mathrm{tr}(x'_\nu x_\nu)$ is positive definite on $F \otimes_{\sigma_\nu} \mathbb{R}$ for all $1 \le \nu \le e$.

Step III: *Either $F \otimes_{\sigma_\nu} \mathbb{R} = \mathrm{M}_2(\mathbb{R})$ for all $1 \le \nu \le e$ or $F \otimes_{\sigma_\nu} \mathbb{R} = \mathbb{H}$ for all $1 \le \nu \le e$. Moreover in the second case $(F, ')$ is of type (c).*

To see this, note that according to Lemma 2.6.3 we can identify $F \otimes_{\sigma_\nu} \mathbb{R}$ with $\mathrm{M}_2(\mathbb{R})$, respectively $\mathbb{H}$, in such a way that

$$x'_\nu = x^*_\nu$$

for all $x_\nu \in F \otimes_{\sigma_\nu} \mathbb{R}$ and every $1 \le \nu \le e$. For a and λ as in Step II we get

$$x^*_\nu = a_\nu^{-1}\overline{x}_\nu a_\nu \quad \text{and} \quad \overline{a}_\nu = \lambda a_\nu. \tag{2.11}$$

Suppose $F \otimes_{\sigma_\mu} \mathbb{R} = \mathbb{H}$ for some $1 \le \mu \le e$. By (2.10) the two canonical anti-involutions agree; that is, $x^*_\mu = \overline{x}_\mu$ for all $x_\mu \in F \otimes_{\sigma_\mu} \mathbb{R}$. Hence $a_\mu \in \mathbb{R}$, implying $\lambda = +1$, and thus $a \in K$. It follows that

$$x' = a^{-1}\overline{x}a = \overline{x}.$$

Assuming $F \otimes_{\sigma_\nu} \mathbb{R} = \mathrm{M}_2(\mathbb{R})$ for some $\nu \neq \mu$, so $x_\nu = \begin{pmatrix} \alpha & \gamma \\ -\gamma & \alpha \end{pmatrix}$, we get

$$\operatorname{tr}(x'_\nu x_\nu) = \operatorname{tr}(\overline{x}_\nu x_\nu) = 2\det(x_\nu). \tag{2.12}$$

Using the Approximation Theorem again, this easily gives a contradiction.

Step IV: *If $F \otimes_{\sigma_\nu} \mathbb{R} = \mathrm{M}_2(\mathbb{R})$ for $1 \leq \nu \leq e$, then a^2 is totally negative.*

In this case $\lambda = -1$, since the canonical involution $x_\nu \mapsto \overline{x}_\nu$ is not positive as we saw in (2.12). So we have $\overline{a}_\nu = -a_\nu$ for every $1 \leq \nu \leq e$. Using (2.11) we get $a_\nu^* = -a_\nu$ and thus

$$a_\nu = \begin{pmatrix} 0 & \alpha_\nu \\ -\alpha_\nu & 0 \end{pmatrix}$$

for some $\alpha_\nu \in \mathbb{R}$. This implies that $a_\nu^2 = -\alpha_\nu^2 \mathbf{1}_2$ for $\nu = 1, \ldots, e$. So a^2 is totally negative.

Step V: *The algebras $(F, \,')$ of type (a), (b) and (c) are of finite dimension over $\mathbb{Q}$ with positive anti-involution of the first kind.*

The assertion is obvious in the cases (a) and (c). To verify this for the case (b) we have to show that for any $a \in F$ with $a^2 \in K$ totally negative, the anti-involution $x \mapsto x' = a^{-1}\overline{x}a$ is positive.

Let $\sigma_1, \ldots, \sigma_e \colon K \to \mathbb{R}$ denote the real embeddings. For $x \in F$ we denote again by x_ν its image in $F \otimes_{\sigma_\nu} \mathbb{R} = \mathrm{M}_2(\mathbb{R})$. By assumption we have $a_\nu^2 = k_\nu \mathbf{1}_2$ with $k_\nu < 0$. An elementary matrix calculation shows that this means

$$a_\nu = \begin{pmatrix} \alpha & \beta \\ \gamma & -\alpha \end{pmatrix} \quad \text{with} \quad \alpha^2 + \beta\gamma = k_\nu < 0.$$

By definition $x_\nu^* = \begin{pmatrix} 0 & 1 \\ -1 & 0 \end{pmatrix} \overline{x}_\nu \begin{pmatrix} 0 & 1 \\ -1 & 0 \end{pmatrix}^{-1}$ and thus

$$x'_\nu = \left(\begin{pmatrix} 0 & 1 \\ -1 & 0 \end{pmatrix} a_\nu \right)^{-1} x_\nu^* \begin{pmatrix} 0 & 1 \\ -1 & 0 \end{pmatrix} a_\nu .$$

But $\begin{pmatrix} 0 & 1 \\ -1 & 0 \end{pmatrix} a_\nu = \begin{pmatrix} \gamma & -\alpha \\ -\alpha & -\beta \end{pmatrix}$ is symmetric and either positive definite or negative definite, as $\det\left(\begin{pmatrix} 0 & 1 \\ -1 & 0 \end{pmatrix} a_\nu \right) = -\alpha^2 - \beta\gamma = -k_\nu > 0$. It follows that

$$\operatorname{tr}(x'_\nu x_\nu) = \operatorname{tr}\left(\left(\begin{pmatrix} 0 & 1 \\ -1 & 0 \end{pmatrix} a_\nu \right)^{-1} x_\nu^* \begin{pmatrix} 0 & 1 \\ -1 & 0 \end{pmatrix} a_\nu x_\nu \right) > 0,$$

since the product of a positive definite symmetric matrix with a nonzero positive semidefinite symmetric matrix in $\mathrm{M}_2(\mathbb{R})$ has positive trace. This completes the proof of the theorem. □

Suppose now that $(F, \,')$ is of the second kind; that is, the anti-involution $x \mapsto x'$ does not act trivially on the centre K of F. According to Lemma 2.6.4 its fixed field K_0 is totally real. Moreover we have:

Lemma 2.6.6 *If $(F, ')$ is a skew field of finite dimension over $\mathbb{Q}$ with positive anti-involution of the second kind, then its centre K is totally complex; that is, no embedding $K \hookrightarrow \mathbb{C}$ factors via $\mathbb{R}$, and the restriction of the anti-involution to K is complex conjugation.*

Proof Assume K is not totally complex. Then there is an embedding $\sigma_1 : K \hookrightarrow \mathbb{R}$. Let $\sigma_2 : K \hookrightarrow \mathbb{R}$ denote the embedding defined by

$$\sigma_2(x) = \sigma_1(x')$$

for all $x \in K$ and denote the remaining embeddings by $\sigma_3, \dots, \sigma_e : K \hookrightarrow \mathbb{C}$. According to the Approximation Theorem (see van der Waerden [136, II, p. 234]) for any $\varepsilon > 0$ there is an $x \in K$ with $|\sigma_1(x) + 1| < \varepsilon$, $|\sigma_2(x) - 1| < \varepsilon$, and $|\sigma_\nu(x)| < \varepsilon$ for $3 \leq \nu \leq e$. For small ε the dominant term of $\mathrm{tr}_{K|\mathbb{Q}}(x'x) = \sum_{\nu=1}^{e} \sigma_\nu(x'x)$ is $\sigma_1(x'x) + \sigma_2(x'x) = 2\sigma_1(x)\sigma_2(x) \approx -2$. Hence

$$\mathrm{tr}_{F|\mathbb{Q}}(x'x) = [F : K] \ \mathrm{tr}_{K|\mathbb{Q}}(x'x) < 0,$$

contradicting the positivity of the anti-involution $x \mapsto x'$. Hence K is totally complex.

Moreover by Lemma 2.6.4 complex conjugation induces an involution on K with fixed field K_0, implying that it coincides with the involution $x \mapsto x'$ on K. □

Denoting by $\overline{F}$ the complex conjugate K-algebra of F and by $\overline{F}^{\mathrm{op}}$ the K-algebra opposite to $\overline{F}$, we may consider the anti-involution $x \mapsto x'$ as an isomorphism $F \to \overline{F}^{\mathrm{op}}$ of K-algebras. However, not every such isomorphism corresponds to an anti-involution. In other words, a necessary condition for the existence of an anti-involution of the second kind on F is that F is isomorphic to $\overline{F}^{\mathrm{op}}$.

Conversely, suppose $\tau : F \to \overline{F}^{\mathrm{op}}$ is any isomorphism. Since τ^2 is an automorphism of F over K, by the Skolem–Noether Theorem there is a $c \in F$ such that

$$\tau^2(x) = c^{-1}xc$$

for all $x \in F$. The following proposition gives a criterion for the existence of an anti-involution on F in terms of c.

Proposition 2.6.7 *For a skew field F of finite dimension over $\mathbb{Q}$ with centre a totally complex quadratic extension K of a totally real number field K_0 the following conditions are equivalent:*

(i) *There exists an anti-involution of the second kind on F.*
(ii) *There exists an isomorphism $\tau : F \to \overline{F}^{\mathrm{op}}$ such that $\tau^2(x) = c^{-1}xc$ for all $x \in F$ and some $c \in F$ implies $c\tau(c) \in N_{K|K_0}(K^*)$.*

Proof By what we said above we have only to show the implication (ii) $\Rightarrow$ (i). We may assume that $c \notin K$, since otherwise τ is already an anti-involution of the second kind. By assumption there is a $\lambda \in K$ such that $c\tau(c) = \bar{\lambda}\lambda$. Define a map

$$\tilde{\ } : F \to F, \qquad x \mapsto \tilde{x} = (\lambda + c)\tau(x)(\lambda + c)^{-1}.$$

This is an anti-involution, since

$$\begin{aligned}\tilde{\tilde{x}} &= (\lambda + c)\ \tau(\lambda + c)^{-1}\tau^2(x)\ \tau(\lambda + c)\ (\lambda + c)^{-1}\\ &= (\lambda + c)\ \tau(\lambda + c)^{-1}c^{-1}xc\ \tau(\lambda + c)\ (\lambda + c)^{-1}\\ &= (\lambda + c)\ (\lambda + c)^{-1}\bar{\lambda}^{-1}x\bar{\lambda}(\lambda + c)\ (\lambda + c)^{-1} = x.\end{aligned}$$

□

The next theorem shows that there is a positive anti-involution on F whenever there is any anti-involution of the second kind. Furthermore it classifies all positive anti-involutions on F.

Theorem 2.6.8 *Let F be a skew field of finite dimension over $\mathbb{Q}$ with centre a totally complex quadratic extension K of a totally real number field K_0. Moreover, suppose that F admits an anti-involution $x \mapsto \tilde{x}$ of the second kind.*

Then there exists a positive anti-involution $x \mapsto x'$ of the second kind and for every embedding $\sigma: K \hookrightarrow \mathbb{C}$ an isomorphism

$$\varphi: F \otimes_\sigma \mathbb{C} \xrightarrow{\sim} \mathrm{M}_d(\mathbb{C})$$

such that $x \mapsto x'$ extends via φ to the canonical anti-involution $X \mapsto {}^t\overline{X}$ on $\mathrm{M}_d(\mathbb{C})$.

Any other positive anti-involution on F is of the form

$$x \mapsto ax'a^{-1}$$

with $a \in F$, $a' = a$ and such that $\varphi(a \otimes 1)$ is a positive definite hermitian matrix in $\mathrm{M}_d(\mathbb{C})$ for every embedding $\sigma: K \to \mathbb{C}$.

Proof **Step I:** Denote by σ_0 the restriction of σ to K_0. Then we can identify $F \otimes_\sigma \mathbb{C} = F \otimes_\sigma (K \otimes_{\sigma_0} \mathbb{R}) = F \otimes_{\sigma_0} \mathbb{R}$ in such a way that the anti-involution $x \otimes \alpha \mapsto \tilde{x} \otimes \bar{\alpha}$ on $F \otimes_\sigma \mathbb{C}$ translates to the anti-involution $x \otimes r \mapsto \tilde{x} \otimes r$ on $F \otimes_{\sigma_0} \mathbb{R}$. Hence there is an isomorphism

$$\psi: F \otimes_{\sigma_0} \mathbb{R} \xrightarrow{\sim} \mathrm{M}_d(\mathbb{C})$$

such that the anti-involution $x \mapsto \tilde{x}$ on F extends via ψ to an anti-involution on $\mathrm{M}_d(\mathbb{C})$, which by the Skolem–Noether Theorem is of the form

$$X \mapsto A^{-1}\,{}^t\overline{X}A$$

for some $A \in \mathrm{GL}_d(\mathbb{C})$. From the proof of Lemma 2.6.3 we see that we may assume ${}^t\overline{A} = A$.

Hence A is contained in the set $\{B \in \mathrm{M}_d(\mathbb{C}) \mid A^{-1}\,{}^t\overline{B}A = B\}$. On the other hand, if U denotes the K_0-vector space $U = \{b \in F \mid \tilde{b} = b\}$, we have

$$\psi(U \otimes_{\sigma_0} \mathbb{R}) = \{B \in \mathrm{M}_d(\mathbb{C}) \mid A^{-1}\,{}^t\overline{B}A = B\}.$$

Since U is dense in $U \otimes_{\sigma_0} \mathbb{R}$, there is an $u \in U$ such that $\psi(u \otimes 1)$ is arbitrarily close to A. The map

$$x \mapsto x' = a\tilde{x}a^{-1}$$

is again an anti-involution, since $\tilde{a} = a$. Its extension to $\mathrm{M}_d(\mathbb{C})$ is

$$X \mapsto \psi(a \otimes 1)A^{-1}\, {}^t\overline{X} A \psi(a \otimes 1)^{-1}.$$

This is a positive anti-involution, since $X \mapsto {}^t\overline{X}$ is a positive anti-involution on $\mathrm{M}_d(\mathbb{C})$ and $\psi(a \otimes 1)A^{-1}$ is arbitrarily close to $\mathbf{1}_d$. Thus we have shown that $x \mapsto x'$ is a positive anti-involution on F. According to Lemma 2.6.3 there is an isomorphism $\varphi\colon F \otimes_\sigma \mathbb{C} \xrightarrow{\sim} \mathrm{M}_d(\mathbb{C})$ as claimed.

Step II: By the Skolem–Noether Theorem any positive anti-involution on F is of the form

$$x \mapsto ax'a^{-1}$$

with $a \in F$. As in the proof of Lemma 2.6.3 we see that $a' = \lambda a$ for some $\lambda \in K$ with $\bar{\lambda}\lambda = 1$. Applying Hilbert's Satz 90 (see Jacobson [73, I, Theorem 4.31]) there is a $\mu \in K$ such that $\lambda = \bar{\mu}\mu^{-1}$. Replacing a by $\mu^{-1}a$ we see that we may assume $a' = a$.

Hence for $A = \varphi(a \otimes 1)$ we have ${}^t\overline{A} = A$ and it remains to show that A is positive definite. For the hermitian matrix A there is a unitary matrix $T \in \mathrm{M}_d(\mathbb{C})$ such that ${}^t\overline{T}AT = \mathrm{diag}(r_1, \dots, r_d)$ for some $r_\nu \in \mathbb{R}$, $1 \le \nu \le d$. But for all matrices $X = (x_{ij}) \in \mathrm{M}_d(\mathbb{C})$, $X \neq 0$, we have

$$\begin{aligned}
0 &< \mathrm{tr}\big(A\, {}^t\overline{(TX\,{}^t\overline{T})} A^{-1} TX\,{}^t\overline{T}\big) \\
&= \mathrm{tr}\big(({}^t\overline{T}AT)\, {}^t\overline{X} ({}^t\overline{T}AT)^{-1} X\big) \\
&= \mathrm{tr}\big(\mathrm{diag}(r_1, \dots, r_d)\, {}^t\overline{X}\ \mathrm{diag}(r_1, \dots, r_d)^{-1} X\big) \\
&= \sum_{i,j=1}^{d} |x_{ij}|^2 \frac{r_j}{r_i}.
\end{aligned}$$

Hence A or $-A$ is positive definite. Since we may replace a by $-a$, this completes the proof. □

2.6.3 Exercises

(1) For a square-free integer $d \ge 1$ consider the imaginary quadratic field $\mathbb{Q}(\sqrt{-d})$ with maximal order with usual basis $\{1, \omega\}$ and f a positive integer, then $\Omega_f = \mathbb{Z} \oplus f\omega\mathbb{Z}$ is a lattice in $\mathbb{C}$ and $E_f = \mathbb{C}/\Omega_f$ is an elliptic curve.
Show that

$$\mathrm{End}(E_f) = \Omega_f.$$

In particular, if $f \ge 2$, then $\mathrm{End}(E_f)$ is not a maximal order in $\mathbb{Q}(\sqrt{-d})$.

(2) Let X be an abelian variety of dimension g. Recall from Exercise 1.3.4 (10) that $\rho(X) \leq g^2$. The following conditions are equivalent

(i) $\rho(X) = g^2$.
(ii) X is isogenous to E^g with an elliptic curve E with complex multiplication.
(iii) X admits a period matrix $\Pi \in \mathrm{M}(g \times 2g, K)$ with K an imaginary quadratic field.
(iv) X is isomorphic to a product $E_1 \times \cdots \times E_g$ with pairwise isogenous elliptic curves with complex multiplication.

(Hint: For (iii) $\Leftrightarrow$ (iv) use Exercise 5.1.5 (15).)

(3) Suppose X is an abelian variety of dimension g with $\mathrm{End}_{\mathbb{Q}}(X)$ a commutative field. Let K_0 be the maximal totally real subfield and $m = \frac{g}{[K_0:\mathbb{Q}]}$. Show that any non-trivial line bundle L on X is non-degenerate of index $i(L) = \nu m$ for some integer $0 \leq \nu \leq [K_0 : \mathbb{Q}]$. Moreover, for any of these values there is a line bundle of this index. In particular, if $\mathrm{End}(X) = \mathbb{Z}$, any non-trivial line bundle is of index 0 or g.

Let K be a totally complex quadratic extension of a totally real number field of degree g over $\mathbb{Q}$. A *CM-type of K* is a set $\Phi = \{\sigma_1, \ldots, \sigma_g\}$ of pairwise non-complex conjugate embeddings $K \hookrightarrow \mathbb{C}$. An abelian variety $X = \mathbb{C}^g/\Lambda$ is said to be of *CM-type* (K, Φ) if there is an embedding $\rho : K \hookrightarrow \mathrm{End}_{\mathbb{Q}}(X)$ such that

$$\rho_a \circ \rho \simeq \mathrm{diag}(\sigma_1, \ldots, \sigma_g) : K \to \mathrm{M}_g(\mathbb{C}).$$

The next exercise shows that to every CM-type Φ of K one can associate an abelian variety in a canonical way:

(4) The tensor product $K \otimes_{\mathbb{Q}} \mathbb{R}$ is an $\mathbb{R}$-vector space of dimension $2g$. The CM-type $\Phi = \{\sigma_1, \ldots, \sigma_g\}$ induces a complex structure on $K \otimes_{\mathbb{Q}} \mathbb{R}$ via the $\mathbb{R}$-linear isomorphism

$$(\sigma_1, \ldots, \sigma_g) \otimes \mathbf{1}_{\mathbb{R}} : K \otimes_{\mathbb{Q}} \mathbb{R} \xrightarrow{\sim} \mathbb{C}^g.$$

The ring of integers $\mathcal{O}$ of K is a lattice of rank $2g$ in $K \otimes_{\mathbb{Q}} \mathbb{R}$. Hence the quotient $X(K, \Phi) := K \otimes_{\mathbb{Q}} \mathbb{R}/\mathcal{O}$ is a complex torus of dimension g.
Show that $X(K, \Phi)$ is an abelian variety.

2.7 The Commutator Map Associated to a Theta Group

To every line bundle L on an abelian variety one can associate a group, its theta group. The main result of this section is that the commutator map of the theta group of L coincides with the Weil pairing on $K(L)$.

2.7.1 The Weil Pairing on $K(L)$

Let $X = V/\Lambda$ be an abelian variety and let $L \in \operatorname{Pic}(X)$ with $H = c_1(L)$ considered as a hermitian form. In Section 2.3.3 we introduced the Weil pairing e^L on X_2. Here we consider a map $\varepsilon^L : K(L) \times K(L) \to \mathbb{C}^*$, which will be called the Weil pairing on $K(L)$.

Recall $K(L) = \Lambda(L)/\Lambda$. Then the map

$$\varepsilon^L : K(L) \times K(L) \to \mathbb{C}^*, \qquad \varepsilon(\overline{v}, \overline{w}) := \mathbf{e}(-2\pi i \operatorname{Im} H(v, w)) \tag{2.13}$$

is well defined according to Exercise 2.7.4 (1). For the minus sign in the definition, see Theorem 2.7.7 below.

Lemma 2.7.1

(i) *The map $\varepsilon^L : K(L) \times K(L) \to \mathbb{C}^*$ is bimultiplicative in the following sense: for all $v, v_1, v_2, w \in \Lambda(L)$ we have*

$$\varepsilon^L(\overline{v}_1 + \overline{v}_2, \overline{w}) = \varepsilon^L(\overline{v}_1, \overline{w})\varepsilon^L(\overline{v}_2, \overline{w}) \qquad \textit{and} \qquad \varepsilon^L(\overline{v}, \overline{w}) = \varepsilon^L(\overline{w}, \overline{v})^{-1}.$$

(ii) *For any non-degenerate $L \in \operatorname{Pic}(X)$ the map ε^L is non-degenerate in the following sense:*

$$\textit{If } \varepsilon^L(\overline{v}, \overline{w}) = 1 \textit{ for all } \overline{w} \in K(L), \textit{ then } \overline{v} = 0.$$

For the proof, see Exercise 2.7.4 (2). Lemma 2.7.1 means that for any non-degenerate line bundle on X the map is a pairing on the group $K(L)$. It is called the *Weil paring on $K(L)$*.

Lemma 2.7.1 implies that for a non-degenerate line bundle L and any subgroup U of $K(L)$ there is an *orthogonal complement $U^\perp$* defined by

$$U^\perp = \{\overline{w} \in K(L) \mid \varepsilon^L(\overline{v}, \overline{w}) = 1 \text{ for all } v \in U\}.$$

Clearly $U^\perp$ is a subgroup, uniquely determined by U (and L). Recall that a subgroup $U \subset K(L)$ is called *isotropic* if with respect to ε^L if $\varepsilon^L(\overline{v}, \overline{w}) = 1$ for all $\overline{v}, \overline{w} \in U$.

Proposition 2.7.2 *For an isogeny $f : X \to Y$ of abelian varieties and a non-degenerate line bundle L on X the following conditions are equivalent:*

(i) $L = f^*M$ *for some* $M \in \operatorname{Pic}(Y)$;
(ii) $\operatorname{Ker} f$ *is an isotropic subgroup of $K(L)$ with respect to ε^L.*

Proof Note first that it follows from the fact that $c_1(L) = c_1(f^*K) = f^*c_1(M)$ that $\mathrm{e}^{f^*(M)}(x, y) = \mathrm{e}^M(f(x), f(y))$. Hence according to Corollary 1.4.4 it suffices to show that $\varepsilon^L(\operatorname{Ker} f, \operatorname{Ker} f) = 1$. But this means just that $\operatorname{Ker} f$ is isotropic with respect to ε^L. □

Corollary 2.7.3 *Let $L \in \operatorname{Pic}(X)$ be ample.*

(a) *If K_1 is a maximal isotropic subgroup of $K(L)$ with respect to ε^L and $\pi : X \to Y = X/K_1$ the canonical map, then there is an $M \in \operatorname{Pic}(Y)$ defining a principal polarization on Y such that $L = \pi^* M$.*

(b) *If L is of type $(d_1, \dots, d_g)$, then there is an $N \in \operatorname{Pic}(X)$ such that $L = N^{d_1}$; N is of type $(1, \frac{d_2}{d_1}, \dots, \frac{d_g}{d_1})$.*

Proof Let $X = V/\Lambda$ and $p : V \to X$ be the canonical projection. Suppose L is of type $(d_1, \dots, d_g)$ and let $V_1 = p^{-1}K_1$. It is easy to see that there is a subspace $V_2 \subset V$ such that $V = V_1 \oplus V_2$ is a decomposition for L. If $\Lambda(L)_i = \Lambda(L) \cap V_i$ for $i = 1$ and 2, then according to Exercise 1.5.5 (5) the subgroup $K_1 = \Lambda(L)_1$ is isomorphic to $\oplus_{i=1}^g \mathbb{Z}/d_i\mathbb{Z}$.

Let $Y = X/K_1$ with projection $\pi : X \to Y$. Since K_1 is isotropic with respect to ε^L, Proposition 2.7.2 implies that there is an $M \in \operatorname{Pic}(Y)$ with $L = \pi^* M$. Since $\deg \pi = d_1 \cdots d_g$ and K_1 is maximal isotropic, M defines a principal polarization, which completes the proof of (a).

For the proof of (b) one uses the fact that all d_1-division points of X are contained in K_1. □

2.7.2 The Theta Group of a Line Bundle

Let $L \in \operatorname{Pic}(X)$ and $X = V/\Lambda$. In the next section we give another interpretation of the Weil pairing in terms of a certain central extension of the group $K(L)$ by the group $\mathbb{C}^*$, namely the theta group of L. It will be introduced in this section.

Let $x \in X$ be a point. A biholomorphic map $\varphi : L \to L$ is called an *automorphism over* x if the diagram

$$\begin{array}{ccc} L & \xrightarrow{\varphi} & L \\ \downarrow & & \downarrow \\ X & \xrightarrow{t_x} & X \end{array}$$

commutes and for every $y \in X$ the induced map on the fibres $\varphi(x) : L(y) \to L(x+y)$ is a $\mathbb{C}$-vector space homomorphism. Denote by $\mathcal{G}(L)$ the set of pairs

$$\mathcal{G}(L) := \{(\varphi, x) \mid x \in X,\ \varphi : L \to L \text{ an automorphism over } x\}.$$

Of course in a pair (φ, x) the map φ determines the point x, but it is more convenient to denote the element in this way. Clearly the composition of pairs

$$(\varphi_1, x_1)(\varphi_2, x_2) := (\varphi_1\varphi_2, x_1 + x_2)$$

defines a group structure on $\mathcal{G}(L)$. The group $\mathcal{G}(L)$ is called the *theta group of L*. For a slightly different definition of $\mathcal{G}(L)$ compare Exercise 2.7.4 (5).

Recall that $K(L)$ denotes the group of all $x \in X$ with $t_x^*L \simeq L$ over X.

Proposition 2.7.4 *The sequence*

$$1 \longrightarrow \mathbb{C}^* \xrightarrow{\iota} \mathcal{G}(L) \xrightarrow{p} K(L) \longrightarrow 0$$

with $\iota(\alpha) = (\alpha, 0)$ *and* $p(\varphi, x) = x$ *is exact. Moreover,* $\mathcal{G}(L)$ *is a central extension of* $K(L)$ *by* $\mathbb{C}^*$.

Proof Let $(\varphi, x) \in \mathcal{G}(L)$. By definition $t_x^*L = X \times_X L$ is the fibre product of $t_x : X \to X$ with the bundle projection $L \to X$. According to the universal property of the fibre product there is a unique isomorphism $\widetilde{\varphi} : L \to t_x^*L$ of line bundles over X such that the following diagram commutes

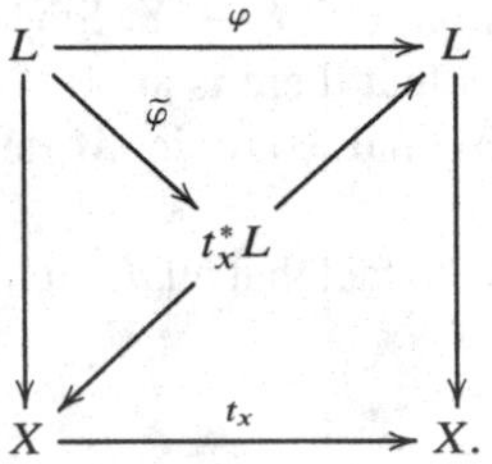

In particular $t_x L \simeq L$, so $x \in K(L)$.

By definition of $K(L)$, for every $x \in K(L)$ there is an isomorphism $\widetilde{\varphi} : L \to t_x^*L$ over X. The composition of $\widetilde{\varphi}$ with the projection $t_x^*L = X \times_X L \to L$ is an isomorphism over x. So p is surjective.

For the exactness it remains to show that $\operatorname{Im} \iota = \operatorname{Ker} p$. But this is an immediate consequence of the fact that every automorphism of L over X is multiplication by a nonzero constant. Obviously every automorphism $(\alpha, 0)$ commutes with every automorphism over x. Hence $\mathcal{G}(L)$ is a central extension of $K(L)$ by $\mathbb{C}^*$. □

In the next section we need the pullback of the exact sequence of Proposition 2.7.4 via $\pi : \Lambda(L) \to K(L)$, where $\pi : V \to X$ denotes the natural projection. For this recall from Lemma 1.2.1 that $\pi^*L \simeq V \times \mathbb{C}$, the trivial line bundle on V.

Let $L = L(H, \chi) \in \operatorname{Pic}(X)$. For every $\alpha \in \mathbb{C}^*$ and $w \in \Lambda(L)$ define the holomorphic map

$$[\alpha, w] : V \times \mathbb{C} \to V \times \mathbb{C}, \qquad (v, t) \mapsto (v + w, \alpha\, \mathbf{e}(\pi H(v, w)t)). \tag{2.14}$$

For the proof of the facts that $[\alpha, w]$ is a linear automorphism on the trivial line bundle $V \times \mathbb{C}$ over the translation $t_w : V \to V$ and the set

$$G(L) := \{[\alpha, w] \mid \alpha \in \mathbb{C}^*,\ w \in \Lambda(L)\}$$

forms a group under the composition

$$[\alpha_1, w_1][\alpha_2, w_2] = [\alpha_1\alpha_2\, \mathbf{e}\big(\pi H(w_2, w_1)\big), w_1 + w_2], \tag{2.15}$$

note that $[1,0]$ is the unit in $G(L)$ and $[\alpha,w]^{-1} = [\alpha^{-1}\,\mathbf{e}(\pi H(w,w)), -w]$ (see Exercise 2.7.4 (5)). The following sequence is exact,

$$1 \longrightarrow \mathbb{C}^* \xrightarrow{\ j\ } G(L) \xrightarrow{\ q\ } \Lambda(L) \longrightarrow 0, \tag{2.16}$$

where $j(\alpha) = [\alpha, 0]$ and $q(\alpha, w] = w$ shows that $G(L)$ is a central extension of $\Lambda(L)$ by $\mathbb{C}^*$ (see also Exercise 2.7.4 (5)).

Lemma 2.7.5 *Let a_L denote the canonical factor of $L = L(H,\chi) \in \operatorname{Pic}(X)$. The map*

$$s_L : \lambda \to G(L) \qquad \lambda \mapsto [a_L(\lambda, 0], \lambda]$$

is a section of $q : G(L) \to \Lambda(L)$ over Λ.

Proof Using the cocycle relation and Lemma 1.5.3 (a), or just by an immediate computation, one checks that s_L is an injective group homomorphism. It remains to show that $s_L(\Lambda)$ is contained in the centre of $G(L)$.

For this note that by the definition of $\Lambda(L)$ (see Section 1.4.2) we have

$$\mathbf{e}\big(\pi H(\lambda, w) - \pi H(w,\lambda)\big) = \mathbf{e}\big(2\pi i \operatorname{Im} H(\lambda, w)\big) = 1 \quad \text{for all} \quad w \in \Lambda(L),\ \lambda \in \Lambda.$$

Hence

$$[\alpha, w][a_L(\lambda, 0), \lambda] = [\alpha a_L(\lambda,0)\,\mathbf{e}(\pi H(\lambda, w)), \lambda + w] = [a_L(\lambda, 0), \lambda][\alpha, w]$$

for all $[\alpha, w] \in G(L)$ and $\lambda \in \Lambda$. This gives the assertion. □

Proposition 2.7.6 *There is a canonical isomorphism of exact sequences*

$$\begin{array}{ccccccccc} 1 & \longrightarrow & \mathbb{C}^* & \longrightarrow & G(L)/s_L(\Lambda) & \longrightarrow & \Lambda(L)/\Lambda & \longrightarrow & 0 \\ & & \| & & \downarrow{\scriptstyle\overline{\sigma}} & & \| & & \\ 1 & \longrightarrow & \mathbb{C}^* & \longrightarrow & \mathcal{G}(L) & \longrightarrow & K(L) & \longrightarrow & 0. \end{array}$$

Proof In Lemma 1.3.1 we saw that $L \simeq V \times \mathbb{C}/\Lambda$, where Λ acts on $V \times \mathbb{C}$ by the canonical factor a_L of L. This action is given by

$$\lambda(v,t) = \big(v + \lambda, a_L(\lambda, v)t\big) = [a_L(\lambda, 0), \lambda](v,t)$$

for all $\lambda \in \Lambda$ and $(v,t) \in V \times \mathbb{C}$ and hence coincides with the action of the subgroup $s_L(\Lambda)$ of $G(L)$ on $V \times \mathbb{C}$. So $L \simeq V \times \mathbb{C}/s_L(\Lambda)$.

Define a map $\sigma : G(L) \to \mathcal{G}(L)$ as follows: Every $[\alpha, w]$ defines an automorphism $\varphi_{[\alpha,w]}$ of L, since $s_L(\Lambda)$ is contained in the centre of $G(L)$. It is obvious that $\varphi_{[\alpha,w]}$ is an automorphism over $\overline{w} \in \pi(\Lambda(L)) = K(L)$, so $\varphi_{[\alpha,w]} \in \mathcal{G}(L)$. Moreover, the map $\sigma : G(L) \to \mathcal{G}(L)$, $[\alpha, w] \mapsto \varphi_{[\alpha,w]}$ is a homomorphism of groups and the following diagram commutes

$$\begin{array}{ccccccccc} 1 & \longrightarrow & \mathbb{C}^* & \longrightarrow & G(L) & \longrightarrow & \Lambda(L) & \longrightarrow & 0 \\ & & \| & & \downarrow{\scriptstyle\sigma} & & \downarrow{\scriptstyle\pi} & & \\ 1 & \longrightarrow & \mathbb{C}^* & \longrightarrow & \mathcal{G}(L) & \longrightarrow & K(L) & \longrightarrow & 0. \end{array} \tag{2.17}$$

By construction $\sigma : G(L) \to \mathcal{G}(L)$ factorizes via $G(L)/s_L(\Lambda)$ and the assertion follows from the snake lemma. □

2.7.3 The Commutator Map

Let $X = V/\Lambda$ be an abelian variety and $L = L(H, \chi) \in \operatorname{Pic}(X)$. The groups $\mathcal{G}(L)$ and $G(L)$ are non-commutative in general. In this subsection we study the corresponding commutator map. The main result is that it coincides with the Weil pairing on $K(L)$. This has some consequences for the Weil pairing.

The group $G(L)$ is a central extension of abelian groups, so its commutator map induces a map

$$\widetilde{\mathrm{e}}^L : \Lambda(L)\times\Lambda(L) \longrightarrow \mathbb{C}^*, \qquad (w_1, w_2) \mapsto [\alpha_1, w_1][\alpha_2, w_2][\alpha_1, w_1]^{-1}[\alpha_2, w_2]^{-1}, \tag{2.18}$$

where $\alpha_1, \alpha_2 \in \mathbb{C}^*$ are any constants. Note that this definition does not depend on the chosen α_i. By diagram (2.17) it makes sense to define

$$\widetilde{\mathrm{e}}^L : K(L) \times K(L) \qquad (\overline{w}_1, \overline{w}_2) \mapsto \widetilde{\mathrm{e}}^L(w_1, w_2).$$

Theorem 2.7.7 *The commutator map $\widetilde{\mathrm{e}}^L$ coincides with the Weil pairing ε^L. In other words, for all $w_1, w_2 \in \Lambda(L)$,*

$$\widetilde{\mathrm{e}}^L(w_1, w_2) = \mathbf{e}(-2\pi i \operatorname{Im} H(w_1, w_2)) = \varepsilon^L(\overline{w}_1, \overline{w}_2).$$

Proof Using (2.18) and the group structure of $G(L)$, we get

$$\begin{aligned} \widetilde{\mathrm{e}}^L(w_1, w_2) &= \big([\alpha_1, w_1][\alpha_2, w_2]\big) \cdot \big([\alpha_2, w_2][\alpha_1, w_1]\big)^{-1} \\ &= [\alpha_1\alpha_2\, \mathbf{e}(\pi H(w_2, w_1), w_1 + w_2] \cdot [\alpha_1\alpha_2\, \mathbf{e}(\pi H(w_1, w_2), w_1 + w_2]^{-1} \\ &= [\alpha_1\alpha_2\, \mathbf{e}(\pi H(w_2, w_1), w_1 + w_2] \\ &\quad \cdot [\alpha_1^{-1}\alpha_2^{-1}\, \mathbf{e}(-\pi H(w_1, w_2)\, \mathbf{e}(\pi H(w_1 + w_2, w_1 + w_2), -w_1 - w_2] \\ &= [\mathbf{e}(\pi H(w_2, w_1) - \pi H(w_1, w_2), 0] \\ &= \mathbf{e}(-2\pi i \operatorname{Im} H(w_1, w_2)), \end{aligned}$$

which gives the first equation. The second equation follows from the definition (2.13). □

For some properties of the commutator map, see Exercise 2.7.4 (6).

Proposition 2.7.8 *For $L \in \mathrm{Pic}(X)$ the following statements are equivalent:*

(i) *L is non-degenerate;*
(ii) *$\widetilde{e}^L : K(L) \times K(L) \to \mathbb{C}^*$ is non-degenerate;*
(iii) *$\mathbb{C}^*$ is the centre of $\mathcal{G}(L)$;*
(iv) *there is a decomposition $K(L) = K_1 \oplus K_2$ with subgroups K_1 and K_2, isotropic with respect to $\widetilde{e}^L$, such that the map*

$$K_2 \to \widehat{K_1} = \mathrm{Hom}(K_1, \mathbb{C}^*), \quad x \mapsto \widetilde{e}^L(\cdot, x)$$

is an isomorphism.

Proof The equivalence (ii) $\Leftrightarrow$ (iii) and the implication (iv) $\Rightarrow$ (ii) are trivial. Moreover, (i) implies (iv) by Exercise 1.5.5 (5). Hence it remains to show (ii) $\Rightarrow$ (i).

Suppose L is degenerate, that is, the group $K(L)$ is infinite. We have to show that $\widetilde{e}^L$ is degenerate. Consider the homomorphism $p : X \to X/K(L)^0$, where as usual $K(L)^0$ denotes the connected component of $K(L)$ containing 0.

We claim that we may assume $L|_{K(L)^0}$ is trivial. Since the canonical map $\mathrm{Pic}^0(X) \to \mathrm{Pic}^0(K(L)^0)$ is surjective by Proposition 1.4.2, there is a $P \in \mathrm{Pic}^0(X)$ with $P|_{K(L)^0} = L|_{K(L)^0}$. So $L \otimes P^{-1}$ is trivial on $K(L)^0$. Since L and $L \otimes P^{-1}$ are algebraically equivalent, we may replace L by $L \otimes P^{-1}$.

By Lemma 1.5.10 there is a line bundle $\overline{L}$ on the abelian variety $X/K(L)^0$ with $p^*\overline{L} = L$. According to Exercise 2.7.4 (6)(c) we have $\widetilde{e}^L = p^*\widetilde{e}^{\overline{L}}$, since $p^{-1}(K(\overline{L})) = K(L)$ by Exercise 2.7.4 (7). But $p^*\widetilde{e}^{\overline{L}}$ is certainly degenerate, since $K(L)^0 \neq 0$. □

2.7.4 Exercises

(1) Show that the map $\varepsilon^L : K(L) \times K(L) \to \mathbb{C}^*$ of equation (2.13) is well defined.

(2) For $L \in \mathrm{Pic}(X)$ show that for all $\overline{v}, \overline{v}_1, \overline{v}_2, \overline{w} \in K(L)$:

(i) $\varepsilon^L(\overline{v}_1 + \overline{v}_2, \overline{w}) = \varepsilon^L(\overline{v}_1, \overline{w})\varepsilon^L(\overline{v}_2, \overline{w})$ and $\varepsilon^L(\overline{v}, \overline{w})^{-1} = \varepsilon^L(\overline{w}, \overline{v})$;
(ii) if L is non-degenerate and $\varepsilon^L(\overline{v}, \overline{w}) = 1$ for all $\overline{w} \in K(L)$, then $\overline{v} = 0$.

(3) Let L be a symmetric line bundle on an abelian variety $X = V/\Lambda$ with $L^n = \mathcal{O}_X$ for some $n \in \mathbb{Z}$. Suppose D is a divisor of X with $\mathcal{O}_X(D) = L$. Then there is a rational function g on X such that $(g) = n^*D$.

(a) Show that for any $x \in X_n$

$$q_L^{(n)}(x) := \frac{g(x+y)}{g(y)}$$

is an n-th root of unity independent of the choice of D and of the point $y \in X$.

(b) Suppose $L = L(0, \chi)$. Show that

$$q_L^{(n)}(\overline{v}) = \chi(nv)$$

for all $\overline{v} \in X_n$.

(c) The map

$$q^{(n)} : X_n \times \widehat{X}_n \to \mu_n, \quad (x, L) \mapsto q_L^{(n)}(x)$$

is a non-degenerate pairing. In particular $q_L^{(2)}$ coincides with the Weil form q_L on X_2 (see Lemma 2.3.10).

(4) Show that the set of all linear isomorphisms $\widetilde{\varphi} : L \to t_x^* L$ over X is a group $\widetilde{\mathcal{G}}(L)$ with respect to the composition

$$\widetilde{\varphi}_1 \cdot \widetilde{\varphi}_2 := (t_{x_2}^* \widetilde{\varphi}_1)\widetilde{\varphi}_2$$

for linear isomorphisms $\widetilde{\varphi}_i : L \to t_{x_i}^* L,\ i = 1, 2$. Moreover, the map

$$\mathcal{G}(L) \to \widetilde{\mathcal{G}}(L), \qquad \varphi \mapsto \widetilde{\varphi}$$

is a group isomorphism.
(Hint: Compare the proof of Proposition 2.7.4.)

(5) Let $L = L(H, \chi) \in \mathrm{Pic}(X)$. Give a proof of the following facts:

(i) the map $[\alpha, w] : V \times \mathbb{C}^* \to V \times \mathbb{C}^*$ of equation (2.14) is an automorphism of the trivial line bundle over V over the translation $t_w : V \to V$;

(ii) the set $G(L) = \{[\alpha, w] \mid \alpha \in \mathbb{C}^*,\ w \in \Lambda(L)\}$ is a group under the composition (2.15);

(iii) the sequence (2.16) is a central extension of $\Lambda(L)$ by $\mathbb{C}^*$.

(Hint: Use the canonical factor $a(\lambda, v)$ of $L(H, \chi)$.)

(6) Let $L, L_1, L_2 \in \mathrm{Pic}(X)$ and $x, x_1, x_2 \in K(L)$. Show that

(a) $\widetilde{\mathrm{e}}^L(x_1 + x_2, x) = \widetilde{\mathrm{e}}^L(x_1, x)\widetilde{\mathrm{e}}^L(x_2, x)$,
$\widetilde{\mathrm{e}}^L(x_1, x_2) = \widetilde{\mathrm{e}}^L(x_2, x_1)^{-1}$ and
$\widetilde{\mathrm{e}}^L(x, x) = 1$.

(b) $\widetilde{\mathrm{e}}^{L_1 \otimes L_2} = \widetilde{\mathrm{e}}^{L_1}\widetilde{\mathrm{e}}^{L_2}$ on $K(L_1) \cap K(L_2)$,
$\widetilde{\mathrm{e}}^{L_1} = \widetilde{\mathrm{e}}^{L_2}$ if L_1 and L_2 are algebraically equivalent;

(c) for any surjective homomorphism $f : Y \to X$ of abelian varieties,

$$\widetilde{\mathrm{e}}^{f^*L}(x, y) = \widetilde{\mathrm{e}}^L(f(x), f(y)) \quad \text{for all} \quad x, y \in f^{-1}(K(L)).$$

(7) Let $f : Y = W/\Gamma \to X = V/\Lambda$ be a surjective homomorphism of abelian varieties with connected kernel, analytic representation F and $L \in \mathrm{Pic}(X)$. Show that

(a) $f^{-1}(K(L)) = K(f^*L)$ and $F^{-1}(\Lambda(L)) = \Gamma(f^*L)$;

(b) the sequence

$$0 \longrightarrow \operatorname{Ker} F \stackrel{\iota}{\longrightarrow} G(f^*L) \stackrel{\widetilde{F}}{\longrightarrow} G(L) \longrightarrow 0$$

is exact (here $\iota(w) = [1, w]$ and $\widetilde{F}[\alpha, w] = [\alpha, F(w)]$);

(c) the sequence of (b) induces an exact sequence

$$1 \longrightarrow \ker f \longrightarrow \mathcal{G}(f^*L) \longrightarrow \mathcal{G}(L) \longrightarrow 0.$$

(8) (*Theorem of Serre–Rosenlicht*) Let X be an abelian variety. Any extension of algebraic groups of X by $\mathbb{C}^*$ is of the form $1 \to \mathbb{C}^* \to \mathcal{G}(L) \to X \to 0$ for a uniquely determined $L \in \operatorname{Pic}(X)$. To be more precise, there is a canonical isomorphism

$$\operatorname{Ext}^1(X, \mathbb{C}^*) \simeq \operatorname{Pic}^0(X)$$

(see Serre [123, Chapter 7]).

(9) Let X be an abelian variety and $L \in \operatorname{Pic}(X)$ semipositive such that $L|_{K(L)^0}$ is trivial (see Proposition 1.6.11). Show that:

(a) The map

$$\mathcal{G}(L) \times H^0(L) \to H^0(L) \to H^0(L), \qquad ((\varphi, x), s) \mapsto \varphi s t_{-x}$$

defines an action of $\mathcal{G}(L)$ on $H^0(L)$. The corresponding representation $\widetilde{\rho} : \mathcal{G}(L) \to \operatorname{GL}(H^0(L))$ is called the *canonical representation of the theta group* $\mathcal{G}(L)$.

(b) $\widetilde{\rho}$ induces a projective representation $\rho : K(L) \to \operatorname{PGL}(H^0(L))$ such that the following diagram commutes

$$\begin{array}{ccccccccc} 1 & \longrightarrow & \mathbb{C}^* & \longrightarrow & \mathcal{G}(L) & \longrightarrow & K(L) & \longrightarrow & 0 \\ & & \| & & \downarrow{\scriptstyle \widetilde{\rho}} & & \downarrow{\scriptstyle \rho} & & \\ 1 & \longrightarrow & \mathbb{C}^* & \longrightarrow & \operatorname{GL}(H^0(L)) & \longrightarrow & \operatorname{PGL}(H^0(L)) & \longrightarrow & 1. \end{array}$$

(c) The canonical representation is irreducible.

(Hint: use a decomposition of $\Lambda(L)$ and express the action with canonical factors.)

(10) Let L be a line bundle on the abelian variety X. Let s be the number of negative eigenvalues of the hermitian form $c_1(L)$. Show that $H^s(L)$ is an irreducible representation of the theta group $\mathcal{G}(L)$ such that the subgroup $\mathbb{C}^*$ acts by multiplication.

Chapter 3
Moduli Spaces

In this chapter several moduli spaces of polarized abelian varieties with an additional structure are constructed. The first thing to notice is that the notion "moduli space" is considered in a slightly naive way: A *moduli space for a set of abelian varieties with some additional structure* means a complex analytic space whose points are in some natural one to one correspondence with the elements of the set. In many case we show that the spaces are manifolds or algebraic varieties. The uniqueness and the functorial properties of these spaces will be totally ignored.

Given a type $D = \operatorname{diag}(d_1, \ldots, d_g)$, there is a polarized abelian variety of this type if and only if $d_g > 0$. If in this chapter a type occurs, this is always assumed.

The starting point for studying the corresponding moduli spaces is the Siegel upper half space $\mathfrak{H}_g$ of complex symmetric $(g \times g)$-matrices with positive definite imaginary part. It parametrizes the set of polarized abelian varieties of type D with a symplectic basis. The corresponding symplectic group G_D acts on $\mathfrak{H}_g$ in a natural way and the quotient $\mathcal{A}_g = \mathfrak{H}_g / G_D$ is a moduli space for polarized abelian varieties of type D (Theorem 3.1.12). In Section 3.2 several level structures are introduced. The corresponding moduli spaces are quotients of $\mathfrak{H}_g$ by the corresponding subgroup of G_D.

The last part of the chapter is devoted to Theorem 3.5.4, due to Igusa [70], which provides an analytic embedding of the moduli space of polarized abelian varieties $\mathcal{A}_D(D)_0$ with orthogonal level D-structure into projective space. It is not difficult to conclude that $\mathcal{A}_D(D)_0$ and hence also $\mathcal{A}_D$ are actually algebraic varieties.

Here is a short outline of the proof of Igusa's Theorem: First a universal family of abelian varieties $\mathcal{X}_D \to \mathfrak{H}_g$ parametrizing all polarized abelian varieties of type D with symplectic basis is constructed. Since the classical factor of automorphy is holomorphic in $Z \in \mathfrak{H}_g$, it extends to a factor on $\mathcal{X}_D$ and this defines a line bundle $\mathcal{L}$ on $\mathcal{X}_D$. Composing the zero section of $\mathcal{X}_D \to \mathfrak{H}_g$ with the map $\mathcal{X}_D \to \mathbb{P}^N$ associated to a certain sublinear system of $|\mathcal{L}|$, we get a holomorphic map $\psi_D : \mathfrak{H}_g \to \mathbb{P}^N$. Now the classical Theta Transformation Formula 3.3.9 implies that ψ_D factorizes via the quotient $\mathcal{A}_D(D)_0$ of $\mathfrak{H}_g$, to give a holomorphic map $\overline{\psi}_D : \mathcal{A}_D(D)_0 \to \mathbb{P}^N$, which turns out to be an embedding.

H. Lange, *Abelian Varieties over the Complex Numbers*, Grundlehren Text Editions,
https://doi.org/10.1007/978-3-031-25570-0_3

Finally a word about period matrices: in almost every book they appear in different forms. The following two requirements seem natural:

1. period matrices should be $(g \times 2g)$-matrices rather that $(2g \times g)$-matrices;
2. a symplectic $\begin{pmatrix} \alpha & \beta \\ \gamma & \delta \end{pmatrix} \in \mathrm{Sp}_{2g}(\mathbb{R})$ should act on $\mathfrak{H}_g$ by $Z \mapsto (\alpha Z + \beta)(\gamma Z + \delta)^{-1}$.

Under these conditions the period matrices are necessarily of the form (Z, D) with $Z \in \mathfrak{H}_g$ and type D. There is a slight disadvantage: there is a transposition coming up. Namely, if R is the rational representation of an isomorphism of polarized abelian varieties, the corresponding action on $\mathfrak{H}_g$ is given by ${}^tR$.

3.1 The Moduli Spaces of Polarized Abelian Varieties

This section contains the construction of the moduli space of polarized abelian varieties of a given type. In the first part the Siegel upper half space $\mathfrak{H}_g$ is introduced. It is shown that it parametrizes all polarized abelian varieties of a given type D with a symplectic basis (Theorem 3.1.2). Then the group G_D is introduced and its action on $\mathfrak{H}_g$ is worked out. Finally, the moduli space of polarized abelian varieties of type D is constructed as a normal complex analytic space (Theorem 3.1.12).

3.1.1 The Siegel Upper Half Space

Given any type D, we introduce in this section the Siegel upper half space $\mathfrak{H}_g$ and show that it parametrizes the set of polarized abelian varieties of this type with a symplectic basis. Moreover, we work out what it means for two points of $\mathfrak{H}_g$ that the associated polarized abelian varieties are isomorphic.

Let $X = V/\Lambda$ be an abelian variety of dimension g and $H \in \mathrm{NS}(X)$ a hermitian form on V defining a polarization of type $D = \mathrm{diag}(d_1, \dots, d_g)$. Let $\lambda_1, \dots, \lambda_g, \mu_1, \dots, \mu_g$ denote a symplectic basis of Λ for H; that is, the alternating form of $\mathrm{Im}\, H$ is given by the matrix $\begin{pmatrix} 0 & D \\ -D & 0 \end{pmatrix}$ with respect to this basis. Define $e_\nu = \frac{1}{d_\nu}\mu_\nu$ for $\nu = 1, \dots, g$. By Lemma 1.5.4 the vectors $e_1, \dots, e_g$ form a basis of the $\mathbb{C}$-vector space V. With respect to these bases the period matrix of X is of the form

$$\Pi = (Z, D)$$

for some $Z \in \mathrm{M}_g(\mathbb{C})$. The matrix Z has the following properties:

Proposition 3.1.1

(a) ${}^tZ = Z$ *and* $\mathrm{Im}\, Z > 0$.
(b) $(\mathrm{Im}\, Z)^{-1}$ *is the matrix of the hermitian form* H *with respect to the basis* $e_1, \dots, e_g$.

Proof Assertion (a) is just a repetition of Corollary 2.1.20. According to Lemma 2.1.22 the matrix of H is $2i\left(\overline{\Pi}\begin{pmatrix} 0 & D \\ -D & 0\end{pmatrix}^{-1}{}^t\Pi\right)^{-1} = (\operatorname{Im} Z)^{-1}$. □

Define a *polarized abelian variety of type D with symplectic basis* to be a triplet

$$(X, H, (\lambda_1, \dots, \lambda_g, \mu_1, \dots, \mu_g))$$

with $X = V/\Lambda$ an abelian variety, H a polarization of type D on X and $(\lambda_1, \dots, \lambda_g, \mu_1, \dots, \mu_g)$ a symplectic basis of Λ for H. The set

$$\mathfrak{H}_g := \{Z \in \mathrm{M}_g(\mathbb{C}) \mid {}^tZ = Z,\ \operatorname{Im} Z > 0\}$$

is called the *Siegel upper half space*. It is a $\frac{1}{2}g(g+1)$-dimensional open submanifold of the vector space of symmetric matrices in $\mathrm{M}_g(\mathbb{C})$.

With our loose notion of a moduli space we have the following theorem.

Theorem 3.1.2 *Given a type D, the Siegel upper half space $\mathfrak{H}_g$ can be considered as a moduli space of polarized abelian varieties of type D with symplectic basis.*

Proof We have seen that a polarized abelian variety of type D with symplectic basis determines a point Z of $\mathfrak{H}_g$. Conversely, given a type D and any point $Z \in \mathfrak{H}_g$, we have to associate a polarized abelian variety of type D with symplectic basis in a canonical way.

Since $\Lambda_Z := (Z, D)\mathbb{Z}^{2g}$ is a lattice in $V = \mathbb{C}^g$, the quotient

$$X_Z := \mathbb{C}^g/\Lambda_Z$$

is a complex torus. Define a hermitian form H_Z by the matrix $(\operatorname{Im} Z)^{-1}$ with respect to the canonical basis of $\mathbb{C}^g$.

We claim that H_Z is a polarization of type D on X_Z. To see this, consider the $\mathbb{R}$-linear isomorphism

$$(Z, D) : \mathbb{R}^{2g} \to \mathbb{C}^g.$$

By definition the columns of the matrix (Z, D) are just the images, say $\lambda_1, \dots, \lambda_g, \mu_1, \dots, \mu_g$, of the standard basis of $\mathbb{R}^{2g}$. With respect to this basis of Λ_Z, $\operatorname{Im} H_Z|_{\Lambda_Z \times \Lambda_Z}$ is given by the matrix

$$\operatorname{Im}\left({}^t(Z, D)(\operatorname{Im} Z)^{-1}\overline{(Z, D)}\right) = \begin{pmatrix} 0 & D \\ -D & 0\end{pmatrix}. \tag{3.1}$$

Since $\operatorname{Im} Z$ is positive definite by definition, this completes the proof of the claim.

Summing up, we get: the assignment

$$Z \mapsto (X_Z, H_Z, (\text{colums of}(Z, D)))$$

is a canonical bijection between $\mathfrak{H}_g$ and the set of (isomorphism classes of) polarized abelian varieties of type D with symplectic basis. □

3.1.2 Action of the Group G_D on $\mathfrak{H}_g$

Let $Z, Z' \in \mathfrak{H}_g$ such that there is an isomorphism $\varphi : (X_{Z'}, H_{Z'}) \to (X_Z, H_Z)$. Let $A \in \mathrm{GL}_g(\mathbb{C})$ and $R \in \mathrm{GL}_{2g}(\mathbb{Z})$ denote the matrices of the analytic and rational representation of φ with respect to the standard basis of $\mathbb{C}^g$ and the symplectic bases of $\Lambda_{Z'}$ and Λ_Z respectively. According to equation (1.2) the matrices A and R are related by $A(Z', D) = (Z, D)R$. Define

$$N := \begin{pmatrix} \mathbf{1}_g & 0 \\ 0 & D \end{pmatrix} R \begin{pmatrix} \mathbf{1}_g & 0 \\ 0 & D \end{pmatrix}^{-1} = {}^t\begin{pmatrix} \alpha & \beta \\ \gamma & \delta \end{pmatrix} \tag{3.2}$$

with $(g \times g)$-blocks $\alpha, \beta, \gamma, \delta$. Then the above equation is equivalent to

$$AZ' = Z^t\alpha +^t \beta \qquad \text{and} \qquad A = Z^t\gamma +^t \delta. \tag{3.3}$$

Since φ is an isomorphism, the matrix ${}^tA = \gamma Z + \delta$ is invertible. Thus we can write Z' in terms of Z and $N = {}^t\left(\begin{smallmatrix} \alpha & \beta \\ \gamma & \delta \end{smallmatrix}\right)$ as follows

$$Z' =^t Z' =^t (Z^t\alpha +^t \beta)^t A^{-1} = (\alpha Z + \beta)(\gamma Z + \delta)^{-1}.$$

Moreover, taking imaginary parts of the equation $\varphi^* H_Z = H_{Z'}$ gives

$${}^tR \begin{pmatrix} 0 & D \\ -D & 0 \end{pmatrix} R = \begin{pmatrix} 0 & D \\ -D & 0 \end{pmatrix}.$$

In terms of N this is

$${}^tN \begin{pmatrix} 0 & \mathbf{1}_g \\ -\mathbf{1}_g & 0 \end{pmatrix} N = \begin{pmatrix} 0 & \mathbf{1}_g \\ -\mathbf{1}_g & 0 \end{pmatrix}. \tag{3.4}$$

Before we go on, let us recall that for any commutative ring $\mathcal{R}$ (with 1 and of characteristic 0) the *symplectic group* $\mathrm{Sp}_{2g}(\mathcal{R})$ is the group of matrices N in $\mathrm{M}_{2g}(\mathcal{R})$ satisfying (3.4).

Lemma 3.1.3

(a) *The group* $\mathrm{Sp}_{2g}(\mathcal{R})$ *is closed under transposition.*

(b) *For any* $M = \left(\begin{smallmatrix} \alpha & \beta \\ \gamma & \delta \end{smallmatrix}\right) \in \mathrm{M}_{2g}(\mathcal{R})$ *the following statements are equivalent:*

(i) $M \in \mathrm{Sp}_{2g}(\mathcal{R})$;
(ii) ${}^t\alpha\gamma$ *and* ${}^t\beta\delta$ *are symmetric and* ${}^t\alpha\delta -^t \gamma\beta = \mathbf{1}_g$;
(iii) $\alpha^t\beta$ *and* $\gamma^t\delta$ *are symmetric and* $\alpha^t\delta - \beta^t\gamma = \mathbf{1}_g$.

Proof For the proof of (a) see Exercise 3.1.5 (1). Statement (b) follows directly from the definition and (a). □

So by definition, $N \in \mathrm{Sp}_{2g}(\mathbb{Q})$. Moreover, for

$$\Lambda_D := \begin{pmatrix} \mathbf{1}_g & 0 \\ 0 & D \end{pmatrix} \mathbb{Z}^{2g}$$

equation (3.2) means $N\Lambda_D \subseteq \Lambda_D$, since $R \in \mathrm{M}_{2g}(\mathbb{Z})$ (see Equation 3.2).

Noting that by Lemma 3.1.3 (a) $\mathrm{Sp}_{2g}(\mathbb{Q})$ is invariant under transposition, we saw that the matrix $M :={}^t N$ is an element of the group

$$G_D := \{M \in \mathrm{Sp}_{2g}(\mathbb{Q}) \mid {}^t M\Lambda_D \subseteq \Lambda_D\}. \tag{3.5}$$

For $M = \begin{pmatrix} \alpha & \beta \\ \gamma & \delta \end{pmatrix} \in G_D$ and $Z \in \mathfrak{H}_g$ define

$$M(Z) = (\alpha Z + \beta)(\gamma Z + \delta)^{-1}.$$

Summing up, we have proved the implication (i) $\Rightarrow$ (ii) of the following proposition.

Proposition 3.1.4 *For any $Z, Z' \in \mathfrak{H}_g$ the following statements are equivalent:*

(i) *the polarized abelian varieties (X_Z, H_Z) and $(X_{Z'}, H_{Z'})$ are isomorphic;*
(ii) *$Z' = M(Z)$ for some $M \in G_D$.*

Proof It remains to show (ii) $\Rightarrow$ (i). Suppose $Z' = M(Z)$ for some $M \in G_D$. From the arguments above we see that the matrix $\begin{pmatrix} \mathbf{1}_g & 0 \\ 0 & D \end{pmatrix}^{-1} {}^t M \begin{pmatrix} \mathbf{1}_g & 0 \\ 0 & D \end{pmatrix}$ is the rational representation of an isomorphism $(X_{Z'}, H_{Z'}) \to (X_Z, H_Z)$ with respect to the symplectic bases determined by Z and Z'. □

For later use we note the following corollary. For the proof we refer to Exercise 3.1.5 (2).

Corollary 3.1.5 *For any $Z \in \mathfrak{H}_g$ and $M = \begin{pmatrix} \alpha & \beta \\ \gamma & \delta \end{pmatrix} \in G_D$ the isomorphism $X_{M(Z)} \to X_Z$ is given by the equation*

$$A(M(Z), \mathbf{1}_g) = (Z, \mathbf{1}_g){}^t M.$$

Here $A ={}^t (\gamma Z + \delta)$ is the matrix of the corresponding analytical representation and the matrix $\begin{pmatrix} \mathbf{1}_g & 0 \\ 0 & D \end{pmatrix}^{-1} {}^t M \begin{pmatrix} \mathbf{1}_g & 0 \\ 0 & D \end{pmatrix}$ describes the rational representation with respect to the chosen bases.

3.1.3 The Action of $\mathrm{Sp}_{2g}(\mathbb{R})$ on $\mathfrak{H}_g$

Proposition 3.1.6 *The group* $\mathrm{Sp}_{2g}(\mathbb{R})$ *acts biholomorphically (from the left) on* $\mathfrak{H}_g$ *by*

$$Z \mapsto M(Z) = (\alpha Z + \beta)(\gamma Z + \delta)^{-1}$$

for all $M = \begin{pmatrix} \alpha & \beta \\ \gamma & \delta \end{pmatrix} \in \mathrm{Sp}_{2g}(\mathbb{R})$.

Proof To see that the matrix $(\gamma Z + \delta)$ is invertible, apply Lemma 3.1.3 (ii) to get

$$\overline{{}^t(\gamma Z + \delta)}(\alpha Z + \beta) - {}^t\overline{(\alpha Z + \beta)}(\gamma Z + \delta) = Z - \overline{Z} = 2i \operatorname{Im} Z. \tag{3.6}$$

Suppose $(\gamma Z + \delta)v = 0$ for some $v \in \mathbb{C}^g$. Then (3.6) implies ${}^t\overline{v}(\operatorname{Im} Z)v = 0$ and thus $v = 0$, since $\operatorname{Im} Z > 0$. So $(\gamma Z + \delta)$ is invertible and hence $M(Z)$ is well defined. Similarly one obtains

$${}^t(\gamma Z + \delta)\big(M(Z) - {}^tM(Z)\big)(\gamma Z + \delta) = Z - {}^tZ = 0.$$

This implies that $M(Z)$ is symmetric. By (3.6) and the symmetry of $M(Z)$,

$$\overline{{}^t(\gamma Z + \delta)} \operatorname{Im} M(Z)(\gamma Z + \delta) = \operatorname{Im} Z > 0.$$

Together this gives $M(Z) \in \mathfrak{H}_g$. It remains to show that $M_1(M_2(Z)) = (M_1M_2)(Z)$ for all $M_1, M_2 \in \mathfrak{H}_g$. But this is an immediate computation which will be omitted. □

The next three propositions give some properties of the action of $\mathrm{Sp}_{2g}(\mathbb{R})$ on $\mathfrak{H}_g$.

Proposition 3.1.7

(a) *The group* $\mathrm{Sp}_{2g}(\mathbb{R})$ *acts transitively on* $\mathfrak{H}_g$.
(b) *The stabilizer of* $i\mathbf{1}_g \in \mathfrak{H}_g$ *is the compact group*

$$\mathrm{Sp}_{2g}(\mathbb{R}) \cap O_{2g}(\mathbb{R}) = \left\{ \begin{pmatrix} \alpha & \beta \\ -\beta & \alpha \end{pmatrix} \in \mathrm{M}_{2g}(\mathbb{R}) \;\middle|\; \begin{array}{c} \alpha^t\beta = \beta^t\alpha \\ \alpha^t\alpha + \beta^t\beta = \mathbf{1}_g \end{array} \right\}.$$

Proof (a): Let $Z = X + iY \in \mathfrak{H}_g$. Since Y is positive definite and symmetric, there is an $\alpha \in \mathrm{GL}_g(\mathbb{R})$ with $Y = \alpha^t\alpha$. One checks that the matrix $N = \begin{pmatrix} \alpha & X^t\alpha^{-1} \\ 0 & {}^t\alpha^{-1} \end{pmatrix}$ is symplectic and satisfies $N(i\mathbf{1}_g) = Z$. This proves the transitivity.

(b): This follows by an immediate computation using that ${}^tM^{-1} = \begin{pmatrix} \delta & -\gamma \\ -\beta & \alpha \end{pmatrix}$ for any symplectic matrix $M = \begin{pmatrix} \alpha & \beta \\ \gamma & \delta \end{pmatrix}$. □

A consequence of Proposition 3.1.7 is that the map

$$h : \mathrm{Sp}_{2g}(\mathbb{R}) \to \mathfrak{H}_g, \qquad M \mapsto M(i\mathbf{1}_g)$$

is surjective and all its fibres are of the form $M \cdot \big(\mathrm{Sp}_{2g}(\mathbb{R}) \cap O_{2g}(\mathbb{R})\big)$ for some $M \in \mathrm{Sp}_{2g}(\mathbb{R})$. In particular the fibres are compact. Moreover, we have:

Proposition 3.1.8 *The map $h : \mathrm{Sp}_{2g}(\mathbb{R}) \to \mathfrak{H}_g$ is proper; that is, for any compact set $K \subset \mathfrak{H}_g$ the preimage $h^{-1}(K)$ is compact.*

Proof For the proof we apply the Iwasawa decomposition (see Exercise 3.1.5 (3)). It gives a diffeomorphism $\mathrm{Sp}_{2g}(\mathbb{R}) \xrightarrow{\simeq} N \times \Delta \times O$ with $O = \mathrm{Sp}_{2g}(\mathbb{R}) \cap O_{2g}(\mathbb{R})$ and under this isomorphism the map h corresponds to the projection $N \times \Delta \times O \to N \times \Delta$, which obviously is proper. For an elementary proof, see Exercise 3.1.5 (4). □

More important than the action of $\mathrm{Sp}_{2g}(\mathbb{R})$ on $\mathfrak{H}_g$ are the induced actions of certain subgroups. Recall that a subgroup $G \subseteq \mathrm{Sp}_{2g}(\mathbb{R})$ acts properly and discontinuously on $\mathfrak{H}_g$ if for any pair of compact subsets K_1, K_2 of $\mathfrak{H}_g$ the set $\{g \in G \mid gK_1 \cap K_2 \neq \emptyset\}$ is finite.

Proposition 3.1.9 *Any discrete subgroup $G \subset \mathrm{Sp}_{2g}(\mathbb{R})$ acts properly and discontinuously on $\mathfrak{H}_g$.*

Proof Let $K_1, K_2 \subseteq \mathfrak{H}_g$ any two compact subsets. We have to show that there are only finitely many $M \in G$ such that $M(K_1) \cap K_2 \neq \emptyset$. From the definition of the map h it follows that $M(K_1) \cap K_2 \neq \emptyset$ if and only if

$$M \in h^{-1}(K_2)(h^{-1}(K_1))^{-1} = \{M_2 M_1^{-1} \mid M_\nu \in h^{-1}(K_\nu), \nu = 1, 2\}.$$

Hence it suffices to show that $h^{-1}(K_2)(h^{-1}(K_1))^{-1}$ is compact in $\mathrm{Sp}_{2g}(\mathbb{R})$. But $h^{-1}(K_i)$ is compact for $i = 1, 2$, since h is proper. The assertion follows, since $h^{-1}(K_2)(h^{-1}(K_1))^{-1}$ is the image of the compact set $h^{-1}(K_1) \times h^{-1}(K_2)$ under the continuous map $(M_1, M_2) \mapsto M_2 M_1^{-1}$. □

Our main aim is the construction of the moduli space of polarized abelian varieties of type D. Although the approach via matrix calculation makes the subject very accessible, there is a danger that the reader might get lost in the formulas. So perhaps the following remark might be useful.

Remark 3.1.10 (1): Consider the group $K := \mathrm{Sp}_{2g}(\mathbb{R}) \cap O_{2g}(\mathbb{R})$ of Proposition 3.1.7. It is a maximal compact subgroup of $\mathrm{Sp}_{2g}(\mathbb{R})$ and the cited proposition identifies

$$\mathfrak{H}_g \simeq \mathrm{Sp}_{2g}(\mathbb{R})/K.$$

Using the definition of the moduli space $\mathcal{A}_D$ of Section 3.1.4, we get an isomorphism of topological spaces

$$\mathcal{A}_D \simeq G_D \backslash \mathrm{Sp}_{2g}(\mathbb{R})/K.$$

(2): Given D, one can always find D' such that $G_{D'} \subseteq G_D$, of finite index, which acts freely on $\mathfrak{H}_g$. Therefore $\mathcal{A}_{D'}$ is a complex manifold and $\mathcal{A}_D$ is a finite quotient of it.

3.1.4 The Moduli Space of Polarized Abelian Varieties of Type D

Let G be a group acting as a group of isomorphisms on a complex manifold X. The quotient X/G, endowed with the quotient topology, admits the structure of a ringed space in a natural way: denote by $\pi : X \to X/G$ the canonical projection. Then by definition, for $U \subseteq X/G$ open, $\mathcal{O}_{X/G}(U)$ is the set of functions $f : U \to \mathbb{C}$ for which $f\pi$ is holomorphic on $\pi^{-1}(U)$. We will use without proof the following theorem due to Cartan (see [30], see also Theorem 2.3.17).

Theorem 3.1.11 *Let G be a group acting properly and discontinuously on a complex manifold X. Then the quotient X/G is a normal complex analytic space.*

According to Proposition 3.1.9 the group G_D defined in equation (3.5) acts properly and discontinuously on $\mathfrak{H}_g$. So by Theorem 3.1.11 the quotient

$$\mathcal{A}_D := \mathfrak{H}_g / G_D$$

is a normal complex analytic space of dimension $\frac{1}{2}g(g+1)$. According to Theorem 3.1.2 and Proposition 3.1.4 the elements of $\mathcal{A}_D$ correspond one to one to the isomorphism classes of polarized abelian varieties of type D. Hence with our loose notion of moduli spaces we have proven:

Theorem 3.1.12 *For any type D the normal complex analytic space $\mathcal{A}_D = \mathfrak{H}_g/G_D$ is a moduli space of polarized abelian varieties of type D.*

We will see later that $\mathcal{A}_D$ is a normal algebraic variety.

For some purposes the following equivalent construction of the moduli $\mathcal{A}_D$ is more convenient: For any commutative ring $\mathcal{R}$ (with 1 and of characteristic 0) define the group

$$\mathrm{Sp}_{2g}^D(\mathcal{R}) = \left\{ R \in \mathrm{M}_{2g}(\mathcal{R}) \;\middle|\; R \begin{pmatrix} 0 & D \\ -D & 0 \end{pmatrix} {}^tR = \begin{pmatrix} 0 & D \\ -D & 0 \end{pmatrix} \right\}. \tag{3.7}$$

The map

$$\sigma_D : \mathrm{Sp}_{2g}^D(\mathbb{R}) \to \mathrm{Sp}_{2g}(\mathbb{R}), \qquad R \mapsto \begin{pmatrix} \mathbf{1}_g & 0 \\ 0 & D \end{pmatrix}^{-1} R \begin{pmatrix} \mathbf{1}_g & 0 \\ 0 & D \end{pmatrix} \tag{3.8}$$

is an isomorphism of groups, since $\mathrm{Sp}_{2g}(\mathbb{R})$ is invariant under transposition (for this apply the defining equation of $\mathrm{Sp}_{2g}^D(\mathbb{R})$ to the inverse map). The action of $\mathrm{Sp}_{2g}(\mathbb{R})$ on $\mathfrak{H}_g$ induces an action of $\mathrm{Sp}_{2g}^D(\mathbb{R})$ on $\mathfrak{H}_g$ via σ_D, namely

$$R(Z) = (aZ + bD)(D^{-1}cZ + D^{-1}dD)^{-1} \tag{3.9}$$

for all $R = \left(\begin{smallmatrix} a & b \\ c & d \end{smallmatrix}\right) \in \mathrm{Sp}_{2g}^D(\mathbb{R})$ and $Z \in \mathfrak{H}_g$. Note that $\sigma_D(\mathrm{Sp}_{2g}^D(\mathbb{Z}))$ is just the group

G_D defined in equation (3.5). For later use we define

$$\Gamma_D := \mathrm{Sp}_{2g}^D(\mathbb{Z}).$$

According to Theorem 3.1.11 the quotient

$$\widetilde{A}_D := \mathfrak{H}_g/\Gamma_D$$

is a normal complex analytic space and the identity on $\mathfrak{H}_g$ induces an isomorphism $\widetilde{\mathcal{A}}_D \xrightarrow{\simeq} \mathcal{A}_D$. Hence Theorem 3.1.12 implies:

Corollary 3.1.13 *For any type D, the normal complex analytic space $\widetilde{\mathcal{A}}_D = \mathfrak{H}_g/\Gamma_D$ is a moduli space of polarized abelian varieties of type D.*

Note that for a principal polarization $G_{1_g} = \mathrm{Sp}_{2g}^{1_g}(Z)$ and hence $\widetilde{\mathcal{A}}_{1_g} = \mathcal{A}_{1_g}$. For any D we have the following interpretation of the elements of Γ_D.

Remark 3.1.14 For $Z \in \mathfrak{H}_g$ and $R \in \Gamma_D$ the polarized abelian varieties (X_Z, H_Z) and $(X_{R(Z)}, H_{R(Z)})$ are isomorphic. Then ${}^tR$ is the rational representation of the corresponding isomorphism $X_{R(Z)} \longrightarrow X_Z$ with respect to the symplectic bases determined by Z and $R(Z)$. For a proof, see Exercise 3.1.5 (5) below.

Sometimes the second approach is more convenient, since the group Γ_D has integer coefficients. The advantage of the first approach is that the action of G_D on $\mathfrak{H}_g$ is more familiar.

3.1.5 Exercises

(1) Show that the symplectic group $\mathrm{Sp}_{2g}(\mathcal{R})$ is closed under transposition for any commutative ring $\mathcal{R}$ with 1 and of characteristic 0.

(2) Give a proof of Corollary 3.1.5.

(3) (*Iwasawa decomposition of* $\mathrm{Sp}_{2g}(\mathbb{R})$) Consider the group $\mathrm{O} = \mathrm{O}_{2g}(\mathbb{R}) \cap \mathrm{Sp}_{2g}(\mathbb{R})$ and the sets $\Delta = \left\{\begin{pmatrix} C & 0 \\ 0 & C^{-1}\end{pmatrix} \in \mathrm{SL}_{2g}(\mathbb{R}) \mid C = \mathrm{diag}(c_1, \dots, c_g)\right\}$ and

$$N = \left\{\begin{pmatrix} A & 0 \\ 0 & {}^tA^{-1}\end{pmatrix}\begin{pmatrix} 1_g & B \\ 0 & 1_g\end{pmatrix} \in \mathrm{SL}_{2g} \;\middle|\; \begin{matrix} A = \text{unipotent upper triangular} \\ B \text{ symmetric}\end{matrix}\right\}$$

(a) Show that Δ and N are subgroups of $\mathrm{Sp}_{2g}(\mathbb{R})$.
(b) The canonical map $N \times \Delta \times O \longrightarrow \mathrm{Sp}_{2g}(\mathbb{R})$ is a diffeomorphism. So any symplectic matrix M can be written uniquely as a product $M = M_1 M_2 M_3$ with $M_1 \in N, M_2 \in \Delta$ and $M_3 \in O$. This is the *Iwasawa decomposition* of M.

(4) Give an elementary proof of Proposition 3.1.8.
(Hint: Show that any sequence of matrices of $\mathrm{Sp}_{2g}(\mathbb{R})$, for which the sequence of images in $\mathfrak{H}_g$ under the map h converges, admits a convergent subsequence. For this use Proposition 3.1.7 (b).)

(5) Suppose $Z \in \mathfrak{H}_g$ and $R \in \Gamma_D$. Show that the polarized abelian varieties (X_Z, H_Z) and $(X_{R(Z)}, H_{R(Z)})$ are isomorphic.

(6) (a) Show that there are at most countably many proper analytic subvarieties A_i of the moduli space $\mathcal{A}_D$ such that every $(X, L) \in \mathcal{A}_D - \cup_i A_i$ has endomorphism ring $\mathbb{Z}$.
(Hint: Consider for any $\left(\begin{smallmatrix} a & b \\ c & d \end{smallmatrix}\right) \in \mathrm{M}_{2g}(\mathbb{Z})$ the equation $ZD^{-1}(cZ + dD) = aZ + bD$ in $\mathcal{H}_g$.)
(b) Deduce that $\mathrm{NS}(X) = \mathbb{Z}$ for a general $(X, L) \in \mathcal{A}_D$.

3.2 Level Structures

Given a type D, we saw in Theorem 3.1.2 that the Siegel upper half space $\mathfrak{H}_g$ can be considered as the moduli space of polarized abelian varieties of type D with a symplectic basis. A symplectic basis cannot be defined in algebraic terms. A *level structure* on a polarized abelian variety is roughly speaking an algebraic replacement of the notion of a symplectic basis or only some properties of it. The corresponding moduli spaces are quotients of $\mathfrak{H}_g$ by suitable subgroups of G_D (respectively Γ_D) and hence are situated between $\mathfrak{H}_g$ and $\mathcal{A}_D$ (respectively $\widetilde{\mathcal{A}}_D$).

Moduli spaces of polarized abelian varieties with level structures have various applications in arithmetic and algebraic geometry. In this section we present the most important examples. For other examples, see Section 3.5.1 and Exercises 3.2.3 (3) to (6).

3.2.1 Level D-structures

Let $(X = V/\Lambda, H)$ be a polarized abelian variety of type D. Recall from Section 2.7.1 the finite group $K(H) = \Lambda(H)/\Lambda$ with the Weil pairing

$$\varepsilon^H : K(H) \times K(H) \to \mathbb{C}^*, \qquad (\overline{v}, \overline{w}) \mapsto \mathbf{e}\big(-2\pi i \operatorname{Im} H(v, w)\big).$$

If $D = (d_1, \dots, d_g)$, then according to Exercise 1.5.5 (5) the group $K(D)$ has the following structure

$$K(H) \simeq K(D) := \mathbb{Z}^g/D\mathbb{Z}^g \oplus \mathbb{Z}^g/D\mathbb{Z}^g \quad \text{with} \quad \mathbb{Z}^g/D\mathbb{Z}^g = \prod_{i=1}^{g} \mathbb{Z}/d_i\mathbb{Z}.$$

Let $f_1, \dots f_{2g}$ denote standard generators of $K(D)$. Then the group $K(D)$ admits the following (multiplicative) alternating pairing:

$$\varepsilon^D : K(D) \times K(D) \longrightarrow \mathbb{C}^*, \qquad (f_\mu, f_\nu) \mapsto \begin{cases} \mathbf{e}(-\frac{2\pi i}{d_\mu}) & \mu = g + \nu, \\ \mathbf{e}(\frac{2\pi i}{d_\mu}) & \text{if } \nu = g + \mu, \\ 1 & \text{otherwise.} \end{cases}$$

A *level D-structure on* $K(H)$ is by definition a symplectic isomorphism

$$\beta : K(H) \to K(D).$$

Here a symplectic isomorphism means a group isomorphism respecting the pairings.

Recall that the group Γ_D acts on $\mathfrak{H}_g$ by equation (3.9). So its subgroup

$$\Gamma_D(D) := \left\{ \begin{pmatrix} a & b \\ c & d \end{pmatrix} \in \Gamma_D \;\middle|\; a - \mathbf{1}_g \equiv b \equiv c \equiv d - \mathbf{1}_g \equiv 0 \bmod D \right\}$$

acts in the same way. Here we write $a \equiv 0 \bmod D$ if $a \in D \cdot \mathrm{M}_g(\mathbb{Z})$.

Theorem 3.2.1 *The quotient*

$$\widetilde{\mathfrak{A}}_D(D) := \mathfrak{H}_g / \Gamma_D(D)$$

is a normal complex analytic space. It is a moduli space of polarized abelian varieties of type D with level D-structure. The embedding $\Gamma_D(D) \hookrightarrow \Gamma_D$ induces a surjective holomorphic map $\widetilde{\mathfrak{A}}_D(D) \to \widetilde{\mathfrak{A}}_D$ of finite degree.

Proof Let $f_1, \dots, f_{2g}$ denote the standard basis of $K(D)$ and let $Z \in \mathfrak{H}_g$. So if $\lambda_1, \dots, \lambda_g, \mu_1, \dots, \mu_g$ denotes the symplectic basis of Λ_Z of Theorem 3.1.2, then

$$\overline{\frac{1}{d_1}\lambda_1}, \dots, \overline{\frac{1}{d_g}\lambda_g}, \overline{\frac{1}{d_{g+1}}\mu_1}, \dots, \overline{\frac{1}{d_{2g}}\mu_g}$$

is a symplectic basis of $K(H_Z)$. Associating $\overline{\frac{1}{d_i}\lambda_i} \mapsto f_i$ and $\overline{\frac{1}{d_{g+i}}\mu_i} \mapsto f_{g+i}$ determines a symplectic isomorphism $\beta_Z : K(H_Z) \to K(D)$. In this way we associate to every $Z \in \mathfrak{H}_g$ a unique polarized abelian variety of type D with level D-structure.

$$Z \mapsto (X_Z, H_Z, \beta_Z).$$

Clearly every polarized abelian variety of type D with level D-structure is isomorphic to one of these. We have to analyse when two of them are isomorphic.

So suppose $Z, Z' \in \mathfrak{H}_g$ such that

$$\varphi : (X_{Z'}, H_{Z'}, \beta_{Z'}) \to (X_Z, H_Z, \beta_Z)$$

is an isomorphism. Then

(i) $\varphi : (X_{Z'}, H_{Z'}) \to (X_Z, H_Z)$ is an isomorphism of polarized abelian varieties and
(ii) $\varphi(\overline{\frac{1}{d_i}\lambda_i'}) = \overline{\frac{1}{d_i}\lambda_i}$ and $\varphi(\overline{\frac{1}{d_i}\mu_i'}) = \overline{\frac{1}{d_i}\mu_i}$ for $i = 1, \dots, g$.

Condition (i) is equivalent to

$$A(Z', D) = (Z, D)^t R$$

with $R \in \Gamma_D$ according to Remark 3.1.14. In terms of matrices condition (ii) reads

$$A(Z'D^{-1}, \mathbf{1}_g) \equiv (ZD^{-1}, \mathbf{1}_g) \bmod \Lambda_Z = (Z, D)\mathbb{Z}^{2g}.$$

In other words,

$$(Z, D)({}^tR - \mathbf{1}_{2g}) = A(Z', D) - (Z, D) \in (Z, D)\mathrm{M}_{2g}(\mathbb{Z}) \begin{pmatrix} D & 0 \\ 0 & D \end{pmatrix}.$$

This means

$$R - \mathbf{1}_{2g} \in \begin{pmatrix} D & 0 \\ 0 & D \end{pmatrix} \mathrm{M}_{2g}(\mathbb{Z}).$$

Summing up, we have shown that Z and Z' determine isomorphic polarized abelian varieties of type D if and only if $Z' = RZ$, where R is an elements of the group $\Gamma_D(D)$. Hence $\widetilde{\mathcal{A}}_D(D)$ is a moduli space as claimed.

As a subgroup of Γ_D, the group $\Gamma_D(D)$ acts properly and discontinuously on $\mathcal{H}_g$. So by Theorem 3.1.11, $\widetilde{\mathcal{A}}_D(D)$ is a normal complex analytic space.

For the last assertion it suffices to show that $\Gamma_D(D)$ is of finite index in Γ_D. But this is easy to see using the definition of $\Gamma_D(D)$. □

3.2.2 Generalized Level n-structures

A *level n-structure* on a principally polarized abelian variety (X, H) is by definition a level $(n\mathbf{1}_g)$-structure on the polarized abelian variety (X, nH) in the sense of the previous section. So a level D-structure is a generalization of this notion. In this section a different generalization is given.

Let $(X = V/\Lambda, H)$ be a polarized abelian variety of (arbitrary) type D and n a positive integer. A symplectic basis $\lambda_1, \dots, \lambda_g, \mu_1, \dots \mu_g$ of Λ for H determines a basis for the n-division points X_n in X, namely $\overline{\frac{1}{n}\lambda_1}, \dots, \overline{\frac{1}{n}\lambda_g}, \overline{\frac{1}{n}\mu_1}, \dots, \overline{\frac{1}{n}\mu_g}$. A *generalized level n-structure* for (X, H) is by definition a basis of X_n coming from a symplectic basis in this way.

For any $n > 1$ the *principal congruence subgroup* $\Gamma_D(n)$ of Γ_D is defined to be

$$\Gamma_D(n) := \{R \in \Gamma_D \mid R \equiv \mathbf{1}_{2g} \bmod n\}.$$

Note that $\Gamma_D(n)$ is a normal subgroup of Γ_D.

Theorem 3.2.2 *The quotient*

$$\overline{\mathcal{A}}_D(n) := \mathfrak{H}_g/\Gamma_D(n)$$

is a normal complex analytic space. It is a moduli space for polarized abelian varieties of type D with generalized level n-structure. The embedding $\Gamma_D(n) \hookrightarrow \Gamma_D$ *induces a holomorphic map of finite degree.*

The proof is completely analogous to the proof of Theorem 3.2.1. So we leave it to the reader as Exercise 3.2.3 (2).

3.2.3 Exercises

(1) Suppose $D = \operatorname{diag}(d_1, \ldots, d_g)$ is the type of a polarization and $M = \begin{pmatrix} \alpha & \beta \\ \gamma & \delta \end{pmatrix} \in \operatorname{Sp}_{2g}(\mathbb{Q})$. Show that $M \in G_D$ if and only if $\alpha, D\gamma, \beta D^{-1}$ and $D\delta D^{-1}$ are all contained in $\operatorname{M}_g(\mathbb{Z})$.

In particular, if $g = 1$ and $D = (d)$, then $G_D = \operatorname{Sp}_2(\mathbb{Q}) \cap \begin{pmatrix} \mathbb{Z} & d\mathbb{Z} \\ \frac{1}{d}\mathbb{Z} & \mathbb{Z} \end{pmatrix}$, and if $g = 2$ and $D = \begin{pmatrix} 1 & 0 \\ 0 & d \end{pmatrix}$, then $G_D = \operatorname{Sp}_4(\mathbb{Q}) \cap \begin{pmatrix} \mathbb{Z} & \mathbb{Z} & \mathbb{Z} & d\mathbb{Z} \\ \mathbb{Z} & \mathbb{Z} & \mathbb{Z} & d\mathbb{Z} \\ \mathbb{Z} & \mathbb{Z} & \mathbb{Z} & d\mathbb{Z} \\ \frac{1}{d}\mathbb{Z} & \frac{1}{d}\mathbb{Z} & \frac{1}{d}\mathbb{Z} & \mathbb{Z} \end{pmatrix}$.

(2) Show that the quotient $\overline{\mathcal{A}}_D(n) := \mathfrak{H}_g/\Gamma_n(n)$ is a normal complex analytic space. It is a moduli space for polarized abelian varieties of type D with generalized level n-structure. The embedding $\Gamma_D(n) \hookrightarrow \Gamma_D$ induces a holomorphic map of finite degree.

(3) (*Moduli space of Polarized Abelian Varieties with Isogeny of Type D*) Let (X, L) be a polarized abelian variety of type D. An isogeny $p : (X, L) \to (Y, M)$ onto a principally polarized abelian variety (Y, M) is called *of type D* if $\operatorname{Ker} p \simeq \mathbb{Z}^g/D\mathbb{Z}^g$. The triplet (X, L, p) is called a *polarized abelian variety with isogeny of type D*. Show that:

(a) The normal complex analytic space

$$\mathcal{A}_D^0 := \mathfrak{H}_g/G_D \cap G_{1_g}$$

is a moduli space for the (isomorphism classes of) polarized abelian varieties with isogeny of type D.

(b) There is a canonical Galois cover $\mathcal{A}_D(D) \to \mathcal{A}_D^0$ of finite degree.

(In the case $g = 2$, $D = \operatorname{diag}(1, d)$ an isogeny of type D is sometimes called a *root*.)

(4) (*Moduli Spaces of Elliptic Curves with a Cyclic Subgroup*) For any integer $n \geq 1$ consider the set of (isomorphism classes of) elliptic curves E with a cyclic subgroup K of order n. Show that

(a) the space $\mathcal{A}^0_{(n)} = \mathfrak{H}_1/G_{(n)} \cap \mathrm{SL}_2(\mathbb{Z})$ (see Exercise 3.2.3 (3)) is a moduli space for elliptic curves with cyclic subgroup of order n;

(b) $G_{(n)} \cap \mathrm{SL}_2(\mathbb{Z}) = \Gamma^0(n) := \left\{\begin{pmatrix} a & b \\ c & d \end{pmatrix} \in \mathrm{SL}_2(\mathbb{Z}) \mid b \equiv 0 \bmod n\right\}$.

(5) (*Moduli Space of Elliptic Curves with n-division Point*) For an integer $n \geq 1$ consider the set of (isomorphism classes of) elliptic curves E with a point $x \in E$ of order n. Define

$$\Gamma^{1,0}(n) := \left\{\begin{pmatrix} a & b \\ c & d \end{pmatrix} \in \mathrm{SL}_2(\mathbb{Z}) \mid a \equiv 1, b \equiv 0 \bmod n\right\}.$$

Show that $\mathcal{A}^{1,0}(n) := \mathfrak{H}_1/\Gamma^{1,0}(n)$ is a moduli space for elliptic curves with n-division point.

(6) Consider the set of triplets (X, L, A_p) with (X, L) a principally polarized abelian surface and A_p a subgroup of order p^2 of the group of p-division points X_p (p a prime), non-isotropic for the Weil form ε^{L^p} on X_p (defined in Section 2.7.1). Define

$$\Gamma^0_{1_2}(p) := \mathrm{Sp}_4(\mathbb{Z}) \cap \begin{pmatrix} \mathbb{Z} & p\mathbb{Z} & \mathbb{Z} & p\mathbb{Z} \\ \mathbb{Z} & \mathbb{Z} & \mathbb{Z} & \mathbb{Z} \\ \mathbb{Z} & p\mathbb{Z} & \mathbb{Z} & p\mathbb{Z} \\ \mathbb{Z} & \mathbb{Z} & \mathbb{Z} & \mathbb{Z} \end{pmatrix}.$$

(a) Show that the normal complex analytic space $\mathcal{A}^0_{1_2}(p) := \mathfrak{H}_2/\Gamma^0_{1_2}(p)$ is a moduli space of isomorphism classes of triplets as above.

(b) Show that there is a canonical isomorphism of moduli spaces $\mathcal{A}^0_{1_2}(p) \simeq \mathcal{A}_D$ with $D = \mathrm{diag}(1, p^2)$.

(7) Let $(X = V/\Lambda, H)$ be a polarized abelian variety of dimension $g > 1$ and type D. According to equation (1.15) a direct sum $\Lambda = \Lambda_1 \oplus \Lambda_2$ is a decomposition of Λ for H if Λ_i is isotropic with respect to $\mathrm{Im}\, H$ for $i = 1, 2$. Let

$$\Delta_D := \left\{\begin{pmatrix} a & b \\ c & d \end{pmatrix} \in \Gamma_D \mid b = c = 0\right\}.$$

(a) The quotient $\widetilde{\mathcal{A}}^{\Delta}_D := \mathfrak{H}_g/\Delta_D$ is a moduli space for abelian varieties with polarization H of type D with a decomposition of Λ for H.

(b) The embedding $\Delta_D \hookrightarrow \Gamma_D$ induces holomorphic maps $\mathfrak{H}_g \xrightarrow{\pi_1} \widetilde{\mathcal{A}}^{\Delta}_D \xrightarrow{\pi_2} \widetilde{\mathcal{A}}_D$ and both π_1 and π_2 have infinite fibres.

3.3 The Theta Transformation Formula

Given a type D, theta functions on $\mathbb{C}^g$ with respect to a lattice $\Lambda_Z = (Z, D)\mathbb{Z}^{2g}$ (for $Z \in \mathfrak{H}_g$) are holomorphic functions on $\mathbb{C}^g$ with a certain functional behaviour with respect to translations with elements of Λ_Z. Varying Z within $\mathfrak{H}_g$ by the action of the symplectic group on $\mathfrak{H}_g$, one may ask how the corresponding theta functions are related. The answer is given by the classical Theta Transformation Formula.

3.3.1 Preliminary Version

Let $D = \operatorname{diag}(d_1, \dots, d_g)$ be a type. Suppose $Z \in \mathfrak{H}_g$, $M = \left(\begin{smallmatrix} \alpha & \beta \\ \gamma & \delta \end{smallmatrix}\right) \in G_D$ (defined in equation (3.5)) and $Z' = M(Z) = (\alpha Z + \beta)(\gamma Z + \delta)^{-1}$. Then the matrix M induces an isomorphism

$$\varphi : (X_{Z'}, H_{Z'}) \tilde{\rightarrow} (X_Z, H_Z)$$

of polarized abelian varieties of type D corresponding to Z' and Z. According to Corollary 3.1.5 the isomorphism φ is given by the equation $A(M(Z), \mathbf{1}_g) = (Z, \mathbf{1}_g)^t M$. Here the matrix

$$A := {}^t(\gamma Z + \delta)$$

is the analytic representation $\rho_a(\varphi)$ with respect to the standard basis of $\mathbb{C}^g$.

Recall the decompositions $\Lambda_Z = Z\mathbb{Z}^g \oplus D\mathbb{Z}^g$ for H_Z and $\Lambda_{Z'} = Z'\mathbb{Z}^g \oplus D\mathbb{Z}^g$ for $H_{Z'}$ and let $L = L(H_Z, \chi)$ denote the line bundle with characteristic $c \in \mathbb{C}^g$ with respect to the decomposition of Λ_Z. The next lemma computes the characteristic of $L' = \varphi^* L$ in terms of c and M. For this, for any $S = (s_{ij}) \in M_g(\mathbb{R})$ we denote by $(S)_0$ the vector

$$(S)_0 = {}^t(s_{11}, \dots, s_{gg}) \in \mathbb{R}^g.$$

Lemma 3.3.1

(a) *The line bundle $\varphi^* L$ is of characteristic*

$$M[c] := A^{-1}c + \frac{1}{2}(Z', \mathbf{1}_g)\begin{pmatrix} D(\gamma{}^t\delta)_0 \\ (\alpha{}^t\beta)_0 \end{pmatrix}$$

with respect to the decomposition $\Lambda_{Z'} = Z'\mathbb{Z}^g \oplus D\mathbb{Z}^g$.

(b) *If $c = Zc^1 + c^2$ with $c^1, c^2 \in \mathbb{R}^g$ and $M[c] = M(Z)M[c]^1 + M[c]^2$, then*

$$M[c]^1 = \delta c^1 - \gamma c^2 + \frac{1}{2}D(\gamma\, {}^t\delta)_0 \quad \text{and} \quad M[c]^2 = -\beta c^1 + \alpha c^2 + \frac{1}{2}(\alpha\, {}^t\beta)_0.$$

Proof (a): According to Lemma 1.3.6 the semicharacter of L' is $\rho_r(\varphi)^*\chi$. Suppose $\mu = Z'\mu^1 + \mu^2 \in \Lambda_{Z'}$ and $\lambda = \rho_r(\varphi)\mu = Z\lambda^1 + \lambda^2 \in \Lambda_Z$. In terms of matrices this reads

$$\begin{pmatrix} \lambda^1 \\ \lambda^2 \end{pmatrix} = {}^tM \begin{pmatrix} \mu^1 \\ \mu^2 \end{pmatrix}.$$

Since $\left(\begin{smallmatrix} 0 & \mathbf{1}_g \\ -\mathbf{1}_g & 0 \end{smallmatrix}\right)$ is the matrix of $\operatorname{Im} H_Z$ (respectively $\operatorname{Im} H_{Z'}$) with respect to the $\mathbb{R}$-basis of $\mathbb{C}^g$ given by the columns of $(Z, \mathbf{1}_g)$ (respectively $(Z', \mathbf{1}_g)$), we have by Corollary 1.5.2,

$$\begin{aligned}
\rho_r(\varphi)^*\chi(\mu) &= \chi(\lambda) = \mathbf{e}\big(\pi i \operatorname{Im} H_Z(Z\lambda^1, \lambda^2) + 2\pi i \operatorname{Im} H_Z(c, \lambda)\big) \\
&= \mathbf{e}\left(\pi i\, {}^t\lambda^1\lambda^2 + 2\pi i\, {}^t\begin{pmatrix} c^1 \\ c^2 \end{pmatrix} \begin{pmatrix} 0 & \mathbf{1}_g \\ -\mathbf{1}_g & 0 \end{pmatrix} \begin{pmatrix} \lambda^1 \\ \lambda^2 \end{pmatrix}\right) \\
&= \mathbf{e}\Big(\pi i\, {}^t\mu^1(\alpha\, {}^t\delta - \beta\, {}^t\gamma)\mu^2 + \pi i\, {}^t\mu^2\gamma\, {}^t\delta\mu^2 - \pi i\, {}^t\mu^1\alpha\, {}^t\beta\mu^1 \\
&\qquad + 2\pi i\, {}^t\begin{pmatrix} c^1 \\ c^2 \end{pmatrix} M^{-1} \begin{pmatrix} 0 & \mathbf{1}_g \\ -\mathbf{1}_g & 0 \end{pmatrix} \begin{pmatrix} \mu^1 \\ \mu^2 \end{pmatrix}\Big),
\end{aligned}$$

where for the last equation we used that ${}^t\mu^1\beta\, {}^t\gamma\mu^2$ and ${}^t\mu^1\alpha\, {}^t\beta\mu^1$ are integers and that $\left(\begin{smallmatrix} 0 & \mathbf{1}_g \\ -\mathbf{1}_g & 0 \end{smallmatrix}\right) {}^tM = M^{-1} \left(\begin{smallmatrix} 0 & \mathbf{1}_g \\ -\mathbf{1}_g & 0 \end{smallmatrix}\right)$.

Note that for any $\ell \in \mathbb{Z}^g$ and any symmetric matrix $S = (s_{ij}) \in \mathrm{M}_g(\mathbb{Z})$,

$${}^t\ell S\ell = \sum_{i=1}^{g} s_{ii}\ell_i^2 + 2 \sum_{1 \le i < j \le g} s_{ij}\ell_i\ell_j \equiv {}^t(S)_0\ell \bmod 2.$$

For this we also use that ℓ_i is odd if and only if ℓ_i^2 is odd. Hence using Lemma 3.1.3 (b) we get

$$\begin{aligned}
\rho_r(\varphi)^*\chi(\mu) &= \mathbf{e}\left(\pi i\, {}^t\mu^1\mu^2 + 2\pi i\, {}^t\left[{}^tM^{-1}\begin{pmatrix} c^1 \\ c^2 \end{pmatrix} + \frac{1}{2}\begin{pmatrix} D(\gamma\, {}^t\delta)_0 \\ (\alpha{}^t\beta)_0 \end{pmatrix}\right] \begin{pmatrix} 0 & \mathbf{1}_g \\ -\mathbf{1}_g & 0 \end{pmatrix} \begin{pmatrix} \mu^1 \\ \mu^2 \end{pmatrix}\right) \\
&= \mathbf{e}\left(\pi i \operatorname{Im} H_{Z'}(Z'\mu^1, \mu^2) + 2\pi i \operatorname{Im} H_{Z'}(M[c], \mu)\right),
\end{aligned}$$

since

$$M[c] = A^{-1}c + \frac{1}{2}(Z', \mathbf{1}_g)\begin{pmatrix} D(\gamma\, {}^t\delta)_0 \\ (\alpha{}^t\beta)_0 \end{pmatrix} = (Z', \mathbf{1}_g)\left({}^tM^{-1}\begin{pmatrix} c^1 \\ c^2 \end{pmatrix} + \frac{1}{2}\begin{pmatrix} D(\gamma\, {}^t\delta)_0 \\ (\alpha{}^t\beta)_0 \end{pmatrix}\right).$$

Together this gives (a). As for (b):

$$\begin{aligned}
M[c] &= A^{-1}c + \frac{1}{2}(M(Z), \mathbf{1}_g)\begin{pmatrix} D(\gamma\, {}^t\delta)_0 \\ (\alpha{}^t\beta)_0 \end{pmatrix} \\
&= (M(Z), \mathbf{1}_g)\left({}^tM^{-1}\begin{pmatrix} c^1 \\ c^2 \end{pmatrix} + \frac{1}{2}\begin{pmatrix} D(\gamma\, {}^t\delta)_0 \\ (\alpha{}^t\beta)_0 \end{pmatrix}\right)
\end{aligned}$$

$$= M(Z)\left(\delta c^1 - \gamma c^2 + \frac{1}{2}D(\gamma\,{}^t\delta)_0\right) + \left(-\beta c^1 + \alpha c^2 + \frac{1}{2}(\alpha\,{}^t\beta)_0\right).$$

This completes the proof of the lemma. □

From now on in this section consider the special case of principal polarizations; that is, $D = \mathbf{1}_g$ and $G_{\mathbf{1}_g} = \mathrm{Sp}_{2g}^{\mathbf{1}_g}(\mathbb{Z})$.

Let $Z \in \mathfrak{H}_g$ and $L \in \mathrm{Pic}(X_Z)$, defining a principal polarization H and of characteristic c with respect to the decomposition $\Lambda_Z = Z\mathbb{Z}^g \oplus \mathbb{Z}^g$. Let $\varphi : (X_{Z'}, H) \to (X_Z, H)$ be an isomorphism of polarized abelian varieties. So $Z' = M(Z)$ with $M \in G_{\mathbf{1}_g}$.

According to Riemann–Roch and Section 1.5.3 there is a unique canonical theta function ϑ_Z^c given by Exercise 1.5.5 (10) and generating $H^0(X_Z, L)$. According to Lemma 3.3.1, $\varphi^* L$ is of characteristic $M[c]$. The canonical theta function of $\vartheta_{Z'}^{M[c]}$ is a basis of $H^0(X_{Z'}, \varphi^* L)$. Moreover, the map

$$A^* : H^0(X_Z, L) \to H^0(Z_{Z'}, \varphi^* L) \qquad \text{with} \qquad A = \rho_a(\varphi)$$

is an isomorphism. This implies that, up to a multiplicative constant, the canonical theta functions ϑ_Z^c and $\vartheta_{Z'}^{M[c]}$ coincide and we have proved the following proposition.

Proposition 3.3.2 (Preliminary version of the Theta Transformation Formula) *With the above notation there is a constant $C(Z, M, c)$ depending only on Z, M and c such that*

$$A^* \vartheta_Z^c = C(Z, M, c)\vartheta_{Z'}^{M[c]}.$$

For the final version of the theta transformation formula it remains to compute the constant $C(Z, M, c)$. This will be done in Section 3.3.3. As a first step we show

Lemma 3.3.3 $C(Z, M, c) = C(Z, M, 0)\ \mathbf{e}(\pi i \operatorname{Im} H_{Z'}(M[0], A^{-1}c))$.

Proof Let $L_0 \in \mathrm{Pic}(X_Z)$ be of characteristic 0 in the class of L and ϑ_Z^0 the corresponding canonical theta function. According to Lemma 3.3.1 the line bundle $\varphi^* L_0$ is of characteristic $M[0]$. Let $\vartheta_{Z'}^{M[0]}$ denote the corresponding canonical theta function. According to Exercise 1.5.5 (3) these functions are related by

$$\vartheta_Z^c = \tau \cdot t_c^* \vartheta_Z^0 \qquad \text{and} \qquad \vartheta_{Z'}^{M[0]} = \tau' \cdot t_{A^{-1}c}^* \vartheta_{Z'}^{M[c]}$$

with

$$\tau = \mathbf{e}\left(-\pi H_Z(\cdot, c) - \frac{\pi}{2} H_Z(c, c)\right) \text{ and}$$
$$\tau' = \mathbf{e}\left(\pi i \operatorname{Im} H_{Z'}(M[0], A^{-1}c) + \pi H_{Z'}(\cdot, A^{-1}c) + \frac{\pi}{2} H_{Z'}(A^{-1}c, A^{-1}c)\right).$$

This gives, using Proposition 3.3.2 and Lemma 3.3.1,

$$\begin{aligned} A^*\vartheta_Z^c &= A^*(\tau \cdot t_c^*\vartheta_Z^0) = A^*\tau \cdot t_{A^{-1}c}^* A^*\vartheta_Z^0 \\ &= A^*\tau \cdot t_{A^{-1}c}^*\left(C(Z,M,0)\vartheta_{Z'}^{M[0]}\right) \\ &= C(Z,M,0)A^*\tau \cdot t_{A^{-1}c}^*\tau' \cdot \vartheta_{Z'}^{M[c]}. \end{aligned}$$

So

$$\begin{aligned} C(Z,M,c) &= C(Z,M,0)A^*\tau \cdot t_{A^{-1}c}^*\tau' \\ &= \mathbf{e}\big(\pi i \operatorname{Im} H_{Z'}(M[0], A^{-1}c)\big)C(Z,M,0), \end{aligned}$$

where the last equation follows by an immediate computation using $A^*H_Z = H_{Z'}$.□

3.3.2 Classical Theta Functions

In Section 1.5.2 we introduced classical theta functions. For the proof of the final version of the theta transformation formula we need some more of their properties. For example, that they depend holomorphically on $Z \in \mathfrak{H}_g$. We start by introducing the relation between the canonical theta functions ϑ_Z^c and the classical Riemann theta functions with (real) characteristics.

Let $(X_Z = \mathbb{C}^g/\Lambda_Z, H = H_Z)$ denote the principally polarized abelian variety given by $Z \in \mathfrak{H}_g$. Then $\Lambda_Z = Z\mathbb{Z}^g \oplus \mathbb{Z}^g$ is a decomposition for H. It induces a decomposition $\mathbb{C}^g = V_1 \oplus V_2$ with real vector spaces $V_1 = Z\mathbb{R}^g$ and $V_2 = \mathbb{R}^g$. So we can write every $v \in \mathbb{C}^g$ uniquely as

$$v = Zv^1 + v^2 \qquad \text{with} \qquad v^1, v^2 \in \mathbb{R}^g.$$

As in Section 1.5.2 denote by B the $\mathbb{C}$-bilinear extension of the symmetric bilinear form $H|_{V_2 \times V_2}$.

Lemma 3.3.4 *For all $v, w \in \mathbb{C}^g$ we have*

(a) $B(v,w) = {}^t v(\operatorname{Im} Z)^{-1}w$;
(b) $(H - B)(v,w) = -2i\, {}^t v w^1$.

Proof (a) is a consequence of the definition and Proposition 3.1.1. Using (a) we get

$$(H-B)(v,w) = {}^t v(\operatorname{Im} Z)^{-1}(\overline{w} - w) = {}^t v(\operatorname{Im} Z)^{-1}(\overline{Z} - Z)w^1 = -2i\, {}^t v w^1. \quad \square$$

Let $L = L(H,\chi)$ be the line bundle on X_Z with characteristic $c = Zc_1 + c_2$ and, as in the last section, let ϑ_Z^c be the canonical theta function generating $H^0(X_Z, L)$ (for an explicit formula see Exercise 1.5.5 (10)).

The classical *Riemann theta function with (real) characteristic* $\begin{bmatrix} c^1 \\ c^2 \end{bmatrix}$, $\vartheta \begin{bmatrix} c^1 \\ c^2 \end{bmatrix} : \mathbb{C}^g \times \mathfrak{H}_g \to \mathbb{C}$ is defined by

$$\vartheta \begin{bmatrix} c^1 \\ c^2 \end{bmatrix} (v, Z) = \sum_{\ell \in \mathbb{Z}^g} \mathbf{e}\left(\pi i \, {}^t(\ell + c^1) Z (\ell + c^1) + 2\pi i \, {}^t(v + c^2)(\ell + c^1)\right). \tag{3.10}$$

Note that this notation coincides with the classical notation (see Krazer [77, page 30]).

Lemma 3.3.5 *The canonical theta function* ϑ_Z^c *and* $\vartheta \begin{bmatrix} c^1 \\ c^2 \end{bmatrix}$ *are related by*

$$\vartheta_Z^c = \mathbf{e}\left(\frac{\pi}{2} B(\cdot, \cdot) - \pi i \, {}^t c^1 c^2\right) \vartheta \begin{bmatrix} c^1 \\ c^2 \end{bmatrix}.$$

This shows in particular that in the principally polarized case our definition of characteristics of theta functions coincides with the classical one. Hence our notion of characteristics of non-degenerate line bundles of arbitrary type is a natural generalization of this.

Proof Write $c = Zc^1 + c^2$. By Exercise 1.5.5 (10) we have for all $v \in \mathbb{C}^g$,

$$\begin{aligned}
\vartheta^{Zc^1+c^2}(v) &= \mathbf{e}(-\pi H(v, c) - \frac{\pi}{2} H(c, c) + \frac{\pi}{2} B(v + c, v + c)) \\
&\quad \cdot \sum_{\lambda \in \Lambda_1} \mathbf{e}\left(\pi (H - B)(v + c, \lambda) - \frac{\pi}{2} (H - B)(\lambda, \lambda)\right) \\
&= \mathbf{e}\left(\frac{\pi}{2} B(v, v)\right) \sum_{\lambda \in \Lambda_1} \mathbf{e}\left(-\frac{\pi}{2} (H - B)(\lambda, \lambda) - \frac{\pi}{2} (H - B)(c, c)\right. \\
&\qquad \left. - \pi (H - B)(c, \lambda) - \pi (H - B)(v, \lambda + c)\right) \\
&\qquad\qquad \text{(where we replaced } \lambda \text{ by } -\lambda) \\
&= \mathbf{e}\left(\frac{\pi}{2} B(v, v)\right) \sum_{\lambda^1 \in \mathbb{Z}^g} \mathbf{e}\left(\pi i \, {}^t\lambda^1 Z \lambda^1 + \pi i \, {}^t(Zc^1 + c^2) c^1\right. \\
&\qquad \left. + 2\pi i \, {}^t(Zc^1 + c^2)\lambda^1 + 2\pi i \, {}^t v (\lambda^1 + c^1)\right) \\
&\qquad\qquad \text{(by Lemma 3.3.4 (b))} \\
&= \mathbf{e}\left(\frac{\pi}{2} B(v, v) - \pi i \, {}^t c^1 c^2\right) \\
&\quad \cdot \sum_{\lambda^1 \in \mathbb{Z}^g} \mathbf{e}(\pi i \, {}^t(\lambda^1 + c^1) Z (\lambda^1 + c^1) + 2\pi i \, {}^t(v + c^2)(\lambda^1 + c^1)),
\end{aligned}$$

which is the assertion. □

Using Lemma 3.3.5 one can translate all properties of canonical theta functions into terms of the classical theta functions. For some of them, see Exercises 3.3.4 (1) to (5).

Proposition 3.3.6 *The classical theta function* $\vartheta\begin{bmatrix} c^1 \\ c^2 \end{bmatrix}$ *is holomorphic on the complex manifold* $\mathbb{C}^g \times \mathfrak{H}_g$ *for every* $c^1, c^2 \in \mathbb{R}^g$.

Proof It suffices to show that the series (3.10) converges absolutely and uniformly on every compact set in $\mathbb{C}^g \times \mathfrak{H}_g$. Let $K \subset \mathbb{C}^g \times \mathfrak{H}_g$ be compact. There is a point $(v_0, Z_0) \in K$ such that $\operatorname{Im} Z \geq \operatorname{Im} Z_0$ for all $(v, Z) \in K$. Then for every $(v, Z) \in K$,

$$\sum_{\ell \in \mathbb{Z}^g} \left|\mathbf{e}(\pi i\, {}^t(\ell + c^1)Z(\ell + c^1) + 2\pi i\, {}^t(v + c^2)(\ell + c^1))\right|$$

$$\leq \sum_{\ell \in \mathbb{Z}^g} \mathbf{e}(-\pi i\, {}^t(\ell + c^1) \operatorname{Im} Z_0(\ell + c^1) - 2\pi\, {}^t\operatorname{Im}(v)(\ell + c^1)).$$

According to Lemma 1.5.7 and 3.3.5 the series of $\vartheta\begin{bmatrix} c^1 \\ c^2 \end{bmatrix}(v, Z_0)$ converges absolutely and uniformly on compact sets in $\mathbb{C}^g$. This means that the series on the right-hand side converges uniformly on the image of K under the natural projection $p : \mathbb{C}^g \times \mathfrak{H}_g \to \mathbb{C}^g$. This completes the proof. □

As it is a holomorphic function, we can differentiate the series term by term and immediately get the following proposition.

Proposition 3.3.7 (Heat Equation) *For every symmetric matrix* $(s_{ij}) \in \mathrm{M}_g(\mathbb{C})$ *and all vectors* $c^1, c^2 \in \mathbb{R}^g$,

$$\sum_{i,j=1}^{g} s_{ij} \frac{\partial^2 \vartheta\begin{bmatrix} c^1 \\ c^2 \end{bmatrix}}{\partial v_i \partial v_j} = 4\pi i \sum_{1 \leq i \leq j \leq g} s_{ij} \frac{\partial \vartheta\begin{bmatrix} c^1 \\ c^2 \end{bmatrix}}{\partial c_{ij}}.$$

For the proof see Exercise 3.3.4 (6). For a suitable choice of (s_{ij}) and certain restrictions on the values of Z this equation describes the conduction of heat in physics.

In Section 1.6.1, equation (1.24), we introduced the hermitian metric $\mathbf{e}(-\pi H(\cdot, \cdot))$ on the line bundle $L = L(H, \chi)$. In particular, for all canonical theta functions $\vartheta_1, \vartheta_2 \in H^0(L)$ the function

$$\langle \vartheta_1, \vartheta_2 \rangle : \mathbb{C}^g \to \mathbb{C}, \qquad v \mapsto \vartheta_1(v)\overline{\vartheta_2(v)}\,\mathbf{e}(-\pi H(v, v))$$

is C^∞ on $\mathbb{C}^g$ and periodic with respect to the lattice Λ_Z. Denote by $\mathrm{d}v$ the volume element of $\mathbb{C}^g$ (respectively X_Z corresponding to the symplectic basis of Λ_Z given by the columns of the matrix $(Z, \mathbf{1}_g)$, written as $\mathrm{d}v = \mathrm{d}v_1 \wedge \cdots \wedge \mathrm{d}v_{2g}$. Then

$$(\vartheta_1, \vartheta_2) := \int_{\mathbb{C}^g/\Lambda_Z} \langle \vartheta_1, \vartheta_2 \rangle \mathrm{d}v$$

defines the inner product on the vector space of canonical theta functions $H^0(L)$. The norm $||\vartheta_Z^c||$ of ϑ_Z^c with respect to this inner product is given as follows.

Proposition 3.3.8 $||\vartheta_Z^c|| = (\det 2\operatorname{Im} Z)^{-\frac{1}{4}}$.

Proof Note first that by Exercise 1.5.5 (11) the functions ϑ_Z^c and ϑ_Z^0 are related as follows

$$\vartheta_Z^c(v) = \mathbf{e}\left(-\pi H(v,c) - \frac{\pi}{2}H(c,c)\right)\vartheta_Z^0(v+c).$$

This gives $\langle\vartheta_Z^c, \vartheta_Z^c\rangle(v) = \langle\vartheta_Z^0, \vartheta_Z^0\rangle(v+c)$. Hence it suffices to prove the assertion for $c = 0$. Using Lemmas 3.3.4 and 3.3.5 we get

$$\langle\vartheta_Z^0, \vartheta_Z^0\rangle(v) = \vartheta\begin{bmatrix}0\\0\end{bmatrix}(v,Z)\,\overline{\vartheta\begin{bmatrix}0\\0\end{bmatrix}(v,Z)}\cdot\mathbf{e}\left(-\frac{\pi}{2}(H-B)(v,v) - \frac{\pi}{2}\overline{(H-B)(v,v)}\right)$$

$$= \sum_{\ell\in\mathbb{Z}^g}\mathbf{e}(\pi i\,{}^t\ell Z\ell + 2\pi i\,{}^tv\ell)\sum_{m\in\mathbb{Z}^g}\mathbf{e}(-\pi i\,{}^tm\overline{Z}m - 2\pi i\,{}^t\overline{v}m - 2\pi\,{}^tv^1\operatorname{Im} Zv^1)$$

$$= \sum_{\ell,m\in\mathbb{Z}^g}\mathbf{e}\Big(\pi i\,{}^t\ell Z\ell - \pi i\,{}^tm\overline{Z}m + 2\pi i\,{}^tv^1(Z\ell - \overline{Z}m)$$

$$- 2\pi\,{}^tv^1\operatorname{Im} Zv^1 + 2\pi i\,{}^tv^2(\ell - m)\Big).$$

Since this series converges absolutely and uniformly on every compact set in $\mathbb{C}^g$, we get, writing $dv^1 = dv_1\wedge\cdots\wedge dv_g$ and $dv^2 = dv_{g+1}\wedge\cdots\wedge dv_{2g}$,

$$||\vartheta^0||^2$$

$$= \sum_{\ell,m\in\mathbb{Z}^g}\int_{\mathbb{R}^g/\mathbb{Z}^g}\mathbf{e}(\pi i\,{}^t\ell Z\ell - \pi i\,{}^tm\overline{Z}m + 2\pi i\,{}^tv^1(Z\ell - \overline{Z}m) - 2\pi\,{}^tv^1\operatorname{Im} Zv^1)dv^1$$

$$\cdot\int_{\mathbb{R}^g/\mathbb{Z}^g}\mathbf{e}(2\pi i\,{}^tv^2(\ell - m))dv^2.$$

But $\int_{\mathbb{R}^g/\mathbb{Z}^g}\mathbf{e}(2\pi i\,{}^tv^2(\ell-m))dv^2 = \begin{cases}1 \text{ if } \ell = m\\ 0 \text{ if } \ell \neq m\end{cases}$ such that

$$||\vartheta^0||^2 = \sum_{\ell\in\mathbb{Z}^g}\int_{\mathbb{R}^g/\mathbb{Z}^g}\mathbf{e}(\pi i\,{}^t\ell(Z-\overline{Z})\ell + 2\pi i\,{}^tv^1(Z-\overline{Z})\ell - 2\pi\,{}^tv^1\operatorname{Im} Zv^1)dv^1$$

$$= \sum_{\ell\in\mathbb{Z}^g}\int_{\mathbb{R}^g/\mathbb{Z}^g}\mathbf{e}(-\,{}^t(\ell+v^1)(2\pi\operatorname{Im} Z)(\ell+v^1))dv^1$$

$$= \int_{\mathbb{R}^g}\mathbf{e}(-\,{}^tv^1(2\pi\operatorname{Im} Z)v^1)dv^1 = (\det 2\operatorname{Im} Z)^{-\frac{1}{2}}.$$

The last equation is easily deduced from the well-known formula $\int_{\infty}^{+\infty}\mathbf{e}(-x^2) = \pi^{\frac{1}{2}}$. □

3.3.3 The Theta Transformation Formula, Final Version

Proposition 3.3.2 contains a preliminary version of the Theta Transformation Formula. According to Lemma 3.3.3 it remains to compute the factor $C(Z, M, 0)$.

Theorem 3.3.9 (Theta Transformation Formula) *For all* $(v, Z) \in \mathbb{C}^g \times \mathfrak{H}_g$, *characteristics* $c = Zc^1 + c^2$, $c^1, c^2 \in \mathbb{R}^g$ *and matrices* $M = \left(\begin{smallmatrix} \alpha & \beta \\ \gamma & \delta \end{smallmatrix}\right) \in G_{1_g} = \mathrm{Sp}_{2g}(\mathbb{Z})$,

$$\vartheta \begin{bmatrix} M[c]^1 \\ M[c]^2 \end{bmatrix} ({}^t(\gamma Z + \delta)^{-1} v, M(z))$$
$$= \kappa(M) \det(\gamma Z + \delta)^{1/2} \mathbf{e}(\pi i\, k(M, c^1, c^2) + \pi i\, {}^t v(\gamma Z + \delta)^{-1} \gamma v) \vartheta \begin{bmatrix} c^1 \\ c^2 \end{bmatrix} (v, Z),$$

where $k(M, c^1, c^2) = {}^t(\delta c^1 - \gamma c^2)(-\beta c^1 + \alpha c^2 + (\alpha\, {}^t\beta)_0) -{}^t c^1 c^2$ *and* $\kappa(M) \in \mathbb{C}_1$ *is a constant with the same sign ambiguity as* $\det(\gamma Z + \delta)^{1/2}$.

Concerning the constant $\kappa(M)$, one can say more (see Exercise 3.3.4 (7)). In particular $\kappa(M)$ is an 8-th root of unity for every $M \in \mathrm{Sp}_{2g}(\mathbb{Z})$.

Proof **Step I**: *Translation of Proposition 3.3.2 into terms of classical theta function.* Let ϑ^c_Z and $\vartheta^{M[c]}_{M(Z)}$ be the canonical theta functions corresponding to the classical theta functions $\vartheta \left[\begin{smallmatrix} c_1 \\ c^2 \end{smallmatrix}\right] (\cdot, Z)$ and $\vartheta \left[\begin{smallmatrix} M[c]^1 \\ M[c]^2 \end{smallmatrix}\right] (\cdot, M(Z))$ respectively. According to Lemmas 3.3.5 and 3.3.4 (a) for all $v \in \mathbb{C}^g$,

$$\vartheta^c_Z(v) = \mathbf{e}\left(\frac{\pi}{2}\, {}^t v(\operatorname{Im} Z)^{-1} v - \pi i\, {}^t c^1 c^2\right) \vartheta \begin{bmatrix} c^1 \\ c^2 \end{bmatrix} (v, Z),$$
$$\vartheta^{M[c]}_{M(Z)}(v) = \mathbf{e}\left(\frac{\pi}{2}\, {}^t v(\operatorname{Im} M(Z))^{-1} v - \pi i\, {}^t M[c]^1 M[c]^2\right) \vartheta \begin{bmatrix} M[c]^1 \\ M[c]^2 \end{bmatrix} (v, M(Z)).$$

Since $A = {}^t(\gamma Z + \delta)$ by Corollary 3.1.5, Proposition 3.3.2 gives

$$\vartheta^{M[c]}_{M(Z)}({}^t(\gamma Z + \delta)^{-1} v) = C(Z, M, c)^{-1} \vartheta^c_Z(v).$$

Replacing canonical theta functions by classical ones, we get using $A^* H_Z = H_{M(Z)}$, which reads in terms of matrices $(\operatorname{Im} M(Z))^{-1} = {}^t A(\operatorname{Im} Z)^{-1} \overline{A}$,

$$\vartheta \begin{bmatrix} M[c]^1 \\ M[c]^2 \end{bmatrix} ({}^t(\gamma Z + \delta)^{-1} v, M(Z))$$
$$= C(Z, M, c)^{-1} \mathbf{e}\Big(\pi i\, {}^t M[c]^1 M[c]^2 - \pi i\, {}^t c^1 c^2 - \frac{\pi}{2}\, {}^t v\, {}^t A^{-1} (\operatorname{Im} M(Z))^{-1} A^{-1} v$$
$$+ \frac{\pi}{2}\, {}^t v(\operatorname{Im} Z)^{-1} v\Big) \vartheta \begin{bmatrix} c^1 \\ c^2 \end{bmatrix} (v, Z)$$

$$= C(Z,M,c)^{-1}\,\mathbf{e}\Big(\pi i\,{}^tM[c]^1M[c]^2 - \pi i\,{}^tc^1c^2 - \frac{\pi}{2}\,{}^tv(\operatorname{Im} Z)^{-1}(\overline{A}A^{-1} - \mathbf{1}_g)v\Big)$$
$$\cdot\,\vartheta\begin{bmatrix} c^1 \\ c^2 \end{bmatrix}(v,Z)$$
$$= C(Z,M,c)^{-1}\,\mathbf{e}\Big(\pi i\,{}^tM[c]^1M[c]^2 - \pi i\,{}^tc^1c^2 + \pi i\,{}^tv(\gamma Z+\delta)^{-1}\gamma v\Big)\vartheta\begin{bmatrix} c^1 \\ c^2 \end{bmatrix}(v,Z).$$

For the last equation we used that

$$(\operatorname{Im} Z)^{-1}(\overline{A}A^{-1} - \mathbf{1}_g) = (\operatorname{Im} Z)^{-1}(\,{}^t(\gamma\overline{Z}+\delta) - {}^t(\gamma Z+\delta))\,{}^t(\gamma Z+\delta)^{-1}$$
$$= -2i\,{}^t\gamma\,{}^t(\gamma Z+\delta)^{-1} = -2i(\gamma Z+\delta)^{-1}\gamma.$$

Here we used Lemma 3.1.3 (b) (iii).

Now write $\binom{d^1}{d^2} = {}^tM^{-1}\binom{c^1}{c^2}$. So $d^1 = \delta c^1 - \gamma c^2$ and $d^2 = -\beta c^1 + \alpha c^2$. Then by Lemma 3.3.1 (b) we have $M[c]^1 = d^1 + \frac{1}{2}(\gamma\,{}^t\delta)_0$ and $M[c]^2 = d^2 + \frac{1}{2}(\alpha\,{}^t\beta)_0$. Together with Lemma 3.3.3 this implies

$$C(Z,M,c)^{-1}\,\mathbf{e}(\pi i\,{}^tM[c]^1M[c]^2 - \pi i\,{}^tc^1c^2)$$
$$= C(Z,M,0)^{-1}\,\mathbf{e}\Big(-\frac{1}{2}\pi i\,{}^t\begin{pmatrix}(\gamma\,{}^t\delta)_0 \\ (\alpha\,{}^t\beta)_0\end{pmatrix}\begin{pmatrix}0 & \mathbf{1}_g \\ -\mathbf{1}_g & 0\end{pmatrix}\begin{pmatrix}d^1 \\ d^2\end{pmatrix}$$
$$+\,\pi i\,{}^t(d^1 + \frac{1}{2}(\gamma\,{}^t\delta)_0)(d^2 + \frac{1}{2}(\alpha\,{}^t\beta)_0) - \pi i\,{}^tc^1c^2\Big)$$
$$= C(Z,M,0)^{-1}\,\mathbf{e}\left(\frac{1}{4}\pi i\,{}^t(\gamma\,{}^t\delta)_0(\alpha\,{}^t\beta)_0 + \pi i\,{}^td^1(d^2 + (\alpha\,{}^t\beta)_0) - \pi i\,{}^tc^1c^2\right).$$

Finally we obtain with $k(M,c^1,c^2)$ as in the theorem,

$$\vartheta\begin{bmatrix} M[c]^1 \\ M[c]^2 \end{bmatrix}(\,{}^t(\gamma Z+\delta)^{-1}v, M(Z)) = C(Z,M,0)^{-1}\,\mathbf{e}\Big(\frac{1}{4}\pi i\,{}^t(\gamma\,{}^t\delta)_0(\alpha\,{}^t\beta)_0$$
$$+\pi i(k(M,c^1,c^2) + \pi i\,{}^tv(\gamma Z+\delta)^{-1}\gamma v\Big)\vartheta\begin{bmatrix} c^1 \\ c^2 \end{bmatrix}(v,Z).$$

Step II: *The factor* $C(Z,M,0)^{-1}\,\mathbf{e}(\frac{1}{4}\pi i\,{}^t(\gamma\,{}^t\delta)_0(\alpha\,{}^t\beta)_0)$.
According to Proposition 3.3.2 we have

$$|C(Z,M,0)^{-1}|^2 = ||A^*\vartheta_Z^0||^{-2}\cdot||\vartheta_{Z'}^{M[0]}||^2.$$

But $||A^*\vartheta_Z^0|| = ||\vartheta_Z^0||$, since the corresponding change of variables of $\mathbb{R}^{2g}$ is given by the matrix ${}^tM$, which has determinant 1. Applying Proposition 3.3.8, we get

$$|C(Z,M,0)^{-1}|^2 = \det(2\operatorname{Im} Z)^{\frac{1}{2}}\cdot\det(2\operatorname{Im} M(Z))^{-\frac{1}{2}}$$
$$= \det(2\operatorname{Im} Z)^{\frac{1}{2}}\cdot\det(2\overline{A}^{-1}\operatorname{Im} Z\,{}^tA^{-1})^{-\frac{1}{2}}$$
$$= |\det A| = |\det(\gamma Z+\delta|.$$

Now the last equation of Step I implies that $C(Z,M,0)^{-1}$ depends holomorphically on Z such that $C(Z,M,0)^{-1}$ and $\det(\gamma Z+\delta)^{\frac{1}{2}}$ differ only by a constant. Hence we obtain

$$C(Z,M,0)^{-1}\,\mathbf{e}\left(\frac{1}{4}\pi i\,{}^t(\gamma\,{}^t\delta)_0(\alpha\,{}^t\beta)_0\right)=\kappa(M)\det(\gamma Z+\delta)^{\frac{1}{2}},$$

where the factor $\kappa(M)\in\mathbb{C}_1$ depends only on M and the chosen root of $\det(\gamma Z+\delta)$. Together with Step I this completes the proof of the theorem. □

3.3.4 Exercises

(1) Let $Z\in\mathfrak{H}_g$. For every $c^1,c^2\in\mathbb{R}^g$ the function $\vartheta\begin{bmatrix}c^1\\c^2\end{bmatrix}(\cdot,Z)$ is a theta function with respect to the lattice $Z\mathbb{Z}^g\oplus\mathbb{Z}^g$ with functional equation

$$\begin{aligned}&\vartheta\begin{bmatrix}c^1\\c^2\end{bmatrix}(v+Z\lambda^1+\lambda^2,Z)\\&=\mathbf{e}\left(2\pi i({}^tc^1\lambda^2-{}^tc^2\lambda^1)-\pi i\,{}^t\lambda^1Z\lambda^1-2\pi i\,{}^tv\lambda^1\right)\vartheta\begin{bmatrix}c^1\\c^2\end{bmatrix}(v,Z)\end{aligned}$$

for all $v\in\mathbb{C}^g$, $\lambda^1,\lambda^2\in\mathbb{Z}^g$.

(2) For $Z\in\mathfrak{H}_g$ and all $c^1,c^2\in\mathbb{R}^g$ the lattice $Z\mathbb{Z}^g\oplus\mathbb{Z}^g$ is a maximal lattice for which $\vartheta\begin{bmatrix}c^1\\c^2\end{bmatrix}(\cdot,Z)$ satisfies a functional equation.

(Hint: Use the fact that the line bundle given by the function $\vartheta\begin{bmatrix}c^1\\c^2\end{bmatrix}$ defines a principal polarization.)

(3) For $Z\in\mathfrak{H}_g$ and all $c^1,c^2\in\mathbb{R}^g$ show that

$$\vartheta\begin{bmatrix}c^1+\ell^1\\c^2+\ell^2\end{bmatrix}=\vartheta\begin{bmatrix}c^1\\c^2\end{bmatrix}\quad\Leftrightarrow\quad\ell^1,\ell^2\in\mathbb{Z}^g.$$

This reflects the fact that in the principally polarized case the characteristic $c\in\mathbb{C}^g$ is uniquely determined modulo the lattice $Z\mathbb{Z}^g\oplus\mathbb{Z}^g$.

(4) Define for $Z\in\mathfrak{H}_g$, for all $\lambda^1,\lambda^2\in\mathbb{Z}^g$ and $v\in\mathbb{C}^g$ the function

$$e\begin{bmatrix}0\\0\end{bmatrix}:\Lambda_Z\times\mathbb{C}^g\to\mathbb{C}^*,\qquad(Z\lambda^1+\lambda^2,v)\mapsto\mathbf{e}(-\pi i\,{}^t\lambda^1Z\lambda^1-2\pi i\,{}^tv\lambda^1).$$

The notation $\begin{bmatrix}0\\0\end{bmatrix}$ indicates that $e\begin{bmatrix}0\\0\end{bmatrix}$ is the classical factor of automorphy for the line bundle of characteristic 0 in $\mathrm{Pic}^{H_Z}(X_Z)$ with respect to the decomposition $\Lambda_Z=Z\mathbb{Z}^g\oplus\mathbb{Z}^g$.

(i) Let $D = \operatorname{diag}(d_1, \ldots, d_g)$ be a type. Show that $\vartheta \begin{bmatrix} c^1 \\ c^2 \end{bmatrix}$ is a theta function with factor $e \begin{bmatrix} 0 \\ 0 \end{bmatrix}$ with respect to the lattice $Z\mathbb{Z}^g \oplus D\mathbb{Z}^g$ if and only if $\begin{pmatrix} c^1 \\ c^2 \end{pmatrix} \in D^{-1}\mathbb{Z}^g \oplus \mathbb{Z}^g$.

(ii) If $c_0, \ldots, c_N$ is a set of representatives of $D^{-1}\mathbb{Z}^g/\mathbb{Z}^g$, show that the functions $\vartheta \begin{bmatrix} c_0 \\ 0 \end{bmatrix}, \ldots, \vartheta \begin{bmatrix} c_N \\ 0 \end{bmatrix}$ are a basis of the space of classical theta functions for the line bundle on $X_{(Z,D)} = \mathbb{C}^g/(Z\mathbb{Z}^g \oplus \mathbb{Z}^g)$ determined by the factor $e \begin{bmatrix} 0 \\ 0 \end{bmatrix}$.

(Hint: For (ii) use the proof of Theorem 1.5.9.)

(5) Show that for $Z \in \mathfrak{H}_g$ and all $c^1, c^2 \in \mathbb{R}^g, v \in \mathbb{C}^g$,

$$\vartheta \begin{bmatrix} c_1 \\ c^2 \end{bmatrix} (v, Z) = \mathbf{e}\big(\pi i\, {}^t c^1 Z c^1 + 2\pi i\, {}^t c^1 (v + c^2)\big) \vartheta \begin{bmatrix} 0 \\ 0 \end{bmatrix} (v + Zc^1 + c^2, Z).$$

(This is a translation of Exercise 1.5.5 (11).)

(6) (Heat equation) Show that for every symmetric matrix $(s_{ij}) \in \mathrm{M}_g(\mathbb{C})$ and all vectors $c^1, c^2 \in \mathbb{R}^g$,

$$\sum_{i,j=1}^{g} s_{ij} \frac{\partial^2 \vartheta \begin{bmatrix} c^1 \\ c^2 \end{bmatrix}}{\partial v_i \partial v_j} = 4\pi i \sum_{1 \le i \le j \le g} s_{ij} \frac{\partial \vartheta \begin{bmatrix} c^1 \\ c^2 \end{bmatrix}}{\partial c_{ij}}.$$

(7) For $M = \begin{pmatrix} \alpha & \beta \\ \gamma & \delta \end{pmatrix} \in \mathrm{Sp}_{2g}(\mathbb{Z})$ let $\kappa(M)$ denote the constant in the theta transformation formula, Theorem 3.3.9.

(a) Show that $\kappa^2 : \mathrm{Sp}_{2g}(\mathbb{Z}) \to \mathbb{C}_1^*$ is a homomorphism of groups.

(b) The matrices $\begin{pmatrix} 0 & \mathbf{1}_g \\ -\mathbf{1}_g & 0 \end{pmatrix}$, $\begin{pmatrix} \mathbf{1}_g & \beta \\ 0 & \mathbf{1}_g \end{pmatrix}$ and $\begin{pmatrix} \alpha & 0 \\ 0 & {}^t\alpha^{-1} \end{pmatrix}$ with $\beta \in \mathrm{M}_g(\mathbb{Z})$ symmetric and $\alpha \in \mathrm{GL}_g(\mathbb{Z})$ generate the group $\mathrm{Sp}_{2g}(\mathbb{Z})$.

(c) Show that $\kappa^2 \begin{pmatrix} 0 & \mathbf{1}_g \\ -\mathbf{1}_g & 0 \end{pmatrix} = (-i)^g$, $\kappa^2 \begin{pmatrix} \mathbf{1}_g & \beta \\ 0 & \mathbf{1}_g \end{pmatrix} = 1$, $\kappa^2 \begin{pmatrix} \alpha & 0 \\ 0 & {}^t\alpha^{-1} \end{pmatrix} = \det(\alpha) = \pm 1$.

3.4 The Universal Family

Given a type D, a family $p : \mathfrak{X}_D \to \mathfrak{H}_g$ is constructed which parametrizes all abelian varieties X_Z with $Z \in \mathfrak{H}_g$, polarization of type D and symplectic basis. Recall that a *family of abelian varieties* (of type D and with symplectic basis) is a holomorphic map of analytic varieties $q : \mathcal{X} \to T$ such that every fibre $q^{-1}(t)$ is an abelian variety (with polarization of type D and symplectic basis depending holomorphically on t). The family $p : \mathfrak{X}_D \to \mathfrak{H}_g$ is *universal* in the following sense: Given any family q as above, there is a unique holomorphic map $f : T \to \mathfrak{H}_g$ such that $f^*(p) = q$. We will not prove the universality of the family p, but see Exercise 3.4.5 (7).

The variety $\mathfrak{X}_D$ admits a line bundle $\mathfrak{L}$ (defined in Section 3.3.2) providing a holomorphic map $\varphi_D : \mathfrak{X}_D \to \mathbb{P}^N$ which finally induces a holomorphic map of the Siegel upper half space $\mathfrak{H}_g$ into projective space $\mathbb{P}^N$.

3.4.1 Construction of the Family

Fix a type $D = \operatorname{diag}(d_1, \dots, d_g)$. In Section 3.1.1 the Siegel upper space $\mathfrak{H}_g$ was introduced and it was shown that it parametrizes the abelian varieties with polarization of type D together with a symplectic basis. In this subsection we put them together and construct a family.

For any $Z \in \mathfrak{H}_g$ consider the isomorphism of $\mathbb{R}$-vector spaces

$$j_Z : \mathbb{R}^{2g} \to \mathbb{C}^g, \qquad x \mapsto (Z, \mathbf{1}_g)x. \tag{3.11}$$

If Λ_D denotes the lattice

$$\Lambda_D = \begin{pmatrix} \mathbf{1}_g & 0 \\ 0 & D \end{pmatrix} \mathbb{Z}^{2g} \quad \text{in} \quad \mathbb{R}^{2g},$$

then $j_Z(\Lambda_D)$ is just the lattice $\Lambda_Z = (Z, D)\mathbb{Z}^{2g}$ in $\mathbb{C}^g$ determined by Z. In other words, if $f_1, \dots f_{2g}$ denotes the standard basis of $\mathbb{R}^{2g}$ and $\lambda_i, \dots, \lambda_g, \mu_1, \dots, \mu_g$ the symplectic basis of Λ_Z associated to Z, then we have

$$\lambda_i = j_Z(f_i) \quad \text{and} \quad \mu_i = j_Z(d_i f_{g+i}) \qquad \text{for} \quad 1 \le i \le g.$$

The lattice Λ_D acts freely and properly discontinuously on the manifold $\mathbb{C}^g \times \mathfrak{H}_g$ by

$$\ell(v, Z) = (v + j_Z(\ell), Z) \qquad \text{for all} \quad \ell \in \Lambda_D, \ (v, Z) \in \mathbb{C}^g \times \mathfrak{H}_g.$$

In Theorem 3.1.11 Cartan's theorem was recalled, which says that the quotient of a complex manifold by a group acting properly and discontinuously is a normal complex analytic space. Clearly the singularities of the quotient correspond to points with non-trivial stabilizer. Hence if the group in addition acts freely, the quotient is a manifold. So we immediately obtain from this and the definitions the first part of the following proposition.

Proposition 3.4.1 *The quotient*

$$\mathfrak{X}_D := (\mathbb{C}^g \times \mathfrak{H}_g)/\Lambda_D$$

is a complex manifold with the following property:

Let $p : \mathfrak{X}_D \to \mathfrak{H}_g$ be the canonical projection. For every $Z \in \mathfrak{H}_g$ the fibre is

$$p^{-1}(Z) = \mathbb{C}^g / j_Z(\Lambda_D) = X_Z,$$

the abelian variety associated to Z with symplectic basis.

It remains to show that X_Z is an abelian variety with polarization of type D and symplectic basis. This is done at the end of the next section, since for this we need a line bundle defining the polarization.

3.4.2 The Line Bundle $\mathfrak{L}$ on $\mathfrak{X}_D$

Define a map

$$e_{\Lambda_D} : \Lambda_D \times (\mathbb{C}^g \times \mathfrak{H}_g) \to \mathbb{C}^*, \qquad (\ell, (v, Z)) \mapsto \mathbf{e}(-\pi i\, {}^t\ell^1 Z \ell^1 - 2\pi i\, {}^t v \ell^1),$$

where for $\ell \in \mathbb{R}^{2g}$ we denote by ℓ^1 the vector of its first g components.

Lemma 3.4.2 *The map e_{Λ_D} is a cocycle in $Z^1(\Lambda_D, H^0(\mathcal{O}^*_{\mathbb{C}^g \times \mathfrak{H}_g}))$.*

For the proof, see Exercise 3.4.5 (1). According to Proposition 1.2.2 the cocycle e_{Λ_D} defines a line bundle $\mathfrak{L}$ in $\operatorname{Pic}(\mathfrak{X}_D)$.

Lemma 3.4.3 *For any $Z \in \mathfrak{H}_g$ the restriction of the line bundle $\mathfrak{L}$ on $\mathfrak{X}_D$ to the fibre X_Z defines the polarization H_Z. To be more precise,*

$$\mathfrak{L}|_{X_Z} \simeq L(H_Z, \chi_0),$$

the line bundle of characteristic 0 with respect to the decomposition $\Lambda_Z = Z\mathbb{Z}^g \oplus D\mathbb{Z}^g$.

For the proof, see Exercise 3.4.5 (2).

Recall that $f_1, \dots, f_{2g}$ denote the standard basis of $\mathbb{R}^{2g}$ and let $j_Z : \mathbb{R}^{2g} \to \mathbb{C}^g$ be as in Section 3.4.1. Define for $1 \le i \le g$ holomorphic maps $\lambda_i, \mu_i : \mathfrak{H}^g \to \mathbb{C}^g$ by

$$\lambda_i(Z) = j_Z(f_i) \qquad \text{and} \qquad \mu_i(Z) = j_Z(d_i f_{g+i}).$$

For every $Z \in \mathfrak{H}_g$ the elements $\lambda_1(Z), \dots, \mu_g(Z)$ are a symplectic basis of Λ_Z for H_Z, clearly depending holomorphically on $Z \in \mathfrak{H}_g$. Summarizing, this gives:

Theorem 3.4.4 *For any type D,*

$$\left(p : \mathfrak{X}_D \to \mathfrak{H}_g, \mathfrak{L}, (\lambda_1, \dots, \mu_g)\right)$$

is a holomorphic family parametrizing the set of polarized abelian varieties with polarization of type D and symplectic basis.

3.4.3 The Map $\varphi_D : \mathfrak{X}_D \to \mathbb{P}^N$

Every sublinear system of $|\mathfrak{L}|$ induces a rational map of $\mathfrak{X}_D$ into some projective space in the usual way. In this section we want to single out a sublinear system $|U_D|$ with the property that the restriction $|U_D|_{X_Z}|$ coincides with the complete linear system $|\mathfrak{L}|_{X_Z}| = |L(H_Z, \chi_0)|$ for every fibre $X_Z = p^{-1}(Z)$.

Consider $H^0(\mathfrak{L})$ as the vector space of holomorphic functions $f : \mathbb{C}^g \times \mathfrak{H}_g \to \mathbb{C}$ satisfying the functional equation $f(v + j_Z(\ell), Z) = e_{\Lambda_D}(\ell, (v, Z))f(v, Z)$ for all $\ell \in \Lambda_D$ and all $(v, Z) \in \mathbb{C}^g \times \mathfrak{H}_g$. Proposition 3.3.6 suggests that the classical Riemann theta function might be a global section of $\mathfrak{L}$.

Define an alternating form J on $\mathbb{R}^{2g}$ by the matrix $\begin{pmatrix} 0 & \mathbf{1}_g \\ -\mathbf{1}_g & 0 \end{pmatrix}$ with respect to the standard basis. Observe that for every $Z \in \mathfrak{H}_g$ the isomorphism j_Z is defined in such a way that

$$J = j_Z^*(\operatorname{Im} H_Z). \tag{3.12}$$

The orthogonal complement $\Lambda_D^\perp$ of the lattice Λ_D with respect to the form $\mathbf{e}(2\pi i J)$ is given by

$$\Lambda_D^\perp = \begin{pmatrix} D^{-1} & 0 \\ 0 & \mathbf{1}_g \end{pmatrix} \mathbb{Z}^{2g}. \tag{3.13}$$

Lemma 3.4.5 $\vartheta \begin{bmatrix} c^1 \\ c^2 \end{bmatrix} \in H^0(\mathfrak{L})$ *for every* $\begin{pmatrix} c^1 \\ c^2 \end{pmatrix} \in \Lambda_D^\perp$.

Proof It suffices to check the functional equation for $\vartheta \begin{bmatrix} c^1 \\ c^2 \end{bmatrix}$ with respect to Λ_D, which is immediate. This also follows from Exercise 3.3.4 (4). □

Define

$$U_D := \left\langle \vartheta \begin{bmatrix} c^1 \\ c^2 \end{bmatrix} \;\middle|\; \begin{pmatrix} c^1 \\ c^2 \end{pmatrix} \in \Lambda_D^\perp \right\rangle,$$

the $\mathbb{C}$-vector space spanned by the functions $\vartheta \begin{bmatrix} c^1 \\ c^2 \end{bmatrix}$ with $\begin{pmatrix} c^1 \\ c^2 \end{pmatrix} \in \Lambda_D^\perp$. Since the characteristic of $\vartheta \begin{bmatrix} c^1 \\ c^2 \end{bmatrix}$ is defined modulo $\mathbb{Z}^{2g}$, U_D is a subvector space of $H^0(\mathfrak{L})$.

Proposition 3.4.6

(a) *If* $\{c_0, \ldots, c_N\}$ *is a set of representatives of* $D^{-1}\mathbb{Z}^g/\mathbb{Z}^g$, *then* $\left\{\vartheta \begin{bmatrix} c_i \\ 0 \end{bmatrix}, 0 \le i \le N\right\}$ *is a basis of* U_D.

(b) *For every* $Z \in \mathfrak{H}_g$ *the restriction of the linear system* $|U_D|$ *to the fibre* X_Z *coincides with the complete linear system* $|L(H_Z, \chi_0)|$.

Proof (a) and (b) follow from Exercises 3.3.4 (3) and (4) respectively. □

Denote by

$$\varphi_D : \mathfrak{X}_D \to \mathbb{P}^N, \qquad (\bar{v}, Z) \mapsto \left(\vartheta \begin{bmatrix} c_0 \\ 0 \end{bmatrix}(v, Z), \dots, \vartheta \begin{bmatrix} c_N \\ 0 \end{bmatrix}(v, Z)\right)$$

the meromorphic map associated with the linear system U_D. According to Proposition 3.4.6 (b), for every $Z \in \mathfrak{H}_g$ the restriction of φ_D to the fibre X_Z is just the rational map $\varphi_{L(X_Z,\chi_0)} : X_Z \to \mathbb{P}^N$ associated to the line bundle $L(H_Z, \chi_0)$.

By Proposition 2.1.5, for $d_1 \geq 2$ the map $\varphi_{L(X_Z,\chi_0)}$ is holomorphic for every $Z \in \mathfrak{H}_g$. Consequently $\varphi_D : \mathfrak{X}_D \to \mathbb{P}^N$ is a holomorphic map in this case.

Now let

$$s_0 : \mathfrak{H}_g \to \mathfrak{X}_D, \qquad Z \mapsto (0, Z)$$

denote the zero section and define the meromorphic map $\psi_D := \varphi_D s_0$. Then the following diagram commutes

$$\begin{array}{ccc} \mathfrak{X}_D & & \\ s_0 \uparrow\downarrow p & \searrow^{\varphi_D} & \\ \mathfrak{H}_g & \xrightarrow[\psi_D]{} & \mathbb{P}^N \end{array} \tag{3.14}$$

One says that the map ψ_D is given by *theta null values*:

$$\psi_Z(Z) = \left(\vartheta \begin{bmatrix} c_0 \\ 0 \end{bmatrix}(0, Z), \dots, \vartheta \begin{bmatrix} c_N \\ 0 \end{bmatrix}(0, Z)\right).$$

3.4.4 The Action of the Symplectic Group

In Section 3.1.3 the action of the symplectic group on $\mathfrak{H}_g$ was introduced, namely $Z \mapsto M(Z) = (\alpha Z + \beta)(\gamma Z + \delta)^{-1}$ for $M = \begin{pmatrix} \alpha & \beta \\ \gamma & \delta \end{pmatrix} \in \mathrm{Sp}_{2g}(\mathbb{R})$. We will see that for suitable subgroups this action extends to an action on the manifold $\mathfrak{X}_D$.

Recall the group $G_D \subset \mathrm{Sp}_{2g}(\mathbb{Q})$ defined in equation (3.5). According to Proposition 3.1.4 and Corollary (3.1.5) there is for every $Z \in \mathfrak{H}_g$ and $M \in G_D$ an isomorphism $X_{M(Z)} \to X_Z$ given by the equation

$${}^t(\gamma Z + \delta)(M(Z), \mathbf{1}_g) = (Z, \mathbf{1}_g)\, {}^tM.$$

Define

$$M_Z := {}^t(\gamma Z + \delta)^{-1},$$

which is the analytic representation of the inverse map $X_Z \to X_{M(Z)}$. The following diagram commutes

$$\begin{array}{ccc} \mathbb{R}^{2g} & \xrightarrow{j_Z} & \mathbb{C}^g \\ {}^t M^{-1} \downarrow & & \downarrow M_Z \\ \mathbb{R}^{2g} & \xrightarrow{j_{M(Z)}} & \mathbb{C}^g. \end{array} \tag{3.15}$$

This immediately implies that

$$M(v, Z) := (M_Z v, M(Z)) \tag{3.16}$$

defines an action of G_D on $\mathbb{C}^g \times \mathfrak{H}_g$. Actually, (3.16) gives an action of the whole group $\mathrm{Sp}_{2g}(\mathbb{R})$, but we do not need this fact.

Lemma 3.4.7 *The action of the group G_D on $\mathbb{C}^g \times \mathfrak{H}_g$ descends to an action on the family of abelian varieties $p : \mathfrak{X}_D \to \mathfrak{H}_g$.*

Here an *action on the family of abelian varieties* $p : \mathfrak{X}_D \to \mathfrak{H}_g$ means an action $\tau : G_D \times \mathfrak{X}_D \to \mathfrak{X}_D$ in such a way that the restriction $\tau|_{[M]\times p^{-1}(Z)}$ is an isomorphism $X_Z \to X_{M(Z)}$ of abelian varieties for every $M \in G_D$ and $Z \in \mathfrak{H}_g$. For the proof of the lemma, see Exercise 3.4.5 (5).

Any subgroup G of G_D acts properly and discontinuously on $\mathfrak{H}_g$ and thus on $\mathfrak{X}_D$. Hence $\mathfrak{X}_D/G$ is a normal complex analytic space and p induces a holomorphic map

$$\overline{p} : \mathfrak{X}_D/G \to \mathfrak{H}_g/G.$$

Since the action of the whole group G_D on $\mathfrak{H}_g$ has fixed points, $\overline{p} : \mathfrak{X}_D/G \to \mathfrak{H}_g/G$ is not necessarily a family of abelian varieties. In fact, the fibre of $\overline{p}$ over $Z \in \mathfrak{H}_g$ is the quotient of X_Z modulo the isotropy subgroup $(G_D)_Z$ of G_D in Z. We will see that for suitable subgroups the corresponding quotient is a family of abelian varieties.

Recall the group defined in Section 3.2.1,

$$\Gamma_D(D) = \left\{ \begin{pmatrix} a & b \\ c & d \end{pmatrix} \in \Gamma_D = \mathrm{Sp}^D_{2g}(\mathbb{R}) \;\middle|\; a - \mathbf{1}_g \equiv b \equiv c \equiv d - \mathbf{1}_g \equiv 0 \bmod D \right\}$$

representing isomorphisms of polarized abelian varieties with level D-structure. Denote by $G_D(D)$ the image of $\Gamma_D(D)$ under the isomorphism $\sigma_D : \Gamma_D \to \mathrm{Sp}_{2g}(\mathbb{R})$ defined in (3.8). We have

$$G_D(D) = \left\{ \begin{pmatrix} \mathbf{1}_g + D\overline{a} & D\overline{b}D \\ \overline{c} & \mathbf{1}_g + \overline{d}D \end{pmatrix} \in G_D \;\middle|\; \overline{a}, \overline{b}, \overline{c}, \overline{d} \in \mathrm{M}_g(\mathbb{Z}) \right\}. \tag{3.17}$$

Note that $G_D(D)$ is a subgroup of $\mathrm{M}_{2g}(\mathbb{Z}) \cap G_D \subseteq G_{\mathbf{1}_g} = \mathrm{Sp}_{2g}(\mathbb{Z})$. This observation will turn out to be important. According to Theorem 3.2.1 the quotient

$$\mathcal{A}_D(D) := \mathfrak{H}_g/G_D(D)$$

is a moduli space of polarized abelian varieties with level D-structure.

Proposition 3.4.8 *Suppose* $D = \operatorname{diag}(d_1, \ldots, d_g)$ *with* $d_1 \geq 3$. *Then*

$$\overline{p} : \mathfrak{X}_D / G_D(D) \to \mathcal{A}_D(D)$$

is a family of abelian varieties.

Proof It suffices to show that $G_D(D)$ acts fixed point free of $\mathfrak{H}_g$. Suppose $M(Z) = Z$ for some $M \in G_D(D)$ and $Z \in \mathfrak{H}_g$. According to Theorem 3.2.1 the corresponding automorphism φ_M of X_Z restricts to the identity on the group $K(H_Z)$. But $K(H_Z)$ contains the group of d_1-division points in X_Z. Since $d_1 \geq 3$, Corollary 2.4.11 implies $\varphi_M = \mathbf{1}_{X_Z}$. □

3.4.5 Exercises

(1) Show that the map $e_{\Lambda_D} : \Lambda_D \times (\mathbb{C}^g \times \mathfrak{H}_g) \to \mathbb{C}^*$ defined at the beginning of Section 3.4.2 is a cocycle in $Z^1(\Lambda_D, H^0(\mathcal{O}^*_{\mathbb{C}^g \times \mathfrak{H}_g}))$.

(2) Show that for any $Z \in \mathfrak{H}_g$ we have $\mathfrak{L}|_{X_Z} \simeq L(H_Z, \chi_0)$, where H_Z is a polarization of type D and $L(H_Z, \chi_0)$ the line bundle of characteristic 0 with respect to the decomposition $\Lambda_Z = Z\mathbb{Z}^g \oplus D\mathbb{Z}^g$.
(Hint: use Exercise 3.3.4 (4).)

(3) If $\{c_0, \ldots, c_N\}$ is a set of representatives of $D^{-1}\mathbb{Z}^g/\mathbb{Z}^g$, then $\left\{\vartheta\left[\begin{smallmatrix} c_i \\ 0 \end{smallmatrix}\right], 0 \leq i \leq N\right\}$ is a basis of U_D.
(Hint: use Exercise 3.3.4 (3).)

(4) Show that for every $Z \in \mathfrak{H}_g$ the restriction of the linear system $|U_D|$ to the fibre X_Z coincides with the complete linear system $|L(H_Z, \chi_0)|$.
(Hint: use Exercise 3.3.4 (4).)

(5) Show that the action of the group G_D on $\mathbb{C}^g \times \mathfrak{H}_g$ descends to an action on the family of abelian varieties $p : \mathfrak{X}_D \to \mathfrak{H}_g$.
(Hint: use diagram 3.15.)

(6) For any $Z = X + iY \in \mathfrak{H}_g$ consider the $\mathbb{R}$-linear isomorphism $j_Z : \mathbb{R}^{2g} \to \mathbb{C}^g$ given in equation (3.11). The complex structure on $\mathbb{C}^g$ induces a complex structure $J \in \mathrm{M}_{2g}(\mathbb{R})$ on $\mathbb{R}^{2g}$, meaning that $ij_Z = j_Z J$. Show that

$$J = \begin{pmatrix} Y^{-1}X & Y^{-1} \\ -Y - XY^{-1}X & -XY^{-1} \end{pmatrix}.$$

(7) Show that the family $p : \mathcal{X}_D \to \mathfrak{H}_g$ of Proposition 3.4.1 is a universal family of abelian varieties of type D with symplectic basis.

3.5 Projective Embeddings of Moduli Spaces

We saw in Section 3.4.3 that the map $\varphi_D : \mathfrak{X}_D \to \mathbb{P}^N$ given by a certain sublinear system of the line bundle $\mathfrak{L}$ on $\mathfrak{X}_D$ is holomorphic if only $d_1 \geq 2$. If G is any subgroup of finite index in $G_D(D)$, we saw in Proposition 3.4.8 that the quotient $\mathfrak{X}_D/G$ admits the structure of a family of abelian varieties, if $d_1 \geq 3$. The main result of this section is that, if G is the group of orthogonal level D-structures and $d_1 \geq 4$, then the map $\varphi_D : \mathfrak{X}_D \to \mathbb{P}^n$ goes down to a holomorphic embedding of the corresponding moduli space of abelian varieties.

3.5.1 Orthogonal Level D-structures

In this section the subgroup $G_D(D)_0$ of $G_D(D)$ of orthogonal level D-structures is introduced.

Assume that D is a type with $d_1 \geq 2$ so that $\varphi_D : \mathfrak{X}_D \to \mathbb{P}^N$ is a holomorphic map as seen at the end of Section 3.4.3. Hence the map $\psi_D : \mathfrak{H}_g \to \mathbb{P}^N$ of diagram (3.14) is holomorphic. Recall the lattice $\Lambda_D^\perp = \begin{pmatrix} D^{-1} & 0 \\ 0 & \mathbf{1}_g \end{pmatrix} \mathbb{Z}^{2g}$ from equation (3.13). By definition, the group G_D and thus also $G_D(D)$ acts from the right on $\Lambda_D^\perp$ by

$$\left(M, \begin{pmatrix} c^1 \\ c^2 \end{pmatrix}\right) \mapsto {}^tM \begin{pmatrix} c_1 \\ c_2 \end{pmatrix}.$$

Consider the quadratic form

$$Q : \Lambda_D^\perp \to \mathbb{C}_1, \qquad \begin{pmatrix} c^1 \\ c^2 \end{pmatrix} \mapsto \mathbf{e}(\pi i \, {}^tc^1 c^2).$$

The group $G_D(D)_0$ is defined to be the subgroup of $G_D(D)$ preserving Q. Roughly speaking, the quotient

$$\mathcal{A}(D)_0 := \mathfrak{H}_g / G_D(D)_0$$

is the space of polarized abelian varieties with *orthogonal level D-structure*. In this section we show that the holomorphic map $\psi_D : \mathfrak{H}_g \to \mathbb{P}^N$ factorizes via $\mathfrak{H}_g \to \mathcal{A}_D(D)_0$. We need the following characterization of $G_D(D)_0$.

Lemma 3.5.1

(a) *For any* $M = \begin{pmatrix} \alpha & \beta \\ \gamma & \delta \end{pmatrix} \in G_D(D)$ *the following conditions are equivalent*

(i) $M \in G_D(D)_0$;
(ii) $(D^{-1}\alpha \, {}^t\beta D^{-1})_0 \equiv (\gamma \, {}^t\delta)_0 \equiv 0 \mod 2$;
(iii) $(D^{-1} \, {}^t\delta\beta D^{-1})_0 \equiv ({}^t\gamma\alpha)_0 \equiv 0 \mod 2$.

(b) $G_D(D)_0$ *is of finite index in* G_D.

Proof (a): For every $\ell^1, \ell^2 \in \mathbb{Z}^g$ we have

$$Q\left({}^tM\begin{pmatrix} D^{-1}\ell^1 \\ \ell^2\end{pmatrix}\right)Q\begin{pmatrix} D^{-1}\ell^1 \\ \ell^2\end{pmatrix}^{-1}$$
$$= \mathbf{e}\left(\pi i\, {}^t\ell^1 D^{-1}(\alpha\, {}^t\delta + \beta\, {}^t\gamma - \mathbf{1}_g)\ell^2 + \pi i\, {}^t\ell^1 D^{-1}\alpha\, {}^t\beta D^{-1}\ell^1 + \pi i\, {}^t\ell^2\gamma\, {}^t\delta\ell^2\right)$$
$$= \mathbf{e}\left(\pi i\, {}^t\ell^1 D^{-1}\alpha\, {}^t\beta D^{-1}\ell^1 + \pi i\, {}^t\ell^2\gamma\, {}^t\delta\ell^2\right)$$
$$\text{(by Lemma 3.1.3 (iii), since } {}^t\ell^1 D^{-1}\beta\, {}^t\gamma\ell^2 \in \mathbb{Z} \text{ by equation 3.17)}$$
$$= \mathbf{e}\left(\pi i\, {}^t(D^{-1}\alpha\, {}^t\beta D^{-1})_0\ell^1 + \pi i\, {}^t(\gamma\, {}^t\delta)_0\ell^2\right).$$

For the last equation we used that $D^{-1}\alpha\, {}^t\beta D^{-1} \in \mathrm{M}_g(\mathbb{Z})$ according to (3.17) and moreover that ${}^t\ell S\ell \equiv {}^t(S)_0\ell \mod 2$ for any $\ell \in \mathbb{Z}^g$ and symmetric $S \in \mathrm{M}_g(\mathbb{Z})$. This implies (i) $\Leftrightarrow$ (ii).

The equivalence (i) $\Leftrightarrow$ (iii) is proven in the same way replacing only M by M^{-1} (see Exercise 3.5.3 (1)).

(b): Recall that $D = \mathrm{diag}(d_1, \ldots, d_g)$ with $d_i | d_{i+1}, 1 \leq i \leq g-1$. Since the principal congruence subgroup $\Gamma(2d_g)$ is of finite index in Γ_D, it suffices to show that its image under the isomorphism $\sigma_D : \Gamma_D \to G_D$ is contained in $G_D(D)_0$. But

$$\sigma_D(\Gamma(2d_g)) = \left\{ \begin{pmatrix} \mathbf{1}_g + 2d_g a & 2d_g bD \\ 2d_g D^{-1}c & \mathbf{1}_g + 2d_g D^{-1}dD \end{pmatrix} \in G_D \;\middle|\; a, b, c, d \in \mathrm{M}_g(\mathbb{Z}) \right\}$$

and one easily checks using (a) that this is a subgroup of $G_D(D)_0$. □

Proposition 3.5.2 *If $d_1 \geq 2$, there is a holomorphic map $\overline{\psi}_D : \mathcal{A}_D(D)_0 \to \mathbb{P}^N$ such that the following diagram commutes*

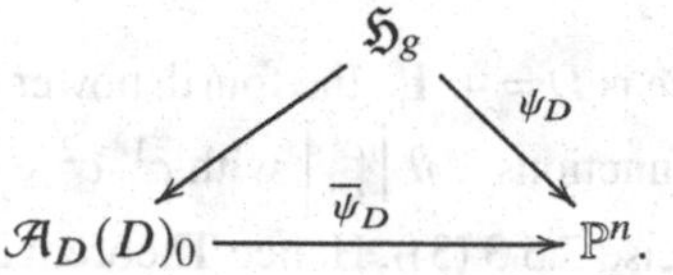

Proof According to the definition of ψ_D (see diagram 3.14), it suffices to show that for every $M \in G_D(D)_0$ there is a holomorphic map $\tau_M : \mathfrak{H}_g \to \mathbb{C}^*$ such that for all $Z \in \mathfrak{H}_g$ and $\ell \in \mathbb{Z}^g$,

$$\vartheta\begin{bmatrix} D^{-1}\ell \\ 0 \end{bmatrix}(0, M(Z)) = \tau_M(Z)\vartheta\begin{bmatrix} D^{-1}\ell \\ 0 \end{bmatrix}(0, Z).$$

The essential observation is that we can apply the Theta Transformation Formula Theorem 3.3.9, since $G_D(D)_0$ is a subgroup of $G_1 = \mathrm{Sp}_{2g}(\mathbb{Z})$. It gives for every $M = \left(\begin{smallmatrix} \alpha & \beta \\ \gamma & \delta \end{smallmatrix}\right) \in G_D(D)_0$ and every characteristic $c = ZD^{-1}\ell$,

$$\vartheta\begin{bmatrix}\delta D^{-1}\ell+\frac{1}{2}D(\gamma\,{}^t\delta)_0\\ -\beta D^{-1}\ell+\frac{1}{2}(\alpha\,{}^t\beta)_0\end{bmatrix}(0,M(Z))$$
$$=\kappa(M)\det(\gamma Z+\delta)^{\frac{1}{2}}\,\mathbf{e}(\pi ik(M,D^{-1}\ell,0))\vartheta\begin{bmatrix}D^{-1}\ell\\ 0\end{bmatrix}(0,Z).$$

Hence using Exercise 3.3.4 (3), it suffices to show:

(i) $\delta D^{-1}\ell+\frac{1}{2}D(\gamma\,{}^t\delta)_0\equiv D^{-1}\ell \mod \mathbb{Z}$;
(ii) $-\beta D^{-1}\ell+\frac{1}{2}(\alpha\,{}^t\beta)_0\equiv 0 \mod \mathbb{Z}$; and
(iii) $k(M,D^{-1}\ell,0)={}^t\ell D^{-1}\,{}^t\delta\big(-\beta D^{-1}\ell+(\alpha\,{}^t\beta)_0\big)\equiv 0 \mod 2$.

But this is an immediate computation to be done in Exercise 3.5.3 (2). □

Remark 3.5.3 A slight modification of the above proof shows that the map $\varphi_D:\mathfrak{X}_D\to\mathbb{P}^N$ also factorizes via $\mathfrak{X}_D\to\mathfrak{X}_D/G_D(D)_0$.

3.5.2 Projective Embedding of $\mathcal{A}_D(D)_0$

Let $D=\operatorname{diag}(d_1,\ldots,d_g)$ be a type with $d_1\geq 2$. In this section we show that under some additional hypotheses the holomorphic map $\overline{\psi}_D$ of Proposition 3.5.2 is an embedding. The main result is the following theorem due to Igusa [70].

Theorem 3.5.4 *If d_1 is an even number ≥ 4, then $\overline{\psi}_D:\mathcal{A}_D(D)_0\hookrightarrow\mathbb{P}^N$ is an analytic embedding.*

Using Proposition 3.4.6 and the definition of $\overline{\psi}_D$, the theorem can also be interpreted as saying that the theta null values $\left\{\vartheta\left[\begin{smallmatrix}c_i\\0\end{smallmatrix}\right](0,Z)\,\middle|\,\overline{c}_i\in D^{-1}\mathbb{Z}^g/\mathbb{Z}^g\right\}$ embed the moduli space.

The most important case is $D=4\cdot\mathbf{1_g}$, the fourth power of a principal polarization L_Z. In this case the theta functions $2^*\vartheta\left[\begin{smallmatrix}c^1\\c^2\end{smallmatrix}\right]$ with $c^1,c^2\in\frac{1}{2}\mathbb{Z}^g/\mathbb{Z}^g$ also form a basis for $H^0(X_Z,L_Z^4)$ (see Exercise 3.5.3 (3)). Hence Theorem 3.5.4 implies that the theta null values $\{\vartheta\left[\begin{smallmatrix}c^1\\c^2\end{smallmatrix}\right](0,Z)\mid\overline{c^1},\overline{c^2}\in\frac{1}{2}\mathbb{Z}^g/\mathbb{Z}^g\}$ give an analytic embedding of the moduli space $\mathcal{A}_{4\cdot\mathbf{1}_g}(4\cdot\mathbf{1}_g)_0$ into $\mathbb{P}^{2^{2g}-1}$.

The theorem is a consequence of the following two propositions.

Proposition 3.5.5 *If d_1 is an even number ≥ 4. then $\overline{\psi}_D:\mathcal{A}_D(D)_0\hookrightarrow\mathbb{P}^N$ is an injective holomorphic map.*

Proof According to Proposition 3.5.2 the map $\overline{\psi}_D$ is holomorphic. Let $Z,Z'\in\mathfrak{H}_g$ with $\psi_D(Z)=\psi_D(Z')$. We have to show that there is an $M\in G_D(D)_0$ with $Z'=MZ$.

Step I: *There is an $M \in G_D$ with $Z' = M(Z)$.*

By Definition of ψ_D and Exercise 3.3.4 (3) there is a constant $\tau \in \mathbb{C}^*$ such that for all $\binom{c^1}{c^2} \in \Lambda_D^{\perp}$,

$$\vartheta\begin{bmatrix} c^1 \\ c^2 \end{bmatrix}(0, Z') = \tau\,\vartheta\begin{bmatrix} c^1 \\ c^2 \end{bmatrix}(0, Z).$$

Now we use the fact that the theta null values determine the abelian varieties $X_{Z'}$ and X_Z. To be more precise, the images of $X_{Z'}$ and X_Z under the embeddings given by the linear systems $|L(H_{Z'}, \chi_0')|$ and $|L(X_Z, \chi_0)|$ coincide. In other words, there is an isomorphism $f : X_{Z'} \to X_Z$ such that the following diagram commutes

$$\begin{array}{ccc} X_{Z'} & \xrightarrow{\quad f \quad} & X_Z \\ & {\scriptstyle \varphi_{L(H_{Z'},\chi_0')}} \searrow \quad \swarrow {\scriptstyle \varphi_{L(H_Z,\chi_0)}} & \\ & \mathbb{P}^N & \end{array} \tag{3.18}$$

The fact that the theta null values determine the diagram is a consequence of Riemann's equations, the proof of which will not be given here, because this would require a large part of another chapter (Birkenhake–Lange [24, Chapter 7]). We only remark that here the assumption of d_1 is required (see [24, Riemann's Equations 7.5.2]).

Given diagram 3.18, by Proposition 3.1.4 there is an $M \in G_D$ with $Z' = M(Z)$ and f is the isomorphism defined by the equation $M_Z^{-1}(M(Z), \mathbf{1}_g) = (Z, \mathbf{1}_g)\,{}^tM$ (see Section 3.4.4). Here M_Z is the analytic representation of f^{-1}. Then $f^*\varphi_{L(H_Z,\chi_0)} = \varphi_{L(H_{Z'},\chi_0')}$ translates to

$$\vartheta\begin{bmatrix} c^1 \\ c^2 \end{bmatrix}(M_Z v, M(Z)) = \tau(v)\,\vartheta\begin{bmatrix} c^1 \\ c^2 \end{bmatrix}(v, Z) \tag{3.19}$$

for all $v \in \mathbb{C}^g$, $\binom{c^1}{c^2} \in \Lambda_D^{\perp}$, with some holomorphic function $\tau : \mathbb{C}^g \to \mathbb{C}^*$, not depending on $\binom{c^1}{c^2}$.

Step II: $\vartheta\left[\begin{smallmatrix} c^1 \\ c^2 \end{smallmatrix}\right](v, Z) = \mathbf{e}(\pi i\,{}^tc^1c^2 - \pi i\,{}^td^1d^2)\vartheta\left[\begin{smallmatrix} d^1 \\ d^2 \end{smallmatrix}\right](v, Z)$ for all $\binom{c^1}{c^2} \in \Lambda_D^{\perp}$ and $\binom{d^1}{d^2} := {}^tM^{-1}\binom{c^1}{c^2}$.

Recall the isomorphism $j_Z : \mathbb{R}^{2g} \to \mathbb{C}^g$ and note that

$$j_Z(\Lambda_D^{\perp}) = \{v \in \mathbb{C}^g \mid \operatorname{Im} H_Z(v, j_Z(\Lambda_D)) \subset \mathbb{Z}\} = \Lambda(H_Z))$$

(see Section 1.4.1). Hence $Zc^1 + c^2 \in \Lambda(H_Z)$ and $\vartheta_Z^{Zc^1+c^2}$ is a canonical theta function for $L(H_Z, \chi_0)$ of characteristic 0 on $X_Z = \mathbb{C}^g/(Z\mathbb{Z}^g \oplus D\mathbb{Z}^g)$. By Lemmas 3.3.1 and 3.3.5 we have $\vartheta_Z^{Zc^1+c^2}(v) = \mathbf{e}\left(\frac{\pi}{2}\,{}^tv(\operatorname{Im} Z)^{-1}v - \pi i\,{}^tc^1c^2\right)\vartheta\left[\begin{smallmatrix} c^1 \\ c^2 \end{smallmatrix}\right](v, Z)$, and (3.19) translates to

$$\vartheta_{M(Z)}^{M(Z)c^1+c^2}(M(Z)v) = \tilde{\tau}(v)\vartheta_Z^{Zc^1+c^2}(v) \tag{3.20}$$

for some holomorphic $\tilde{\tau}: \mathbb{C}^g \to \mathbb{C}^*$ and all $v \in \mathbb{C}^g$, $\binom{c^1}{c^2} \in \Lambda_D^\perp$. But the canonical factors $a_{L(H_Z,\chi_0)}$ of $\vartheta_Z^{Zc^1+c^2}$ and $a_{L(H_{M(Z)},\chi_0')}$ of $\vartheta_{M(Z)}^{M(Z)c^1+c^2}$ are related by

$$M_Z^* a_{L(H_{M(Z)},\chi_0')} = a_{L(H_Z,\chi_0)},$$

since M_Z is the analytic representation of the isomorphism f^{-1} of Step I and $f^{-1*}L(H_{M(Z)},\chi_0') = L(H_Z,\chi_0)$. This implies that $\tilde{\tau}$ is periodic with respect to the lattice Λ_Z and thus constant.

Using Exercise 1.5.5 (3) twice as well as the equations $H_Z = M_Z^* H_{M(Z)}$ and $M_Z j_Z \binom{c^1}{c^2} = j_{M(Z)}({}^tM^{-1}\binom{c^1}{c^2}) = j_{M(Z)}\binom{d^1}{d^2}$, equation (3.20) gives

$$\vartheta_Z^{Zc^1+c^2}(v)$$

$$= \mathbf{e}\left(-\pi H_Z\left(v, j_Z\binom{c^1}{c^2}\right) - \frac{\pi}{2}H_Z\left(j_Z\binom{c^1}{c^2}, j_Z\binom{c^1}{c^2}\right)\right)\vartheta_Z^0\left(v + j_Z\binom{c^1}{c^2}\right)$$

$$= \tilde{\tau}(0)^{-1}\mathbf{e}\left(-\pi M_Z^* H_{M(Z)}\left(v, j_Z\binom{c^1}{c^2}\right) - \frac{\pi}{2}M_Z^* H_{M(Z)}\left(j_Z\binom{c^1}{c^2}, j_Z\binom{c^1}{c^2}\right)\right)$$

$$\cdot \vartheta_{M(Z)}^0\left(M_Z v + M_Z j_Z\binom{c^1}{c^2}\right)$$

$$= \tilde{\tau}(0)^{-1}\vartheta_{M(Z)}^{M(Z)d^1+d^2}(M_Z v) = \vartheta_Z^{Zd^1+d^2}(v).$$

For the last equation note that $\binom{d_1}{d^2} \in \Lambda_D^\perp$. Translating this back into terms of classical theta functions gives the assertion of Step II.

Step III: $M \in G_D(D)_0$.

The functions $\vartheta\begin{bmatrix} c^1 \\ c^2 \end{bmatrix}(\cdot, Z)$ and $\vartheta\begin{bmatrix} d^1 \\ d^2 \end{bmatrix}(\cdot, Z)$ differ only by a constant for all $\binom{c^1}{c^2} \in \Lambda_D^\perp$ according to Step II. Hence their factors of automorphy with respect to the lattice $Z\mathbb{Z}^g \oplus \mathbb{Z}^g$ coincide. This gives (see Exercise 3.3.4 (3))

$${}^tM^{-1}\binom{c^1}{c^2} = \binom{d^1}{d^2} \equiv \binom{c^1}{c^2} \bmod \mathbb{Z} \tag{3.21}$$

for all $\binom{c^1}{c^2} \in \Lambda_D^\perp = \begin{pmatrix} D^{-1} & o \\ 0 & 1_g \end{pmatrix}\mathbb{Z}^{2g}$, or equivalently

$$(1_{2g} - {}^tM^{-1}) \in M_{2g}(\mathbb{Z}) \cdot \begin{pmatrix} D & 0 \\ 0 & 1_g \end{pmatrix}.$$

Using the fact that this relation holds also for ${}^tM$ instead of ${}^tM^{-1}$, we derive that M is of the form

$$M = \begin{pmatrix} \alpha & \beta \\ \gamma & \delta \end{pmatrix} = \begin{pmatrix} \mathbf{1}_g + Da & bD \\ c & \mathbf{1}_g + dD \end{pmatrix}$$

for some $a, b, c, d \in \mathrm{M}_g(\mathbb{Z})$. On the other hand, combining (3.21) with the assertion of Step II gives

$$\vartheta \begin{bmatrix} c^1 \\ c^2 \end{bmatrix} (v, Z) = \mathbf{e}(\pi i\, {}^tc^1 c^2 - \pi i\, {}^td^1 d^2) \vartheta \begin{bmatrix} c^1 \\ c^2 \end{bmatrix} (v, Z).$$

Inserting $\binom{d^1}{d^2} = \binom{\delta c^1 - \gamma c^2}{-\beta c^1 + \alpha c^2}$, this implies

$${}^tc^1 c^2 - {}^td_1 d_2 = {}^tc^1(\mathbf{1}_g - {}^t\delta\alpha - {}^t\beta\gamma)c^2 + {}^tc^1\, {}^t\delta\beta c^1 + {}^tc^2\, {}^t\gamma\alpha c^2 \equiv 0 \bmod 2.$$

Since M is symplectic, Lemma 3.1.3 (b) gives $\mathbf{1}_g - {}^t\delta\alpha + {}^t\beta\gamma = 0$. We may replace c^1 by $D^{-1}\ell^1$ and c^2 by ℓ^2 for some $\ell^1, \ell^2 \in \mathbb{Z}^g$. Then the above equation reads

$${}^t\ell^1 D^{-1}\, {}^t\delta\beta D^{-1}\ell^1 + {}^t\ell^2\, {}^t\gamma\alpha\ell^2 \equiv {}^t(D^{-1}\, {}^t\delta\beta D^{-1})_0\ell^1 + {}^t({}^t\gamma\alpha)_0\ell^2 \equiv 0 \bmod 2$$

for all $\ell^1, \ell^2 \in \mathbb{Z}^g$. So according to Lemma 3.5.1 and equation 3.17 it remains to show that $\beta = D\widetilde{b}D$ for some $\widetilde{b} \in \mathrm{M}_g(\mathbb{Z})$. Using $\beta = bD$ and $\delta = \mathbf{1}_g + dD$ this follows from the subsequent computation

$$D^{-1}b \equiv D^{-1}b + {}^tdb \equiv D^{-1}\, {}^t(\mathbf{1}_g + dD)b = D^{-1}\, {}^t\delta\beta D^{-1} \equiv 0 \bmod \mathbb{Z}. \qquad \square$$

Proposition 3.5.6 *For $d_1 \geq 4$ the differential $\mathrm{d}\overline{\psi}_{D,Z}$ is injective at any point $Z \in \mathcal{A}_D(D)_0$.*

Proof The projection $\mathfrak{H}_g \to \mathcal{A}_D(D)_0$ is an unramified map according to the proof of Proposition 3.4.8. Hence it suffices to show that $\mathrm{d}\psi_{D,Z}$ is injective for all $Z \in \mathfrak{H}_g$.

Let $Z = (z_{ij}) \in \mathfrak{H}_g$. There is a $c \in \{c_0, \ldots, c_N\}$, a set of representatives of $\Lambda_D^\perp$ in $\Lambda_D^\perp/\Lambda_D$, such that $\vartheta \left[\begin{smallmatrix} c \\ 0 \end{smallmatrix}\right] (0, Z) \neq 0$. We have to show

$$\mathrm{rk} \left(\frac{\partial \vartheta \left[\begin{smallmatrix} c_\nu \\ 0 \end{smallmatrix}\right] / \vartheta \left[\begin{smallmatrix} c \\ 0 \end{smallmatrix}\right]}{\partial z_{11}}, \frac{\partial \vartheta \left[\begin{smallmatrix} c_\nu \\ 0 \end{smallmatrix}\right] / \vartheta \left[\begin{smallmatrix} c \\ 0 \end{smallmatrix}\right]}{\partial z_{12}}, \ldots, \frac{\partial \vartheta \left[\begin{smallmatrix} c_\nu \\ 0 \end{smallmatrix}\right] / \vartheta \left[\begin{smallmatrix} c \\ 0 \end{smallmatrix}\right]}{\partial z_{gg}} \right)_{0 \leq \nu \leq N} (0, Z)$$
$$= \frac{1}{2} g(g+1).$$

For this it suffices to show that

$$\mathrm{rk} \left(\vartheta \begin{bmatrix} c_\nu \\ 0 \end{bmatrix}, \frac{\partial \vartheta \left[\begin{smallmatrix} c_\nu \\ 0 \end{smallmatrix}\right]}{\partial z_{11}}, \frac{\partial \vartheta \left[\begin{smallmatrix} c_\nu \\ 0 \end{smallmatrix}\right]}{\partial z_{12}}, \ldots, \frac{\partial \vartheta \left[\begin{smallmatrix} c_\nu \\ 0 \end{smallmatrix}\right]}{\partial z_{gg}} \right)_{0 \leq \nu \leq N} (0, Z) = \frac{1}{2} g(g+1) + 1. \tag{3.22}$$

Assume that (3.22) does not hold. In other words, assume that the columns of the matrix in (3.22) are linearly dependent. Since the functions

$$\vartheta \begin{bmatrix} c_0 \\ 0 \end{bmatrix} (\cdot, Z), \ldots, \vartheta \begin{bmatrix} c_N \\ 0 \end{bmatrix} (\cdot, Z)$$

span the vector space $H^0(L(H_Z, \chi_0))$, this means that for all (classical theta functions) $\vartheta = \vartheta(\cdot, Z) \in H^0(L(H_Z, \chi_0))$

$$s\,\vartheta(0) = \sum_{1 \le i \le j \le g} s_{ij} \frac{\partial \vartheta}{\partial z_{ij}}(0) \tag{3.23}$$

for some constants $s, s_{ij} \in \mathbb{C}$, not all zero and not depending on ϑ.

Defining $s_{ji} = s_{ij}$ for $i < j$, we can apply the Heat Equation Proposition 3.3.7 for the matrix $S = (s_{ij})$ to get

$$4\pi i\; s\,\vartheta(0) = \sum_{i,j=1}^{g} s_{ij} \frac{\partial^2 \vartheta}{\partial v_i \partial v_j}(0) \tag{3.24}$$

for all $\vartheta \in H^0(L(H_Z, \chi_0))$.

According to Lemma 1.4.14 there is an ample line bundle $M \in \operatorname{Pic}(X_Z)$ such that

$$L(H_Z, \chi_0) \simeq M^{d_1}.$$

Choose a classical theta function $\theta \neq 0$ in $H^0(M)$ and points a_1, a_2 in its divisor (θ). Since $d_1 \geq 4$, there are $a_3, \ldots, a_{d_1} \neq (\theta)$ with $\sum_{i=1}^{d_1} a_i = 0$. According to Lemma 2.1.4 the function $\vartheta_0 := \prod_{i=1}^{d_1} \theta(\cdot + a_i)$ is a theta function for $L(H_Z, \chi_0)$. Equation (3.24) gives for $\vartheta = \vartheta_0$:

$$0 = 4\pi i s \vartheta_0(0) = 2\theta(a_3) \cdots \theta(a_{d_1}) \sum_{i,j=1}^{g} s_{ij} \frac{\partial \theta}{\partial v_i}(a_1) \frac{\partial \theta}{\partial v_j}(a_2)$$

and equivalently

$$\left(\frac{\partial \theta}{\partial v_1}(a_1), \ldots, \frac{\partial \theta}{\partial v_g}(a_1) \right) \cdot S \cdot {}^t\!\left(\frac{\partial \theta}{\partial v_1}(a_2), \ldots, \frac{\partial \theta}{\partial v_g}(a_2) \right) = 0. \tag{3.25}$$

Note that $\left(\frac{\partial \theta}{\partial v_1}(a_i), \ldots, \frac{\partial \theta}{\partial v_g}(a_i)\right) \in \mathbb{P}^{g-1}$ is just the image of the point $\overline{a}_i \in X_Z$ under the Gauss map for the divisor (θ). Now by Proposition 2.1.9, the image of the Gauss map spans $\mathbb{P}^{g-1}$. So varying a_1 and a_2 within (θ), equation (3.25) implies that $s_{ij} = 0$ for all i, j. Inserting this into (3.23), it follows that either $s = 0$ or 0 is a base point of the linear system $|L(H_Z, \chi_0)|$, a contradiction in both cases. □

Corollary 3.5.7 *The space $\mathcal{A}_D(D)_0$ is a quasi-projective variety of dimension $\frac{1}{2}g(g+1)$.*

Proof Denote for a moment by X the image of the analytic embedding $\overline{\psi}_D : \mathcal{A}_D(D)_0 \hookrightarrow \mathbb{P}^N$ of Theorem 3.5.4 and let $\overline{X}$ be its closure in $\mathbb{P}^N$ with respect to the euclidean topology. Using the theory of modular forms, it is not difficult to show that $\overline{X}$ coincides with the closure of X with respect to the Zariski topology (see Igusa [70, Theorem V.8]). So $\overline{X}$ is a projective algebraic variety by Chow's Theorem 2.1.16. Moreover, by Igusa [70, Remark page 224], the variety X is Zariski open in $\overline{X}$. This implies the assertion, since $\dim X = \dim \mathfrak{H}_g = \frac{1}{2}g(g+1)$. □

Corollary 3.5.8 *Let G be a subgroup of finite index in G_D. Then $\mathcal{A}_G := \mathfrak{H}_g/G$ is an algebraic variety. In particular, most of the moduli space occurring above are algebraic.*

Proof Since $G_D(D)_0$ is of finite index in G_D, so is the intersection of all its conjugates, say $\widetilde{G}_D(D)_0$ in G_D. Let

$$\widetilde{\mathcal{A}}_D(D)_0 = \mathfrak{H}_g/\widetilde{G}_D(D)_0.$$

Then the natural map $\widetilde{\mathcal{A}}_D(D)_0 \to \mathcal{A}_D(D)_0$ is finite. Since $\mathcal{A}_D(D)_0$ is algebraic by Corollary 3.5.7 and a finite cover of an algebraic variety, ramified at most in an algebraic set, is algebraic, $\widetilde{\mathcal{A}}_D(D)_0$ is algebraic.

Now $\widetilde{\mathcal{A}}_D(D)_0$ being normal in G_D, the quotient $G_D/\widetilde{\mathcal{A}}_D(D)_0$ is a finite group, acting on $\widetilde{\mathcal{A}}_D(D)_0$ with quotient $\mathcal{A}_D$. Since the quotient of an algebraic variety by a finite group is algebraic, this gives that $\mathcal{A}_D$ is algebraic.

Finally, as a finite covering of the algebraic variety $\mathcal{A}_D$ with at most algebraic ramification, the variety $\mathcal{A}_G$ is algebraic. □

Remark 3.5.9 The moduli space $\mathcal{A}_g := \mathcal{A}_{1_g}$ of principally polarized abelian varieties of dimension g is not compact. There are several toroidal compactifications of $\mathcal{A}_g$, the most prominent being the second Voronoi compactification $\overline{\mathcal{A}_g}$. For these compactifications (as well as for the other moduli spaces) consider the book of Ash, Mumford, Rapoport and Tai [10].

3.5.3 Exercises

(1) Show that $M = \begin{pmatrix} \alpha & \beta \\ \gamma & \delta \end{pmatrix} \in G_D(D)$ is contained in $G_D(D)_0$ if and only if $(D^{-1}\,{}^t\delta\beta D^{-1})_0 \equiv ({}^t\gamma\alpha)_0 \equiv 0 \bmod 2$.
(Hint: Replace M by M^{-1} in the proof of Lemma 3.5.1 (a).)

(2) Show that for $M = \left(\begin{smallmatrix}\alpha & \beta \\ \gamma & \delta\end{smallmatrix}\right) \in G_D(d)_0$, we have for all $\ell \in \mathbb{Z}^g$:

(i) $\delta D^{-1}\ell + \frac{1}{2}D(\gamma\, {}^t\delta)_0 \equiv D^{-1}\ell \mod \mathbb{Z}$;
(ii) $-\beta D^{-1}\ell + \frac{1}{2}(\alpha\, {}^t\beta)_0 \equiv 0 \mod \mathbb{Z}$; and
(iii) $k(M, D^{-1}\ell, 0) = {}^t\ell D^{-1}\, {}^t\delta\big(-\beta D^{-1}\ell + (\alpha\, {}^t\beta)_0\big) \equiv 0 \mod 2$.

(Hint: Use the special form of M given in (3.17) and Lemma 3.5.1 (a).)

(3) Let (X, L) be a principally polarized abelian variety. For any $x \in X_2$ denote by ϑ_x the basis of $H^0(t_x^*L)$ given in the proof of Theorem 1.5.9. Show that $\{2^*\vartheta_x \mid x \in X_2\}$ is a basis of $H^0(L^4)$.
(Hint: Use the canonical representation of the theta group $\mathcal{G}(L^4)$.)

Chapter 4
Jacobian Varieties

A curve in this chapter means a complex smooth projective curve. To every curve C one can associate in a natural way a principally polarized abelian variety, its *Jacobian* $J(C)$. As we mentioned already in the introduction, the theory of abelian varieties originated with the investigation of Jacobians. They are not only the most important, but also the best-known examples of abelian varieties. Much more can be said about them than about a general principally polarized abelian variety. In fact, presenting the theory of Jacobian varieties in a satisfactory way would require a whole volume for itself. This chapter contains, apart from the basic definitions and constructions, only some selected topics on Jacobian varieties.

Section 4.1 contains the basic definitions of the Jacobian $J(C)$ of a curve C. Moreover, it is shown that the Abel–Jacobi map $C \to J(C)$ is an embedding. Without proof we state the Abel–Jacobi Theorem, the proof of which is contained in almost every book on compact Riemann surfaces or algebraic curves. In the second section the theta divisor of a Jacobian variety is studied. We prove Poincaré's formula and Riemann's theorem and state without proof Riemann's Singularity Theorem. As a consequence we can compute the number of even and odd theta characteristics, as well as the dimension of the singularity locus of the theta divisor. Section 4.3 contains a proof of the Torelli Theorem. In Sections 4.4 and 4.5 we construct the Poincaré bundles of a curve C out of the Poincaré bundle for $J(C)$ (constructed in Section 1.4.4) and derive the universal property of the Jacobian. In the sixth section it is shown that the endomorphism ring of the Jacobian $J(C)$ may be interpreted as the ring of equivalence classes of correspondences on C, a result which is probably due to Hurwitz [68]. Proposition 4.6.12, expressing the rational trace of an endomorphism of $J(C)$ in terms of an associated correspondence of C, is due to Weil [137]. In Section 4.7 we prove an improvement of a criterion of Matsusaka [91] for a principally polarized abelian variety (in terms of a 1-cycle) to be the Jacobian of a curve, the improvement due to Ran [109]. The proof given here is due to Collino [32].

H. Lange, *Abelian Varieties over the Complex Numbers*, Grundlehren Text Editions,
https://doi.org/10.1007/978-3-031-25570-0_4

It is in general difficult to compute the period matrix of the Jacobian of a curve which is given in terms of equations. Under the supervision of F. Klein, Bolza developed in his thesis [26] a method for doing this, provided that the curve admits a sufficiently large group of automorphisms. We explain the method and give an example in Section 4.8.

We use some basic results of the theory of algebraic curves, such as the Riemann–Roch Theorem, the Plücker Formulas, and the description of finite coverings. By g^r_d we denote a linear system of dimension r and degree d on a curve C. Finally, we use Bertini's Theorem and the inequality for the dimension of the non-empty intersection of two closed subvarieties of a smooth projective variety.

4.1 Definitions and Basic Results

This section contains the definition of the Jacobian $J(C)$ of a curve C as well as the canonical principal polarization of $J(C)$. Moreover, the Abel–Jacobi map is investigated.

4.1.1 First Definition of the Jacobian

Let C be a smooth projective curve of genus g over the field of complex numbers.

Recall the g-dimensional $\mathbb{C}$-vector space $H^0(\omega_C)$ of holomorphic 1-forms on C. The homology group $H_1(C,\mathbb{Z})$ is a free abelian group of rank $2g$. For convenience we use the same letter for (topological) 1-cycles on C and their corresponding classes in $H_1(C,\mathbb{Z})$. By Stokes' theorem any element $\gamma \in H_1(C,\mathbb{Z})$ yields in a canonical way a linear form on the vector space $H^0(\omega_C)$, which we also denote by γ:

$$\gamma\colon H^0(\omega_C) \to \mathbb{C}, \qquad \omega \mapsto \int_\gamma \omega.$$

Lemma 4.1.1 *The canonical map*

$$H_1(C,\mathbb{Z}) \to H^0(\omega_C)^* = \mathrm{Hom}(H^0(\omega_C),\mathbb{C})$$

is injective.

Proof By the universal coefficient theorem the canonical map $H_1(C,\mathbb{Z}) \hookrightarrow H_1(C,\mathbb{C}) = H^1_{\mathrm{dR}}(C)^*$, $\gamma \mapsto \{\omega \mapsto \int_\gamma \omega\}$ is injective. Recall the Hodge decomposition $H^1_{\mathrm{dR}}(C) = H^0(\omega_C) \oplus \overline{H^0(\omega_C)}$. Clearly the canonical map in question is the composition

$$H_1(C,\mathbb{Z}) \longrightarrow H^1_{\mathrm{dR}}(C)^* = H^0(\omega_C)^* \oplus \overline{H^0(\omega_C)^*} \xrightarrow{\ p\ } H^0(\omega_C)^*,$$

where p denotes the projection. Since the image of any $\gamma \in H_1(C,\mathbb{Z})$ in $H^0(\omega_C)^* \oplus \overline{H^0(\omega_C)^*}$ is invariant under complex conjugation, it is necessarily of the form $l + \bar{l}$ with $l \in H^0(\omega_C)^*$. This implies the assertion. □

It follows that $H_1(C,\mathbb{Z})$ is a lattice in $H^0(\omega_C)^*$ and the quotient

$$J(C) := H^0(\omega_C)^*/H_1(C,\mathbb{Z})$$

is a complex torus of dimension g, called the *Jacobian variety* or simply the *Jacobian* of C. Note that $J(C) = 0$ for $g = 0$. In the sequel we often assume $g \geq 1$ in order to avoid trivialities.

In order to describe $J(C)$ in terms of period matrices, choose bases $\lambda_1, \ldots, \lambda_{2g}$ of $H_1(C,\mathbb{Z})$ and $\omega_1, \ldots, \omega_g$ of $H^0(\omega_C)$. Let $\ell_1, \ldots, \ell_g$ denote the basis of $H^0(\omega_C)^*$ dual to $\omega_1, \ldots, \omega_g$; that is, $\ell_i(\omega_j) = \delta_{ij}$. Considering λ_i as a linear form on $H^0(\omega_C)$ as above, we have $\lambda_i = \sum_{j=1}^{g}(\int_{\lambda_i} \omega_j)\ell_j$ for $i = 1, \ldots, 2g$. Hence the matrix

$$\Pi = \begin{pmatrix} \int_{\lambda_1} \omega_1 & \cdots & \cdots & \int_{\lambda_{2g}} \omega_1 \\ \vdots & & & \vdots \\ \int_{\lambda_1} \omega_g & \cdots & \cdots & \int_{\lambda_{2g}} \omega_g \end{pmatrix} \tag{4.1}$$

describes the periods of $J(C)$ and is therefore called the *period matrix* for $J(C)$ with respect to these bases.

4.1.2 The Canonical Polarization of $J(C)$

The complex torus $J(C)$ turns out to be an abelian variety. In fact, there is a canonical principal polarization on $J(C)$, which will be introduced now.

Fix a homology basis $\lambda_1, \ldots, \lambda_{2g}$ of $H_1(C,\mathbb{Z})$ with intersection matrix $\left(\begin{smallmatrix} 0 & -1_g \\ 1_g & 0 \end{smallmatrix}\right)$ as indicated in the following picture.

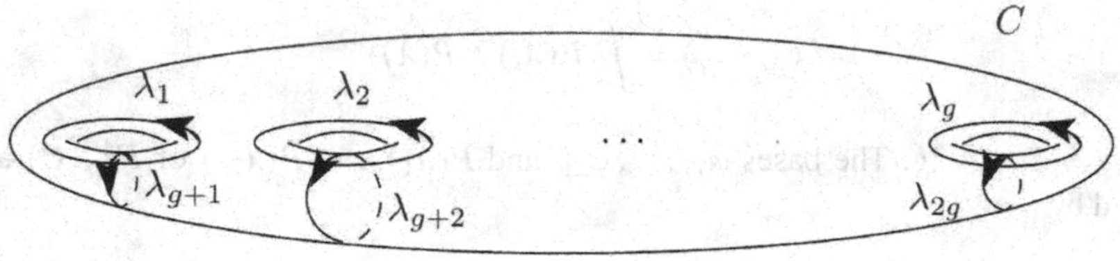

By what we have said above, $\lambda_1, \ldots, \lambda_{2g}$ is a basis of $H^0(\omega_C)^*$ considered as an $\mathbb{R}$-vector space. Denote by E the alternating form on $H^0(\omega_C)^*$ with matrix $\begin{pmatrix} 0 & \mathbf{1}_g \\ -\mathbf{1}_g & 0 \end{pmatrix}$ with respect to the basis $\lambda_1, \ldots, \lambda_{2g}$ of $H^0(\omega_C)^*$ and define

$$H: H^0(\omega_C)^* \times H^0(\omega_C)^* \to \mathbb{C} \qquad (u, v) \mapsto E(iu, v) + iE(u, v).$$

Proposition 4.1.2 *The map H is a hermitian form H defining a principal polarization on $J(C)$. It does not depend on the choice of the bases.*

The polarization H is called the *canonical polarization of* $J(C)$. Any divisor Θ on $J(C)$ such that the line bundle $O_{J(C)}(\Theta)$ defines the canonical polarization is called a *theta divisor* of the Jacobian $J(C)$. We often write $(J(C), \Theta)$ for the canonically polarized Jacobian.

Proof By definition of E it suffices to show that H is a positive definite hermitian form. According to Theorem 2.1.18 this is the case if and only if the period matrix Π of $J(C)$ (with respect to the chosen bases) satisfies the Riemann Relations:

$$\Pi \begin{pmatrix} 0 & -\mathbf{1}_g \\ \mathbf{1}_g & 0 \end{pmatrix} {}^t\Pi = 0 \quad \text{and} \quad i\Pi \begin{pmatrix} 0 & -\mathbf{1}_g \\ \mathbf{1}_g & 0 \end{pmatrix} {}^t\overline{\Pi} > 0.$$

We will check only the inequality, the proof of the equality being very similar (see Exercise 4.1.4 (1)). Recall that Π is the matrix of the $\mathbb{C}$-linear map $H^0(\omega_C)^* \to H^1_{\text{dR}}(C)^*$ with respect to the bases $\ell_1, \ldots, \ell_g$ of $H^0(\omega_C)^*$, dual to $\omega_1, \ldots, \omega_g$, and $\lambda_1, \ldots, \lambda_{2g}$ of $H^1_{\text{dR}}(C)^* = H_1(C, \mathbb{C})$. Let $\varphi_1, \ldots, \varphi_{2g}$ denote the basis of $H^1_{\text{dR}}(C)$, dual to $\lambda_1, \ldots, \lambda_{2g}$. The dual map $H^1_{\text{dR}}(C) \to H^0(\omega_C)$ is given by ${}^t\Pi$ with respect to the bases $\varphi_1, \ldots, \varphi_{2g}$ and $\omega_1, \ldots, \omega_g$, so

$$\omega_s = \sum_{t=1}^{2g} \left(\int_{\lambda_t} \omega_s \right) \varphi_t \quad \text{for } s = 1, \ldots, g.$$

Recall the well-known fact (see Griffiths–Harris [55, p. 59]) that the intersection product in $H_1(C, \mathbb{Z})$ is Poincaré dual to the cup product in $H^1_{\text{dR}}(C)$. Denoting by $P: H_1(C, \mathbb{C}) \to H^1_{\text{dR}}(C)$ the Poincaré duality isomorphism, this means

$$(\lambda_i \cdot \lambda_j) = \int_C P(\lambda_i) \wedge P(\lambda_j)$$

for $i, j = 1, \ldots, 2g$. The bases $\varphi_1, \ldots, \varphi_{2g}$ and $P(\lambda_1), \ldots, P(\lambda_{2g})$ of $H^1_{\text{dR}}(C)$ are related by

$$P(\lambda_j) = \sum_{i=1}^{2g} \begin{pmatrix} 0 & -\mathbf{1}_g \\ \mathbf{1}_g & 0 \end{pmatrix}_{ij} \varphi_i \qquad \text{for} \quad i, j = 1, \ldots, 2g.$$

To see this, write $P(\lambda_j) = \sum_{i=1}^{2g} a_{ij}\varphi_i$. Then using the definition of the Poincaré duality,

$$a_{ij} = \sum_{\nu=1}^{2g} a_{\nu j} \int_{\lambda_i} \varphi_\nu = \int_{\lambda_i} P(\lambda_j) = \int_C P(\lambda_i) \wedge P(\lambda_j) = (\lambda_i \cdot \lambda_j) = \begin{pmatrix} 0 & -\mathbf{1}_g \\ \mathbf{1}_g & 0 \end{pmatrix}_{ij}.$$

Now an immediate matrix computation gives

$$\int_C \varphi_s \wedge \varphi_t = \begin{pmatrix} 0 & -\mathbf{1}_g \\ \mathbf{1}_g & 0 \end{pmatrix}_{st} \quad \text{for } s, t = 1, \ldots, 2g. \tag{4.2}$$

Using this we get

$$\begin{aligned} i \int_C \omega_\mu \wedge \overline{\omega}_\nu &= i \sum_{s,t=1}^{2g} \left(\int_{\lambda_s} \omega_\mu\right)\left(\int_{\lambda_t} \overline{\omega}_\nu\right) \int_C \varphi_s \wedge \varphi_t \\ &= i \sum_{s,t=1}^{2g} \Pi_{\mu s} \begin{pmatrix} 0 & -\mathbf{1}_g \\ \mathbf{1}_g & 0 \end{pmatrix}_{st} \overline{\Pi}_{\nu t} = i \left(\Pi \begin{pmatrix} 0 & -\mathbf{1}_g \\ \mathbf{1}_g & 0 \end{pmatrix} {}^t\overline{\Pi}\right)_{\mu\nu} \end{aligned}$$

for $\mu, \nu = 1, \ldots, g$. So for any holomorphic 1-form $\omega = \sum \alpha_\mu \omega_\mu$

$$i \int \omega \wedge \overline{\omega} = i(\alpha_1, \ldots, \alpha_g)\Pi \begin{pmatrix} 0 & -\mathbf{1}_g \\ \mathbf{1}_g & 0 \end{pmatrix} {}^t\overline{\Pi} \begin{pmatrix} \overline{\alpha}_1 \\ \vdots \\ \overline{\alpha}_g \end{pmatrix}.$$

Since $i \int \omega \wedge \overline{\omega} > 0$ for every $\omega \neq 0$, this implies that $i\Pi \begin{pmatrix} 0 & -\mathbf{1}_g \\ \mathbf{1}_g & 0 \end{pmatrix} {}^t\overline{\Pi}$ is positive definite.

The last assertion of the proposition follows from the fact that changing the bases leads to equivalent hermitian and alternating forms. □

Corollary 4.1.3 *If A is the matrix of the intersection product of cycles on C with respect to some basis of $H_1(C, \mathbb{Z})$, then A^{-1} is the matrix of the alternating form defining the canonical polarization with respect to the same basis.*

Proof This is an immediate consequence of the proof of Proposition 4.1.2, using the fact that if the cycles $\lambda_1, \ldots, \lambda_{2g}$ form a symplectic basis of the lattice $H_1(C, \mathbb{Z})$ in $H^0(\omega_C)^*$ for H, then $\begin{pmatrix} 0 & -\mathbf{1}_g \\ \mathbf{1}_g & 0 \end{pmatrix}^{-1} = \begin{pmatrix} 0 & \mathbf{1}_g \\ -\mathbf{1}_g & 0 \end{pmatrix} = {}^t\begin{pmatrix} 0 & -\mathbf{1}_g \\ \mathbf{1}_g & 0 \end{pmatrix}$. □

4.1.3 The Abel–Jacobi Map

There is another approach to defining the Jacobian variety of C, namely as the group $\mathrm{Pic}^0(C)$ of line bundles of degree zero on C using the Abel–Jacobi Theorem, which will be explained now.

Recall that $\mathrm{Pic}^0(C)$ is the quotient of the group $\mathrm{Div}^0(C)$ of divisors of degree zero on C modulo the subgroup of principal divisors. Define a canonical map

$$\mathrm{Div}^0(C) \longrightarrow J(C) = H^0(\omega_C)^*/H_1(C,\mathbb{Z})$$

as follows: any divisor $D \in \mathrm{Div}^0(C)$ can be written as a finite sum $D = \sum_{\nu=1}^N (p_\nu - q_\nu)$ for some points $p_\nu, q_\nu \in C$. The class of the linear form $\omega \mapsto \sum_{\nu=1}^N \int_{q_\nu}^{p_\nu} \omega$ in $H^0(\omega_C)^*/H_1(C,\mathbb{Z})$ depends only on the divisor D, but not on its special representation as a sum of differences of points. So

$$D \mapsto \left\{\omega \mapsto \sum_{\nu=1}^N \int_{q_\nu}^{p_\nu} \omega\right\} \bmod H_1(C,\mathbb{Z})$$

gives a well-defined map $\mathrm{Div}^0(C) \longrightarrow J(C)$. Obviously this is a homomorphism of groups. It is called the *Abel–Jacobi map* of C.

Jacobi's Inversion Theorem states that the Abel–Jacobi map is surjective (see Griffiths–Harris [55, p. 235]). On the other hand, by a theorem of Abel its kernel is the subgroup of principal divisors in $\mathrm{Div}^0(C)$ (see Griffith-Harris [55, p. 235]). Combining both theorems we obtain the following result.

Theorem 4.1.4 (Abel–Jacobi Theorem) *The Abel–Jacobi map induces a canonical isomorphism*

$$\mathrm{Pic}^0(C) \xrightarrow{\sim} J(C).$$

Via this isomorphism $\mathrm{Pic}^0(C)$ inherits the structure of a principally polarized abelian variety. For the proof of the Abel–Jacobi Theorem we refer to the standard books on Riemann surfaces or algebraic curves. In the sequel we identify

$$J(C) = \mathrm{Pic}^0(C)$$

via the canonical isomorphism. We work with both interpretations without further notice: sometimes we consider elements of $J(C)$ as points and write $+$ for the group law and sometimes as line bundles on C and write $\otimes$ for the group law. The respective meaning will be clear from the context.

For any $n \in \mathbb{Z}$ denote by $\mathrm{Div}^n(C)$ the set of divisors of degree n on C. It is a principal homogeneous space for the group $\mathrm{Div}^0(C)$. This suggests that one can define more generally an Abel–Jacobi map from $\mathrm{Div}^n(C)$ to $J(C)$. Certainly, for $n \neq 0$ this map will not be canonical, but depends on the choice of a divisor in $\mathrm{Div}^n(C)$. To be more precise, fix a divisor $D_n \in \mathrm{Div}^n(C)$ and define

$$\mathrm{Div}^n(C) \longrightarrow J(C)\,,\; D \mapsto \mathcal{O}_C(D - D_n). \tag{4.3}$$

The most important case is $D_n = nc$ for some point $c \in C$. In this case the map $\mathrm{Div}^n(C) \to J(C)$ can also be written as

$$D = \sum r_\nu p_\nu \mapsto \left\{\omega \mapsto \sum r_\nu \int_c^{p_\nu} \omega\right\} \bmod H_1(C, \mathbb{Z}).$$

For $n \geq 1$ let $C^{(n)}$ denote the n-fold symmetric product of C. Recall that $C^{(n)}$ is the quotient of the cartesian product C^n by the natural action of the symmetric group of degree n. As such it is a smooth projective variety of dimension n (see Griffiths–Harris [55, p. 236]). The elements of $C^{(n)}$ can be considered as effective divisors of degree n on C. In this way $C^{(n)}$ is a subset of $\mathrm{Div}^n(C)$ and we denote the restriction to $C^{(n)}$ of the map (4.3) by

$$\alpha_{D_n} : C^{(n)} \to J(C).$$

The map α_{D_n} is also called the *Abel–Jacobi map*.

Let $\mathrm{Pic}^n(C)$ denote the set of line bundles of degree n on C. It is a principal homogeneous space for the group $\mathrm{Pic}^0(C)$: given a line bundle L_n of degree n on C, the map

$$\alpha_{L_n} : \mathrm{Pic}^n(C) \to J(C)\,,\; L \mapsto L \otimes L_n^{-1}$$

is bijective. Finally, consider the canonical map $\rho \colon C^{(n)} \to \mathrm{Pic}^n(C)$ sending an effective divisor D in $C^{(n)}$ to its class $\mathcal{O}_C(D)$ in $\mathrm{Pic}^n(C)$. These maps fit into the following commutative diagram

$$\begin{array}{ccc} & & \mathrm{Pic}^n(C) \\ & \overset{\rho}{\nearrow} & \\ C^{(n)} & & \downarrow \alpha_{\mathcal{O}(D_n)} \\ & \underset{\alpha_{D_n}}{\searrow} & \\ & & J(C). \end{array} \tag{4.4}$$

For any line bundle $L \in \mathrm{Pic}^n(C)$ the fibre $\rho^{-1}(L)$ is by definition the complete linear system $|L|$ on C. This implies the following lemma.

Lemma 4.1.5 *For any $M \in \mathrm{Pic}^0(C)$ the fibre $\alpha_{D_n}^{-1}(M)$ is the complete linear system $|M \otimes \mathcal{O}_C(D_n)|$.*

Suppose now $g \geq 1$ and fix a point $c \in C$. The Abel–Jacobi map

$$\alpha = \alpha_c : C \to J = J(C)$$

is of special interest. In order to show that α is an embedding, we first study its differential $d\alpha$. Recall that $d\alpha$ is a holomorphic map from the tangent bundle $\mathcal{T}_C$ of C to the tangent bundle $\mathcal{T}_J$ of J. According to Lemma 1.1.22 the tangent bundle of J is trivial: $\mathcal{T}_J = J \times \mathbb{C}^g$. The projectivization of the composed map $\mathcal{T}_C \to \mathcal{T}_J \simeq J \times \mathbb{C}^g \to \mathbb{C}^g$ is a priori a rational map $C \to \mathbb{P}^{g-1}$ called the *projectivized differential* of α.

Proposition 4.1.6 *The projectivized differential of the Abel–Jacobi map $\alpha: C \to J$ is the canonical map $\varphi_{\omega_C}: C \to \mathbb{P}^{g-1}$.*

Proof For $x \in J$ consider the tangent space $T_x J$ at the point x. The canonical isomorphisms $T_x J = T_0 J = H^0(\omega_C)^*$ yield an isomorphism $\mathcal{T}_J = J \times H^0(\omega_C)^*$. Choose a basis $\omega_1, \dots, \omega_g$ of $H^0(\omega_C)$ and identify $H^0(\omega_C)^* = \mathbb{C}^g$. Then the Abel–Jacobi map $\alpha: C \to \mathbb{C}^g / H_1(C, \mathbb{Z})$ is given by $\alpha(p) = {}^t(\int_c^p \omega_1, \dots, \int_c^p \omega_g)$ mod $H_1(C, \mathbb{Z})$. Hence by the fundamental theorem of calculus, the projectivization of the composed map $\mathcal{T}_C \xrightarrow{d\alpha} \mathcal{T}_J \simeq J \times \mathbb{C}^g \to \mathbb{C}^g$ is given by $p \mapsto (\omega_1(p) : \cdots : \omega_g(p))$. But this is just the canonical map. □

Corollary 4.1.7 *For any $g \geq 1$ the Abel–Jacobi map $\alpha: C \to J(C)$ is an embedding.*

Proof The map α is injective, since for every line bundle L of degree 1 on a curve of genus ≥ 1 we have $h^0(L) \leq 1$. From Proposition 4.1.6 we conclude that the differential of α is injective at every point $p \in C$, the canonical line bundle ω_C on C being base point free. □

As a second corollary we get a statement which will be applied later.

Corollary 4.1.8 *Suppose C is a curve of genus 2. Let x and y be different points on $J = J(C)$ and $t \in T_x J$ a tangent vector.*

(a) *There is a point $z \in J$ such that the translated curve $t_z^* \alpha(C)$ passes through x and y.*
(b) *There is a point $z \in J$ such that the translated curve $t_z^* \alpha(C)$ passes through x and t is tangential to $t_z^* \alpha(C)$ at x.*

In fact, one can be more precise (see Exercise 4.1.4 (2)): there are exactly two translates of $\alpha(C)$ passing through x and y.

Proof (a): let ι denote the hyperelliptic involution of C. The map $C^{(2)} \to J, (p, q) \mapsto O_J(p - \iota q)$ is surjective. Hence there are points p and q on C such that $x - y = O_J(p - \iota q)$. Then the point $z = \alpha(p) - x$ satisfies the assertion: $x = \alpha(p) - z$ and $y = \alpha(\iota q) - z$.

(b): according to Proposition 4.1.6 the projectivized differential of $\alpha: C \to J$ is a double covering $C \to \mathbb{P}_1$. Hence there is a point $p \in C$ such that t is tangent to C in $\alpha(p)$. Then the point $z = \alpha(p) - x$ satisfies the assertion. □

4.1.4 Exercises

(1) Give a proof of the equality of the Riemann Relations in the proof of Proposition 4.1.2.

(2) Let C be a curve of genus 2 and $\alpha = \alpha_c: C \to J(C)$ the Abel–Jacobi embedding with respect to the point $c \in C$.

(a) Show that for any distinct points x and y in $J(C)$ there are exactly two translates of $\alpha(C)$ passing through x and y.
(b) For any $x \in J$ and any tangent vector $t \neq 0$ of $J(C)$ at x, there are either one or two translates of $\alpha(C)$ passing through x and touching t. There are exactly 6 tangent directions such that there is only one such translate.

(3) (*Dual Jacobian Variety*) Let C be a smooth projective curve of genus g.

(a) Show that the composed map

$$H^1(C,\mathbb{Z}) \hookrightarrow H^1(C,\mathbb{C}) = H^0(\omega_C) \oplus \overline{H^0(\omega_C)} \to \overline{H^0(\omega_C)}$$

is injective. So $\overline{H^0(\omega_C)}/H^1(C,\mathbb{Z})$ is a complex torus.
Show that $\overline{H^0(\omega_C)}/H^1(C,\mathbb{Z}) = \widehat{J(C)}$, the dual Jacobian variety.
(b) The canonical principal polarization on $\widehat{J(C)}$ is given by the hermitian form

$$\overline{H^0(\omega_C)} \times \overline{H^0(\omega_C)} \to \mathbb{C}, \quad (\omega_1,\omega_2) \mapsto 2i \int_C \omega_1 \wedge \overline{\omega_2}.$$

(4) Let C be a smooth projective curve of genus g and $\rho : C^{(g)} \to \operatorname{Pic}^g(C)$ the canonical map of diagram (4.4).
(a) If $g = 2$, then $\rho : C^{(2)} \to \operatorname{Pic}^2(C)$ is the blow up of $\operatorname{Pic}^2(C)$ in the canonical point $\omega_C \in \operatorname{Pic}^2(C)$.
(b) If $g = 3$, then $\rho : C^{(3)} \to \operatorname{Pic}^3(C)$ is the blow up of $\operatorname{Pic}^3(C)$ along the curve $-C + \omega_C = \{\omega_C(-p) \mid p \in C\} \subset \operatorname{Pic}^3(C)$.

(5) Show that for a general smooth projective curve C of genus g we have $\operatorname{End} J(C) \simeq \mathbb{Z}$.
(See Koizumi [76].)

4.2 The Theta Divisor

Let (J,Θ) be the Jacobian of a smooth projective curve C of genus $g \geq 1$. In this section we study the geometry of the theta divisors Θ. In particular we will see that some properties of Θ reflect geometrical properties of the curve itself.

4.2.1 Poincaré's Formula

This section contains a proof of Poincaré's formula which relates the homology classes of the subset of $\operatorname{Pic}^n(C)$ of line bundles with non-empty linear system and an intersection power of Θ.

As we saw in the last section, the varieties $\mathrm{Pic}^n(C)$ are principal homogenous spaces for $J = \mathrm{Pic}^0(C)$. We will see next that there is an intrinsic way of defining a theta divisor in $\mathrm{Pic}^{g-1}(C)$. Recall the canonical map $\rho\colon C^{(n)} \longrightarrow \mathrm{Pic}^n(C)$ for $n \geq 1$. Its image

$$W_n := \rho\big(C^{(n)}\big) \subseteq \mathrm{Pic}^n(C)$$

is the subset of $\mathrm{Pic}^n(C)$ of line bundles with non-empty linear system.

Lemma 4.2.1

(i) *For* $n \geq g$, $W_n = \mathrm{Pic}^n(C)$ *and for* $n \geq g+1$, $\rho\colon C^{(n)} \longrightarrow \mathrm{Pic}^n(C)$ *is a* $\mathbb{P}^{n-g}$*-bundle.*
(ii) *For* $n \leq g-1$, W_n *is irreducible and closed in* $\mathrm{Pic}^n(C)$ *and the map* $\rho\colon C^{(n)} \rightarrow W_n$ *is birational.*
(iii) *For* $n = g-1$, *the image* W_{g-1} *is a theta divisor in* $\mathrm{Pic}^{g-1}(C)$.

Proof (i): For $n \geq g$ Riemann–Roch for curves implies $h^0(L) = n - g + 1$ for all line bundles of degree n. This implies the assertion.

(ii) and (iii): For $1 \leq n \leq g-1$ it is well known (see Griffith-Harris [55, p. 338]) that $h^0(\mathcal{O}_C(D)) = 1$ for a general divisor $D \in C^{(n)}$. This means that the map $\rho\colon C^{(n)} \longrightarrow \mathrm{Pic}^n(C)$ is birational onto its image W_n. Since moreover ρ is a proper morphism, W_n is an irreducible closed subvariety of $\mathrm{Pic}^n(C)$ of dimension n. In particular, W_{g-1} is a divisor in $\mathrm{Pic}^{g-1}(C)$ and hence a theta divisor, intrinsically defined. □

We want to study the relation between the varieties W_n and the theta divisor Θ: fix a point $c \in C$ and denote by $\widetilde{W}_n$ the image of W_n in J under the bijection $\alpha_{\mathcal{O}(nc)}\colon \mathrm{Pic}^n(C) \to J$,

$$\widetilde{W}_n = \alpha_{\mathcal{O}_C(nc)}(W_n) = \alpha_{nc}\big(C^{(n)}\big) \subseteq J.$$

The next theorem compares the fundamental classes of Θ and $\widetilde{W}_n$. For this recall that the *fundamental class* $[Y]$ of an n-dimensional subvariety Y of a smooth projective variety X of dimension g is by definition the element in $H^{2g-2n}(X, \mathbb{Z})$, Poincaré dual to the homology class $\{Y\}$ of Y in $H_{2n}(X, \mathbb{Z})$.

Theorem 4.2.2 (Poincaré's Formula) $[\widetilde{W}_n] = \frac{1}{(g-n)!} \bigwedge^{g-n}[\Theta]$ *for any* $1 \leq n \leq g$.

Notice that the formula also holds for $n = 0$, if we define $\widetilde{W}_0$ to be a point. By definition

$$[\Theta] = c_1(\Theta),$$

the first Chern class of Θ (see Griffiths–Harris [55, p. 141]). Thus $\bigwedge^g[\Theta]$ equals the intersection number (Θ^g) times the class of a point (see Section 1.7.2). So for $n = 0$ the formula is equivalent to $(\Theta^g) = g!$, which is a consequence of Riemann–Roch.

Proof We will prove the Poincaré dual of the above formula, namely

$$\{\widetilde{W}_n\} = \tfrac{1}{(g-n)!}\{\Theta\}^{g-n}, \tag{4.5}$$

using the Pontryagin product $\star$.

First we claim $\{\widetilde{W}_n\} = \frac{1}{n!} \star_{i=1}^{n} \{\widetilde{W}_1\}$ for all $1 \leq n \leq g$.

The proof proceeds by induction on n. For $n = 1$ there is nothing to show. So suppose the formula is valid for some $n \geq 1$. Denote by $p \colon C^{(n)} \times C \to C^{(n+1)}$ the natural map and by μ the addition map. Then the commutativity of the diagram

$$\begin{array}{ccc} C^{(n)} \times C & \xrightarrow{\alpha_{nc} \times \alpha_c} & J \times J \\ \Big\downarrow p & & \Big\downarrow \mu \\ C^{(n+1)} & \xrightarrow{\alpha_{(n+1)c}} & J \end{array}$$

and the induction hypothesis imply

$$\begin{aligned} \{\widetilde{W}_{n+1}\} &= \frac{1}{n+1}\,(\alpha_{(n+1)c})_* p_* \{C^{(n)} \times C\} \\ &= \frac{1}{n+1}\,\mu_*(\alpha_{nc} \times \alpha_c)_*\{C^{(n)} \times C\} \\ &= \frac{1}{n+1}\,\mu_*\big(\alpha_{nc*}\{C^{(n)}\} \times \alpha_{c*}\{C\}\big) \\ &= \frac{1}{n+1}\,\{\widetilde{W}_n\} \star \{\widetilde{W}_1\} = \frac{1}{(n+1)!} \star_{i=1}^{n+1} \{\widetilde{W}_1\}. \end{aligned}$$

This completes the proof of the claim.

Now suppose $\lambda_1, \ldots, \lambda_{2g}$ is a symplectic basis of $H_1(C,\mathbb{Z}) = H_1(J,\mathbb{Z})$ with corresponding real coordinate functions $x_1, \ldots, x_{2g}$ of $H^0(\omega_C)^*$. By construction, the bases $\mathrm{d}x_1, \ldots, \mathrm{d}x_{2g}$ of $H^1_{\mathrm{dR}}(J)$ and $\lambda_1, \ldots, \lambda_{2g}$ of $H_1(J,\mathbb{Z})$ are dual to each other. So under the identification $H_1(C,\mathbb{Z}) = H_1(J,\mathbb{Z})$ also the basis $\alpha_c^*\mathrm{d}x_1, \ldots, \alpha_c^*\mathrm{d}x_{2g}$ of $H^1_{\mathrm{dR}}(C)$ is dual to $\lambda_1, \ldots, \lambda_{2g}$. Arguing as in the proof of Proposition 4.1.2 (see equation (4.2)) we obtain

$$\int_{\widetilde{W}_1} \mathrm{d}x_i \wedge \mathrm{d}x_j = \int_C \alpha_c^*\mathrm{d}x_i \wedge \alpha_c^*\mathrm{d}x_j = (\lambda_i \cdot \lambda_j) = -\delta_{g+i,j} = \int_{-\Sigma_\nu \lambda_\nu \star \lambda_{g+\nu}} \mathrm{d}x_i \wedge \mathrm{d}x_j.$$

For the last equation we used the fact that the basis $\{\mathrm{d}x_i \wedge \mathrm{d}x_j \mid i < j\}$ of $H^2(J,\mathbb{Z})$ is dual to the basis $\{\lambda_i \star \lambda_j | i < j\}$ of $H_2(J,\mathbb{Z})$ according to Lemma 2.5.12 and Proposition 2.5.13. This implies $\{\widetilde{W}_1\} = -\sum_{\nu=1}^{g} \lambda_\nu \star \lambda_{g+\nu}$. Hence using Theorem 2.5.16,

$$\{W_n\} = \tfrac{1}{n!} \star_{i=1}^{n} \left(- \sum_{\nu=1}^{g} \lambda_\nu \star \lambda_{g+\nu} \right) = \tfrac{1}{(g-n)!}\{\Theta\}^{g-n}. \qquad \square$$

We want to work out Poincaré's Formula in the cases $n = 1$ and $g - 1$. According to Corollary 4.1.7 the Abel–Jacobi map $\alpha\colon C \to J$ is an embedding. Identifying C with its image $\widetilde{W}_1 = \alpha(C)$, Poincaré's formula for $n = 1$ gives

Corollary 4.2.3 $(C \cdot \Theta) = g$.

Proof According to Poincaré's Formula we have

$$[C] \wedge [\Theta] = \frac{1}{(g-1)!} \bigwedge^{g-1} [\Theta] \wedge [\Theta] = \tfrac{1}{(g-1)!} \bigwedge^{g} [\Theta].$$

So $(C \cdot \Theta) = \frac{1}{(g-1)!}(\Theta^g) = g$ by Riemann–Roch and the definition of the intersection number. □

In case $n = g - 1$ we get:

Corollary 4.2.4 *There is an* $\eta \in \operatorname{Pic}^{g-1}(C)$ *such that* $W_{g-1} = \alpha_\eta^* \Theta$.

Proof By Poincaré's Formula $[\widetilde{W}_{g-1}] = [\Theta]$, so the first Chern classes of the corresponding line bundles coincide. Hence according to Corollary 1.4.13 there is an $x \in J = \operatorname{Pic}^0(C)$ such that $\widetilde{W}_{g-1} = t_x^* \Theta$. This implies

$$W_{g-1} = \alpha^*_{\mathcal{O}\left((g-1)c\right)} \widetilde{W}_{g-1} = \alpha^*_\eta \Theta$$

with $\eta = \mathcal{O}\big((g-1)c\big) \otimes x^{-1}$. □

4.2.2 Riemann's Theorem

There is no canonical way to distinguish a theta divisor Θ in $J(C)$ within its algebraic equivalence class. On the other hand, the divisor W_{g-1} in $\operatorname{Pic}^{g-1}(C)$ is intrinsic. So the line bundle η in Corollary 4.2.4 depends on the choice of Θ. If Θ is one of the 2^{2g} symmetric theta divisors (see Lemma 2.3.5), then we can say more about the line bundle η. For this recall that a *theta characteristic* on C is a line bundle κ on C with $\kappa^2 = \omega_C$.

Theorem 4.2.5 (Riemann's Theorem) *For any symmetric theta divisor* Θ *there is a theta characteristic* κ *on* C *such that*

$$W_{g-1} = \alpha_\kappa^* \Theta.$$

In classical terminology Riemann's Theorem reads:

$$W_{g-1} - \kappa = \Theta.$$

In view of Riemann's Theorem, it makes sense to call W_{g-1} the *canonical theta divisor* of C. Note that in particular the theta divisor Θ corresponding to a symplectic basis is symmetric. To be more precise, let $\lambda_1, \dots, \lambda_{2g}$ be a symplectic basis for the canonical polarization and L_0 the corresponding line bundle of characteristic zero on J. Then the unique divisor Θ in the linear system $|L_0|$ is symmetric. In fact, Θ may be considered as the zero divisor of the classical Riemann theta function. In this case the theta characteristic κ in the theorem is called *Riemann's constant*.

Proof We have to show that $\kappa^2 = \omega_C$. For this consider the involution ι on $\mathrm{Pic}^{g-1}(C)$, sending a line bundle L on C to $\omega_C \otimes L^{-1}$. Since by Riemann–Roch and Serre duality $h^0(L) = h^0(\omega_C \otimes L^{-1})$ for every $L \in \mathrm{Pic}^{g-1}(C)$, the divisor W_{g-1} is invariant under ι: $\iota^* W_{g-1} \simeq W_{g-1}$. Using this and the fact that Θ is symmetric, we have

$$\iota^* \alpha_\kappa^* (-1)^* \Theta \simeq \iota^* \alpha_\kappa^* \Theta \simeq \iota^* W_{g-1} \simeq W_{g-1} \simeq \alpha_\kappa^* \Theta.$$

Now $(-1)\alpha_\kappa \iota = \alpha_{\omega_C \otimes \kappa^{-1}}$ yields $\alpha_\kappa^* \Theta \simeq \alpha^*_{\omega_C \otimes \kappa^{-1}} \Theta$. This implies the assertion, since Θ defines a principal polarization. □

Consider again the canonical map $\rho \colon C^{(g-1)} \to W_{g-1} \subset \mathrm{Pic}^{g-1}(C)$, $D \mapsto \mathcal{O}_C(D)$. It blows down the whole linear system $|D| \subset C^{(g-1)}$ to the point $\mathcal{O}_C(D)$ of W_{g-1}. This suggests that for positive-dimensional linear systems the corresponding point is a singular point of W_{g-1}. In fact, this is a consequence of the following theorem.

Theorem 4.2.6 (Riemann's Singularity Theorem) *For every $L \in \mathrm{Pic}^{(g-1)}(C)$*

$$\mathrm{mult}_L\, W_{g-1} = h^0(L).$$

For the proof, which we do not repeat here, we refer to Arbarello et al [8]. It belongs more to curve theory. However, we want to give some applications.

4.2.3 Theta Characteristics

Riemann's Theorem 4.2.5 implies in particular that there is a theta characteristic κ on C. Using this κ we have:

Lemma 4.2.7 *The map $\alpha_\kappa \colon \mathrm{Pic}^{g-1}(C) \to J$, $L \mapsto L \otimes \kappa^{-1}$ induces a bijection between the set of theta characteristics on C and the set of 2-division points J_2 of J. In particular the curve C admits exactly 2^{2g} theta characteristics.*

Proof In fact, $x = \alpha_\kappa(\eta) = \eta \otimes \kappa^{-1}$ is a 2-division point if and only if $\eta^2 = \omega_C$. □

A theta characteristic η is called *even* (respectively *odd*) if $h^0(\eta) \equiv 0 \pmod 2$ (respectively $h^0(\eta) \equiv 1 \bmod 2$).

Proposition 4.2.8 *Let Θ be a symmetric theta divisor on J and κ the theta characteristic with $W_{g-1} = \alpha_\kappa^* \Theta$. Then κ is even (respectively odd) if and only if Θ is even (respectively odd).*

In particular Riemann's constant is an even theta characteristic, since the classical Riemann theta function with characteristic $\begin{bmatrix}0\\0\end{bmatrix}$ is even.

Proof Recall from Section 2.3.4 that Θ is even (respectively odd) if $\operatorname{mult}_0 \Theta \equiv 0 \bmod 2$ (respectively $\operatorname{mult}_0 \Theta \equiv 1 \bmod 2$). By Riemann's Theorem 4.2.5, α_κ induces a biholomorphic map $W_{g-1} \to \Theta$. So Riemann's Singularity Theorem 4.2.6 gives $h^0(\kappa) = \operatorname{mult}_{\alpha_\kappa(\kappa)}(\Theta) = \operatorname{mult}_0(\Theta)$ which implies the assertion. □

Fix an even symmetric theta divisor Θ and let κ be the corresponding theta characteristic. As in the proof of Proposition 4.2.8 one has $h^0(\eta) = \operatorname{mult}_{\alpha_\kappa(\eta)}(\Theta)$ for every theta characteristic η. So Proposition 2.3.15 implies:

Proposition 4.2.9 *The curve C admits exactly* $2^{g-1}(2^g+1)$ *(respectively* $2^{g-1}(2^g-1)$*) even (respectively odd) theta characteristics.*

4.2.4 The Singularity Locus of Θ

The singularity locus $\operatorname{sing}\Theta$ of a theta divisor Θ on the Jacobian $J(C)$ is a closed algebraic subset. Using Riemann's Singularity Theorem 4.2.6 one can compute its dimension. For this we assume $g \geq 4$, the cases of smaller genus being trivial.

Proposition 4.2.10
$$\dim \operatorname{sing}\Theta = \begin{cases} g-4 & \text{if } C \text{ is not hyperelliptic,}\\ g-3 & \text{if } C \text{ is hyperelliptic.}\end{cases}$$

For the proof we need some preliminaries: Let $\varphi_{\omega_C} : C \to \mathbb{P}^{g-1}$ denote the canonical map, that is, the map given by the canonical line bundle ω_C on C. For any effective divisor D on C denote by $\overline{\varphi_{\omega_C}(D)}$ the intersection of the hyperplanes $H \subset \mathbb{P}^{g-1}$ such that either $\varphi_{\omega_C}(C) \subset H$ or $D \leq \varphi^*_{\omega_C}(H)$. If D is a sum of different points, then $\overline{\varphi_{\omega_C}(D)}$ is just the usual span of these points. In general one has to take the appropriate osculating spaces into account.

Then the following lemma is called the *Geometric version of the Riemann–Roch Theorem* for curves (see Arbarello et al [8, p. 12]).

Lemma 4.2.11 *For any effective divisor D on C,*

$$h^0(\mathcal{O}_C(D)) = \deg D - \dim \overline{\varphi_{\omega_C}(D)}. \tag{4.6}$$

Moreover, we need the classical general position theorem as stated in Arbarello et al [8, p. 109].

Theorem 4.2.12 (General Position Theorem) *Let* $C \subset \mathbb{P}^r$, $r \geq 2$, *be an irreducible non-degenerate, possibly singular, curve of degree d. Then a general hyperplane meets C in d points, any r of which are linear independent.*

Proof (of Proposition 4.2.10) According to Riemann's Theorem 4.2.5 we have to compute $\dim \operatorname{sing} W_{g-1}$. By Riemann's Singularity Theorem 4.2.6 a line bundle L is contained in $\operatorname{sing} W_{g-1}$ if and only if $h^0(L) \geq 2$.

Suppose first that C is hyperelliptic. Let ι denote the hyperelliptic involution on C and $L_0 = \mathcal{O}_C(p + \iota p)$ the unique line bundle on C with $h^0(L_0) = \deg L_0 = 2$. Consider the map

$$\phi\colon C^{(g-3)} \to \operatorname{sing} W_{g-1}, \quad p_1 + \cdots + p_{g-3} \mapsto \mathcal{O}_C(p_1 + \cdots + p_{g-3}) \otimes L_0.$$

It suffices to show that ϕ is birational and surjective.

In order to show that ϕ is surjective, suppose $\mathcal{O}_C(D) \in \operatorname{sing} W_{g-1}$. Then $2 \leq h^0(\mathcal{O}_C(D)) = g - 1 - \dim \overline{\varphi_{\omega_C}(D)}$ implies $\dim \overline{\varphi_{\omega_C}(D)} \leq g-3$. Using the well-known facts that $\varphi_{\omega_C}(C)$ is the rational normal curve in $\mathbb{P}_{g-1}$ (see Hartshorne [61, p. 343]) and that any $g-1$ pairwise different points on it span a $\mathbb{P}_{g-2}$, we conclude that D is of the form $D = p_1 + \cdots + p_{g-3} + p + \iota p$.

For hyperelliptic C it remains to show that ϕ is generically injective. But, if $p_1, \ldots, p_{g-3}$ are points in C, no two of which correspond to each other under the involution ι, then $h^0(\mathcal{O}_C(p_1 + \cdots + p_{g-3}) \otimes L_0) = 2$, again by (4.6). This implies that $p_1 + \ldots + p_{g-3}$ is uniquely determined by $\mathcal{O}_C(p_1 + \cdots + p_{g-3}) \otimes L_0$.

Finally, suppose that C is not hyperelliptic. Then the canonical map $\varphi_{\omega_C}\colon C \to \mathbb{P}^{g-1}$ is an embedding. Consider the natural map $\rho\colon C^{(g-1)} \to W_{g-1} \subset \operatorname{Pic}^{g-1}(C)$.

We claim: $\dim \rho^{-1}(\operatorname{sing} W_{g-1}) \leq g-3$.

For the proof fix a general divisor $D = p_1 + \cdots + p_{g-3}$. According to the General Position Theorem 4.2.12, $\dim \overline{\varphi_{\omega_C}(D)} = g-4$ and $C \cap \operatorname{span} \varphi_{\omega_C}(D) = \varphi_{\omega_C}(D)$ and moreover the linear projection with centre $\overline{\varphi_{\omega_C}(D)}$ leads to a birational morphism $p\colon C \to \overline{C} \subset \mathbb{P}^2$. Since $g \geq 4$ and $\deg(\overline{C}) = \deg C - g + 3 = g+1$, the curve $\overline{C}$ is not smooth by Plückers's formula for plane curves. Let q be a singular point of $\overline{C}$ and consider the linear projection $\mathbb{P}^2 \to \mathbb{P}^1$ with centre q. Let ν denote the multiplicity of the singular point q of $\overline{C}$ and $p^{-1}(q) = \{q_1, \ldots, q_\nu\}$ counted with multiplicities. The composed map $C \to \mathbb{P}^1$ is given by a linear system $g^1_{g+1-\nu}$, the corresponding line bundle of which is $\omega_C(-D - \sum_{i=1}^{\nu} q_i)$. By Riemann–Roch and Serre duality this implies

$$2 \leq h^0\Big(\omega_C\big(-D - \sum_{i=1}^{\nu} q_i\big)\Big) \leq h^0\big(\omega_C(-D - q_1 - q_2)\big) = h^0\big(\mathcal{O}_C(D + q_1 + q_2)\big).$$

Thus $D + q_1 + q_2 \in \rho^{-1}(\operatorname{sing} W_{g-1})$ by Riemann's Singularity Theorem.

Let F_D denote the surface of divisors in $C^{(g-1)}$ containing D. Since $\overline{C}$ contains only finitely many singularities, the argument above shows that F_D intersects $\rho^{-1}(\operatorname{sing} W_{g-1})$ at most in finitely many points. So we get

$$\begin{aligned} 0 &= \dim(\rho^{-1}(\operatorname{sing} W_{g-1}) \cap F_D) \\ &\geq \dim \rho^{-1}(\operatorname{sing} W_{g-1}) + \dim F_D - \dim C^{(g-1)} \\ &= \dim \rho^{-1}(\operatorname{sing} W_{g-1}) - g + 3\,, \end{aligned}$$

which completes the proof of the claim.

On the other hand, varying the divisor D within an open dense subset of $C^{(g-3)}$, the construction of above shows that the dimension of a component of $\rho^{-1}(\operatorname{sing} W_{g-1})$ is $\geq g-3$. By what we have said above, this gives $\dim \rho^{-1}(\operatorname{sing} W_{g-1}) = g-3$.

The fibre of ρ over $L \in W_{g-1}$ is just the linear system $|L|$, and $h^0(L) \geq 2$ for all $L \in \operatorname{sing} W_{g-1}$. Moreover, it is easy to see using (4.6) that every component of sing W_{g-1} contains a line bundle L with $h^0(L) = 2$. So the general fibre of ρ over any component is of dimension one and thus $\dim \operatorname{sing} W_{g-1} = g-4$. □

Note that the proof shows that sing W_{g-1} is irreducible for hyperelliptic curves C. On the other hand, together with the existence theorem of Brill–Noether theory (see Arbarello et al [8, p. 206]), the proof gives that sing W_{g-1} is equidimensional for non-hyperelliptic C.

4.2.5 Exercises

(1) Let C be a hyperelliptic curve of genus 3. Show that $W_2 \subset \operatorname{Pic}^2(C)$ has an ordinary double point at the unique line bundle $\ell \in \operatorname{Pic}^2(C)$ with $h^0(\ell) = \deg \ell = 2$.

(2) Let $\kappa \in \operatorname{Pic}^{g-1}(C)$ and Θ be the theta divisor on $J(C)$ with $\alpha_\kappa^* \Theta = W_{g-1}$. Since Θ and $(-1)^*\Theta$ are algebraically equivalent, there is an $x \in J(C)$ such that $(-1)^*\Theta = t_x^*\Theta$. Show that $x = \omega_C \otimes \kappa^{-2}$.

(3) Let C be a smooth algebraic curve, $\alpha_c : C \to J(C)$ the embedding with respect to the point $c \in C$ and Θ the theta divisor on $J(C)$ defined by $\alpha_L^*\Theta = W_{g-1}$ with $L = \omega_C \otimes \mathcal{O}_C\big((1-g)c\big)$. Show that $\alpha_c^* \mathcal{O}_{J(C)}(\Theta) = \mathcal{O}_C(g \cdot c)$.
(Hint: use Lemma 4.4.4 below.)

(4) Show that any curve C of genus $g \geq 1$ admits a theta characteristic κ with $h^0(\kappa) = 1$. (Hint: use Exercise 3.5.3 (3) to show that the theta divisor Θ of $J(C)$ contains a 2-division point x with $\operatorname{mult}_x(\Theta) = 1$.)

(5) Let C be a curve of genus g with Jacobian (J, Θ). Denote by L the line bundle of degree $g-1$ defined by $W_{g-1} - L = \Theta$. For a point $c \in C$ and any $n > 0$ consider the map $\alpha_{n,C} : C \to J,\ p \mapsto \mathcal{O}_C(np - nc)$. Show that $\alpha_{n,C}^* \mathcal{O}_J(\Theta) = (L(np))^n \otimes \omega_C^{\frac{n(n-1)}{2}}$.

4.3 The Torelli Theorem

There are many proofs of the Torelli Theorem, for example given in Torelli [133] (1913), Comessatti [34] (1914–15), Weil [139] (1957), Andreotti [5] (1958), Matsusaka [90] (1958), Martens [88] (1963) and Beilinson–Polishuk [19] (2001). Although proofs can be found in almost every book on compact Riemann surfaces or algebraic curves, I think a proof should be contained in an introductory book on abelian varieties.

The proof given here is due to Andreotti [5] with its improvement in Arbarello et al [8], which is probably the easiest proof. Moreover it fits best into this volume, because it uses the Gauss map which we used already for other purposes.

4.3.1 Statement of the Theorem

To every smooth projective curve C we associated its principally polarized Jacobian $(J(C), \Theta)$. The Torelli Theorem says that conversely a principally polarized Jacobian (J, Θ) determines the curve. In other words:

Theorem 4.3.1 (Torelli Theorem) *The map from the set of isomorphism classes of smooth projective curves of genus g to the moduli space of principally polarized abelian varieties of dimension g*

$$C \mapsto (J(C), \Theta)$$

is injective.

The proof uses the Gauss map, which was already introduced in Section 2.1.2. However, here we need a little more.

4.3.2 The Gauss Map of a Canonically Polarized Jacobian

Let C be a smooth curve of genus $g \geq 2$ with Jacobian (J, Θ). As above let Θ_s denote the smooth part of Θ. According to Section 2.1.2 and Exercise 2.1.6 (6), the Gauss map

$$G : \Theta_s \to \mathbb{P}^*_{g-1}$$

is defined by associating to every point $x \in \Theta_s$ the translate to the origin of the projectivized tangent hyperplane $P(T_x\Theta)$. Since G does not change under a translation, we assume that $J = \operatorname{Pic}^{g-1}(C)$ and $\Theta = W_{g-1}$, the theta-divisor of Lemma 4.2.1, that is, the image under the canonical map

$$\rho : C^{(g-1)} \to \operatorname{Pic}^{g-1}(C).$$

Recall from Section 4.2.4 that for any $D \in C^{(g-1)}$ we denote by $\overline{\varphi_{\omega_C}(D)}$ the intersection of hyperplanes $H \in \mathbb{P}^{g-1}$ containing $\varphi_{\omega_C}(D)$.

Lemma 4.3.2 *For any $D \in C^{(g-1)}$,*

$$h^0(\mathcal{O}_C(D)) \geq 2 \Leftrightarrow \dim \overline{\varphi_{\omega_C}(D)} \leq g-3.$$

In particular Θ_s is exactly the set of $\rho(D)$ with $D \in C^{(g-1)}$ such that $\overline{\varphi_{\omega_C}(D)}$ is a hyperplane.

Proof The first assertion is a consequence of Lemma 4.2.11. Using Riemann's Singularity Theorem 4.2.6 this implies the second assertion. □

In Proposition 2.1.9 it is shown that Im G is not contained in a hyperplane. Here we have more precisely:

Lemma 4.3.3 *Let $\Gamma_s \subset \Theta_s \times \mathbb{P}^*_{g-1}$ denote the graph of the Gauss map G and $\Gamma \subset \Theta \times \mathbb{P}^*_{g-1}$ its closure. Let $\widetilde{\Gamma}$ be the normalization of Γ. The induced map $\gamma : \widetilde{\Gamma} \to \mathbb{P}^*_{g-1}$ is a finite morphism.*

Proof According to Lemma 4.2.1, for any $x \in \Theta_s$ there is exactly one divisor D in $C^{(g-1)}$ such that $\rho(D) = x$. On the other hand, by the choice of J and Θ, the space $\mathbb{P}^*_{g-1}$ in the definition of G is just the dual projectivized space of the canonical space $H^0(\omega_C)$. This implies that, if

$$\varphi_\omega : C \to P(H^0(\omega_C))^* = \mathbb{P}^*_{g-1}$$

denotes the canonical map of C, the projectivized tangent space of Θ at x is just the linear span of the divisor $\rho(D)$, that is,

$$G(\rho(D)) = \overline{\varphi_\omega(D)}.$$

Now if C is not hyperelliptic, then the closure of the graph of G in $\Theta \times \mathbb{P}^*_{g-1}$ is

$$\Gamma = \{(\rho(D), H) \mid \overline{\varphi_\omega(D)} \subset H\} \subseteq \Theta \times \mathbb{P}^*_{g-1}.$$

This is easy to see using the fact that by the General Position Theorem 4.2.12, any point in Γ is the limit of points $(\rho(D), \gamma(\rho(D)))$ with $\rho(D) \in \Theta_s$, that is, D a sum of different points.

Then clearly the projection $\Gamma \to \mathbb{P}^*_{g-1}$ is a finite morphism. This implies that the map γ of the assertion is also finite.

Suppose now that C is hyperelliptic. Then we claim that the Gauss map $G : \Theta_s \to \mathbb{P}^*_{g-1}$ extends to a morphism

$$\overline{G} : C^{(g-1)} \to \mathbb{P}^*_{g-1}.$$

To see this note that in this case it is known (see Hartshorne [61, Proposition 4.5.3]) that the canonical map φ_ω factorizes as

$$\varphi_\omega : C \xrightarrow{h} \mathbb{P}^1 \xrightarrow{\overline{\varphi}} \mathbb{P}^*_{g-1},$$

where h is the hyperelliptic double covering and $\overline{\varphi}$ the $(g-1)$-uple embedding of $\mathbb{P}^1$ in $\mathbb{P}^*_{g-1}$. On the other hand, it follows from Riemann–Roch on the curve $\mathbb{P}^1$ that $\overline{\varphi}$ has the property that any $g-1$ points of $\mathbb{P}_1$ determine a unique hyperplane in $\mathbb{P}^*_{g-1}$. It follows that

$$\overline{G} : C^{(g-1)} \longrightarrow \mathbb{P}^*_{g-1}, \qquad D \mapsto \overline{\varphi_{\omega_C}(D)}$$

is an extension of G.

If $\delta : C^{(g-1)} \to \Gamma$ is given by $D \mapsto (\rho(D), \overline{G}(D))$, the following diagram commutes

$$\begin{array}{ccc} C^{(g-1)} & \xrightarrow{\quad \overline{G} \quad} & \mathbb{P}^*_{g-1} \\ & {\scriptstyle\delta}\searrow \quad \nearrow{\scriptstyle p_2} & \\ & \Gamma. & \end{array}$$

Clearly δ is finite and generically of degree 1. It follows that we may identify $\widetilde{\Gamma} = C^{(g-1)}$ and $\overline{G} = \gamma$. Since p_2 is clearly finite, this implies the finiteness of γ in the hyperelliptic case. □

For any projective variety $X \subset \mathbb{P}^n$ the closure in the dual projective space $\mathbb{P}^*_n$ of the locus of hyperplanes containing the tangent plane X at a smooth point is again a variety, called the *dual variety of X* and denoted by X^* (see Harris [60, Example 15.22]). Moreover, denote by $B_\gamma \subset \mathbb{P}^*_{g-1}$ the branch locus of the finite morphism $\gamma : \widetilde{\Gamma} \to \mathbb{P}^*_{g-1}$. With these notations we have:

Proposition 4.3.4 *If C is not hyperelliptic, then*

$$B_\gamma = C^*.$$

Proof It follows from Lemma 4.3.2 that on a dense open set of $C^{(g-1)}$ the space $\overline{\varphi_\omega(D)}$ is a hyperplane in $\mathbb{P}^{g-1}$ and using the properties of the map ρ, that

$$\Gamma' := \{(D, H) \subset C^{(g-1)} \times \mathbb{P}^*_{g-1} \mid \overline{\varphi_\omega(D)} \subseteq H\}$$

is an irreducible variety. Moreover, the map

$$(\rho, \mathbf{1}_{\mathbb{P}^*_{g-1}}) : \Gamma' \to \Gamma \subset \operatorname{Pic}^{g-1} \times \mathbb{P}^*_{g-1}$$

is finite onto Γ. Since $\widetilde{\Gamma}$ is the normalization of Γ, we get a commutative diagram

$$\begin{array}{ccc} \widetilde{\Gamma} & \xrightarrow{\gamma} & \mathbb{P}^*_{g-1} \\ & \searrow \quad \nearrow_{p_2} & \\ & \Gamma' & \end{array}$$

where p_2 is the second projection.

Now, since the canonical line bundle ω is of degree $2g-2$, the map p_2 is finite of degree $\binom{2g-2}{g-1}$ and its branch locus is the locus of the tangent hyperplanes of $\varphi_\omega(C)$ that is the dual variety C^* of C. This implies $B_\gamma \subset C^*$.

On the other hand, according to Lemma 4.3.2 the ramification locus R_{p_2} is

$$R_{p_2} = \{(D, H) \in \Gamma' \mid \overline{\varphi_\omega(D)} \text{ contains multiple points}\}.$$

Hence, if $p_1 : \Gamma' \to C^{(g-1)}$ denotes the first projection, we get

$$p_1(R_{p_2}) = \{2q_1 + q_2 + \cdots + q_{g-2} \mid q_i \in C \text{ for } i = 1, \ldots, g-2\},$$

which is an irreducible subvariety of codimension 1 in $C^{(g-1)}$.

For any $D \in C^{(g-1)}$, the linear span $\overline{\varphi_\omega(D)}$ is a hyperplane if and only if $D \not\subset p_1(R_{p_2})$. Hence p_2 is biregular outside of R_{p_2} and, according to Lemma 4.3.2, has positive-dimensional fibres over $p_2(R_{p_2})$, $p_1(R_{p_2})$ has codimension at least 2 in $C^{(g-1)}$ and a general point of R_{p_2} is a smooth point of Γ'. Hence the map $\widetilde{\Gamma} \to \Gamma'$ is an isomorphism. This implies that the branch locus B_γ contains an open set of C^*. Since C^* is irreducible, we obtain $B_\gamma = C^*$. □

Now let C be hyperelliptic of genus $g \geq 2$. Let $C_0 = \varphi_{\omega_C} \subset \mathbb{P}^*_{g-1}$. Clearly $C_0 \simeq \mathbb{P}^1$. Denote by $p_1, \ldots, p_{2g+2}$ the branch points of the double cover φ_{ω_C}. Note that the dual variety p_i^* of the point p_i is just the space of hyperplanes containing the point p_i.

Proposition 4.3.5 *If C is hyperelliptic, then*

$$B_\gamma = C_0^* \cup p_1^* \cup \cdots \cup p_{2g+2}^*.$$

Proof We saw already at the end of the proof of Lemma 4.3.3 that we may identify $C^{(g-1)} = \widetilde{\Gamma}$ and $\widetilde{G} : C^{(g-1)} \to \mathbb{P}^*_{g-1}$ with $\gamma : \widetilde{\Gamma} \to \mathbb{P}^*_{g-1}$. So by Lemma 4.3.3 the map $\overline{G}$ is finite and clearly of degree 2^{g-1}. It is obvious that its branch locus is the locus of tangent hyperplanes of C_0 plus the locus of hyperplanes passing through one of the p_i. □

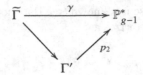

4.3.3 Proof of the Torelli Theorem

Given the principally polarized abelian variety $(J(C), \Theta)$, we have to show that one can reconstruct the curve C from it. But it is a classical theorem that the double dual of any projective variety is the variety itself:

$$(X^*)^* = X$$

(see Harris [60, Theorem 15.24]). Using this, in the non-hyperelliptic case, the Torelli Theorem follows from Proposition 4.3.4. In the hyperelliptic case it follows from Proposition 4.3.5, since every p_i^* determines of course p_i, which is a branch point of the double cover on C_0. This gives the hyperelliptic double cover C of C_0 and thus completes the proof.

4.3.4 Exercises

For the following exercises let C be a curve of genus $g \geq 2$ with Jacobian (J, Θ). We may assume that $\mathcal{O}_J(\Theta)$ is of characteristic zero with respect to some symplectic basis of $H_1(C, \mathbb{Z})$. For any $p, q \in C$ let $p - q$ denote the image of $(p, q) \in C \times C$ under the difference map

$$\delta : C \times C \to G, \quad (p, q) \mapsto \alpha_c(p) - \alpha_c(q).$$

Note that this definition is independent of the choice of $c \in C$.

(1) Show that $\Theta \cap t_{p-q}^* \Theta \subset t_{p-r}^* \Theta \cup t_{s-r}^* \Theta$ for all $p \neq q, r, s \in C$.

(2) Given $x \in J, x \neq 0$, show that the following conditions are equivalent:

(a) there exist y and z in J, distinct from $0, x$ such that $\Theta \cup t_x^* \Theta \subset t_y^* \Theta \cup t_z^* \Theta$;
(b) $x = p - q$ for some $p, q \in C$.

(Hint: see Mumford [100, Lemma, p. 76].)

(3) Let C be a non-hyperelliptic C curve.

(a) For any $x \in J$ such that $\Theta \cap t_x^* \Theta \subset t_y^* \Theta \cup t_z^* \Theta$ for some $y, z \neq 0, x$ show that the intersection $X := \Theta \cap t_x^* \Theta$ consists of two irreducible components X_1 and X_2.
(b) The curve $\delta(C)$ is up to sign a translate of the locus $\{x \in J \mid X_1 \subseteq \Theta + x\}$.

(Hint: see [8, Theorem p. 267].)

(4) Conclude from Exercise (3) the Torelli Theorem for non-hyperelliptic curves. (Hint: see [100, p. 80, lines 1 and 2].)

(5) (*Torelli Problem for Complex Tori of Dimension* 2) Let X be a complex torus of dimension 2 and ω a non-vanishing holomorphic 2-form on X. The map $\pi_X : H_2(X,\mathbb{Z}) \to \mathbb{C}, \gamma \mapsto \int_\gamma \omega$ is called the *period of X (with respect to 2-forms)*. Since ω is unique up to a nonzero constant, so is the period π_X. Given another complex torus Y of dimension 2 with period π_Y, we say that an isomorphism $\varphi : H_2(X,\mathbb{Z}) \to H_2(Y,\mathbb{Z})$ *preserves the periods* if $\pi_X\varphi = c\pi_Y$ for some non-zero constant c. Assume that there exists an isomorphism $\varphi : H_2(X,\mathbb{Z}) \to H_2(Y,\mathbb{Z})$ preserving the intersection form and the periods. Show that either $Y \simeq X$ or $Y \simeq \widehat{X}$.
In particular, for self-dual complex tori of dimension 2 the intersection form and the period determine the complex torus. This is called Torelli's Theorem for self-dual 2-dimensional complex tori.
(See Shioda [125].)

4.4 The Poincaré Bundles for a Curve C

4.4.1 Definition of the Bundle $\mathcal{P}^n_C$

Let C be a smooth curve of genus g and $J = J(C)$ its Jacobian. For any integer n we will construct a universal family of line bundles of degree n on C, the Poincaré bundle of degree n for C.

Fix a point $c \in C$ and let $\alpha_c : C \hookrightarrow J$ be the corresponding Abel–Jacobi map.

Lemma 4.4.1 *The restriction* $\alpha_c^* : \widehat{J} = \mathrm{Pic}^0(J) \to \mathrm{Pic}^0(C)$ *is an isomorphism.*

We will see in Corollary 4.4.5 below that $(\alpha_c^*)^{-1} = -\phi_\Theta$.

Proof The exponential sequences of J and C induce the following commutative diagram

$$\begin{array}{ccccccc}
H^1(J,\mathbb{Z}) & \longrightarrow & H^1(\mathcal{O}_J) & \longrightarrow & \mathrm{Pic}^0(J) & \longrightarrow & 0 \\
\downarrow{\scriptstyle \beta} & & \downarrow{\scriptstyle \gamma} & & \downarrow{\scriptstyle \alpha_c^*} & & \\
H^1(C,\mathbb{Z}) & \longrightarrow & H^1(\mathcal{O}_C) & \longrightarrow & \mathrm{Pic}^0(C) & \longrightarrow & 0.
\end{array}$$

It suffices to show that the restriction maps β and γ are isomorphisms.

For β : we have $H^1(J,\mathbb{Z}) = \mathrm{Hom}(H_1(J,\mathbb{Z}),\mathbb{Z})$, as we saw in Section 1.1.3. Moreover, $H^1(C,\mathbb{Z}) = \mathrm{Hom}(H_1(C,\mathbb{Z}),\mathbb{Z})$ and β is the transposed map of the isomorphism $\alpha_{c*} : H_1(C,\mathbb{Z}) \to H_1(J,\mathbb{Z})$.

For γ: the functoriality of the Hodge duality means that the following diagram is commutative

$$\begin{array}{ccc}
H^1(\Omega_J) & = & H^0(\Omega_J)^* \\
\downarrow{\scriptstyle \gamma} & & \uparrow{\scriptstyle r^*} \\
H^1(\Omega_C) & = & H^0(\Omega_C)^*.
\end{array}$$

Here $r\colon H^0(\Omega_J) \longrightarrow H^0(\Omega_C)$ denotes the restriction map. By definition of the Jacobian, r is an isomorphism, and so is γ. □

Lemma 4.4.1 allows us to use the Poincaré bundle for the Jacobian J in order to construct Poincaré bundles for the curve C itself. Recall that a *Poincaré bundle of degree n for C* (normalized with respect to the point $c \in C$) is a line bundle $\mathcal{P}_C^n$ on $C \times \operatorname{Pic}^n(C)$ satisfying

(i) $\mathcal{P}_C^n|_{C\times\{L\}} \simeq L$ for every $L \in \operatorname{Pic}^n(C)$, and
(ii) $\mathcal{P}_C^n|_{\{c\}\times\operatorname{Pic}^n(C)}$ is trivial.

Proposition 4.4.2 *For every $n \in \mathbb{Z}$ there exists a Poincaré bundle $\mathcal{P}_C^n$ for C, uniquely determined by the choice of the point $c \in C$.*

Proof For any $L \in \operatorname{Pic}^n(C)$ consider the commutative diagram

$$\begin{array}{ccccc} C \times \operatorname{Pic}^n(C) & \xrightarrow{1_C \times \alpha_{\mathcal{O}(nc)}} & C \times \operatorname{Pic}^0(C) & \xrightarrow{\alpha_c \times \alpha_c^{*-1}} & J \times \widehat{J} \\ \uparrow & & \uparrow & & \uparrow \\ C \times \{L\} & \longrightarrow & C \times \{\alpha_{\mathcal{O}(nc)}(L)\} & \longrightarrow & J \times \{\alpha_c^{*-1}(\alpha_{\mathcal{O}(nc)}(L))\}. \end{array}$$

Here the vertical maps are the natural embeddings and the lower horizontal maps are the restrictions of the upper ones. Denote by γ the composed map

$$\gamma = (\alpha_c \times \alpha_c^{*-1})(\mathbf{1}_C \times \alpha_{\mathcal{O}_C(nc)})\colon C \times \operatorname{Pic}^n(C) \to J \times \widehat{J}.$$

If $\mathcal{P}$ is the Poincaré bundle on $J \times \widehat{J}$, define

$$\mathcal{P}_C^n := \gamma^*\mathcal{P}.$$

The commutativity of the diagram above and the corresponding property of $\mathcal{P}$; that is, $\mathcal{P}|_{J\times\{M\}} \simeq M$ for all $M \in \widehat{J}$, implies that $\mathcal{P}_C^n$ satisfies condition (i).

With a similar diagram as above, and using the facts that $\alpha_c\colon C \to J$ maps c to $0 \in J$ and that $\mathcal{P}|_{\{0\}\times\widehat{J}}$ is trivial, one verifies condition (ii). Finally, the uniqueness statement follows from the seesaw theorem, Corollary 1.4.9. □

4.4.2 Universal Property of $\mathcal{P}_C^n$

The Poincaré bundle $\mathcal{P}_C^n$ satisfies the following universal property:

Proposition 4.4.3 *For any normal algebraic variety T and any line bundle $\mathcal{L}$ on $C \times T$ with*
(i) $\mathcal{L}|_{C\times\{t\}} \in \mathrm{Pic}^n(C)$ *for every* $t \in T$, *and*
(ii) $\mathcal{L}|_{\{c\}\times T}$ *is trivial,*
there is a unique morphism $\psi : T \to \mathrm{Pic}^n(C)$ *such that* $\mathcal{L} \simeq (\mathbf{1_C} \times \psi)^*\mathcal{P}_{\mathbf{C}}^{\mathbf{n}}$.

Notice that the underlying set-theoretical map of ψ is $t \mapsto \mathcal{L}|_{C\times\{t\}}$. Moreover, since T is irreducible as an algebraic variety, one can show that it suffices to assume condition (i) for only one point $t_0 \in T$. We omit the proof of Proposition 4.4.3, since it is completely analogous to that of Theorem 1.4.11, (but see Exercise 4.4.3 (1)). For the proof that condition (i) suffices for one point t_0, see Exercise 4.4.3 (2).

As a consequence we show that $(\alpha_c^*)^{-1} = -\phi_\Theta$. For this we need a technical lemma. We prove it in a slightly more general form, since it is useful anyway. Fix a line bundle $\kappa \in \mathrm{Pic}^{g-1}(C)$ (not necessarily a theta characteristic) and define a theta divisor Θ on J by $W_{g-1} = \alpha_\kappa^*\Theta$ $(= \Theta + \kappa \in \mathrm{Pic}^{g-1}(C))$.

Lemma 4.4.4 *For all* $x \in J = \mathrm{Pic}^0(C)$

$$\alpha_c^* O_J(t_x^*(-1)^*\Theta) = x^{-1} \otimes \kappa \otimes O_C(c).$$

Proof **Step I:** Let U denote the subset of J consisting of all points x such that (i) the curve $\alpha_c(C)$ intersects the divisor $t_x^*(-1)^*\Theta$ in g pairwise different points (see Corollary 4.2.3) and (ii) $h^0(x^{-1} \otimes \kappa \otimes O_C(c)) = 1$. According to the Moving Lemma 4.6.4 below and by semicontinuity U is open and dense in J. We claim that the assertion holds for every $x \in U$.

Suppose $x \in U$. There are pairwise different points $p_1, \dots, p_g$ of C, such that $\alpha_c^*(t_x^*(-1)^*\Theta) = p_1 + \cdots + p_g$. We have to show that $O_C(p_1 + \cdots + p_g) = x^{-1} \otimes \kappa \otimes O_C(c)$.

By assumption, we have $\alpha_c(p_i) \in t_x^*(-1)^*\Theta$ for $i = 1, \dots, g$, or equivalently

$$\left(x^{-1} \otimes \kappa \otimes O_C(c)\right) \otimes O_C(-p_i) = \alpha_c(p_i)^{-1} \otimes x^{-1} \otimes \kappa \in \alpha_\kappa^*\Theta = W_{g-1}$$

for $i = 1, \dots, g$. Since $h^0\left(x^{-1} \otimes \kappa \otimes O_C(c)\right) = 1$ and the points p_i are pairwise different, this implies the assertion.

Step II: The proof of the lemma is completed by applying the seesaw principle to a globalized version of the assertion: denoting by p_C and p_J the natural projections of $C \times J$ and by $\mu : J \times J \to J$ the addition map, we claim

$$(\alpha_c \times \mathbf{1}_J)^*\mu^*O_J\left((-1)^*\Theta\right) \otimes p_J^*O_J\left(-(-1)^*\Theta\right) \simeq \mathcal{P}_C^{0\,-1} \otimes p_C^*\left(\kappa \otimes O_C(c)\right). \quad (4.7)$$

But this follows from Step I and Corollary 1.4.9, restricting both sides to $C\times\{x\}$ with $x \in U$ and $\{c\} \times J$. Restricting (4.7) to $C \times \{x\}$ for any $x \in J$ gives the assertion. □

Let Θ be a symmetric theta divisor on $J = J(C)$, inducing an isomorphism $\phi_\Theta : J \to \widehat{J} = \mathrm{Pic}^0(J)$. On the other hand, by Theorem 4.1.4 the Abel–Jacobi map $\alpha_c^* : \mathrm{Pic}^0(J) \to \mathrm{Pic}^0(C)$ is an isomorphism. If we identify $\widehat{J} = \mathrm{Pic}^0(C)$ via α_c^*, we have the following important corollary.

Corollary 4.4.5 $(\alpha_c^*)^{-1} = -\phi_\Theta$.

Proof By Lemma 4.4.4 we have for all $x \in J = \mathrm{Pic}^0(C)$,

$$\begin{aligned}\alpha_c^*\phi_\Theta(x) = \alpha_c^*\phi_{(-1)^*\Theta}(x) &= \alpha_c^* \mathcal{O}_J\left(t_x^*(-1)^*\Theta - (-1)^*\Theta\right)\\ &= x^{-1} \otimes \kappa \otimes \mathcal{O}_C(c) \otimes \left(\kappa \otimes \Omega_C(c)\right)^{-1} = x^{-1}.\end{aligned}$$ □

4.4.3 Exercises

(1) Show that, for any integer n, any normal algebraic variety T and any $\mathcal{L} \in \mathrm{Pic}(C \times T)$ with (i) $\mathcal{L}|_{C\times\{t\}} \in \mathrm{Pic}^n(C)$ for every $t \in T$ and (ii) $\mathcal{L}|_{\{c\}\times T}$ trivial, there is a unique morphism $\psi : T \to \mathrm{Pic}^n(C)$ such that $\mathcal{L} \simeq (\mathbf{1}_C \times \psi)^*\mathcal{P}_{\mathbf{C}}^{\mathbf{n}}$. (Hint: Use Theorem 1.4.11.)

(2) Show that, since T is an irreducible variety in Proposition 4.4.3, it suffices for the existence of the map $\psi : T \to \mathrm{Pic}^n(C)$ to assume condition (i) only for one point $t_0 \in T$.

(3) Let C be a smooth algebraic curve and $\alpha = \alpha_c : C \to J(C)$ the embedding with respect to the point $c \in C$. Suppose $\mathcal{P}_C$ is the Poincaré bundle of degree zero on $C \times J(C)$, normalized with respect to c, and let Δ denote the diagonal in C^2. Show that

$$(\mathbf{1}_C \times (-1)\alpha_c)^*\mathcal{P}_C \simeq \mathcal{O}_{C^2}(\{c\} \times C + C \times \{c\} - \Delta).$$

4.5 The Universal Property of the Jacobian

The Jacobian J of a curve C of genus g admits a universal property: maps from C into abelian varieties factorize via the Jacobian. In this section a proof and some applications will be given.

4.5.1 The Universal Property

Theorem 4.5.1 (Universal Property of J(C)) *Suppose X is an abelian variety and $\varphi\colon C \longrightarrow X$ is a rational map. Then there exists a unique homomorphism $\widetilde{\varphi}\colon J \longrightarrow X$ such that for every $c \in C$ the following diagram is commutative*

$$\begin{array}{ccc} C & \xrightarrow{\ \varphi\ } & X \\ {\scriptstyle \alpha_c}\downarrow & & \downarrow{\scriptstyle t_{-\varphi(c)}} \\ J & \xrightarrow{\ \widetilde{\varphi}\ } & X. \end{array}$$

Proof According to Theorem 2.5.9 the map φ is everywhere defined. Consider the morphism $(t_{-\varphi(c)}\varphi)^{(g)}\colon C^{(g)} \longrightarrow X$ defined by

$$(t_{-\varphi(c)}\varphi)^{(g)}(p_1 + \cdots + p_g) = \varphi(p_1) + \cdots + \varphi(p_g) - g\varphi(c).$$

Since $\alpha_{gc}\colon C^{(g)} \longrightarrow J$ is birational, there is a rational map $\widetilde{\varphi}_c\colon J \longrightarrow X$ such that

$$(t_{-\varphi(c)}\varphi)^{(g)} = \widetilde{\varphi}_c \alpha_{gc} \tag{4.8}$$

on an open dense set of $C^{(g)}$. Again by Theorem 2.5.9 the map $\widetilde{\varphi}_c$ is a morphism, so equation (4.8) holds everywhere. Now $\widetilde{\varphi}_c(0) = (t_{-\varphi(c)}\varphi)^{(g)}(gc) = 0$ implies that $\widetilde{\varphi}_c$ is a homomorphism (see Proposition 1.1.6). Moreover the diagram commutes, since $\alpha_c(p) = \alpha_{gc}(p + (g-1)c)$ and $\varphi(p) - \varphi(c) = (t_{-\varphi(c)}\varphi)^{(g)}(p + (g-1)c)$ for all $p \in C$. The uniqueness of $\widetilde{\varphi}_c$ follows from the fact that $\alpha_c(C)$ generates J as a group.

It remains to show that $\widetilde{\varphi}_c = \widetilde{\varphi}_{c'}$ for all $c, c' \in C$. But

$$\begin{aligned}(\widetilde{\varphi}_c - \widetilde{\varphi}_{c'})\alpha_c(p) &= \widetilde{\varphi}_c\alpha_c(p) - \widetilde{\varphi}_{c'}(\alpha_{c'}(p) - \alpha_{c'}(c)) \\ &= t_{-\varphi(c)}\varphi(p) - t_{-\varphi(c')}\varphi(p) + t_{-\varphi(c')}\varphi(c) = 0\end{aligned}$$

for all $p \in C$. Since $\alpha_c(C)$ generates J, this implies the assertion. □

The dual of the homomorphism $\widetilde{\varphi}$ is

Corollary 4.5.2 $\quad \widehat{\widetilde{\varphi}} = -\phi_\Theta \varphi^* : \widehat{X} \to \widehat{J}.$

Proof Under the identifications $\widehat{X} = \operatorname{Pic}^0(X)$ and $\widehat{J} = \operatorname{Pic}^0(J)$ we have $\widehat{\widetilde{\varphi}} = \widetilde{\varphi}^*$. Moreover, $t^*_{-\varphi(c)}\colon \operatorname{Pic}^0(X) \to \operatorname{Pic}^0(X)$ is the identity according to Exercise 1.4.5 (5) a). So using Corollary 4.4.5 and the Universal Property of the Jacobian we obtain

$$\varphi^* = (t_{-\varphi(c)}\varphi)^* = (\widetilde{\varphi}\alpha_c)^* = \alpha_c^*\widehat{\widetilde{\varphi}} = -\phi_\Theta^{-1}\widehat{\widetilde{\varphi}}. \qquad \square$$

4.5.2 Finite Coverings of Curves

Let $f: C \to C'$ be a finite morphism of smooth projective curves. Denote by J' the Jacobian of C' and consider the composed morphism $\alpha_{f(c)} f: C \to J'$. According to the Universal Property, Theorem 4.5.1, there is a unique homomorphism N_f fitting into the following commutative diagram

$$\begin{array}{ccc} C & \xrightarrow{f} & C' \\ {\scriptstyle \alpha_c}\downarrow & & \downarrow{\scriptstyle \alpha_{f(c)}} \\ J & \xrightarrow{N_f} & J'. \end{array}$$

By definition N_f is just the map $\mathcal{O}_C(\sum r_i p_i) \mapsto \mathcal{O}_{C'}(\sum r_i f(p_i))$, classically called the *norm map* of f. Denote by Θ' a theta divisor on J'. Dualizing the equation $\alpha_{f(c)} f = N_f \alpha_c$ and applying Corollary 4.5.2 gives

$$\widehat{N_f} \phi_{\Theta'} = \phi_\Theta f^*. \tag{4.9}$$

So the investigation of N_f is equivalent to the investigation of f^*.

Proposition 4.5.3 *The homomorphism $f^*: J' \to J$ is not injective if and only if f factorizes via a cyclic étale covering f' of degree $n \geq 2$:*

$$\begin{array}{ccc} C & \xrightarrow{\quad f \quad} & C' \\ & {\scriptstyle f''}\searrow \quad \nearrow{\scriptstyle f'} & \\ & C''. & \end{array}$$

Proof Suppose first that f factorizes via a cyclic étale covering f' of degree $n \geq 2$. It suffices to show that the homomorphism $f'^*: J' \to J'' = J(C'')$ is not injective. To see this, recall that f' is given as follows: there exists a line bundle L on C' of order n in $\operatorname{Pic}^0(C')$ such that C'' is the inverse image of the unit section of $L^n = C' \times \mathbb{C}$ under the n-th power map $L \to L^n$ and $f': C'' \to C'$ is the restriction of $L \to L^n$ to C''.

Denote by $p: L \to C'$ the natural projection. Since the tautological line bundle p^*L is trivial, so is $f'^*L = p^*L|_{C''}$, and thus f'^* is not injective.

Conversely, suppose f^* is not injective. Choose a non-trivial line bundle $L \in \ker f^* \subset \operatorname{Pic}^0(C')$. Necessarily L is of finite order, say $n \geq 2$, since

$$L^{\deg f} = N_f f^* L = N_f \mathcal{O}_C = \mathcal{O}_{C'}.$$

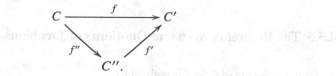

Then the cyclic étale covering $f'\colon C'' \to C'$ associated to L is of degree n. Consider the pullback diagram

$$\begin{array}{ccc} C \times_{C'} C'' & \xrightarrow{q} & C'' \\ {\scriptstyle p}\downarrow & & \downarrow{\scriptstyle f'} \\ C & \xrightarrow{f} & C'. \end{array}$$

The étale covering p is given by the trivial line bundle $f^*L = \mathcal{O}_C$. Hence $C \times_{C'} C''$ is the disjoint union of n copies of C. In particular there exists a section $s\colon C \to C \times_{C'} C''$ and f factorizes as $f = f'qs$. □

From the proof of Proposition 4.5.3 one easily deduces that for the cyclic étale covering $f'\colon C'' \to C'$ the kernel $\ker\{f^*\colon J' \to J''\}$ is generated by the line bundle L defining f'. If $(f'')^*\colon J'' \to J$ is not injective, one can apply the proposition again and factorize f''. Repeating this process we obtain:

Corollary 4.5.4 *For any finite morphism $f\colon C \to C'$ of smooth projective curves C and C' there is a factorization*

$$\begin{array}{ccc} C & \xrightarrow{\quad f \quad} & C' \\ & {\scriptstyle g}\searrow \quad \nearrow{\scriptstyle f_e} & \\ & C_e & \end{array}$$

with f_e étale, $\ker f^* = \ker f_e^*$, *and $g^*\colon J(C_e) \to J$ injective.*

4.5.3 The Difference Map and Quotients of Jacobians

The *difference map* of C in J is defined as

$$\delta\colon C \times C \to J, \qquad \delta(x,y) = \alpha_c(x) - \alpha_c(y).$$

Considering J as $\operatorname{Pic}^0(C)$ as usual, we have $\delta(x,y) = \mathcal{O}_C(x-y)$. In particular δ is independent of the choice of c and vanishes on the diagonal Δ of $C \times C$.

Proposition 4.5.5 *Let X be an abelian variety and $\varphi\colon C \times C \to X$ a rational map with $\varphi(\Delta) = 0$. Then there exists a unique homomorphism $\widetilde{\varphi}\colon J \to X$ such that the following diagram is commutative*

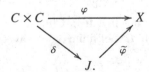

$$\begin{array}{ccc} C \times C & \xrightarrow{\quad \varphi \quad} & X \\ & {\scriptstyle \delta}\searrow \quad \nearrow{\scriptstyle \widetilde{\varphi}} & \\ & J. & \end{array}$$

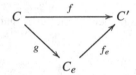

Proof According to Theorem 2.5.9 the map φ is everywhere defined. Since $\varphi(c,c) = 0$, Corollary 2.5.8 provides unique morphisms $\varphi_i\colon C \to X, i = 1,2$ with $\varphi_i(c) = 0$ and $\varphi(x,y) = \varphi_1(x) + \varphi_2(y)$ for all $x, y \in C$. Now $\varphi(\Delta) = 0$ implies $\varphi_2 = -\varphi_1$. By the Universal Property of the Jacobian there is a unique homomorphism $\widetilde{\varphi}\colon J \to X$ such that $\varphi_1 = \widetilde{\varphi}\alpha_c$ and we have for all $x, y \in C$,

$$\widetilde{\varphi}\delta(x,y) = \widetilde{\varphi}\alpha_c(x) - \widetilde{\varphi}\alpha_c(y) = \varphi_1(x) + \varphi_2(y) = \varphi(x,y). \qquad \square$$

Remark 4.5.6 The Universal Property of the Jacobian and Proposition 4.5.5 mean that the Jacobian J is the *Albanese variety* of the curve C (see Section 5.2.1 below).

Finally we show that every abelian variety is a quotient of a Jacobian. This has been very important in the development of the theory of abelian varieties (see Weil [137]). For this we need the following result from Algebraic Geometry.

Lemma 4.5.7 *Suppose $Y \subseteq \mathbb{P}^N$ is a smooth irreducible projective variety of dimension ≥ 2 and $f\colon Z \to Y$ is a finite morphism of a smooth irreducible variety Z onto Y. Then $f^{-1}(Y \cap H)$ is connected for every hyperplane H in $\mathbb{P}_N$.*

For the proof note that $f^{-1}(Y \cap H)$ is the support of an ample divisor, since f is finite. So the statement follows from Hartshorne [61, III Corollary 7.9].

Proposition 4.5.8 *For any abelian variety X there is a smooth projective curve C whose Jacobian $J(C)$ admits a surjective homomorphism*

$$J(C) \to X.$$

Proof Without loss of generality we may assume that $g = \dim X \geq 2$. Choose a projective embedding $X \hookrightarrow \mathbb{P}^N$. According to Bertini's Theorem (see Hartshorne [61, II 8.18]) there is a linear subspace $\mathbb{P}^{N-g-1}$ of $\mathbb{P}^N$ such that $C = X \cap \mathbb{P}^{N-g-1}$ is a smooth irreducible curve. Translating, if necessary, we may assume $0 \in C$. By the Universal Property of the Jacobian the embedding $C \hookrightarrow X$ factorizes via a homomorphism $\widetilde{\varphi}\colon J(C) \to X$.

Assume that $\widetilde{\varphi}$ is not surjective and denote by X_1 the abelian subvariety $\operatorname{Im}\widetilde{\varphi}$. Moreover denote by X_2 the complementary abelian subvariety of X_1 in X with respect to some polarization of X, as defined in Section 2.4.4. By Corollary 2.4.24 the addition map $\mu\colon X_1 \times X_2 \to X$ is an isogeny. Since $C \subset X_1$ and $X_1 \cap X_2$ is finite, so is $C \cap X_2$. Let $f\colon X_1 \times X_2 \to X$ be the composition of $1_{X_1} \times 2_{X_2}\colon X_1 \times X_2 \to X_1 \times X_2$ with μ. Then $f^{-1}(C)$ is not connected. But this contradicts the above fact applied several times. $\square$

4.5.4 Exercises

For the first three exercises let C be a curve of genus g and let

$$\delta^{(n)} : C^{(n)} \times C^{(n)}, \quad (\sum_i x_i, \sum_j y_j) \mapsto \mathcal{O}_C(\sum_i x_i - \sum_j y_j)$$

be the difference map.

(1) Show that for $n < \frac{g}{2}$:

(a) $\delta^{(n)}$ is birational onto its image if C is non-hyperelliptic;
(b) $\delta^{(n)}$ is of degree 2^n onto its image if C is hyperelliptic.

(2) Show that for C non-hyperelliptic, the projectivized tangent cone of $\delta^{(1)}$ at $0 \in J(C)$ is the canonical curve $\varphi_{\omega_C}(C) \subset P(H^0(\omega_C)^*) = \mathbb{P}^{g-1}$.

(3) If C does not admit a covering $C \to \mathbb{P}^1$ of degree $\leq n$, then the multiplicity of $\operatorname{Im} \delta^{(n)}$ at 0 is

$$\operatorname{mult}_0 \operatorname{Im} \delta^{(n)} = \sum_{i=0}^{n} \binom{n-g-1-i}{n-i} \binom{g}{i}.$$

(4) Let C be a smooth projective curve of genus $g \geq 2$ and (J, Θ) its Jacobian. Show that the automorphism groups of C and (J, Θ) are related as follows,

$$\operatorname{Aut}(C) = \begin{cases} \operatorname{Aut}(J, \Theta)/\langle -1_J \rangle & \text{if } C \text{ is non-hyperelliptic,} \\ \operatorname{Aut}(J, \Theta) & \text{if } C \text{ is hyperelliptic.} \end{cases}$$

4.6 Endomorphisms Associated to Curves and Divisors

In this section we prove the criterion of Matsusaka [91] for numerical equivalence of 1-cycles, respectively algebraic equivalence of divisors, in terms of the endomorphisms associated to cycles. The criterion is valid for an arbitrary abelian variety, but the proof uses the theory of Jacobians and in particular some results on correspondences.

4.6.1 Correspondences Between Curves

The main result here is that every homomorphism between Jacobians can be described by correspondences. We associate to every such homomorphism γ a correspondence L_γ which can be used to compute the rational trace of γ.

In the theory of algebraic curves a correspondence between two curves C_1 and C_2 is defined to be a divisor D on the product $C_1 \times C_2$. We will see that to any such correspondence one can associate a homomorphism between the corresponding Jacobians in a canonical way. This homomorphism depends only on the line bundle $L = \mathcal{O}_{C_1 \times C_2}(D)$. So for our purposes it is more convenient to define: a *correspondence* between two smooth projective curves C_1 and C_2 is a line bundle L on $C_1 \times C_2$.

Let J_1 and J_2 denote the Jacobians of C_1 and C_2 respectively. For any correspondence L between C_1 and C_2 and any point $p \in C_1$ define

$$L(p) = L|_{\{p\} \times C_2},$$

considered as a line bundle on C_2. Define the map

$$\gamma_L \colon J_1 \to J_2, \quad \mathcal{O}_{C_1}\Big(\sum_{i=1}^{n} r_i p_i\Big) \mapsto L(p_1)^{r_1} \otimes \cdots \otimes L(p_n)^{r_n}.$$

Note that γ_L is a well defined homomorphism: it is the homomorphism $J_1 \to J_2$ induced by the morphism $C_1 \to J_2$, $p \mapsto L(p) \otimes L(c)^{-1}$ according to the Universal Property of the Jacobian, Theorem 4.5.1.

Two correspondences L and L' between C_1 and C_2 are said to be *equivalent* if there are line bundles L_i on C_i, $i = 1, 2$, such that

$$L' = L \otimes q_1^* L_1 \otimes q_2^* L_2,$$

where q_1 and q_2 denote the canonical projections of $C_1 \times C_2$. This defines an equivalence relation on the set of all correspondences between C_1 and C_2 (Exercise 4.6.4 (1)). Denote by $\operatorname{Corr}(C_1, C_2)$ the $\mathbb{Z}$-module of equivalence classes of correspondences between C_1 and C_2.

Theorem 4.6.1 *The assignment $L \mapsto \gamma_L$ induces a canonical isomorphism of abelian groups*

$$\operatorname{Corr}(C_1, C_2) \to \operatorname{Hom}(J_1, J_2).$$

Proof First note that $\gamma_L = \gamma_{L'}$ for equivalent correspondences L and L', since $(q_1^* L_1)(p) = \mathcal{O}_{C_2}$ and $(q_2^* L_2)(p) = L_2$ for all $p \in C_1$. So the map $\operatorname{Corr}(C_1, C_2) \to \operatorname{Hom}(J_1, J_2)$ is well defined. It is obviously a homomorphism of abelian groups.

In order to show that it is bijective, fix points $c_i \in C_i$ and denote by $\alpha_i \colon C_i \to J_i$ the corresponding embeddings. For $\gamma \in \operatorname{Hom}(J_1, J_2)$ define a correspondence L_γ by

$$L_\gamma = (\gamma\alpha_1 \times \alpha_2)^* \mu^* \mathcal{O}_{J_2}(\Theta_2)^{-1},$$

where $\mu \colon J_2 \times J_2 \to J_2$ is the addition map and Θ_2 is some theta divisor on J_2.

First we claim $\gamma_{L_\gamma} = \gamma$ for every $\gamma \in \mathrm{Hom}(J_1, J_2)$. Since the curve $\alpha_1(C_1)$ generates J_1, this follows from the fact that for all $p \in C_1$,

$$\begin{aligned}\gamma_{L_\gamma}\alpha_1(p) &= \gamma_{L_\gamma}\mathcal{O}_{J_1}(p - c_1) = L_\gamma(p) \otimes L_\gamma(c_1)^{-1} \\ &= \alpha_2^* t^*_{\gamma\alpha_1(p)}\mathcal{O}(\Theta_2)^{-1} \otimes \alpha_2^*\mathcal{O}(\Theta_2) = -\alpha_2^*\phi_{\Theta_2}\gamma\alpha_1(p) = \gamma\alpha_1(p),\end{aligned}$$

where we used Corollary 4.4.5. So $\mathrm{Corr}(C_1, C_2) \to \mathrm{Hom}(J_1, J_2)$ is surjective.

In order to show that it is injective, note that $\gamma_L = 0$ means $\gamma_L\alpha_1(p) = L(p) \otimes L(c_1)^{-1} = \mathcal{O}_{C_2}$ for all $p \in C_1$. By the seesaw theorem, Corollary 1.4.9, this implies $L = q_1^*L_1 \otimes q_2^*L(c_1)$ for some line bundle L_1 on C_1. □

Suppose now $C_1 = C_2 = C$ is a curve of genus g with Jacobian variety J. The ring structure of $\mathrm{Hom}(J, J) = \mathrm{End}(J)$ induces a ring structure on $\mathrm{Corr}(C, C)$, which is easy to work out (see Exercise 4.6.4 (2)). For every $\gamma \in \mathrm{End}(J)$ we can use the associated correspondence L_γ to compute the trace of the rational representation $\mathrm{Tr}_r(\gamma)$. For this define the *bidegree* (d_1, d_2) of a correspondence L on $C \times C$ by

$$d_1 = d_1(L) = \deg L|_{C\times\{p\}} \quad \text{and} \quad d_2 = d_2(L) = \deg L|_{\{p\}\times C}.$$

Certainly this definition is independent of the point $p \in C$. Denoting by Δ the diagonal on $C \times C$ we have:

Proposition 4.6.2 *For every* $\gamma \in \mathrm{End}(J)$,

$$\mathrm{Tr}_r(\gamma) = d_1(L_\gamma) + d_2(L_\gamma) - (\Delta \cdot L_\gamma).$$

Since the right-hand side of the formula is constant on the equivalence classes of correspondences, the proposition implies slightly more generally,

$$\mathrm{Tr}_r(\gamma_L) = d_1(L) + d_2(L) - (\Delta \cdot L)$$

for all line bundles $L \in \mathrm{Pic}(C \times C)$.

Proof Let $\alpha\colon C \to J$ denote the embedding of C with base point $c \in C$. Since $L_\gamma|_{C\times\{p\}} = \alpha^*\gamma^*\mathcal{O}_J(t^*_{\alpha(p)}\Theta)^{-1}$ and $L_\gamma|_{\{p\}\times C} = \alpha^*\mathcal{O}_J(t^*_{\gamma\alpha(p)}\Theta)^{-1}$,

$$d_1 = -\big(\alpha(C) \cdot \gamma^*(\mathcal{O}_J(\Theta)\big) \quad \text{and} \quad d_2 = -\big(\alpha(C) \cdot \mathcal{O}_J(\Theta)\big).$$

Moreover, if $\Delta_C : C \to C \times C$ denotes the diagonal map,

$$\begin{aligned}(\Delta \cdot L_\gamma) = \deg \Delta_C^*(\gamma\alpha \times \alpha)^*\mu^*\mathcal{O}_J(\Theta)^{-1} &= -\deg \alpha^*(\gamma + \mathbf{1}_J)^*\mathcal{O}_J(\Theta) \\ &= -\,(\alpha(C) \cdot (\gamma + \mathbf{1}_J)^*\mathcal{O}_J(\Theta)).\end{aligned}$$

Using this, Proposition 2.4.6 and Poincaré's Formula 4.2.2 imply

$$\begin{aligned}
\operatorname{Tr}_r(\gamma) &= \tfrac{g}{(\Theta^g)}(\Theta^{g-1}\cdot((\gamma+\mathbf{1}_J)^*\mathcal{O}_J(\Theta)\otimes\gamma^*\mathcal{O}_J(\Theta)^{-1}\otimes\mathcal{O}_J(\Theta)^{-1}))\\
&= \tfrac{1}{(g-1)!}\Big(\Theta^{g-1}\cdot((\gamma+\mathbf{1}_J)^*\mathcal{O}_J(\Theta)\otimes\gamma^*\mathcal{O}_J(\Theta)^{-1}\otimes\mathcal{O}_J(\Theta)^{-1})\Big)\\
&= \big(\alpha(C)\cdot(\gamma+\mathbf{1}_J)^*\mathcal{O}_J(\Theta)\big)-\big(\alpha(C)\cdot\gamma^*\mathcal{O}_J(\Theta)\big)-\big(\alpha(C)\cdot\mathcal{O}_J(\Theta)\big)\\
&= -(\Delta\cdot L_\gamma)+d_1+d_2.
\end{aligned}$$

□

Finally, we express the Rosati involution $\gamma\mapsto\gamma'$ on $\operatorname{End}(J)$ with respect to the canonical polarization Θ on J in terms of correspondences on $C\times C$. For this denote by $\tau\colon C\times C\to C\times C$ the canonical involution $(p,q)\mapsto(q,p)$.

Proposition 4.6.3 *$\gamma_L'=\gamma_{\tau^*L}$ for every correspondence L on $C\times C$.*

Proof By Theorem 4.6.1 we may assume $L=(\gamma_L\alpha\times\alpha)^*\mu^*\mathcal{O}_J(\Theta)^{-1}$. Thus $\tau^*L=(\alpha\times\gamma_L\alpha)^*\mu^*\mathcal{O}_J(\Theta)^{-1}$. Moreover, by the Corollary 4.4.5 we have $\alpha^*=-\phi_\theta^{-1}$. Using this, we obtain,

$$\begin{aligned}
\gamma_{\tau^*L}\alpha(p)=(\tau^*L)(p)\otimes(\tau^*L)(c)^{-1}&=\alpha^*\gamma_L^*\mathcal{O}_J(t_{\alpha(p)}^*\Theta)^{-1}\otimes\alpha^*\gamma_L^*\mathcal{O}_J(\Theta)\\
&=-\alpha^*\widehat{\gamma_L}\phi_\Theta\alpha(p)=\phi_\Theta^{-1}\widehat{\gamma_L}\phi_\Theta\alpha(p)=\gamma_L'\alpha(p)
\end{aligned}$$

for every $p\in C$. This implies the assertion, since $\alpha(C)$ generates the group J. □

4.6.2 Endomorphisms Associated to Cycles

In this section let X more generally be an abelian variety of dimension g and denote by V and W algebraic cycles on X of complementary dimension. There is a canonical way to associate to the pair (V,W) an endomorphism $\delta(V,W)$ of X. Following Morikawa [94] and Matsusaka [91], we will show that $\delta(V,W)$ depends only on the algebraic equivalence classes of V and W.

Here we consider only *algebraic cycles* V with coefficients in $\mathbb{Z}$, that is, finite formal sums

$$V=\sum_i r_iV_i$$

with integers r_i and algebraic subvarieties V_i of X, which we assume to be all of the same dimension. If $\dim V_i=p$, then V is also called an *algebraic p-cycle*. Let $W=\sum s_iW_i$ be an algebraic q-cycle on X. The cycles V and W are said to *intersect properly* if $V_i\cap W_j$ is either of pure dimension $p+q-g$ or empty, whenever $r_i\neq0\neq s_j$.

Lemma 4.6.4 (Moving Lemma) *Let V be an algebraic p-cycle and W an algebraic q-cycle on X. There is an open dense subset U in X such that V and t_x^*W intersect properly for all $x\in U$.*

Proof We may assume that V and W are subvarieties of X. Consider the difference map $d: V \times W \to X$, $(v, w) \mapsto w - v$. The fibre of d over any $x \in X$ is

$$d^{-1}(x) \simeq V \cap t_x^* W.$$

Since d is a closed morphism, there is an open dense subset U of X such that $d^{-1}(x)$ is either of dimension $p + q - g$ (if d is surjective) or empty for all $x \in U$ (see Hartshorne [61, Exercise II, 3.22]). □

We need the following version of the Moving Lemma with parameters.

Lemma 4.6.5 *Let T be an algebraic variety and Z an algebraic cycle on $T \times X$ intersecting $\{t\} \times X$ properly for any $t \in T$. Let $Z(t)$ be the cycle on X defined by $\{t\} \times Z(t) = Z \cdot (\{t\} \times X)$. For any algebraic cycle W on X there is an open dense subset $U \subset T \times X$ such that $Z(t)$ and $t_x^* W$ intersect properly for all $(t, x) \in U$.*

Proof The proof is analogous to the proof of the Moving Lemma. Instead of the difference map d one uses the morphism $W \times Z \to T \times X$, $(w, t, x) \mapsto (t, w - x)$. □

Let V and W be algebraic cycles on X of complementary dimension. Suppose V and W intersect properly, then the usual intersection product $V \cdot W$ is a 0-cycle on X; that is, $V \cdot W = \sum_{i=1}^n r_i x_i$ with points x_i on X and integers r_i. Define

$$S(V \cdot W) = r_1 x_1 + \cdots + r_n x_n \in X,$$

where the sum means addition in X. Note that S is symmetric and bilinear; that is, $S(V \cdot W) = S(W \cdot V)$ and $S(V + V', W) = S(V, W) + S(V', W)$ for cycles V and V' both intersecting W properly.

Let now (V, W) be an arbitrary pair of algebraic cycles of complementary dimension on X. The pair (V, W) induces an endomorphism $\delta(V, W)$ of X in the following way.

According to the Moving Lemma 4.6.4 the cycle V intersects $t_x^* W$ properly for all x in an open dense subset of X. So the assignment $x \mapsto S(V \cdot t_x^* W)$ defines a rational map $X \to X$ which according to Theorem 2.5.9 extends to a morphism $S: X \to X$. By Proposition 1.1.6 there is an endomorphism $\delta(V, W)$ of X and a point $c \in X$, both uniquely determined by S, such that $\delta(V, W) = S - c$. So for $\delta(V, W): X \to X$ we have

$$\delta(V, W)(x) = S(V \cdot t_x^* W) - c$$

whenever V intersects $t_x^* W$ properly. The bilinearity of S implies

$$\delta(V + V', W) = \delta(V, W) + \delta(V', W) \quad \text{and}$$
$$\delta(V, W + W') = \delta(V, W) + \delta(V, W')$$

for all algebraic cycles V, V' and W, W' of complementary dimension on X. Note that in the special case when V intersects W properly we have $c = S(V \cdot W)$, that is,

$$\delta(V, W)(x) = S\big(V \cdot (t_x^* W - W)\big)$$

whenever defined. The next proposition shows that we always may assume that V intersects W properly.

Proposition 4.6.6 $\delta(V, W) = \delta(V', W)$ *for any algebraically equivalent algebraic p-cycles V and V' and any algebraic $(g-p)$-cycle W on X.*

Proof Without loss of generality we may assume that V intersects W properly. By the definition of algebraic equivalence we may assume that there is a smooth algebraic variety T and an algebraic cycle Z in $T \times X$ intersecting $\{t\} \times X$ properly for every $t \in T$ such that

$$Z \cdot (\{t_0\} \times X) = \{t_0\} \times V \quad \text{and} \quad Z \cdot (\{t_1\} \times X) = \{t_1\} \times V'$$

for some $t_0, t_1 \in T$. For any $t \in T$ define the p-cycle V_t by $Z \cdot (\{t\} \times X) = \{t\} \times V_t$. According to Lemma 4.6.5 there exists an open dense subset U of $T \times X$ such that V_t intersects $t_x^* W$ properly for every $(t, x) \in U$. Since $V = V_{t_0}$ intersects W properly by assumption, we may assume that $(t_0, 0) \in U$. Passing eventually to a smaller subset, we may assume that V_t also intersects W properly for every $(t, x) \in U$. In other words with $(t, x) \in U$ also $(t, 0) \in U$. So

$$\phi(t, x) := S\big(V_t \cdot (t_x^* W - W)\big),$$

for all $(t, x) \in U$, defines a rational map $\phi\colon T \times X \to X$ which by Theorem 2.5.9 is everywhere defined. We have $\phi(t, 0) = S(V_t \cdot (W - W)) = 0$ for any $(t, 0) \in U$ and thus for all $t \in T$. Hence by Corollary 2.5.7 the morphism ϕ does not depend on $t \in T$. In particular,

$$\delta(V, W) = \phi(t_0, \cdot) = \phi(t_1, \cdot) = \delta(V', W). \qquad \square$$

Recall that for arbitrary algebraic cycles V and W of complementary dimension, $(V \cdot W)$ denotes the intersection number of V and W. If V and $t_x^* W$ intersect properly, then $(V \cdot W)$ is the degree of the 0-cycle $V \cdot t_x^* W$.

Lemma 4.6.7 $\quad \delta(V, W) + \delta(W, V) = -(V \cdot W)\mathbf{1}_X.$

Proof We may assume that V and W intersect properly. Then for all x in an open dense subset of X,

$$\begin{aligned}\delta(V, W)(x) &= S(V \cdot t_x^* W) - S(V \cdot W) = S(t_{-x}^* V \cdot W) - (V \cdot W)x - S(W \cdot V) \\ &= S(W \cdot (t_{-x}^* V - V)) - (V \cdot W)x = -\delta(W, V)(x) - (V \cdot W)x.\end{aligned}$$

So Theorem 2.5.9 implies the assertion. $\square$

Combining Proposition 4.6.6 and Lemma 4.6.7 we obtain:

Corollary 4.6.8 *The homomorphism $\delta(V, W)$ depends only on the algebraic equivalence classes of V and W.*

One can show that $\delta(V, W)$ depends only on the numerical equivalence classes of V and W (see Exercise 4.6.4 (3)), but we do not need this fact.

Lemma 4.6.9 *For algebraic cycles $V_0, \ldots, V_r$ on X with $\sum_{i=0}^{r} \dim V_i = rg$ we have,*

$$\delta(V_0, V_1 \cdot \ldots \cdot V_r) = \sum_{i=1}^{r} \delta(V_0 \cdot V_1 \cdot \ldots \cdot \check{V}_i \cdot \ldots \cdot V_r, V_i).$$

Here the notation $\check{V}_i$ means that the cycle $\check{V}_i$ has to be omitted in the intersection product. Moreover by $V_0 \cdot \ldots \cdot \check{V}_i \cdot \ldots \cdot V_r$ for $i = 0, \ldots, r$ we mean any cycle in the algebraic equivalence class of the corresponding intersection product. The assumption on the dimension implies that the cycles V_i and $V_0 \cdot \ldots \cdot \check{V}_i \cdot \ldots \cdot V_r$ are of complementary dimension for all $0 \le i \le r$. So all endomorphisms in the formula are well defined.

Proof Passing eventually to suitable translations we may assume that V_i and $V_0 \cdot \ldots \check{V}_i \cdot \ldots \cdot V_r$ intersect properly for all i. Suppose first $r = 2$. Then for a general $x \in X$,

$$\begin{aligned}\delta(V_0, V_1 \cdot V_2)(x) &= S(V_0 \cdot (t_x^* V_1 \cdot t_x^* V_2 - V_1 \cdot V_2)) \\ &= S(V_0 \cdot (t_x^* V_1 \cdot t_x^* V_2 - t_x^* V_1 \cdot V_2)) + S(V_0 \cdot (t_x^* V_1 \cdot V_2 - V_1 \cdot V_2)) \\ &= \delta(V_0 \cdot V_1, V_2)(x) + \delta(V_0 \cdot V_2, V_1)(x).\end{aligned}$$

This proves the assertion for $r = 2$. The general case follows by induction. □

Proposition 4.6.10 *For any divisor D on X and $0 \le r \le g$ we have,*

$$\delta(D^r, D^{g-r}) = -\frac{g-r}{g}(D^g)\mathbf{1}_X.$$

Proof Using Lemma 4.6.9 and Lemma 4.6.7 we have,

$$\delta(D, D^{g-1}) = (g-1)\,\delta(D^{g-1}, D) = (g-1)\Big(-\delta(D, D^{g-1}) - (D^g)\mathbf{1}_X\Big)$$

implying the assertion for $r = 1$. Using this and again Lemmas 4.6.9 and 4.6.7 we get for every $0 \le r \le g$,

$$\begin{aligned}\delta(D^r, D^{g-r}) &= (g-r)\,\delta(D^{g-1}, D) = (g-r)\Big(-\delta(D, D^{g-1}) - (D^g)\mathbf{1}_X\Big) \\ &= (g-r)\Big(\frac{g-1}{g} - 1\Big)(D^g)\mathbf{1}_X = -\frac{g-r}{g}(D^g)\mathbf{1}_X.\end{aligned}$$ □

4.6.3 Endomorphisms Associated to Curves and Divisors

Let X be an abelian variety of dimension g. In this section we consider the special case that one of the cycles is of dimension 1 and the other a divisor on X.

Suppose D is a divisor and Γ an algebraic 1-cycle on X. In the previous section we associated to the pair (Γ, D) an endomorphism $\delta(\Gamma, D)$ of X. Since by Corollary 4.6.8 the endomorphism $\delta(\Gamma, D)$ depends only on the divisor D up to linear equivalence, it makes sense to write

$$\delta(\Gamma, L) := \delta(\Gamma, D) \quad \text{for} \quad L = \mathcal{O}_X(D).$$

Our first aim is to deduce a different expression for $\delta(\Gamma, L)$. Since $\delta(\Gamma, L)$ is additive in the first argument, we may assume that Γ is a reduced irreducible curve in X. Let $\varphi\colon C \to \Gamma = \varphi(C)$ be its normalization. Moreover we may assume that $\varphi(c) = 0$ for some $c \in C$. Let J denote the Jacobian of C and $\alpha = \alpha_c\colon C \to J$ the embedding with base point c. According to the universal property of the Jacobian, Theorem 4.5.1, the morphism φ extends to a homomorphism $\widetilde{\varphi}$ fitting into the following commutative diagram

(4.10)

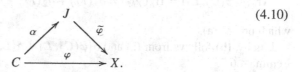

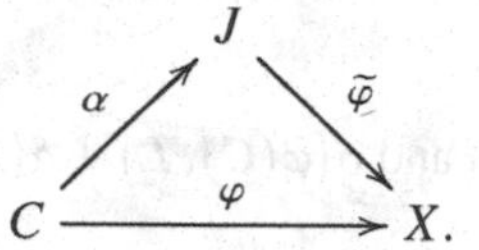

To simplify the notation, we identify J with its dual $\widehat{J}$ via the canonical isomorphism ϕ_Θ. Then we have for any line bundle L on X:

Proposition 4.6.11 $\quad \delta(\varphi(C), L) = -\widetilde{\varphi}\widehat{\widetilde{\varphi}}\phi_L$ *in* $\mathrm{End}(X)$.

Proof For all x in an open dense subset U of X and a suitably chosen divisor D with $L = \mathcal{O}_X(D)$, we have, using Corollary 4.5.2 and the definition of the map S in Section 4.6.2,

$$\begin{aligned} -\widetilde{\varphi}\widehat{\widetilde{\varphi}}\phi_L(x) &= \widetilde{\varphi}\varphi^*(t_x^*L \otimes L^{-1}) = \widetilde{\varphi}\mathcal{O}_C\big(\varphi^*(t_x^*D - D)\big) \\ &= S\big(\varphi(C) \cdot (t_x^*D - D)\big) = \delta\big(\varphi(C), D\big)(x) \end{aligned}$$

for all $x \in U$. This implies the assertion. □

With the preceding notation:

Proposition 4.6.12

(a) $\mathrm{Tr}_r\big(\delta(\varphi(C), L)\big) = -2\big(\varphi(C) \cdot L\big)$.

(b) $\mathrm{Tr}_r\Big(\delta\big(L, \varphi(C)\big)\Big) = -(2g - 2)\big(\varphi(C) \cdot L\big)$.

Proof Note first that $\mathrm{Tr}_r(\delta(\varphi(C), L)) = -\mathrm{Tr}_r(\widetilde{\varphi}\widehat{\widetilde{\varphi}}\phi_L) = -\mathrm{Tr}_r(\widehat{\widetilde{\varphi}}\phi_L\widetilde{\varphi})$ and $-\widehat{\widetilde{\varphi}}\phi_L\widetilde{\varphi}$ is an endomorphism of J, since we identified $J = \widehat{J}$. The idea is to take a correspondence of $C \times C$ associated to $-\widehat{\widetilde{\varphi}}\phi_L\widetilde{\varphi}$ and use Proposition 4.6.2 in order to compute the trace.

Define $M := (\varphi \times \varphi)^*\mu^*L$ (where as usual $\mu\colon X \times X \to X$ denotes the addition map). Then

$$\gamma_M = -\widehat{\widetilde{\varphi}}\phi_L\widetilde{\varphi},$$

since $\gamma_M\alpha(p) = M(p) \otimes M(c)^{-1} = \varphi^*t^*_{\varphi(p)}L \otimes \varphi^*L^{-1} = \varphi^*\phi_L\varphi(p) = -\widehat{\widetilde{\varphi}}\phi_L\widetilde{\varphi}\alpha(p)$ for all $p \in C$ by Corollary 4.5.2.

The bidegree of M and its intersection number with the diagonal are given by

$$d_1(M) = d_2(M) = \deg M|_{\{p\}\times C} = \deg \varphi^*L = (\varphi(C) \cdot L)$$

and

$$(\Delta \cdot M) = \deg \Delta_C^* M = \deg(2\varphi)^*L = 4(\varphi(C) \cdot L).$$

So Proposition 4.6.2 implies

$$\mathrm{Tr}_r(\delta(\varphi(C), L)) = \mathrm{Tr}_r(\gamma_M) = d_1(M) + d_2(M) - (\Delta \cdot M) = -2(\varphi(C) \cdot L),$$

which proves (a).

Finally, (b) follows from (a) and $\delta\big(\varphi(C), L\big) + \delta\big(L, \varphi(C)\big) = -\big(\varphi(C) \cdot L\big)\mathbf{1}_X$ by Lemma 4.6.7. □

For the proof of the following Corollary, see Exercise 4.6.4 (5).

Corollary 4.6.13 $\delta(\varphi(C), L) = 0$ *if and only if* $\delta(L, \varphi(C)) = 0$.

Recall that an algebraic 1-cycle Γ on X is called *numerically equivalent to zero* if $(\Gamma \cdot L) = 0$ for every line bundle L on X. The following criterion is due to Matsusaka [91].

Theorem 4.6.14

(a) *Suppose L is a non-degenerate line bundle and Γ an algebraic 1-cycle on X. If $\delta(\Gamma, L) = 0$, then Γ is numerically equivalent to zero.*
(b) *Suppose $\Gamma \subset X$ is a curve generating X as a group and L a line bundle on X. Then $\delta(\Gamma, L) = 0$ if and only if L is algebraically equivalent to zero.*

Note that in (a) the converse implication is also valid (see Exercise 4.6.4 (3)). We only proved that $\delta(\Gamma, L) = 0$ if Γ is algebraically equivalent to zero (see Proposition 4.6.6).

Proof (a): Suppose $\Gamma = \sum r_i\Gamma_i$ with irreducible reduced curves Γ_i. Let J_i denote the Jacobian of the normalization $\varphi_i\colon C_i \to \Gamma_i \subset X$ and define $\widetilde{\varphi}_i\colon J_i \to X$ as in diagram (4.10). Then $0 = \delta(\Gamma, L) = \sum r_i\delta(\Gamma_i, L) = -\sum r_i\widetilde{\varphi}_i\widehat{\widetilde{\varphi}}_i\phi_L$ by Proposition 4.6.11. Since ϕ_L is an isogeny, this implies $\sum r_i\widetilde{\varphi}_i\widehat{\widetilde{\varphi}}_i = 0$.

Let L' be an arbitrary line bundle on X. We have to show $(\Gamma \cdot L') = 0$. But Proposition 4.6.11 and (4.10) imply $\delta(\Gamma, L') = -\sum_i r_i \widetilde{\varphi}_i \widehat{\widetilde{\varphi}}_i \phi_{L'} = 0$. So from Proposition 4.6.12 we get

$$(\Gamma \cdot L') = \sum_i r_i (\Gamma_i \cdot L') = -\sum_i r_i \tfrac{1}{2} \operatorname{Tr}_r(\delta(\Gamma_i, L')) = -\tfrac{1}{2} \operatorname{Tr}_r(\delta(\Gamma, L')) = 0.$$

(b): By assumption $\widetilde{\varphi}: J \to X$ is surjective and thus $\widehat{\widetilde{\varphi}}: \widehat{X} \to J$ has finite kernel. It follows that $M = \widehat{\widetilde{\varphi}}^* \mathcal{O}_J(\Theta)$ is an ample line bundle on $\widehat{X}$ and $\widetilde{\varphi}\widehat{\widetilde{\varphi}} = \phi_M : \widehat{X} \to X$ (see Corollary 1.4.6) is an isogeny. So $\delta(\Gamma, L) = -\widetilde{\varphi}\widehat{\widetilde{\varphi}}\phi_L = 0$ if and only if $\phi_L = 0$, that is, $L \in \operatorname{Pic}^0(X)$. □

4.6.4 Exercises

(1) Show that the definition of equivalent correspondences gives an equivalence relation on the set of all correspondences.

(2) Let C be a smooth projective curve with Jacobian J. Recall the isomorphism of abelian groups $\operatorname{Corr}(C, C) \to \operatorname{End}(J)$ of Theorem 4.6.1. The ring structure of $\operatorname{End}(J)$ induces a ring structure on $\operatorname{Corr}(C, C)$ as follows. Let $\ell_1, \ell_2 \in \operatorname{Corr}(C, C)$.

(a) Show that there are divisors D_1 and D_2 on $C \times C$ defining ℓ_1 and ℓ_2 such that $C \times D_1$ and $D_2 \times C$ intersect properly in $C \times C \times C$. Moreover the class of the correspondence in $\operatorname{Corr}(C, C)$ defined by the divisor

$$D = p_{13*}((C \times D_1) \cdot (D_2 \times C))$$

does not depend on the choice of the divisors D_1 and D_2.

(b) $\gamma_{\mathcal{O}(D_1)} \gamma_{\mathcal{O}(D_2)} = \gamma_{\mathcal{O}(D)}$.

(3) Two algebraic p-cycles V_1 and V_2 on an abelian variety X are called *numerically equivalent* if the intersection number satisfies $(V_1 \cdot W) = (V_2 \cdot W)$ for all $(g-p)$-cycles W on X. Show that the homomorphism $\delta(V, W)$ depends only on the numerical equivalence classes of V and W.
(See Matsusaka [91].)

(4) Suppose L is a non-degenerate line bundle and Γ an algebraic 1-cycle on X. Show that if Γ is numerically equivalent to zero, then $\delta(\Gamma, L) = 0$.
(Hint: see Matsusaka [91].)

(5) Give a proof of Corollary 4.6.13.

4.7 The Criterion of Matsusaka–Ran

This section contains a proof of the criterion of Matsusaka–Ran (see Ran [109]) for a polarized abelian variety to be a product of Jacobians. We give here a modified version of Collino's proof (see Collino [32]).

4.7.1 Statement of the Criterion

Recall that a curve C on an abelian variety X is said to generate X if X is the smallest abelian variety containing C. More generally, an effective algebraic 1-cycle $\sum r_\nu C_\nu$ on X, $r_\nu > 0$ for all ν, *generates* X if the union of the curves C_ν generates X.

Theorem 4.7.1 (Criterion of Matsusaka–Ran) *Suppose (X, L) is a polarized abelian variety of dimension g and $C = \sum_{\nu=1}^n r_\nu C_\nu$ is an effective 1-cycle generating X with $(C \cdot L) = g$. Then $r_1 = \cdots = r_n = 1$, the curves C_ν are smooth, and (X, L) is isomorphic to the product of the canonically polarized Jacobians of the C_ν's:*

$$(X, L) \simeq (J(C_1), \Theta_1) \times \cdots \times (J(C_n), \Theta_n).$$

In particular, if C is an irreducible curve generating X with $(C \cdot L) = g$, then C is smooth and (X, L) is the Jacobian of C. As a special case we get Matsusaka's criterion.

Corollary 4.7.2 (Matsusaka's Criterion) *Let (X, Θ_X) be a principally polarized abelian variety of dimension g and $C \subset X$ an irreducible curve with*

$$[C] = \frac{1}{(g-1)!} \wedge^{g-1} [\Theta_X] \quad \textit{in} \quad H^{2g-2}(X, \mathbb{Z}).$$

Then C is smooth and (X, Θ_X) is the Jacobian of C.

For a direct proof see Exercise 4.7.3 (2).

Proof It suffices to show that C generates X with $(C \cdot \Theta_X) = g$. But

$$(C \cdot \Theta_X) = \frac{1}{(g-1)!}(\wedge^{g-1}\Theta_X \cdot \Theta_X) = \frac{1}{(g-1)!}(\Theta_X^g) = g,$$

where the last equation follows from the Geometric Riemann–Roch Theorem 1.7.3. This also implies that C generates X. □

Corollary 4.7.3

(a) *A principally polarized abelian surface is either the Jacobian of a smooth curve of genus 2 or the canonically polarized product of two elliptic curves.*
(b) *A principally polarized abelian threefold is either the Jacobian of a smooth curve of genus 3 or the principally polarized product of an abelian surface with an elliptic curve respectively three elliptic curves.*

Proof a) is an immediate consequence of the criterion and Riemann–Roch.

(b): The moduli spaces of curves of genus 3 and of principally polarized abelian threefolds are both irreducible algebraic varieties of dimension 6. Hence by the Torelli Theorem 4.3.1 a general principally polarized abelian threefold is a Jacobian. Since under specialization effective 1-cycles go to effective 1-cycles and the intersection numbers are preserved, Theorem 4.7.1 implies the assertion. □

The argument of the last sentence of the proof of Corollary 4.7.3 yields more generally:

Corollary 4.7.4 *Any specialization of a Jacobian is a product of Jacobians.*

4.7.2 Proof of the Criterion

Step I: First we outline the general set up: for $\nu = 1, \dots, n$ denote by $\widetilde{C}_\nu$ the normalization of C_ν. According to the universal property of the Jacobian, Theorem 4.5.1, the composed map $\iota_\nu \colon \widetilde{C}_\nu \to C_\nu \hookrightarrow X$ factorizes via a unique homomorphism $\psi_\nu \colon J_\nu = J(\widetilde{C}_\nu) \to X$ (of course we may assume that every C_ν passes through the origin in X). Writing $g_\nu = (C_\nu \cdot L)$, we have

$$g = \sum_{\nu=1}^{n} r_\nu g_\nu .$$

Fix a divisor $D \in |L|$. By Proposition 2.1.6 we may assume that D is reduced. By eventually passing to an algebraically equivalent line bundle, we may assume that its pullback $\iota_\nu^* D$ is a divisor and thus a divisor of degree g_ν on $\widetilde{C}_\nu$ for all ν. Recall the Abel–Jacobi map $\alpha_\nu := \alpha_{\iota_\nu^* D} \colon \widetilde{C}_\nu^{(g_\nu)} \to J_\nu$ defined by $\alpha_\nu(\sum x_i) = \mathcal{O}_{\widetilde{C}_\nu}(\sum x_i - \iota_\nu^* D)$.

Define rational maps $h_\nu \colon X \dashrightarrow \widetilde{C}_\nu^{(g_\nu)}$ by $h_\nu(x) = \iota_\nu^* t_x^* D$. Then we have the following diagram

$$\begin{array}{ccccc}
X & \xrightarrow{\;h=(h_1,\dots,h_n)\;} & \widetilde{C}_1^{(g_1)} \times \cdots \times \widetilde{C}_n^{(g_n)} & & \widetilde{C}_1 \cup \cdots \cup \widetilde{C}_n \\
\downarrow{\scriptstyle \phi_L} & & \downarrow{\scriptstyle \alpha=\alpha_1\times\cdots\times\alpha_n} & & \downarrow{\scriptstyle \iota=\iota_1+\cdots+\iota_n} \\
\operatorname{Pic}^0(X) & \xrightarrow{\;\iota^*=(\iota_1^*,\dots,\iota_n^*)\;} & J_1 \times \cdots \times J_n & \xrightarrow{\;\psi=\psi_1+\cdots+\psi_n\;} & X.
\end{array}$$

The diagram is commutative, since for a general $x \in X$ we have

$$\iota_\nu^* \phi_L(x) = \iota_\nu^* \mathcal{O}_X(t_x^* D - D) = \mathcal{O}_{\widetilde{C}_\nu}(\iota_\nu^* t_x^* D - \iota_\nu^* D) = \alpha_\nu h_\nu(x).$$

Step II. *We claim that $r_\nu = 1$ and $\widetilde{C}_\nu$ is a curve of genus g_ν for $\nu = 1, \dots, n$, and that $\psi = \psi_1 + \cdots + \psi_n$ is an isogeny.*

Identifying $J_\nu = \widehat{J}_\nu$ as usual, the map ι^* is dual to $-\psi$ by Exercises 1.4.5 (3) and (4), and Corollary 4.5.2. Since ψ is surjective by assumption, the homomorphism $\iota^* = (\iota_1^*, \dots, \iota_n^*)$ and thus also the composed map $\iota^*\phi_L$ have finite kernel. According to the commutativity of the diagram this gives

$$\dim X = \dim \overline{(\operatorname{Im} \alpha h)} \le \dim \overline{(\operatorname{Im} h)} \le \sum_{\nu=1}^{n} g_\nu \le \sum_{\nu=1}^{n} r_\nu g_\nu = g = \dim X.$$

It follows that $r_1 = \cdots = r_n = 1$. Moreover, since X and $\widetilde{C}_1^{(g_1)} \times \cdots \times \widetilde{C}_n^{(g_n)}$ are both of the same dimension, the map h is dominant. Hence $\iota^*\phi_L(X) = \operatorname{Im} \alpha$.

Since $\operatorname{Im} \alpha$ generates the abelian variety $J_1 \times \cdots \times J_n$, this implies that α is surjective and $g_\nu = \dim J_\nu$ for $\nu = 1, \dots, n$. So $\widetilde{C}_\nu$ is of genus g_ν and ψ is an isogeny.

Step III: *We claim that ψ is an isomorphism and the line bundles ψ^*L and $p_1^*\psi_1^*L \otimes \cdots \otimes p_n^*\psi_n^*L$ define the same principal polarization on $J_1 \times \cdots \times J_n$.*

It suffices to show that ψ^*L and $p_1^*\psi_1^*L \otimes \cdots \otimes p_n^*\psi_n^*L$ are algebraically equivalent line bundles inducing a principal polarization, since then $1 = h^0(\psi^*L) = \deg \psi \cdot h^0(L)$ implies that ψ is an isomorphism.

First we show that ψ_i^*L defines a principal polarization on J_i for $i = 1, \dots, n$. Assume $h^0(\psi_i^*L) \ge 2$. For any $y \in J_i$ consider the exact sequence

$$0 \longrightarrow H^0(t_y^*\psi_i^*L(-\widetilde{C}_i)) \longrightarrow H^0(t_y^*\psi_i^*L) \longrightarrow H^0(t_y^*\psi_i^*L|_{\widetilde{C}_i}).$$

For a general $y \in J_i$ we have $h^0(t_y^*\psi_i^*L(-\widetilde{C}_i)) = 0$ (since not every linear system $|t_y^*L|$ contains $\widetilde{C}_i$) and thus $h^0(t_y^*\psi_i^*L|_{\widetilde{C}_i}) \ge 2$. Together with $\deg t_y^*\psi_i^*L|_{\widetilde{C}_i} = \deg L|_{C_i} = (L \cdot C_i) = g_i$ and Riemann–Roch for curves, this implies that for a general $y \in J_i$ the divisor $t_y^*\psi_i^*D|_{\widetilde{C}_i}$ is special (recall from Step I that $D \in |L|$). On the other hand, the divisors $t_y^*\psi_i^*D|_{\widetilde{C}_i}$ form just the image of the rational map $h_i \colon X \dashrightarrow \widetilde{C}_i^{(g_i)}$. Since $\operatorname{Im} h_i$ is dense in $\widetilde{C}_i^{(g_i)}$, this implies that a general divisor of degree g_i on $\widetilde{C}_i$ is special, a contradiction. Hence $h^0(\psi_i^*L) = 1$ and thus ψ_i^*L defines a principal polarization on J_i.

Next we claim that $\widehat{\psi}_i\phi_L\psi_j = 0$ for all $i \neq j$. Note first that $\psi_i \colon J_i \to X$ is an embedding, since $\phi_{\psi_i^*L} = \widehat{\psi}_i\phi_L\psi_i$ is an isomorphism. Let N_{J_i} denote the corresponding norm-endomorphism (see Section 2.4.3), that is, $N_{J_i} = \psi_i\phi_{\psi_i^*L}^{-1}\widehat{\psi}_i\phi_L$. Since $\psi = \psi_1 + \cdots + \psi_n$ is an isogeny, the subvarieties J_i and J_j of X intersect in finitely many points. This implies

$$0 = N_{J_i}N_{J_j} = (\psi_i\phi_{\psi_i^*L}^{-1})\widehat{\psi}_i\phi_L\psi_j(\phi_{\psi_j^*L}^{-1}\widehat{\psi}_j\phi_L).$$

But $\psi_i\phi_{\psi_i^*L}^{-1}$ is injective and $\operatorname{Im}(\phi_{\psi_j^*L}^{-1}\widehat{\psi}_j\phi_L) = J_j$, which implies the claim.

Using this we obtain

$$\begin{aligned}\phi_{\psi^* L} &= \widehat{\psi}\phi_L\psi \\ &= (\widehat{\psi_1},\dots,\widehat{\psi_n})\phi_L(\psi_1+\cdots+\psi_n) \\ &= (\widehat{\psi_1}\phi_L\psi_1)\times\cdots\times(\widehat{\psi_n}\phi_L\psi_n) \\ &= \phi_{p_1^*\psi_1^*L\otimes\cdots\otimes p_n^*\psi_n^*L}.\end{aligned}$$

Now Proposition 1.4.12 shows that ψ^*L and $p_1^*\psi_1^*L\otimes\cdots\otimes p_n^*\psi_n^*L$ are algebraically equivalent line bundles on X, completing the proof of Step III.

Since ψ is an isomorphism, the curves $C_\nu = \widetilde{C}_\nu$ are smooth. Moreover it follows from Step III that ψ_ν^*L defines a principal polarization on J_ν. It remains to show that this is the canonical polarization of J_ν. So we are reduced to the irreducible case and set $C = C_1 = \widetilde{C}_1$, $J = J_1$ and $\psi = \psi_1$.

Step IV: By what we have seen above, $\alpha = \alpha_1$ is birational and $\iota^* = \iota_1^*$ and ϕ_L are isomorphisms. So $h^{-1} = \phi_L^{-1}\iota^{*-1}\alpha\colon C^{(g)} \to J$ is a birational morphism. It is defined as follows:

For a general $(x_1+\cdots+x_g) \in C^{(g)}$ there is a unique $x \in X$ such that $h(x) = t_x^*D|_C = x_1+\cdots+x_g$. Then $h^{-1}(x_1+\cdots+x_g) = x$.

Define a map $\beta\colon C^{(g-1)}\times C \to J$ by

$$\beta(x_1+\cdots+x_{g-1},x_g) = h^{-1}(x_1+\cdots+x_g)+x_g$$

(the last + means addition in J, which makes sense, since $C \subset J$).

We claim $D = \operatorname{Im}\beta$. In particular, D is an irreducible divisor. For the proof suppose $h^{-1}(x_1+\cdots+x_g) = x$. By definition $x_g \in t_x^*D$ or equivalently $x+x_g \in D$, so $\operatorname{Im}\beta \subseteq D$. Certainly $\operatorname{Im}\beta$ is an irreducible divisor in J.

Assume there is a divisor $D' \neq 0$ such that $D = \operatorname{Im}\beta + D'$. By eventually passing to a translate of C, we may assume that C intersects D' and $\operatorname{Im}\beta$ properly, but not their intersection (note that D is a reduced divisor). So we may write $D|_C = y_1+\cdots+y_g$ with $y_g \in D'$. But then

$$\beta(y_1+\cdots+y_{g-1},y_g) = h^{-1}(y_1+\cdots+y_g)+y_g = y_g \in \operatorname{Im}\beta\,,$$

a contradiction.

Step V: *ψ^*L defines the canonical polarization on J.*

Recall that the canonical polarization on J is defined by the divisor $\Theta = \{x_1+\cdots+x_{g-1} \mid x_\nu \in C \subset J\}$ in J. So it suffices to show that $D = \operatorname{Im}\beta$ is a translate of the divisor $(-1)_J^*\Theta$ in J.

According to Corollary 2.5.8 the morphism β is of the form

$$\beta(x_1+\cdots+x_{g-1},x_g) = \gamma(x_1+\cdots+x_{g-1})+\delta(x_g)$$

with morphisms $\gamma\colon C^{(g-1)} \to J$ and $\delta\colon C \to J$. Since for every $x \in C$ the dimension of $\beta(C^{(g-1)} \times \{x\})$ is $g-1$, we have $\operatorname{Im}\gamma + \delta(x) = \beta(C^{(g-1)} \times \{x\}) = \operatorname{Im}\beta = D$, so δ is constant and we may assume $\delta \equiv 0$. This means

$$\beta(x_1 + \dots + x_{g-1}, x_g) = \gamma(x_1 + \dots + x_{g-1})$$

for all $x_1, \dots, x_g \in C$. Fix $y_1, \dots, y_{g-1} \in C$. For all $x_1, \dots, x_g \in C$ we have

$$\begin{aligned}\beta(x_1 + \dots + x_{g-1}, x_g) &= \beta(x_1 + \dots + x_{g-1}, y_1)\\ &= h^{-1}(x_1 + \dots + x_{g-1} + y_1) + y_1\\ &= h^{-1}(x_2 + \dots + x_{g-1} + y_1 + x_1) + y_1\\ &= \gamma(x_2 + \dots + x_{g-1} + y_1) + y_1 - x_1.\end{aligned}$$

Repeating this process, we finally obtain

$$\beta(x_1 + \dots + x_{g-1}, x_g) = \gamma(y_1 + \dots + y_{g-1}) + y_1 + \dots + y_{g-1} - x_1 - \dots - x_{g-1}.$$

This implies the assertion and completes the proof of the theorem. □

Remark 4.7.5 (The Schottky Problem) Matsusaka's theorem, Corollary 4.7.2, is a good criterion for a principally polarized abelian variety X to be the Jacobian of a curve. However it assumes the existence of suitable curve in X. One would like a criterion without a curve. Moreover, one would like information about where the Jacobians of genus g lie in the moduli space $\mathcal{A}_{1_g}$ of principally polarized abelian varieties of dimension g.

We never spoke about the moduli space of curves of a fixed genus g. Taking for granted that it exists and is an irreducible algebraic variety $\mathcal{M}_g$ of dimension $3g-3$ (see Mumford et al [101], in this remark we assume $g \geq 2$), we conclude from Torelli's Theorem 4.3.1 that the map

$$\mathcal{M}_g \to \mathcal{A}_{1_g}, \qquad C \mapsto J(C)$$

is injective and hence its image is an irreducible subvariety of $\mathcal{A}_{1_g}$ of the same dimension as $\mathcal{M}_g$. According to Corollary 4.7.4 its closure in $\mathcal{A}_{1_g}$, which we denote by $\overline{\mathcal{M}_g}$, is a closed algebraic subvariety of dimension $3g-3$. The *Schottky problem* consists in describing $\overline{\mathcal{M}_g}$ by equations in $\mathcal{A}_{1_g}$ or more generally to give criteria for a Jacobian.

There are a lot of papers concerning this problem of which I mention only a few. First of all, since $\dim \overline{\mathcal{M}_g} = 3g-3$ and $\dim \mathcal{A}_{1_g} = \frac{1}{2}g(g+1)$ we get $\overline{\mathcal{M}_g} = \mathcal{A}_{1_g}$ for $g = 2$ and 3. So we may assume $g \geq 4$.

In 1888 Schottky showed in [120] that a certain polynomial of degree 16 in the theta constants vanishes on $\overline{\mathcal{M}_4}$, but not everywhere on $\mathcal{A}_{1_4}$. Igusa showed in [71] that the zero set of this polynomial equals $\overline{\mathcal{M}_4}$.

In 1909 Schottky and Jung constructed in [121] expressions in the theta constants which vanish on $\overline{\mathcal{M}_g}$ by means of what is now called the theory of Prym varieties (see Section 5.3 below). For the Schottky–Jung relations see [24, 12.10.6]. These expressions define a certain locus $\mathcal{S}_g$ in $\mathcal{A}_{1_g}$, called the *Schottky locus*, which contains $\overline{\mathcal{M}_g}$. It is conjectured that $\mathcal{S}_g = \overline{\mathcal{M}_g}$. Van Geemen showed in [48] that $\overline{\mathcal{M}_g}$ is an irreducible component of $\mathcal{S}_g$.

There are several geometric approaches to the Schottky problem. Let us mention only one. Consider the abelian varieties $X_Z = \mathbb{C}^g/\Lambda_Z$ associated to $Z \in \mathfrak{H}^g$ with theta divisor Θ. According to Theorem 2.3.20 the map $\varphi_{2\Theta}$ induces an embedding of the corresponding Kummer variety K_Z into $\mathbb{P}^{2^g-1}$. If now X_Z is the Jacobian of a curve, then K_Z admits a trisecant in $\mathbb{P}^{2^g-1}$. This is the content of Fay's trisecant identity not proved in this volume (but see [24, 11.10.1]). It was conjectured that if the Kummer variety of $X_Z \in \mathcal{A}_g$ admits a trisecant, then X_Z is a Jacobian. Beauville and Debarre proved in [18] that the existence of a trisecant implies $\dim \operatorname{sing} \Theta \geq 4$ which is a hint that the conjecture might be true. Finally Krichever proved this conjecture in [79]. To be more precise, he proved that if the Kummer variety of an indecomposable $X_Z \in \mathcal{A}_g$ admits a trisecant (which may be degenerate; that is, a trisecant with multiplicities), such that no point of intersection of this line with the Kummer is singular in K_Z, then X_Z is the Jacobian of a curve.

An analytic approach to the Schottky problem is given as follows: there is an infinitesimal version of the trisecant identity, where the three points of K_Z are replaced by three tangent vectors of X_Z at 0 or equivalently by three constant vector fields on X_Z. Consider the theta functions $\theta\left[\begin{smallmatrix}\epsilon\\0\end{smallmatrix}\right](v, Z)$ as defined in equation 4.10 with $\epsilon \in \frac{1}{2}\mathbb{Z}^g/\mathbb{Z}^g$ and let

$$\theta_2[\epsilon](v, Z) := \theta\begin{bmatrix}\epsilon\\0\end{bmatrix}(2v, 2Z).$$

Then if X_Z is the Jacobian of a curve, then there exist constant vector fields D_1, D_2 and D_3 on X_Z and a complex number c such that

$$\left((D_1^4 - D_1D_3 + \frac{3}{4}D_2^2 + c)\ \theta_2[\epsilon]\right)(0, Z) = 0 \text{ for all } \epsilon.$$

This equation is called the KP-equation (after Kadomtsev–Petviashvili). The equation characterizes Jacobians in $\mathcal{A}_{1_g}$. This was conjectured by Novikov and proved by Shiota in [128].

A completely algebraic-geometric proof of both theorems, of Shiota and Krichever, was given by Arbarello, Codogni and Pareschi in [9].

So in a sense both results give a proof of the Schottky problem. But there remain many other interesting approaches to the problem which are open or only partly solved. A good survey is given by Grushevsky in [59].

4.7.3 Exercises

(1) Let (X, Ξ) be a principally polarized abelian variety of dimension g and (J, Θ) the Jacobian variety of a smooth projective curve C. Given a morphism $\varphi\colon C \to X$ and an integer e, show that the following statements are equivalent:

(a) $(\varphi^*)^*\Theta \equiv e\Xi$;

(b) $\varphi_*[C] = \frac{e}{(g-1)!} \bigwedge^{g-1}[\Xi]$ in $H^{2g-2}(X, \mathbb{Z})$.

(2) Use the previous exercise to give a direct proof of Matsusaka's criterion, Corollary 4.7.2.
(Hint: Show that X is contained in $J(C)$ and let Y denote the complement of X in $J(C)$. Use Corollary 5.3.5 below to show that Y is principally polarized and then Lemma 5.3.6 to conclude that it is zero.)

4.8 A Method to Compute the Period Matrix of a Jacobian

Given a smooth projective curve C of genus g, it is difficult in general to compute a period matrix for its Jacobian $J(C) = H^0(\omega_C)^*/H_1(C, \mathbb{Z})$. However, if C admits a sufficiently large group of automorphisms, there is a method, due to Bolza [26], for doing this. We want to explain it, work out an example, and state Bolza's result as well as several other examples in the exercises in Section 4.8.3.

4.8.1 The Method

Suppose $\varphi\colon C \to C$ is an automorphism. The Universal Property of the Jacobian 4.5.1 provides a unique automorphism $\widetilde{\varphi}$ of $J(C)$ such that for any $c \in C$ the following diagram commutes

$$\begin{array}{ccc} C & \xrightarrow{\ \varphi\ } & C \\ {\scriptstyle \alpha_c}\downarrow & & \downarrow{\scriptstyle \alpha_{-\varphi(c)}} \\ J(C) & \xrightarrow{\ \widetilde{\varphi}\ } & J(C). \end{array}$$

Choose a symplectic basis $\lambda_1, \dots, \lambda_g, \mu_1, \dots, \mu_g$ of the lattice $H_1(C, \mathbb{Z})$ and a basis $\omega_1, \dots, \omega_g$ of $H^0(\omega_C)$ with coordinates such that $\int_{\mu_i} \omega_j = \delta_{ij}$. Then according to equation (4.1) the corresponding period matrix is of the form $\Pi = (Z, \mathbf{1}_g)$ for some $Z \in \mathfrak{H}_g$.

Let ${}^tM = {}^t\begin{pmatrix} \alpha & \beta \\ \gamma & \delta \end{pmatrix} \in G_{1_g} = \mathrm{Sp}_{2g}(\mathbb{Z})$ and $A \in \mathrm{M}_g(\mathbb{C})$ denote the rational and analytic representation of $\widetilde{\varphi}$ with respect to these bases. Then, according to Corollary 3.1.5 and the fact that $\widetilde{\varphi}$ is an automorphism, we have $A(Z, \mathbf{1}_g) = (Z, \mathbf{1}_g)\,{}^tM$ or equivalently

$$Z = (\alpha Z + \beta)(\gamma Z + \delta)^{-1} \quad \text{and} \quad A = {}^t(\gamma Z + \delta). \tag{4.11}$$

Suppose now C admits a covering $C \to \mathbb{P}^1$ such that φ descends to an automorphism of $\mathbb{P}^1$. Realizing the covering as a concrete Riemann surface over $\mathbb{P}^1$ with the help of a system of canonical dissections, one can determine the action of φ on the fundamental group $\pi_1(J(C)) = H_1(C, \mathbb{Z})$, in other words, compute the matrix ${}^tM$ of the rational representation of $\widetilde{\varphi}$.

Suppose moreover that C is defined by an equation in $\mathbb{P}^2$ and we are given an explicit basis of $H^0(\omega_C)$ in terms of this equation, such that the analytic representation $\tilde{A}$ of $\widetilde{\varphi}$ with respect to this basis can be computed. Since A and $\tilde{A}$ are equivalent matrices, this gives us the eigenvalues of A and in particular its determinant. All this information gives restrictions on the matrix Z.

In the case that there is only one curve C with a given automorphism group, this procedure may be sufficient to determine a period matrix of $J(C)$, as the following example shows.

4.8.2 An Example

Let C be the smooth curve defined inhomogeneously by the equation

$$y^2 = x^6 - 1.$$

C is a hyperelliptic curve of genus 2 admitting the automorphism

$$\varphi : \begin{cases} x & \mapsto x' = \zeta x \\ y & \mapsto y' = -y \end{cases}$$

with $\zeta = e(\frac{2\pi i}{6})$. The hyperelliptic involution $y \mapsto -y$ induces the double covering $C \to \mathbb{P}^1$, $(x, y) \mapsto x$, ramified in the 6th roots of unity. Consider the following picture

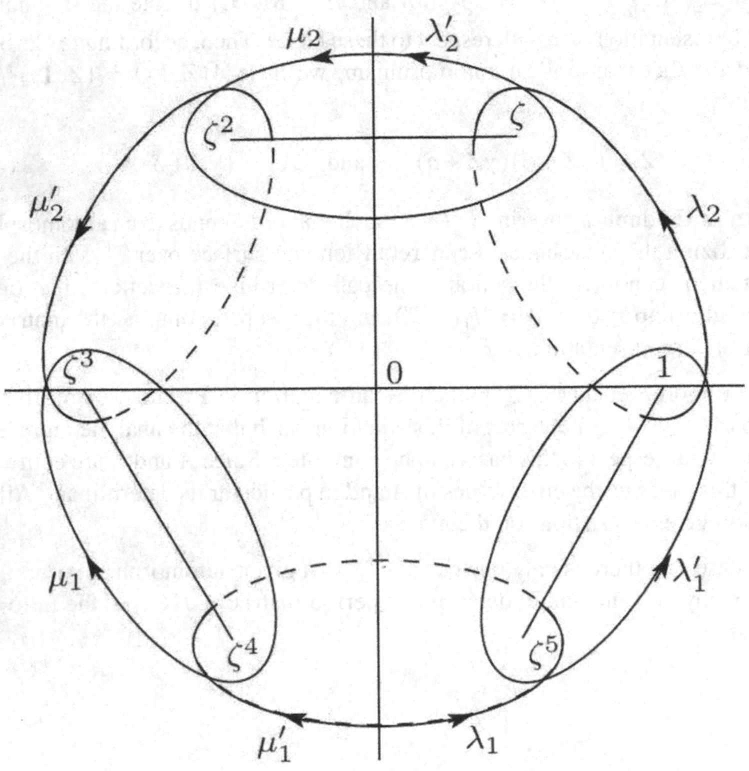

Here C is to be thought of as consisting of two copies of the x-plane $\mathbb{P}^1$, glued together in the usual way via the three straight line dissections joining the ramification points $1, \zeta, \ldots, \zeta^5$. Moreover, $\lambda_i, \mu_i, \lambda_i', \mu_i'$, $i = 1, 2$, indicate 1-cycles on C with dotted segments lying on the lower sheet and full segments lying on the upper sheet. Obviously $\lambda_1, \lambda_2, \mu_1, \mu_2$ is a symplectic basis of $H_1(C, \mathbb{Z})$. The automorphism φ is induced by rotation in the x-plane by $\zeta = e(\frac{2\pi i}{6})$ with centre 0, but exchanges the two sheets over the point 0. Hence $\lambda_1', \lambda_2', \mu_1', \mu_2'$ are the images of the cycles $\lambda_1, \lambda_2, \mu_1, \mu_2$ under φ.

In order to compute ${}^tM$, we have to express the cycles $\lambda_1', \lambda_2', \mu_1'$ and μ_2' in terms of the symplectic basis: obviously $\lambda_2' = \mu_2$ and $\mu_1' = -\lambda_1$. On the other hand, comparing their intersection numbers with the basis elements $\lambda_1, \lambda_2, \mu_1, \mu_2$ yields $\lambda_1' = \mu_1 - \mu_2$ and $\mu_2' = -\lambda_1 - \lambda_2$. Hence

$${}^tM = \rho_r(\widetilde{\varphi}) = \begin{pmatrix} 0 & \left(\begin{smallmatrix} -1 & -1 \\ 0 & -1 \end{smallmatrix}\right) \\ \left(\begin{smallmatrix} 1 & 0 \\ -1 & 1 \end{smallmatrix}\right) & 0 \end{pmatrix}$$

and equations (4.11) read

$$Z = \begin{pmatrix} 1 & -1 \\ 0 & 1 \end{pmatrix} \left(\begin{pmatrix} -1 & 0 \\ -1 & -1 \end{pmatrix} Z \right)^{-1} \quad \text{and} \quad A = {}^t\left(\begin{pmatrix} -1 & 0 \\ -1 & -1 \end{pmatrix} Z \right). \tag{4.12}$$

In order to determine the determinant of A, consider the basis $\frac{dx}{y}, \frac{xdx}{y}$ of $H^0(\omega_C)$ (for the fact that this is a basis of $H^0(\omega_C)$, see Shafarevich [124, III §5.5)]). Recall that $J(C) = H^0(\omega_C)^*/H_1(C,\mathbb{Z})$. Thus the analytic representation of the induced endomorphism $\widetilde{\varphi}$ of $J(C)$ is given by the dual map of $\varphi^* \colon H^0(\omega_C) \to H^0(\omega_C)$. Since $\varphi^*(\frac{dx}{y}) = -\zeta\frac{dx}{y}$ and $\varphi^*(\frac{xdx}{y}) = -\zeta^2\frac{dx}{y}$, it follows that $\widetilde{A} = \operatorname{diag}(-\zeta, -\zeta^2)$ is the matrix of the analytic representation with respect to this basis. We get

$$\det A = \det \rho_a(\widetilde{\varphi}) = \det \widetilde{A} = -1. \tag{4.13}$$

Now (4.12) implies $\det Z = -1$ and writing $Z = \begin{pmatrix} z_1 & z_2 \\ z_3 & z_4 \end{pmatrix}$ we obtain the following equations $z_1 = z_3 = -2z_2$ and $z_1z_3 - z_2^2 = -1$. Since $\operatorname{Im} Z$ is positive definite, this gives $z_1 = z_3 = -2z_2 = \frac{2i}{\sqrt{3}}$ and we finally obtain the following period matrix for $J(C)$

$$\Pi = \begin{pmatrix} \frac{2i}{\sqrt{3}} & -\frac{i}{\sqrt{3}} & 1 & 0 \\ -\frac{i}{\sqrt{3}} & \frac{2i}{\sqrt{3}} & 0 & 1 \end{pmatrix}.$$

If one starts with the basis $\mu_1, \mu_2, \lambda_1, \lambda_2$ instead of the symplectic basis $\lambda_1, \lambda_2, \mu_1, \mu_2$, one ends up with the following period matrix for $J(C)$

$$\widetilde{\Pi} = \begin{pmatrix} \frac{2i}{\sqrt{3}} & \frac{i}{\sqrt{3}} & 1 & 0 \\ \frac{i}{\sqrt{3}} & \frac{2i}{\sqrt{3}} & 0 & 1 \end{pmatrix}.$$

This proves case IV of the result of Bolza [26], given in Exercise 4.8.3 (1).

This method can be applied also to many hyperelliptic curves of higher genus as well as many nonhyperelliptic curves (see Exercises 4.8.3 (2) to (5)).

4.8.3 Exercises

(1) Let $\overline{\operatorname{Aut} C}$ denote the *reduced automorphism group* of a hyperelliptic curve C; that is, the quotient of $\operatorname{Aut} C$ modulo the hyperelliptic involution. Moreover D_n denotes the dihedral group of order $2n$. Use the method of Section 4.8.1 to give a proof of Bolza's complete result, given in the table below.

Note that the list is complete in the sense that every curve C of genus 2 with non-trivial reduced automorphism group appears. Note moreover that the moduli space of curves of type I is of dimension two. Correspondingly the space of period matrices is of dimension two in this case. It is not known which period matrix corresponds to a particular curve of this type. A similar remark is valid also for types II and III.

type	equation	$\overline{\operatorname{Aut} C}$	$\widetilde{\Pi} = (Z, \mathbf{1}_2)$
I	$y^2 = (x^2 - a^2)(x^2 - b^2)(x^2 - 1)$	$\mathbb{Z}/2\mathbb{Z}$	$Z = \begin{pmatrix} z & \frac{1}{2} \\ \frac{1}{2} & z' \end{pmatrix}$
II	$y^2 = x(x^2 - a^2)(x^2 - a^{-2})$	D_2	$Z = \begin{pmatrix} z & \frac{1}{2} \\ \frac{1}{2} & z \end{pmatrix}$
III	$y^2 = (x^3 - a^3)(x^3 - a^{-3})$	D_3	$Z = \begin{pmatrix} 2z & z \\ z & 2z \end{pmatrix}$
IV	$y^2 = x^6 - 1$	D_6	$Z = \begin{pmatrix} \frac{2i}{\sqrt{3}} & \frac{i}{\sqrt{3}} \\ \frac{i}{\sqrt{3}} & \frac{2i}{\sqrt{3}} \end{pmatrix}.$
V	$y^2 = x(x^4 - 1)$	S_4	$Z = \begin{pmatrix} \frac{-1+i\sqrt{2}}{2} & \frac{1}{2} \\ \frac{1}{2} & \frac{-1+i\sqrt{2}}{2} \end{pmatrix}$
VI	$y^2 = x(x^5 - 1)$	$\mathbb{Z}/5\mathbb{Z}$	$Z = \begin{pmatrix} 1-\epsilon^4 & -\epsilon^2-\epsilon^4 \\ -\epsilon^2-\epsilon^4 & \epsilon \end{pmatrix}, \quad \epsilon = \mathbf{e}(\frac{2\pi i}{5})$

(2) Let C be the hyperelliptic curve of genus $g \geq 2$ defined by the affine equation $y^2 = x^{2g+2} - 1$. It is the unique such curve with reduced automorphism group the dihedral group of order $4g + 4$. Show that $\Pi = (Z, \mathbf{1}_g)$ with $Z = (z_{jk})$,

$$z_{jk} = \frac{i}{g+1} \sum_{\nu=1}^{j} \left(\frac{1+\cos \frac{2\nu-1}{g+1}\pi}{\sin \frac{2\nu-1}{g+1}\pi} + \frac{1+\cos \frac{2(k-\nu)+1}{g+1}\pi}{\sin \frac{2(k-\nu)+1}{g+1}\pi} \right)$$

for $j \leq k$, is a period matrix of the Jacobian $J(C)$.

(3) Let C be the hyperelliptic curve of even genus $g \geq 2$ defined by the affine equation $y^2 = x^{2g+2} - x$. It is the unique such curve with reduced automorphism group cyclic of order $2g + 1$. Show that $\Pi = (Z, \mathbf{1}_g)$ with $Z = (z_{jk})$,

$$z_{jk} = 1 - \sigma_1^{-1} \sum_{\nu=1}^{j} \sigma_\nu \sigma_{k-j+\nu} \quad \text{for } 1 \leq j \leq k \leq g \text{ with}$$

$$\sigma_1 = \mathbf{e}\left(-\tfrac{g\pi i}{2g+1}\right) \quad \text{and}$$

$$\sigma_{\nu+1} = \frac{\sigma_1}{1+\mathbf{e}\left(\frac{2\nu\pi i}{2g+1}\right)} \left(1 - \sum_{\mu=2}^{\nu} \mathbf{e}\left((g - \nu + \mu - 1) \tfrac{2\pi i}{2g+1} \right) \sigma_\mu \sigma_{\nu-\mu+2} \right)$$

for $\nu = 1, \ldots, g - 1$, is a period matrix for the Jacobian $J(C)$. An analogous formula is valid for odd g.

(4) Let C be the hyperelliptic curve of genus $g \geq 2$ defined by the affine equation $y^2 = x^{2g+1} - x$. It is the unique such curve with reduced automorphism group the dihedral group of order $4g$ for $g \geq 3$ and the symmetric group S_4 for $g = 2$.

Show that $\Pi = (Z, \mathbf{1_g})$ with $Z = (z_{jk})$,

$$\begin{aligned}
z_{1,k} &= \alpha_k \quad \text{for } k = 1, \ldots, g, \\
z_{j,j} &= -2\alpha_2 \quad \text{for } j = 2, \ldots, g, \\
z_{j,k} &= \alpha_{k-j+1} - \alpha_{k-j+2} \quad \text{for } 2 \le j < k \le g \text{ and} \\
\alpha_j &= \tfrac{2}{g}\, \mathbf{e}\Big((2j-3)\tfrac{\pi i}{2g}\Big)\Big(\mathbf{e}\big((2j-3)\tfrac{\pi i}{g}\big) - 1\Big)^{-1} \quad \text{for } j = 2, \ldots, g \text{ and} \\
\alpha_1 &= \tfrac{1}{2}\Big(-\alpha_2 - \sum_{j=2}^{g} \alpha_j - 1\Big)
\end{aligned}$$

is a period matrix for the Jacobian $J(C)$.
(For the last three exercises see Schindler [116]. For $g = 2$ compare the results with cases IV, V and VI in Bolza's list in Exercise (1). Note that for $g = 2$ the curve in the last exercise is isomorphic to the curve of case V in the list.)

(5) The plane quartic with equation $X_0^3 X_1 + X_1^3 X_2 + X_2^3 X_0 = 0$ is called the *Klein quartic*. It is the unique plane quartic with automorphism group of order 168. Show that $\Pi = (Z, \mathbf{1}_3)$ with

$$Z = \frac{1}{2}\begin{pmatrix} -2 & 3 & -1 \\ 3 & -6 & 2 \\ -1 & 2 & -1 \end{pmatrix} + \frac{\sqrt{7}}{14} i \begin{pmatrix} 6 & -5 & 3 \\ -5 & 10 & -6 \\ 3 & -6 & 5 \end{pmatrix}$$

is a period matrix of the Klein quartic.
(Hint: use a modification of the method of Section 4.8.1. See Schindler [116] with a correction by H. Braden.)

(6) Compute all automorphism groups of curves of genus 3 (hyperelliptic and non-hyperelliptic) and for each of them compute a period matrix.
(Hint: see Schindler [116].)

Chapter 5
Main Examples of Abelian Varieties

In this chapter some of the most prominent examples of abelian varieties will be introduced, namely abelian surfaces, Picard and Albanese varieties, Prym varieties and Intermediate Jacobians.

Section 5.1 deals with abelian surfaces; that is, abelian varieties X of dimension 2. A polarization $L \in \mathrm{Pic}(X)$ is of type (d_1, d_2) for positive integers d_1, d_2 with d_1 dividing d_2. For $d_1 \geq 3$ the map $\varphi_L : X \to P(H^0(L)^*)$ is an embedding according Lefschetz Theorem 2.1.10. If $d_1 = 2$, the behaviour of φ_L is worked out explicitly in Sections 2.2.4 and 2.3.6. It remains to study the behaviour of φ_L in the case $d_1 = 1$. The main result of this section is a proof of a criterion for φ_L to be an embedding, due to Reider [111], Theorem 5.1.6 below.

To every smooth projective curve C one can associate an abelian variety, its Jacobian $J(C)$. The same construction generalizes to give the Picard and Albanese varieties $\mathrm{Pic}^0(M)$ and $\mathrm{Alb}(M)$ for every smooth projective variety M. They are introduced in Section 5.2 and it is shown that they are dual to each other. In the special case of a curve C this gives the self-duality of $J(C)$.

In Section 5.3 we introduce and study the Prym variety of a finite covering $f : C \to C'$ of smooth projective curves. The pullback of line bundle induces an homomorphism $f^* : J(C') \to J(C)$, which is an isogeny onto its image $A \subset J(C)$. The complement $P = P(f)$ of A in $J(C)$ with respect to the canonical polarization of $J(C)$ is called the Prym variety of f if the restriction of the canonical polarization Θ of $J(C)$ to P is a multiple of a principal polarization Ξ on P. Theorem 5.3.9 determines all coverings which lead to Prym varieties. Section 5.3.3 gives a topological construction of Prym varieties.

For a smooth projective variety M of dimension $n \geq 3$ there are several possibilities to associate an abelian variety. The Picard and Albanese varieties use the first cohomology and the $(2n-1)$-th cohomology of M for this. One can also use the intermediate cohomologies in order to associate abelian varieties to M. However there are several ways to do this, which lead to different intermediate Jacobians. In Section 5.4.2 we introduce the Griffiths Intermediate Jacobian, which is a non-

H. Lange, *Abelian Varieties over the Complex Numbers*, Grundlehren Text Editions,
https://doi.org/10.1007/978-3-031-25570-0_5

degenerate complex torus and in Section 5.4.3 the Weil Intermediate Jacobian, which is an abelian variety. Finally we define the Lazzeri Intermediate Jacobian, which in special cases equals an algebraic version of the Griffiths Intermediate Jacobian.

5.1 Abelian Surfaces

An *abelian surface* is by definition an abelian variety of dimension 2. A polarization L on an abelian surface X is of type (d_1, d_2). If $d_1 \geq 3$, the associated map $\varphi_L : X \to \mathbb{P}^N$ is an embedding according to the Theorem of Lefschetz 2.1.10. Theorems 2.2.1, 2.2.8 and 2.3.21 work out the behaviour of φ_L in the case $d_1 = 2$. In this section we prove a criterion, due to Reider, for φ_L to be an embedding in the case $d_1 = 1$.

5.1.1 Preliminaries

Let (X, L) be a polarized abelian surface of type (d_1, d_2). In this section we recall some general results.

For any line bundle L on the abelian surface X the Riemann–Roch Theorem states

$$\chi(L) = \tfrac{1}{2}(L^2).$$

By the Vanishing Theorem 1.6.4 and Proposition 2.1.11 the line bundle L is ample if and only if $h^i(L) = 0$ for $i = 1, 2$ and $(L^2) > 0$. For an ample L the Riemann–Roch Theorem says

$$h^0(L) = \tfrac{1}{2}(L^2) = d_1 d_2.$$

So the line bundle L defines a rational map $\varphi_L : X \dashrightarrow \mathbb{P}^{d_1 d_2 - 1}$.

Effective divisors on a surface can be interpreted as curves (not necessarily smooth or irreducible). We will use here both terms. For any curve C on X the *arithmetic genus* $p_a(C)$ is defined as

$$p_a(C) = 1 - \chi(\mathcal{O}_C).$$

The *adjunction formula* (see Hartshorne [61, Exercise V.1.3 (a)]) says

$$2p_a(C) - 2 = (C^2). \tag{5.1}$$

Here the intersection number of two curves is defined to be the intersection number of the corresponding line bundles. Hence $p_a(C)$ depends only on the line bundle $\mathcal{O}_X(C)$ and not on C itself. So for any curve C in the linear system of the ample line bundle L we obtain

$$p_a(C) = d_1 d_2 + 1. \tag{5.2}$$

Suppose now that L is ample of type $(1, d)$. Then the Decomposition Theorem 2.2.1 reads

Lemma 5.1.1 *L has a fixed component if and only if there are elliptic curves E_1 and E_2 such that $(X, L) \simeq (E_1 \times E_2, p_1^* L_1 \otimes p_2^* L_2)$ with line bundles L_1 of type (1) on E_1 and L_2 of type (d) on E_2.*

From now on assume that L has no fixed component. Then we have

Lemma 5.1.2

(a) *If $d \geq 3$, the line bundle L has no base point.*
(b) *If $d = 2$, the line bundle L has exactly four base points.*

Proof Suppose L has a base point. The group $K(L)$ acts on the base locus of L by translations. Hence L has at least $d^2 = \#K(L)$ base points. On the other hand there are at most $(L^2) = 2d$ base points, implying $d \leq 2$.

If $d = 2$, then φ_L maps the abelian surface X to $\mathbb{P}^1$. Since $(L^2) > 0$, the map $\varphi_L : X \to \mathbb{P}^1$ is not a fibration, so L has a base point. By what we have said above the base locus of L consists exactly of four points. □

According to the Theorem of Bertini (see Griffiths–Harris [55, p. 137]) a general member of $|L|$ is singular at most in the base locus of L. This gives immediately

Proposition 5.1.3

(a) *If $d \geq 2$, the general member of the linear system $|L|$ is smooth.*
(b) *If $D \in |L|$ is an irreducible and reduced divisor and $x \in D$, then the multiplicity $\operatorname{mult}_x D$ satisfies*

$$2d = (L^2) \geq \operatorname{mult}_x D \cdot (\operatorname{mult}_x D - 1) + \deg G,$$

where $G : D_s \longrightarrow \mathbb{P}^1$ is the Gauss map.

Proof (a): If $d \geq 3$ the assertion follows from the previous lemma. If L is of type $(1, 2)$ and $D \in |L|$ is singular in one of the four base points of L, then $(D^2) > 4$, a contradiction.

(b): According to Proposition 2.2.6 the Gauss map G is dominant. Let ϑ be a theta function for D. For every $w \in T_{X,0}$ the derivative $\partial_w \vartheta$ vanishes at x of order $\geq \operatorname{mult}_x D - 1$. On the other hand, the derivatives $\partial_w \vartheta|_D$, with $w \in T_{X,0}$, define the Gauss map (see Lemma 2.2.7). So for a general $w \in T_{X,0}$ the derivative $\partial_w \vartheta$ vanishes at $\deg G$ smooth points of D. According to a Theorem of Noether the local intersection number of D with the divisor $(\partial_w \vartheta)$ at x is $\geq \operatorname{mult}_x D \cdot \operatorname{mult}_x \partial_w \vartheta \geq \operatorname{mult}_x D \cdot (\operatorname{mult}_x D - 1)$ (see Fulton [47, Section 12.4]). Summing up and using the fact that $\partial_w \vartheta|_D$ is a section of $L|_D$ we obtain

$$2d = \deg L|_D \geq \operatorname{mult}_x D \cdot (\operatorname{mult}_x D - 1) + \deg G. \qquad \square$$

5.1.2 Rank-2 Bundles on an Abelian Surface

In the next section we need some properties of rank-2 vector bundles on a polarized abelian surface (X, L).

A vector bundle F of rank 2 on X is called *μ-semistable* (with respect to a polarization L of X) if

$$(c_1(E) \cdot L) \leq \tfrac{1}{2}(c_1(F) \cdot L) \tag{5.3}$$

for every coherent subsheaf E of F of rank 1. It is called *μ-stable* if the inequality is always strict.

Any coherent subsheaf of F is contained in a unique coherent subsheaf of F with a torsion-free quotient sheaf. On the other hand, any coherent subsheaf of rank 1 of F with torsion-free quotient is a line bundle. Hence in the definition of μ-(semi-)stability it suffices to require the inequality (5.3) for all line bundles contained in F.

Finally, recall that any coherent torsion-free sheaf of rank 1 on X is of the form $I_Z \otimes M$, where M is a line bundle on X and I_Z the ideal sheaf of a zero-dimensional subscheme Z of X.

Denote as usual by $\mathcal{E}nd\, F = F^* \otimes F$ the vector bundle of endomorphisms of F.

Lemma 5.1.4 $h^0(\mathcal{E}nd\, F) = 1$ *for any μ-stable vector bundle F of rank 2 on X.*

Proof It suffices to show that any nonzero endomorphism $f \colon F \to F$ is multiplication by a constant. Suppose first that f is of rank 1. Consider the following exact sequences of coherent sheaves

$$0 \longrightarrow \ker f \longrightarrow F \longrightarrow \operatorname{Im} f \longrightarrow 0$$
$$0 \longrightarrow \operatorname{Im} f \longrightarrow F \longrightarrow \operatorname{coker} f \longrightarrow 0.$$

The μ-stability of F implies

$$\tfrac{1}{2}\big(c_1(F)\cdot L\big) < \big(c_1(F)\cdot L\big) - \big(c_1(\ker f)\cdot L\big) = \big(c_1(\operatorname{Im} f)\cdot L\big) < \tfrac{1}{2}\big(c_1(F)\cdot L\big),$$

a contradiction. Hence f is of rank 2. But then $\det f$ is a nonzero endomorphism of the line bundle $\det F$ and as such multiplication by a nonzero constant. This implies that f is an isomorphism. It follows that the $\mathbb{C}$-algebra of endomorphisms of F is a skew field of finite dimension over $\mathbb{C}$ and thus isomorphic to $\mathbb{C}$ itself. □

The main tool for the proof of Theorem 5.1.6 below is the following result due to Bogomolov.

Theorem 5.1.5 (Bogomolov's Inequality) *Any μ-semistable vector bundle F of rank 2 on X satisfies the inequality*

$$\big(c_1(F)^2\big) \leq 4c_2(F).$$

Proof **Step I:** Suppose first F is μ-stable with respect to a polarization L. According to Lemma 5.1.4 and Serre duality we have $h^2(\mathrm{End}\, F) = h^0(\mathrm{End}\, F) = 1$. Moreover, the trace yields a splitting of the natural inclusion $\mathcal{O}_X \hookrightarrow \mathrm{End}\, F$, $\alpha \mapsto \alpha \mathbf{1}_F$. This implies $h^1(\mathrm{End}\, F) \geq h^1(\mathcal{O}_X) = 2$. So

$$\chi(\mathrm{End}\, F) = \sum_{\nu=0}^{2} (-1)^\nu h^\nu(\mathrm{End}\, F) \leq 2 - 2 = 0. \tag{5.4}$$

On the other hand the Riemann–Roch Theorem states

$$\chi(\mathrm{End}\, F) = \mathrm{ch}(\mathrm{End}\, F)_2 \tag{5.5}$$

(see Exercise 1.7.3 (3)). Using the calculus of Chern classes we obtain

$$\mathrm{ch}(\mathrm{End}\, F)_2 = (c_1(F)^2) - 4c_2(F).$$

Combining this with (5.4) and (5.5) gives the assertion in the μ-stable case.

Step II: Suppose F is μ-semistable, but not μ-stable with respect to L. Then there is an exact sequence

$$0 \longrightarrow G_1 \longrightarrow F \longrightarrow I_Z \otimes G_2 \longrightarrow 0$$

with line bundles G_1 and G_2 on X such that $(c_1(G_1) \cdot L) = (c_1(G_2) \cdot L)$, and an ideal sheaf I_Z of a zero-dimensional subscheme Z of X. Tensoring with G_1^{-1}, we obtain the exact sequence

$$0 \longrightarrow \mathcal{O}_X \longrightarrow F \otimes G_1^{-1} \longrightarrow I_Z \otimes G_2 \otimes G_1^{-1} \longrightarrow 0$$

with $\left(c_1(G_2 \otimes G_1^{-1}) \cdot L\right) = 0$. On the other hand, comparing the Chern classes of F and $F \otimes G_1^{-1}$ we get

$$\begin{aligned}(c_1(F)^2) - 4c_2(F) &= \left(c_1(F \otimes G_1^{-1})^2\right) - 4c_2(F \otimes G_1^{-1}) \\ &= \left(c_1(G_2 \otimes G_1^{-1})^2\right) - 4 \deg Z \leq 0.\end{aligned}$$

For the last inequality we used that

$$\left(c_1(G_2 \otimes G_1^{-1})^2\right) \cdot (L^2) \leq \left(c_1(G_2 \otimes G_1^{-1}) \cdot L\right)^2 = 0$$

by the Hodge index theorem (see Hartshorne [61, Ex. V.1.9]). □

5.1.3 Reider's Theorem

Suppose L is an ample line bundle of type $(1, d)$ with $d \geq 5$ on an abelian surface X. The object of this section is to give a criterion for the induced map $\varphi_L \colon X \to \mathbb{P}^{d-1}$ to be an embedding. In order to avoid trivialities, suppose (X, L) does not split; that is, is not isomorphic to a polarized product of elliptic curves $(E_1 \times E_2, p_1^* L_1 \otimes p_2^* L_2)$. The main result of this section is

Theorem 5.1.6 (Reider's Theorem) *Suppose L is an ample line bundle of type $(1, d)$ with $d \geq 5$ on X. Then the morphism $\varphi_L \colon X \to \mathbb{P}^{d-1}$ is an embedding if and only if there is no elliptic curve C on X with $(C \cdot L) = 2$.*

A general abelian surface X with polarization L of type $(1, d)$ does not contain an elliptic curve, since its Néron Severi group $\mathrm{NS}(X)$ is isomorphic to $\mathbb{Z}$ (see Exercise 3.1.5 (6)). So generically the map φ_L is an embedding.

On the other hand, special abelian surfaces X may contain infinitely many elliptic curves. For applications, in particular to moduli problems, it is more useful to have a criterion involving only one curve: according to Exercise 1.5.5 (6) there is a cyclic isogeny of abelian varieties $q \colon X \to Y$ of degree d and a line bundle M on Y with $L = q^* M$. The line bundle M defines a principal polarization on Y. According to Corollary 4.7.3 the unique curve D_Y in $|M|$ is either smooth of genus 2 or a union of two elliptic curves intersecting exactly in one point. Define D_X by the cartesian diagram

$$\begin{array}{ccc} D_X & \hookrightarrow & X \\ {\scriptstyle q}\downarrow & & \downarrow{\scriptstyle q} \\ D_Y & \hookrightarrow & Y \end{array}$$

and recall that an *elliptic involution* on a curve is an involution whose quotient is an elliptic curve.

The following criterion was given by Ramanan [107] in the case of a smooth curve D_Y and by Hulek–Lange [65] in the remaining case. Here we deduce it as a consequence of Reider's Theorem.

Corollary 5.1.7 *For $d \geq 5$ the morphism φ_L is an embedding if and only if the curves D_X and D_Y do not admit elliptic involutions compatible with the covering q.*

Proof It suffices to show that the following two conditions are equivalent:

(i) there is an elliptic curve $C \subset X$ with $(C \cdot L) = 2$; and
(ii) the curves D_X and D_Y admit elliptic involutions compatible with q.

(i) $\Rightarrow$ (ii): Let $C \subset X$ be an elliptic curve with $(C \cdot L) = 2$. We may assume that C is an abelian subvariety of X. Then the quotient X/C is also an elliptic curve. Denote by f the composition $f \colon D_X \hookrightarrow X \to X/C$.

We claim that f is a finite map of degree 2. If D_X is irreducible this is clear, since $\deg f = (C \cdot L) = 2$. Otherwise $D_X = F_1 + F_2$, where F_1 and F_2 are elliptic curves with $(F_1 \cdot F_2) = d$ and $(C \cdot F_1) + (C \cdot F_2) = (C \cdot L) = 2$. Assume $(C \cdot F_1) = 0$. Then $C \equiv F_1$ and thus $2 = (C \cdot F_2) = (F_1 \cdot F_2) = d \geq 5$, a contradiction. By symmetry this gives $(C \cdot F_1) = (C \cdot F_2) = 1$, which implies the assertion.

Define $C' = q(C)$. Since by Proposition 2.5.10 the abelian surface Y does not contain any rational curve, C' and thus also Y/C' are elliptic curves. Let f' denote the composition $f' \colon D_Y \hookrightarrow Y \to Y/C'$. Then we have the following commutative diagram

$$\begin{array}{ccc} D_X & \xrightarrow{f} & X/C \\ {\scriptstyle q}\downarrow & & \downarrow{\scriptstyle \overline{q}} \\ D_Y & \xrightarrow{f'} & Y/C'. \end{array}$$

Let $d' = \deg(q|_C \colon C \to C')$. The composed map $\overline{q} f$ is finite of degree $\frac{2d}{d'}$. Since $q \colon D_X \to D_Y$ is of degree d, this implies that f' is finite of degree $\frac{2}{d'}$. But f' cannot be of degree 1, since Y/C' is an elliptic curve and either D_Y is smooth of genus 2 or reducible. So f and f' provide a pair of elliptic involutions compatible with the covering q.

(ii) $\Rightarrow$ (i): Suppose ι_X and ι_Y are elliptic involutions on D_X and D_Y compatible with q; that is, the following diagram commutes

$$\begin{array}{ccc} D_X & \xrightarrow{\iota_X} & D_X \\ {\scriptstyle q}\downarrow & & \downarrow{\scriptstyle q} \\ D_Y & \xrightarrow{\iota_Y} & D_Y. \end{array}$$

We may assume that $0 \in D_Y$ and $\iota_Y(0) = 0$. This implies that $\ker\{q \colon X \to Y\}$ is contained in D_X and ι_X acts trivially on it.

We first claim that ι_X extends to an involution $\widetilde{\iota_X}$ on X. Since D_X generates X, every $x \in X$ is of the form $x = \sum_{i=1}^n n_i p_i$ with $p_i \in D_X$ and $n_i \in \mathbb{Z}$. Define

$$\widetilde{\iota_X}(x) = \sum_{i=1}^n n_i \iota_X(p_i).$$

We have to show that this definition is independent of the chosen representation of x. It suffices to show that for any representation of the origin $0 = \sum n_i p_i$ we have $\sum n_i \iota_X(p_i) = 0$. According to the Universal Property of the Jacobian 4.5.1 the involution ι_Y on D_Y extends to an involution $\widetilde{\iota_Y}$ on Y. Note that this also holds for reducible D_Y. So

$$q(\sum n_i \iota_X(p_i)) = \sum n_i \iota_Y(q(p_i)) = \widetilde{\iota_Y}(q(\sum n_i p_i)) = 0.$$

This implies $\sum n_i \iota_X(p_i) =: p \in \ker q$. But $p = \iota_X(p) = \sum n_i p_i = 0$, which completes the proof of the claim.

Let $F = \mathrm{Im}(\mathbf{1}_X + \widetilde{\iota}_X)$. It is easy to see that this is an elliptic curve. Moreover, by definition the composed map $f\colon D_X \hookrightarrow X \to F$ is the double covering associated to the involution ι_X. The kernel $\ker(\mathbf{1}_X + \widetilde{\iota}_X)$ is a disjoint union of translates of an elliptic curve C, say $\ker(\mathbf{1}_X + \widetilde{\iota}_X) = \bigcup_{i=1}^n t_{x_i}^* C$. Since $n(C \cdot L) = \deg f = 2$, we obtain $(C \cdot L) = 1$ or 2. If $(C \cdot L) = 1$, the polarized abelian variety (X, L) splits according to Lemma 5.1.9 below, a contradiction. So $(C \cdot L) = 2$. □

5.1.4 Proof of the Theorem

It suffices to show that φ_L is an embedding provided that there is no elliptic curve C in X with $(C \cdot L) = 2$, the converse implication being trivial, since a line bundle of degree 2 on an elliptic curve is never very ample.

The idea of the proof is as follows: assuming that there are points p and q (possibly infinitely near) with $\varphi_L(p) = \varphi_L(q)$ (see Hartshorne [61, p. 392]), one constructs a vector bundle F on X from which a contradiction can be derived.

Recall that L is an ample line bundle of type $(1, d)$ with $d \geq 5$ without fixed components on X. According to Lemma 5.1.2 it is generated by global sections. Our assumption means

$$h^0(I_{p+q} \otimes L) = h^0(L) - 1 = d - 1. \tag{5.6}$$

Here I_{p+q} denotes the ideal of the zero-dimensional subscheme $p + q$ on X. Then we can construct a vector bundle F of rank 2 on X as follows: the exact sequence $0 \longrightarrow I_{p+q} \otimes L \longrightarrow L \longrightarrow L|p + q \longrightarrow 0$ gives the cohomology sequence

$$0 \to H^0(I_{p+q} \otimes L) \to H^0(L) \to H^0(L|p + q) \to H^1(I_{p+q} \otimes L) \to 0.$$

Together with (5.6) we deduce $h^1(I_{p+q} \otimes L) = 1$. According to Serre duality this gives

$$\mathrm{Ext}^1(I_{p+q} \otimes L, \mathcal{O}_X)^* \simeq H^1(I_{p+q} \otimes L) \simeq \mathbb{C}.$$

Hence there exists a non-trivial extension

$$0 \longrightarrow \mathcal{O}_X \longrightarrow F \longrightarrow I_{p+q} \otimes L \longrightarrow 0 \tag{5.7}$$

uniquely determined up to multiplication by a constant. Since $\mathrm{Ext}^1(L, \mathcal{O}_X) \simeq H^1(L)^* = 0$, the extension is not in the image of the canonical map $\mathrm{Ext}^1(L, \mathcal{O}_X) \to \mathrm{Ext}^1(I_{p+q} \otimes L, \mathcal{O}_X)$, implying that F is a vector bundle. For the proof of this fact we refer to Griffiths–Harris [55, p. 727]. With this notation we have

Proposition 5.1.8 *If F is not μ-semistable with respect to L, then there exists a curve C on X with $(C^2) = 0$ and $(C \cdot L) = 1$ or 2.*

Proof The assumption on F implies that there is a diagram

$$\begin{array}{ccccccccc} & & & & 0 & & & & \\ & & & & \downarrow & & & & \\ & & & & G_1 & & & & \\ & & & & \downarrow & & & & \\ 0 & \longrightarrow & \mathcal{O}_X & \longrightarrow & F & \longrightarrow & I_{p+q} \otimes L & \longrightarrow & 0 \\ & & & \searrow^{\sigma} & \downarrow & & & & \\ & & & & I_Z \otimes G_2 & & & & \\ & & & & \downarrow & & & & \\ & & & & 0 & & & & \end{array}$$

where Z is a zero-dimensional subscheme of X and G_1 and G_2 are line bundles on X with $(G_1 \cdot L) > \frac{1}{2}(c_1(F) \cdot L)$ or equivalently

$$(G_1 \cdot L) > (G_2 \cdot L). \tag{5.8}$$

We claim that there is an effective divisor C on X with $G_2 = \mathcal{O}_X(C)$ and $G_1 = L(-C)$. For the proof it suffices to show that the composed map $\sigma \colon \mathcal{O}_X \to F \to I_Z \otimes G_2$ is nonzero. Otherwise there would be a nonzero homomorphism $I_{p+q} \otimes L \to I_Z \otimes G_2$ and thus, taking double duals, a nonzero homomorphism $L \to G_2$. This would imply that $G_2 \otimes L^{-1} = G_1^{-1}$ is effective, contradicting $2(G_1 \cdot L) > (c_1(F) \cdot L) = (L^2) > 0$.

Next we claim:

$$0 \le (C \cdot L(-C)) \le 2. \tag{5.9}$$

The right-hand inequality follows from $2 = c_2(F) = (C \cdot L(-C)) + \deg Z$. For the left-hand inequality consider the diagram above restricted to an irreducible component C_i of C. Since the section σ vanishes on C_i, we obtain an injective homomorphism of sheaves $0 \to \mathcal{O}_{C_i} \to L(-C) \otimes \mathcal{O}_{C_i}$ implying that $(C_i \cdot L(-C)) = \deg L(-C)|_{C_i} \ge 0$. This proves (5.9).

The next point to observe is $(C^2) = 0$. In fact, we have $\left((L(-C) \otimes \mathcal{O}_X(C))^2\right) = (L^2) = 2d \ge 10$. Using the right-hand inequality of (5.9), this gives

$$(L(-C)^2) + (C^2) \ge 6.$$

On the other hand, by (5.8),

$$\begin{aligned}(L(-C)^2) - (C^2) &= \left((L(-C) \otimes \mathcal{O}_X(-C)) \cdot (L(-C) \otimes \mathcal{O}_X(C))\right) \\ &= (L(-C) \cdot L) - (C \cdot L) = (G_1 \cdot L) - (G_2 \cdot L) > 0.\end{aligned}$$

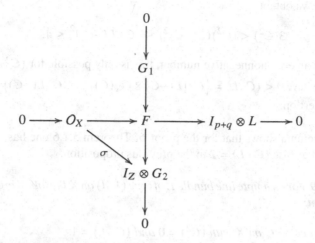

Adding both inequalities gives $\left(L(-C)^2\right) > 3$. So by Hartshorne [61, Exercise V.1.9 (a)] and (5.9) we obtain

$$3(C^2) < (C^2)\left(L(-C)^2\right) \leq \left(C \cdot L(-C)\right)^2 \leq 4.$$

Since (C^2) is an even nonnegative number, this is only possible for $(C^2) = 0$.

Finally we have $0 < (C \cdot L) = \Big(C \cdot \big(L(-C) \otimes \mathcal{O}_X(C)\big)\Big) = \big(C \cdot L(-C)\big) \leq 2$, which is the last assertion. □

The next lemma shows that for the proof of Theorem 5.1.6 one has only to take into account the case $(C \cdot L) = 2$ in the previous proposition.

Lemma 5.1.9 *For an ample line bundle L of type $(1, d)$ on X the following conditions are equivalent:*

(i) *There is a curve C on X with $(C^2) = 0$ and $(C \cdot L) = 1$.*
(ii) *The polarized abelian variety (X, L) is isomorphic to a polarized product of elliptic curves $(E_1 \times E_2, p_1^* L_1 \otimes p_2^* L_2)$.*

Proof (ii)$\Rightarrow$(i): Without loss of generality we may assume that L_1 is of type (1) on E_1 and L_2 is type (d) on E_2. Suppose there is an isomorphism $\varphi\colon (X, L) \to (E_1 \times E_2, p_1^* L_1 \otimes p_2^* L_2)$. Then $C := \varphi^{-1} E_1$ satisfies $(C^2) = 0$ by the adjunction formula (5.1) and moreover

$$(C \cdot L) = (E_1 \cdot p_1^* L_1 \otimes p_2^* L_2) = (E_1 \cdot E_2) + d(E_1^2) = 1.$$

(i)$\Rightarrow$ (ii): Note first that necessarily C is an elliptic curve by the adjunction formula (5.1) and Proposition 2.5.10. According to the Nakai–Moishezon Criterion (Corollary 2.2.3) the line bundle $L\big(-(d-1)C\big)$ is ample. Hence $h^0\Big(L\big(-(d-1)C\big)\Big) = \frac{1}{2}\Big(L\big(-(d-1)C\big)^2\Big) = 1$ and $h^0\Big(L\big(-(d-1)C\big)|_C\Big) = \deg L\big(-(d-1)C\big)|_C = 1$. The exact sequence

$$\begin{aligned} 0 \to H^0\big(L(-dC)\big) \to H^0\big(L(-(d-1)C)\big) &\to H^0\big(L(-(d-1)C)|_C\big) \\ &\to H^1\big(L(-dC)\big) \to 0 \end{aligned}$$

yields $h^0\big(L(-dC)\big) = h^1\big(L(-dC)\big) = 1$ or 0. On the other hand, we have $\chi\big(L(-dC)\big) = \frac{1}{2}\big(L(-dC)^2\big) = 0$. So, twisting eventually $L(-dC)$ by a line bundle of $\operatorname{Pic}^0(X)$, we may assume that $h^0\big(L(-dC)\big) = 1$ (see Theorem 1.6.8).

Finally, let $E_1 = C$ and E_2 the unique curve in $|L(-dC)|$. Since then $(E_1^2) = (E_2^2) = 0$ and $(E_1 \cdot E_2) = 1$, the curves E_1 and E_2 are elliptic and the map $\varphi\colon E_1 \times E_2 \to X$, $(p, q) \mapsto p - q$ is an isomorphism of abelian varieties. Moreover, for $L_1 = \mathcal{O}_{E_1}(0)$ and $L_2 = \mathcal{O}_{E_2}(d \cdot 0)$ we have $(X, L) \simeq (E_1 \times E_2, p_1^* L_1 \otimes p_2^* L_2)$ as polarized abelian varieties. □

Proof (of Reider's Theorem 5.1.6) If φ_L is not an embedding, then there exists the extension (5.7). But $c_1(F)^2 - 4c_2(F) = 2d - 8 > 0$. So F is not μ-semistable by Bogomolov's Inequality 5.1.5. Then Proposition 5.1.8 and Lemma 5.1.9 predict the existence of a curve C on X with $(C^2) = 0$ and $(C \cdot L) = 2$. Finally, C is elliptic by the adjunction formula (5.1) and Proposition 2.5.10. □

5.1.5 Exercises and Further Results

(1) Use Theorem 2.6.2 to see that the endomorphism algebra of an elliptic curve is either $\mathbb{Q}$ or an imaginary quadratic field.

(2) Use Theorem 2.6.2 and the previous exercise to see that only the following endomorphism algebras $\mathrm{End}_{\mathbb{Q}}(X)$ are possible for an abelian surface X:

(a) A non-simple X is isogenous to the product of elliptic curves $X \sim E_1 \times E_2$.

(i) If E_1 and E_2 are non-isogenous, then $\mathrm{End}_{\mathbb{Q}}(X) \simeq \mathrm{End}_{\mathbb{Q}}(E_1) \oplus \mathrm{End}_{\mathbb{Q}}(E_2)$.

(ii) If E_1 and E_2 are isogenous, then $\mathrm{End}_{\mathbb{Q}}(X) \simeq \mathrm{M}_2(\mathrm{End}_Q(E_1))$.

(b) If X is simple, then

(i) $\mathrm{End}_Q(X) \simeq \mathbb{Q}$ or a real quadratic field;

(ii) $\mathrm{End}_Q(X)$ is either a totally definite of a totally indefinite quaternion algebra over $\mathbb{Q}$;

(iii) $\mathrm{End}_Q(X)$ is a totally complex quadratic extension of either $\mathbb{Q}$ or a real quadratic field.

(3) Let X be a complex torus of dimension 2 with period matrix $(iY, \mathbf{1}_2)$, where $Y = (y_{ij}) \in \mathrm{M}_2(\mathbb{R})$.

(a) Use Exercise 1.3.4 (9) to show that the Picard number of X is

$$\rho(x) = 4 - \dim_{\mathbb{Q}}(y_{11}, y_{12}, y_{21}, y_{22}) + \begin{cases} 1 & \text{if } \det Y \in \mathbb{Q}, \\ 0 & \text{if } \det Y \notin \mathbb{Q}. \end{cases}$$

(b) The matrices Y in the following table give examples of complex tori X realizing all possible values for the Picard number $\rho = \rho(X)$ and the algebraic dimension $a = a(X)$ (defined in Chapter 2). Here p, q, r denote pairwise different prime numbers.

ρ \ a	0	1	2
0	$\begin{pmatrix}\sqrt{p} & \sqrt{q}\\ \sqrt{r} & 1\end{pmatrix}$	impossible	impossible
1	$\frac{1}{\sqrt{\sqrt{p}-\sqrt{qr}}}\begin{pmatrix}\sqrt{p} & \sqrt{q}\\ \sqrt{r} & 1\end{pmatrix}$	$\begin{pmatrix}\sqrt{p} & \sqrt{q}\\ 0 & 1\end{pmatrix}$	$\begin{pmatrix}\sqrt{p} & 1\\ 1 & \sqrt{q}\end{pmatrix}$
2	$\begin{pmatrix}1 & -\sqrt[3]{p}\\ \sqrt[3]{p} & 1\end{pmatrix}$	$\begin{pmatrix}\sqrt[3]{p} & 1\\ 0 & \sqrt[3]{p}\end{pmatrix}$	$\begin{pmatrix}\sqrt{p} & 0\\ 0 & 1\end{pmatrix}$
3	$\begin{pmatrix}1 & -\sqrt{p}\\ \sqrt{p} & 1\end{pmatrix}$	$\begin{pmatrix}1 & \sqrt{p}\\ 0 & 1\end{pmatrix}$	$\begin{pmatrix}\sqrt{p} & 1\\ 1 & \sqrt{p}\end{pmatrix}$
4	impossible	impossible	$\begin{pmatrix}1 & 0\\ 0 & 1\end{pmatrix}$

(Hint: for the computation of $a(X)$ use Exercises 1.3.4 (9) and 2.1.6 (10) (a). For the restrictions in the first line use Exercise 2.1.6 (10) (b), for the restrictions in the last line show that $\rho(X)$ being maximal implies that X is abelian (see Exercise 2.6.3 (2).)

(4) Let X be a complex torus of dimension 2 with algebraic dimension $a(X) = 0$. Show that any line bundle on X, not analytically equivalent to zero, is non-degenerate of index 1. For examples, see the previous exercise.

(5) Let X be a complex torus of dimension 2 with Picard number and algebraic dimension $\rho(X) = a(X) = 1$. Show that up to translation X admits exactly one elliptic curve. In particular, Poincaré's Reducibility Theorem for Complex Tori (see Exercise 1.6.4 (6)) is not valid for X.

(6) Let L be an ample line bundle of type $(1, 2)$ on an abelian surface X. Any curve $D \in |L|$ is of one of the following types

(a) D smooth of genus 3, admitting an elliptic involution.
(b) D irreducible of genus 2 with one double point, admitting an elliptic involution.
(c) $D = E_1 + E_2$ with elliptic curves E_1 and E_2 and $(E_1 \cdot E_2) = 2$.
(d) $D = E_0 + E_1 + E_2$ with elliptic curves E_i, such that $(E_0 \cdot E_1) = (E_0 \cdot E_2) = 1$ and $(E_1 \cdot E_2) = 0$.

The linear system $|L|$ always contains singular curves. In case (d) we have $(X, L) \simeq (E_0 \times E_1, p_1^* \mathcal{O}_{E_0}(0) \otimes p_2^* \mathcal{O}_{E_1}(2 \cdot 0))$, where 0 denotes the point 0 of E_0 respectively E_1.

(7) Let L be a symmetric ample line bundle of type $(1, 2)$. Show that the four base points of L are 4-division points.

(8) Show that any polarized abelian surface of type $(1, d)$ contains a curve of genus 2, not necessarily smooth and irreducible.

(9) Let L be an ample line bundle of type $(1, 3)$ on an abelian surface X. Then

(a) the map $\varphi : X \to \mathbb{P}^2$ given by $h^0(L)$ is a covering of degree 6;
(b) $\varphi_* \mathcal{O}_X$ is a vector bundle of rank 6 on $\mathbb{P}^2$ with Chern polynomial $c_t(\varphi \mathcal{O}_X) = 1 - 9t + 33t^2$;
(c) the covering φ is ramified along a smooth and irreducible curve $D \subset X$ with $\mathcal{O}_X(D) \simeq L^3$;
(d) $\varphi|_D : D \to C$ is birational onto the branch locus C of φ.

(10) Let $L \in \operatorname{Pic}(X)$ be of characteristic zero with respect to some decomposition on the abelian surface X. Suppose L is ample of type $(1,4)$ with associated map $\varphi_L : X \to \mathbb{P}^3$.

(a) The variety $\varphi_L(X)$ is a surface of degree 8, 4 or 2 in $\mathbb{P}^3$.
(b) The coordinates can be chosen in such a way that the coordinate points are of multiplicity 4 (2 or 1 respectively) in $\varphi_L(X)$ if $\deg \varphi_L(X) = 8$ (4 or 2, respectively).

The following exercise is more precise.

(11) Suppose L is an ample line bundle of type $(1, 4)$ on an abelian surface X. Let the abelian surface Y and the curves D_X on X and D_Y on Y be as in Corollary 5.1.7 (note that they exist also for type $(1, 4)$).

(a) If D_X and D_Y do not admit elliptic involutions, compatible with the covering q, then $\varphi_L \colon X \to \mathbb{P}^3$ is birational onto its image.
(b) If D_X and D_Y admit elliptic involutions, compatible with the covering q, then $\varphi_L \colon X \to \mathbb{P}^3$ is a double covering of a singular quartic $\overline{X}$, which is birational to an elliptic scroll. Moreover, the coordinates of $\mathbb{P}^3$ can be chosen in such a way that $\overline{X}$ is given by the equation

$$\lambda_1(Y_0^2Y_1^2 + Y_2^2Y_3^2) - \lambda_2(Y_0^2Y_2^2 - Y_1^2Y_3^2) = 0$$

for some $(\lambda_1 : \lambda_2) \in \mathbb{P}^1 - \{(1 : 0), (0 : 1), (1 : i), (1 : -i)\}$. The surface $\overline{X}$ is singular exactly along the coordinate lines $\{Y_0 = Y_3 = 0\}$ and $\{Y_1 = Y_2 = 0\}$.

(12) Let (X, L) denote a principally polarized abelian surface with $L = \mathcal{O}_X(D)$, D a symmetric effective divisor, and $K = \varphi_{L^2}(X) \subset \mathbb{P}^3$ the corresponding Kummer surface.

(a) $\#(D \cap X_2) = 6$; that is D contains exactly 6 two-division points of X.
(b) For any $x \in X_2$ denote by D_x the unique divisor in the linear system $|t_x^* L|$. Show that the curve $C_x = \varphi_{L^2}(D_x)$ is a conic and $2C_x$ is a complete intersection of K with a plane in $\mathbb{P}^3$. These planes are called *singular planes of* K. There are exactly 16 of them.
(c) The double covering $\varphi_{L^2} : X \to K$ maps the 16 2-division points of X to the 16 *singular points* of K in $\mathbb{P}^3$.

(13) (*The* 16_6-*Configuration*) Let (X, L) be a principally polarized abelian surface with the notation of the previous exercise. Show that the 16 singular planes and the 16 singular points of the Kummer surface K form a 16_6-configuration; that is

(a) any singular plane contains exactly 6 singular points,
(b) any singular point is contained in exactly 6 singular planes.

Work out explicitly which singular points lie in a given singular plane etc.

(14) (*Generalization of the* 16_6-*Configuration*) Let X be an abelian variety of dimension g and L a symmetric line bundle on X defining an irreducible principal polarization. According to Theorem 2.3.20 the map $\varphi = \varphi_{L^2} : X \to K \subset \mathbb{P}^{2^g-1}$ is of degree 2. Its image K is called the *Kummer variety* of X. The singular points of K are exactly the images of the 2-division points $x \in X_2$. For $x \in X_2$ denote by D_x the unique divisor in the linear system $|t_x^* L|$. Then $2D_x \in |L^2|$ and thus corresponds to a uniquely determined hyperplane P_x in $\mathbb{P}^{2^g-1}$, called a *singular hyperplane*. Show that the 2^{2g} singular points of K and the 2^{2g} singular hyperplanes form a $(2^{2g})_{2^{g-1}(2^g-1)}$-configuration; that is, any singular hyperplane contains exactly $2^{g-1}(2^g - 1)$ singular points and any singular point lies in exactly $2^{g-1}(2^g - 1)$ singular hyperplanes.

(15) Let X be an abelian variety of dimension g, isogenous to a product $\times_{i=1}^{g} E$ with E an elliptic curve with complex multiplication, then X is isomorphic to a product of elliptic curves.
(Hint: For $g = 2$ see Shioda–Mitani [127] or Ruppert [114]. For $g \geq 3$ see Lange [81] or Schoen [117]. See also Exercise 2.6.3 (2).)

(16) Give an example of elliptic curves E, E_1, E_2 such that $E \times E_1 \simeq E \times E_2$, but E_1 is not isomorphic to E_2.

(17) Let X be a polarized abelian surface of type $(1, n)$ with period matrix $\left(Z, \begin{smallmatrix} 1 & 0 \\ 0 & n \end{smallmatrix}\right)$.
The surface X contains an elliptic curve if and only if there exist integers $a, b, c, d, e, f \in \mathbb{Z}$ satisfying

(a) $na + nez_{11} + (f - nb)z_{12} - dz_{22} + c \det Z = 0$ and
(b) $ac + de - bf = 0$.

(Hint: use Exercise 1.3.4 (9).)

(18) Consider for $z \in \mathbb{C}$, $|z| < 1$, the abelian surface X with period matrix $\left(\begin{smallmatrix} 1 & i & z & iz \\ z & -iz & 1 & -i \end{smallmatrix}\right)$. Show that X is isogenous to a product of elliptic curves. Moreover, X is isomorphic to a product of elliptic curves if and only if iz, $1 - z^2$, $i + iz^2$ are linearly dependent over $\mathbb{Q}$.

(19) Let C be a smooth projective curve of genus 2 with non-trivial reduced automorphism group. According to Exercise 4.8.3 (1) the curve C is isomorphic to one of the 6 types of curves in the list of the exercise. Show that:

(a) If C is of type I, its Jacobian J is isogenous to a product of elliptic curves. In general J is not isomorphic to a product of elliptic curves, for example, if $1, z, z', zz'$ are linearly independent over $\mathbb{Q}$.
(b) If C is of type II, its Jacobian is isomorphic to a product of elliptic curves if and only if z is contained in some imaginary quadratic field.
(c) If C is of type III, IV or V, its Jacobian is isomorphic to a product of elliptic curves.
(d) If C is of type VI, its Jacobian is a simple abelian surface.

(20) Show that a non-elliptic curve on an abelian surface is an ample divisor. (Hint: Use Corollary 2.2.3.)

(21) Show that a principal polarization on an abelian surface X induces a division of the 16 2-division points X_2 into sets of 10 and 6 points. (Hint: Use Proposition 2.3.15.)

5.2 Albanese and Picard Varieties

In this section we generalize the notion of a Jacobian of a smooth projective curve to higher-dimensional varieties. More generally we associate to any compact Kähler manifold M of dimension $n \geq 1$ two complex tori, the Albanese torus $\mathrm{Alb}(M)$ and the Picard torus $\mathrm{Pic}^0(M)$. If M is a smooth projective variety, they are abelian varieties and in fact dual to each other.

5.2.1 The Albanese Torus

Let M be a compact Kähler manifold of dimension n. Recall that $q(M) = h^0(\Omega^1_M)$ is called the *irregularity* of M. The Hodge decomposition $H^1(M, \mathbb{C}) = H^0(\Omega^1_M) \oplus H^1(\mathcal{O}_M)$ with $\overline{H^1(\mathcal{O}_M)} \simeq H^0(\Omega^1_M)$ implies that

$$H_1(M)_{\mathbb{Z}} := H_1(M, \mathbb{Z})/\text{torsion}$$

is a free abelian group of rank $2q$. By Stokes' theorem any element $\gamma \in H_1(M)_{\mathbb{Z}}$ yields in a canonical way a linear form on the vector space $H^0(\Omega_C)$, which we also denote by γ :

$$\gamma : H^0(\Omega^1_M) \to \mathbb{C}, \quad \omega \mapsto \int_\gamma \omega.$$

The same proof as for Lemma 4.1.1 shows that the canonical map $H_1(M)_{\mathbb{Z}} \to H^0(\Omega^1_M)^*$ is injective. It follows that $H_1(M)_{\mathbb{Z}}$ is a lattice in $H^0(\Omega^1_M)^*$ and the quotient

$$\mathrm{Alb}(M) := H^0(\Omega^1_M)^*/H_1(M)_{\mathbb{Z}}$$

is a complex torus of dimension $q(M)$, called the *Albanese torus* of M.

Proposition 5.2.1 *Any complex torus X is the Albanese torus of a manifold, namely of itself:*

$$\mathrm{Alb}(X) = X.$$

Proof If $X = V/\Lambda$, we have $V = H^0(\Omega^1_X)^*$ by Theorem 1.1.21 and $\Lambda = H_1(X, \mathbb{Z})$ by Section 1.1.3. □

The analogue of the Abel–Jacobi map is the Albanese map defined as follows: For a point $p_0 \in M$ the holomorphic map

$$\alpha_{p_0} : M \to \mathrm{Alb}(M), \quad p \mapsto \left\{\omega \mapsto \int_{p_0}^{p} \omega\right\} \mod H_1(M)_{\mathbb{Z}}$$

is called the *Albanese map* of M (with base point p_0). The pair $(\mathrm{Alb}(M), \alpha_{p_0})$ satisfies the following universal property .

Theorem 5.2.2 (Universal Property of the Albanese Torus) *Let $\varphi : M \to X$ be a holomorphic map into a complex torus X. There exists a unique homomorphism $\widetilde{\varphi} : \mathrm{Alb}(M) \to X$ of complex tori such that the following diagram is commutative*

$$\begin{array}{ccc} M & \xrightarrow{\varphi} & X \\ {\scriptstyle \alpha_{p_0}}\downarrow & & \downarrow{\scriptstyle t_{-\varphi(p_0)}} \\ \mathrm{Alb}(M) & \xrightarrow{\widetilde{\varphi}} & X. \end{array}$$

Proof Suppose $X = \mathbb{C}^g/\Lambda$. Let $\widetilde{M}$ denote the universal covering of M. Then $\varphi : M \to X$ lifts to a holomorphic map $\phi = (\phi_1, \ldots, \phi_g) : \widetilde{M} \to \mathbb{C}^g$. Considering the fundamental group $\pi_1(M, p_0)$ as a group of covering transformations on $\widetilde{M}$, the lifting ϕ maps $\pi_1(M, p_0)$ into Λ; that is,

$$\phi \circ \gamma(z) - \phi(z) \in \Lambda \tag{5.10}$$

for all $\gamma \in \pi_1(M, p_0)$ and $z \in \widetilde{M}$. Hence $\mathrm{d}(\phi_i \circ \gamma) = \mathrm{d}\phi_i$ and thus the differentials $\mathrm{d}\phi_i$ may be considered as elements of $H^0(\Omega^1_M)$. Choose a basis $\omega_1, \ldots, \omega_q$, with $q = q(M)$, of $H^0(\Omega^1_M)$ and write

$$\mathrm{d}\phi_i = \sum_{j=1}^{q} a_{ij}\omega_j. \tag{5.11}$$

Considering the ω_i as coordinate functions on $H^0(\Omega^1_M)^*$ the matrix $A = (a_{ij})$ defines a linear map $A : H^0(\Omega^1_M)^* \to \mathbb{C}^g$. Note that $H_1(M)_{\mathbb{Z}}$ is a quotient of the fundamental group $\pi_1(M, p_0)$, hence equation (5.10) implies that $A\big(H_1(M)_{\mathbb{Z}}\big) \subset \Lambda$. So A is the analytic representation of a homomorphism

$$\widetilde{\varphi} : \mathrm{Alb}(M) \to X.$$

Using equation (5.11) we obtain for all $p \in M$

$$\begin{aligned} t_{-\varphi(p_0)}\varphi(p) &= \phi(p) - \phi(p_0) \bmod \Lambda \\ &= {}^t\!\left(\int_{p_0}^{p} \mathrm{d}\phi_1 \cdots \int_{p_0}^{p} \mathrm{d}\phi_g\right) \bmod \Lambda \\ &= A\ {}^t\!\left(\int_{p_0}^{p} \omega_1 \cdots \int_{p_0}^{p} \omega_g\right) \bmod \Lambda = \widetilde{\varphi}\alpha_{p_0}(p). \end{aligned}$$

Thus the diagram commutes. The uniqueness of $\widetilde{\varphi}$ follows from the construction. □

5.2.2 The Picard Torus

In order to define the Picard torus of M note that the composed map

$$\iota : H^1(M, \mathbb{R}) \to H^1(M, \mathbb{C}) = H^0(\Omega^1_M) \oplus \overline{H^0(\Omega^1_M)} \xrightarrow{pr} \overline{H^0(\Omega^1_M)}$$

is injective, since every real differential 1-form is of the form $\alpha + \overline{\alpha}$ with $\alpha \in H^0(\Omega^1_M)$. Denote by

$$H^1_{\mathbb{Z}}(M) \simeq H^1(M, \mathbb{Z})/\text{torsion}$$

the image of $H^1(M, \mathbb{Z})$ in $\overline{H^0(\Omega^1_M)}$. Then the quotient

$$\mathrm{Pic}^0(M) := \overline{H^0(\Omega^1_M)}/H^1_{\mathbb{Z}}(M)$$

is a complex torus, since $\mathrm{rk}\, H^1_{\mathbb{Z}}(M) = \dim_{\mathbb{C}} H^1(M, \mathbb{C}) = 2\dim_{\mathbb{C}} \overline{H^0(\Omega^1_M)}$. $\mathrm{Pic}^0(M)$ is called the *Picard torus* of M. Note that in the special case of a complex torus M the Picard torus $\mathrm{Pic}^0(M)$ coincides with the dual torus $\widehat{X}$ (see Proposition 1.4.1). In particular the old and new notation $\mathrm{Pic}^0(M)$ coincide for a complex torus M. Note moreover that the construction of $\mathrm{Pic}^0(M)$ is functorial: if $f : M_1 \to M_2$ is a holomorphic map of compact Kähler manifolds, the pullback f^* of holomorphic 1-forms induces a homomorphism of complex tori (see Exercise 5.2.5 (1) (ii))

$$f^* : \mathrm{Pic}^0(M_2) \to \mathrm{Pic}^0(M_1).$$

As in the case of a complex torus the Picard torus can be identified with the group of line bundles with vanishing first Chern class. To be more precise:

Proposition 5.2.3 *For any compact Kähler manifold there is a canonical isomorphism*
$$\mathrm{Pic}^0(M) \simeq \ker\{c_1 : H^1(\mathcal{O}_M^*) \to H^2(M,\mathbb{Z})\}.$$

Proof The exponential sequence of M gives the exact sequence
$$\cdots \to H^1(M,\mathbb{Z}) \to H^1(\mathcal{O}_M) \to H^1(\mathcal{O}_M^*) \xrightarrow{c_1} H^2(M,\mathbb{Z}) \to \cdots .$$

Hence using Hodge duality
$$\ker c_1 = H^1(\mathcal{O}_M)/\big(\mathrm{Im}\, H^1(M,\mathbb{Z})\big) = \overline{H^0(\Omega_M^1)}/H^1_{\mathbb{Z}}(M) = \mathrm{Pic}^0(M). \qquad \square$$

5.2.3 The Picard Variety

If M is a smooth projective variety, then we will see that $\mathrm{Pic}^0(M)$ is an abelian variety. In this case $\mathrm{Pic}^0(M)$ is also called the *Picard variety* of M.

Let $\omega \in H^{1,1}(M) \cap H^2(M,\mathbb{Z})$ denote the first Chern class of the line bundle $\mathcal{O}_M(1)$.

Lemma 5.2.4 *The hermitian form*
$$H : \overline{H^0(\Omega_M^1)} \times \overline{H^0(\Omega_M^1)} \to \mathbb{C}, \quad H(\varphi,\psi) := -2i \int_M \bigwedge^{n-1} \omega \wedge \varphi \wedge \overline{\psi}$$

defines a polarization on $\mathrm{Pic}^0(M)$*, called the canonical polarization of* $\mathrm{Pic}^0(M)$.

Proof For $\varphi, \psi \in H^1_{\mathbb{Z}}(M) \subset \overline{H^0(\Omega_M^1)}$ the sums $\varphi + \overline{\varphi}$ and $\psi + \overline{\psi}$ are integral 1-forms in $H^1_{\mathbb{Z}}(M) = H^1(M,\mathbb{Z})/\text{torsion}$. So
$$\begin{aligned}\mathrm{Im}\, H(\varphi,\psi) &= \tfrac{1}{2i}\big(H(\varphi,\psi) - H(\psi,\varphi)\big)\\ &= -\int_M \bigwedge^{n-1} \omega \wedge \varphi \wedge \overline{\psi} + \int_M \bigwedge^{n-1} \omega \wedge \psi \wedge \overline{\varphi}\\ &= -\int_M \bigwedge^{n-1} \omega \wedge (\varphi + \overline{\varphi}) \wedge (\psi + \overline{\psi}) \in \mathbb{Z}.\end{aligned}$$

It remains to show that H is positive definite. For this recall the Hodge star-operator $* : H^{p,q}(M) \longrightarrow H^{n-p,n-q}(M)$ (see Griffiths–Harris [55, page 82]). It is defined in such a way that
$$(\ ,\ ) : H^{p,q}(M) \times H^{p,q}(M) \to \mathbb{C}, \quad (\varphi,\psi) = \int_M \varphi \wedge *\psi$$

is a hermitian inner product. In particular $(\varphi,\varphi) > 0$ for every $\varphi \neq 0$.

Of course one can define the star operator explicitly. We only need this for elements of $\overline{H^0(\Omega^1_M)}$, that is for forms of type $(0,1)$. But for every differential 1-form $\varphi \in \overline{H^0(\Omega^1_M)}$ we have according to Wells [142, V, Theorem 3.16] (applied with r=0, p =1),

$$*\varphi = \frac{-i}{(n-1)!} \bigwedge^{n-1} \omega \wedge \overline{\varphi}.$$

So for any $0 \neq \varphi \in \overline{H^0(\Omega^1_M)}$,

$$H(\varphi,\varphi) = -2i \int_M \bigwedge^{n-1} \omega \wedge \varphi \wedge \overline{\varphi} = 2(n-1)! \int_M \varphi \wedge *\varphi > 0. \qquad \square$$

Suppose (X, L) is a polarized abelian variety. Then Lemma 5.2.4 provides the dual abelian variety $\widehat{X} = \mathrm{Pic}^0(X)$ with a polarization H. On the other hand there is the notion of a dual polarization L_δ of $\widehat{X}$ as defined in Proposition 2.5.1. The next lemma shows that these polarizations are multiples of each other.

Lemma 5.2.5 *Let (X, L) be a polarized abelian variety of dimension g and type $(d_1, \dots, d_g)$ and $(\widehat{X}, L_\delta)$ its dual polarization. Then*

$$H = 4\,(g-1)!\, d_2 \cdots d_{g-1}\, c_1(L_\delta)$$

is the canonical polarization of $\mathrm{Pic}^0(X) = \widehat{X}$.

Proof By definition $\phi_L^* L_\delta \equiv L^{d_1 d_g}$ (see Proposition 2.5.1). So it suffices to check the following identity of hermitian forms

$$\rho_a(\phi_L)^* H = c_1(L^{4(g-1)!d}) = 4\,(g-1)!\, d\, c_1(L)$$

with $d = d_1 \cdots d_g = h^0(L)$.

For this choose a basis $e_1, \dots, e_g$ of $V := H^0(\Omega^1_X)^*$ with respect to which the hermitian form H_L of L is given by the identity matrix. If $v_1, \dots, v_g$ denote the corresponding coordinate functions, then the first Chern class of L, considered as an element of $H^2_{DR}(X)$, is

$$\omega = c_1(L) = \tfrac{i}{2} \sum_{\nu=1}^{g} \mathrm{d}v_\nu \wedge \mathrm{d}\overline{v}_\nu$$

(see Exercise 1.3.4 (8)). Moreover, the differentials $\mathrm{d}\overline{v}_1, \dots, \mathrm{d}\overline{v}_g$ give a basis of the tangent space $\overline{H^0(\Omega^1_X)} = H^{0,1}_{DR}(X)$ of $\mathrm{Pic}^0(X)$. With respect to these coordinates the analytic representation of the isogeny $\phi_L : X \to \widehat{X} = \mathrm{Pic}^0(X)$ is

$$\rho_a(\phi_L) : V \longrightarrow \overline{H^0(\Omega^1_X)}, \quad e_i \mapsto \mathrm{d}\overline{v}_i$$

(see Lemma 1.4.5). So we have by Lemma 5.2.4,

$$
\begin{aligned}
(\rho_a(\phi_L)^*H)(e_i,e_j) &= -2i\int_X \bigwedge^{g-1} \omega \wedge (\rho_a(\phi_L)(e_i)) \wedge \overline{(\rho_a(\phi_L)(e_j))} \\
&= -2i\int_X \bigwedge^{g-1} \omega \wedge d\overline{v}_i \wedge dv_j \\
&= -2i\left(\tfrac{i}{2}\right)^{g-1}(g-1)!\sum_{\nu=1}^{g}\int_X dv_1 \wedge d\overline{v}_1 \wedge \cdots \\
&\qquad \cdots \wedge \overbrace{dv_\nu \wedge d\overline{v}_\nu}^{\vee} \wedge \cdots \wedge dv_g \wedge d\overline{v}_g \wedge d\overline{v}_i \wedge dv_j \\
&= \tfrac{4}{g}\left(\tfrac{i}{2}\right)^g g!\,\delta_{ij}\int_X dv_1 \wedge d\overline{v}_1 \wedge \cdots \wedge dv_g \wedge d\overline{v}_g \\
&= \tfrac{4}{g}\delta_{ij}\int_X \bigwedge^{g} c_1(L) \\
&= \tfrac{4}{g}\delta_{ij}(L^g) && \text{(see Section 1.7.2)} \\
&= \tfrac{4}{g}g!\,d\,\delta_{ij}, && \text{(by the Riemann–Roch Theorem)} \\
&= 4\,(g-1)!\,d\,H(e_i,e_j). && \text{(by the choice of the coordinates)}
\end{aligned}
$$

This implies the assertion. □

5.2.4 Duality of Pic^0 and Alb

For a smooth projective variety M we have:

Proposition 5.2.6 *The dual abelian variety of the Albanese variety is the Picard variety:*

$$\mathrm{Pic}^0(M) = \widehat{\mathrm{Alb}(M)}.$$

Proof Let $\alpha = \alpha_{p_0} : M \to \mathrm{Alb}(M)$ be the Albanese map with respect to some base point $p_0 \in M$. By what we have said above $\mathrm{Pic}^0(\mathrm{Alb}(M)) \simeq \widehat{\mathrm{Alb}(M)}$. Hence it suffices to show that $\alpha^* : \mathrm{Pic}^0(\mathrm{Alb}(M)) \to \mathrm{Pic}^0(M)$ is an isomorphism. But its analytic representation $\alpha^* : \overline{H^0(\Omega^1_{\mathrm{Alb}(M)})} \to \overline{H^0(\Omega^1_M)}$ is an isomorphism by the Hodge Decomposition Theorem 1.1.21 (b) (applied to the complex torus $\mathrm{Alb}(M)$) and the rational representation $\alpha^* : H^1(\mathrm{Alb}(M),\mathbb{Z}) \to H^1_{\mathbb{Z}}(M)$ is an isomorphism, since $H^1(\mathrm{Alb}(M),\mathbb{Z}) = \mathrm{Hom}(H_1(M)_{\mathbb{Z}},\mathbb{Z}) \simeq H^1_{\mathbb{Z}}(M)$ (see Section 1.1.3). □

Corollary 5.2.7 *For any smooth projective variety M the complex torus $\mathrm{Alb}(M)$ is an abelian variety, called the Albanese variety.*

Proposition 5.2.8 *Let M be a smooth projective variety and $p_0 \in M$. There is a positive integer n such that the holomorphic map*

$$\alpha^n_{p_o} : M^n \longrightarrow \mathrm{Alb}(M), \quad (p_1, \ldots, p_n) \mapsto \sum_{i-1}^{n} \alpha_{p_0}(p_i)$$

is surjective. In particular, $\alpha_{p_0}(M)$ generates $\mathrm{Alb}(M)$ as a group.

Proof For every n the subset $A_n := \mathrm{im}(\alpha^n_{p_0})$ is an irreducible closed subvariety of $\mathrm{Alb}(M)$. As $0 \in A_n$ for all n, there is a sequence of embeddings

$$A_1 \subset A_2 \subset A_3 \subset \cdots .$$

Clearly there is an n_0 such that $A_n = A_{n_0}$ for all $n \geq n_0$. We claim that A_{n_0} is an abelian subvariety. By construction A_{n_0} is closed under addition. So it suffices to show that with $x \in A_{n_0}$ also $-x \in A_{n_0}$.

For this consider the universal covering $\pi : \mathbb{C}^q \longrightarrow \mathrm{Alb}(M)$ and let $V_{n_0} \subset \mathbb{C}^q$ denote the irreducible component of $\pi^{-1}(A_{n_0})$ containing 0.

Note first that multiplication by positive integers map A_{n_0} surjectively onto itself. Hence for any $v \in V_{n_0}$ and $k >> 0$ we have $\frac{1}{k}v \in V_{n_0}$. For any $0 \neq v \in V_{n_0}$ denote by $\ell_v \subset \mathbb{C}^q$ the line joining 0 and v. It suffices to show that $\ell_v \subset V_{n_0}$, since the map $\pi : V_{n_0} \longrightarrow A_{n_0}$ is surjective. For any holomorphic function f on $\mathbb{C}^q$ vanishing on V_{n_0} we have $f(\frac{1}{k}v) = 0$ for $k \gg 0$. Hence by the identity theorem for holomorphic functions on $\mathbb{C} = \ell_v$ the function f vanishes on the whole line l_v. This implies that $\ell_v \subset V_{n_0}$. This completes the proof that A_{n_0} is an abelian subvariety of $\mathrm{Alb}(M)$. Applying the Universal Property of the Albanese variety 5.2.2, one concludes that $A_{n_0} = \mathrm{Alb}(M)$. □

As a direct consequence of the Lefschetz Hyperplane Theorem (see Griffiths–Harris [55, p. 156]) we obtain:

Proposition 5.2.9 *Let M be a smooth projective variety of dimension $n \geq 3$ and $N \subset M$ a smooth hyperplane section. The embedding $N \hookrightarrow M$ induces an isomorphism of canonically polarized Picard varieties*

$$(\mathrm{Pic}^0(M), H_M) \xrightarrow{\sim} (\mathrm{Pic}^0(N), H_N).$$

Proof According to the Lefschetz Hyperplane Theorem the restriction maps $\mathrm{res} : \overline{H^0(\Omega^1_M)} \to \overline{H^0(\Omega^1_N)}$ and $\mathrm{res} : H^1_{\mathbb{Z}}(M) \to H^1_{\mathbb{Z}}(N)$ are isomorphisms. This implies that $\mathrm{Pic}^0(M) = \mathrm{Pic}^0(N)$. It remains to show that $\mathrm{res}^* H_N = H_M$.

If $\omega \in H^{1,1}(M) \cap H^2(M,\mathbb{Z})$ denotes the first Chern class of $\mathcal{O}_M(1)$ then clearly $\omega|_N$ is the first Chern class of $\mathcal{O}_N(1)$. So for all $\varphi, \psi \in H^0(\Omega^1_M)$

$$\begin{aligned}(\mathrm{res}^* H_N)(\varphi,\psi) &= -2i \int_N \Big(\bigwedge^{n-2} \omega \wedge \varphi \wedge \overline{\psi}\Big)\Big|_N \\ &= -2i \int_M \bigwedge^{n-1} \omega \wedge \varphi \wedge \overline{\psi} = H_M(\varphi,\psi),\end{aligned}$$

since $\omega|_N$ is the fundamental class of N. □

5.2.5 Exercises

(1) Let $f: M_1 \to M_2$ be a morphism of compact Kähler manifolds.

(i) There is a homomorphism of complex tori $\tilde{f}$ such that for every $p_1 \in M_1$ the following diagram commutes

$$\begin{array}{ccc} M_1 & \xrightarrow{\ f\ } & M_2 \\ \downarrow{\scriptstyle \alpha_{p_1}} & & \downarrow{\scriptstyle \alpha_{f(p_1)}} \\ \mathrm{Alb}(M_1) & \xrightarrow{\ \tilde{f}\ } & \mathrm{Alb}(M_2); \end{array}$$

(ii) the dual homomorphism of $\tilde{f}$ is $\widehat{f}: \mathrm{Pic}(M_2) \to \mathrm{Pic}(M_1)$, given by pullback of line bundles.

(2) Let M and N be compact Kähler manifolds. Then

(a) $\mathrm{Alb}(M \times N) \simeq \mathrm{Alb}(M) \times \mathrm{Alb}(N)$;
(b) $\mathrm{Pic}(M \times N) \simeq \mathrm{Pic}(M) \times \mathrm{Pic}(N)$.

(3) For any polarized abelian variety (X, H) there is an $m \in \mathbb{N}$ and a smooth projective surface S such that (X, mH) is isomorphic to the canonically polarized Picard variety of S:

$$(X, mH) \simeq (\mathrm{Pic}^0(S), H_S).$$

(Hint: Use Proposition 5.2.9 and Bertini's Theorem.)

5.3 Prym Varieties

Given a finite covering $f : C \to C'$ with Jacobians J and J', the complement P of f^*J' in J with respect to the canonical polarization is called a Prym variety if the restriction of the canonical polarization of J to P is a multiple of a principal polarization. We determine all coverings f leading to a Prym variety and give a topological construction of the most important ones. For more on Prym varieties, see Chapter 2 of [83].

5.3.1 Abelian Subvarieties of a Principally Polarized Abelian Variety

In Section 2.4.3 we introduced the notion of complementary abelian subvarieties of a polarized abelian variety (X, L) and studied some first properties. In this section we derive further results on such subvarieties in the special case of a principal polarization $L = \mathcal{O}_X(\Theta)$.

Let (X, Θ) be a principally polarized abelian variety and $\iota = \iota_Y : Y \hookrightarrow X$ an abelian subvariety of X. In order to simplify the notation, we identify X with its dual abelian variety $\widehat{X}$ via the isomorphism $\phi_\Theta : X \to \widehat{X}$, and write $\phi_Y := \phi_{\iota^*\Theta} : Y \to \widehat{Y}$ for the isogeny of the induced polarization. Recall that the exponent $e(Y)$ of Y is defined as the exponent of the finite group $\operatorname{Ker}\phi_Y$. According to Proposition 1.1.15 the map $\psi_Y = e(Y)\phi_Y^{-1}$ is an isogeny. With this notation the norm-endomorphism N_Y and the symmetric idempotent ε_Y of Y are

$$N_Y = \iota\psi_Y\widehat{\iota} \in \operatorname{End}(X) \quad \text{and} \quad \varepsilon_Y = \iota\phi_Y^{-1}\widehat{\iota} \in \operatorname{End}_{\mathbb{Q}}(X). \tag{5.12}$$

As we saw in Theorem 2.4.19, the assignment $Y \mapsto \varepsilon_Y$ gives a bijection between the sets of abelian subvarieties Y of X and symmetric idempotents in $\operatorname{End}_{\mathbb{Q}}(X)$. In this way the involution $\varepsilon \mapsto 1 - \varepsilon$ on the set of symmetric idempotents of $\operatorname{End}_{\mathbb{Q}}(X)$ leads to the notion of complementary abelian subvarieties.

Let Z be the abelian subvariety of X complementary to Y,

$$Z = \operatorname{Im}\big(e(Y)\mathbf{1}_X - N_Y\big).$$

Our first aim is to show that the exponents of Y and Z coincide in the case of a principally polarized X. This is a consequence of the following

Proposition 5.3.1 $e(Y) = \min\{n > 0 \mid n\varepsilon_Y \in \operatorname{End}(X)\}$ *for any abelian subvariety Y of a principally polarized abelian variety (X, Θ).*

Proof Define $e := \min\{n > 0 \mid n\varepsilon_Y \in \operatorname{End}(X)\}$. By definition of the exponent $e(Y)\varepsilon_Y = \iota\psi_Y\widehat{\iota}$ is an endomorphism, so $e \leq e(Y)$.

On the other hand, since ι is a closed immersion, it follows that $e\phi_Y^{-1}\widehat{\iota}$ is a homomorphism. So its dual $\iota(e\phi_Y^{-1})\widehat{} = \iota e\phi_Y^{-1}$ is also a homomorphism. Again, since ι is a closed immersion, $e\phi_Y^{-1} \in \operatorname{Hom}(\widehat{Y}, Y)$. But it follows immediately from the definition of the exponent that $e(Y)$ is the smallest positive integer such that $e(Y)\phi_Y^{-1} \in \operatorname{Hom}(\widehat{Y}, Y)$. Hence $e(Y) \le e$, which completes the proof. □

Since $\varepsilon_Z = 1 - \varepsilon_Y$ in $\operatorname{End}_{\mathbb{Q}}(X)$ for any pair (Y, Z) of complementary abelian subvarieties, the proposition implies:

Corollary 5.3.2 *Complementary abelian subvarieties of a principally polarized abelian variety have the same exponent.*

Note that for an arbitrary polarization Corollary 5.3.2 is not valid. For an example, see Exercise 5.3.4 (1).

In the sequel we denote by e the common exponent of the complementary abelian subvarieties Y and Z. Then Lemma 2.4.22 (4) simplifies to

$$N_Z = e_X - N_Y. \tag{5.13}$$

The following proposition gives further possibilities for expressing Z in terms of Y.

Proposition 5.3.3 $Z = (\operatorname{Ker} N_Y)^0 = \operatorname{Ker} \widehat{\iota} \simeq (X/Y)\widehat{}$.

Proof We have $Z = \operatorname{Im} N_Z \subset (\operatorname{Ker} N_Y)^0$, since $N_Y N_Z = 0$ by Lemma 2.4.22 (3). As Z and $(\operatorname{Ker} N_Y)^0$ are abelian subvarieties of the same dimension, this gives the first equation. Moreover, $(\operatorname{Ker} N_Y)^0 = (\operatorname{Ker} \widehat{\iota})^0$ by equation (5.12), since ι a closed immersion and ψ_Y an isogeny.

In order to show that $(\operatorname{Ker} \widehat{\iota})^0 = \operatorname{Ker} \widehat{\iota}$; that is, that $\operatorname{Ker} \widehat{\iota}$ is connected, consider the exact sequence $0 \longrightarrow Y \hookrightarrow X \longrightarrow X/Y \longrightarrow 0$. By Proposition 1.4.2 the dual sequence is also exact: $0 \longrightarrow (X/Y)\widehat{} \longrightarrow \widehat{X} \xrightarrow{\widehat{\iota}} \widehat{Y} \longrightarrow 0$. So $\operatorname{Ker} \widehat{\iota} \simeq (X/Y)\widehat{}$. In particular, $\operatorname{Ker} \widehat{\iota}$ is connected. □

Corollary 5.3.4 $K(\iota^*\Theta) = \iota^{-1}Z \simeq Y \cap Z$.

Proof $K(\iota^*\Theta) = \operatorname{Ker} \phi_{\iota^*\Theta} = \operatorname{Ker}(\widehat{\iota}\iota) = \iota^{-1} \operatorname{Ker}(\widehat{\iota}) = \iota^{-1}Z \simeq Y \cap Z$. □

By the symmetry of the situation Corollary 5.3.4 implies that $K(\iota_Y^*\Theta)$ and $K(\iota_Z^*\Theta)$ are isomorphic as abelian groups. Thus the types of the induced polarizations are related as follows:

Corollary 5.3.5 *Let (Y, Z) be a pair of complementary abelian subvarieties of a principally polarized abelian variety with $\dim Y \ge \dim Z = r$. If the induced polarization $\iota_Z^*\Theta$ is of type $(d_1, \dots, d_r)$, then $\iota_Y^*\Theta$ is of type $(1, \dots, 1, d_1, \dots, d_r)$.*

By definition the integer d_r is the exponent of the abelian subvariety Z. In particular we see again that the exponents of Y and Z coincide.

According to Corollary 2.4.24 the homomorphism $\mu = \iota_Y + \iota_Z : Y \times Z \to X$ is an isogeny. It is of exponent e, since by equation (5.13),

$$(N_Y, N_Z)(\iota_Y + \iota_Z) = e_X. \tag{5.14}$$

Lemma 5.3.6 *The induced polarization on $Y \times Z$ splits: $\phi_{(\iota_Y+\iota_Z)^*\Theta} = \phi_Y \times \phi_Z$.*

Proof This is a consequence of Corollary 2.4.24. □

5.3.2 Definition of a Prym Variety

Let $f : C \to C'$ be a covering of degree n of smooth projective curves. In order to avoid trivialities, we assume that C' is of genus ≥ 1. Denote by (J, Θ) and (J', Θ') respectively the corresponding Jacobians. Recall from Section 4.5.2 the norm map

$$N_f : J \to J', \qquad O_C\left(\sum r_\nu p_\nu\right) \mapsto O_{C'}\left(\sum r_\nu f(p_\nu)\right).$$

Identifying $J = \widehat{J}$ and $J' = \widehat{J'}$ as usual, the pull back map f^* is a homomorphism of J' into J.

Lemma 5.3.7 $(f^*)^*\Theta \equiv n\Theta'$*; that is, the divisors $(f^*)^*\Theta$ and $n\Theta'$ are algebraically equivalent, so define the same polarization.*

Proof By definition $N_f f^*$ is multiplication by n on J'. So using equation (4.9) we obtain $\phi_{n\Theta'} = n_{J'} = N_f f^* = \widehat{f^*} f^* = \phi_{(f^*)^*\Theta}$, and Proposition 1.4.12 gives the assertion. □

Letting $A = \operatorname{Im} f^*$, the map f^* factorizes into an isogeny j and the canonical embedding ι_A. With $\phi_A = \phi_{\iota_A^*\Theta}$ as above the following diagram commutes

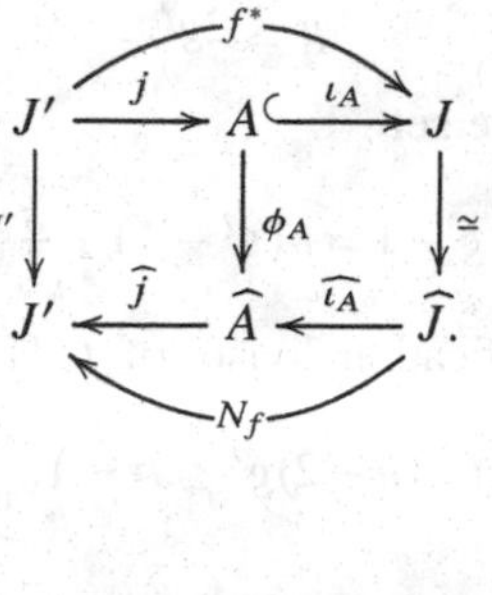

The norm map N_f and the norm-endomorphism N_A are related as follows:

Proposition 5.3.8 $f^*N_f = \frac{n}{e(A)}N_A$.

Proof From the diagram we deduce $\phi_A = n\widehat{j}^{-1}j^{-1}$, since j is an isogeny. So

$$\psi_A = e(A)\phi_A^{-1} = \frac{e(A)}{n} j\,\widehat{j},$$

which implies

$$f^*N_f = \iota_A j\,\widehat{j}\,\widehat{\iota}_A = \frac{n}{e(A)}\iota_A\psi_A\widehat{\iota}_A = \frac{n}{e(A)}N_A. \qquad \square$$

Another important abelian subvariety of J is the complementary abelian subvariety, say P, of A in J with respect to the canonical polarization Θ as defined in Section 2.4.4. It is often called the Prym variety of the cover f (see Lange–Rodriguez [83, Chapter 2]). Here we follow the original definition of Mumford (see [98]):

If $\Theta|_P$ is a multiple of a principle polarization, say Ξ, then (P, Ξ) is called the *Prym variety of the covering* $f\colon C \to C'$.

The following theorem, due to Wirtinger [147] and Mumford [98], gives a list of all coverings determining Prym varieties in this way.

Theorem 5.3.9 *Let $f\colon C \to C'$ be a covering of degree $n \geq 2$ of smooth projective curves of genus g and $g' \geq 1$ with f ramified if $g(C') = 1$. Then the abelian subvariety P of $J(C)$, as defined above, is a Prym variety if and only if f is of one of the following types:*

(a) *f is étale of degree 2;*
(b) *f is of degree 2 and ramified in 2 points;*
(c) *f is non-cyclic étale of degree 3 and $g(C') = 2$.*
(d) $\deg f \geq 2$, *$g(C) = 2$ and $g(C') = 1$.*

From Corollary 5.3.5 one easily deduces that the Prym variety P is of exponent 2 in the cases (a) and (b) and of exponent 3 in case (c).

Proof Suppose P is a Prym variety. Necessarily P is of exponent $e \geq 2$ in J, since otherwise the canonical polarization on J would split by Lemma 5.3.6 and Corollary 5.3.5. Since $\iota_P^*\Theta$ is of type $(e, \dots, e)$, the polarization on A, defined by $\iota_A^*\Theta$, is of type $(1, \dots, 1, e, \dots, e)$ again by Corollary 5.3.5. This implies $g' = \dim A \geq \dim P = g - g'$. So

$$g \leq 2g'. \tag{5.15}$$

Using the Hurwitz formula we get

$$2g' - 1 \geq g - 1 = n(g' - 1) + \tfrac{\delta}{2} \geq n(g' - 1) \tag{5.16}$$

with δ the degree of the ramification divisor of f. Hence

$$(n - 2)g' \leq n - 1. \tag{5.17}$$

We consider the following four cases separately:

Case 1: $n \geq 3$, $g' \geq 3$: On the one hand we have $6 \leq 2n$, on the other hand (5.17) implies $2n \leq 5$, a contradiction.

Case 2: $n \geq 3$, $g' = 2$: Here equation (5.17) gives $n = 3$ implying $\delta = 0$ and $g = 4$ by equation (5.16). So f is étale and $\dim Y = \dim Z = 2$.

Since the exponent e divides $n = 3$ by Proposition 5.3.8 and $e \geq 2$, we have $e = 3$ and the polarization $\iota_A^*\Theta$ is of type $(3, 3)$. Hence f is either cyclic or non-cyclic étale of degree 3. If f were cyclic, Proposition 4.5.3 would imply that f^* would not be injective and hence $\iota_A^*\Theta$ of type $(1, 3)$, a contradiction. So f is non-cyclic and we are in case (c).

Case 3: $n \geq 2$, $g' = 1$: By (5.15) the curve C is of genus $g = 2$ and we are in case (d).

Case 4: $n = 2$, $g' \geq 2$: Inequality (5.16) gives: $2g' - 1 \geq 2g' - 2 + \frac{\delta}{2}$. So $\delta \leq 2$ and we are either in case (a) or (b) of the theorem.

Conversely we have to show that in the cases (a), ..., (d) the abelian subvariety P is a Prym variety. It suffices to show that the induced polarization is of type $(e, \ldots, e)$. This is an easy exercise using Proposition 4.5.3, Corollary 5.3.5 and Lemma 5.3.7. □

Finally we prove a formula relating the theta divisors of J, J' and the Prym variety P in cases (a) and (b) of the theorem.

Proposition 5.3.10 *Suppose $f: C \to C'$ is a double covering, ramified in at most two points, and (P, Ξ) the associated Prym variety. Then*

$$2\Theta \equiv N_f^*\Theta' + \widehat{\iota_P}^*\Xi.$$

Proof In terms of divisors Lemma 5.3.6 reads $(\iota_A + \iota_P)^*\Theta \equiv q_A^*\iota_A^*\Theta + q_P^*\iota_P^*\Theta$ with q_A and q_P the natural projections of $A \times P$. Recall that $N_A + N_P = (\iota_A + \iota_P)(N_A, N_P) = 2_J$. So by Proposition 5.3.8,

$$\begin{aligned} 4\Theta \equiv 2_J^*\Theta &\equiv (N_A, N_P)^*(q_A^*\iota_A^*\Theta + q_P^*\iota_P^*\Theta) \\ &= N_A^*\Theta + N_P^*\Theta = N_f^*(f^*)^*\Theta + \widehat{\iota_P}^*\iota_P^*\Theta. \end{aligned}$$

But $(f^*)^*\Theta \equiv 2\Theta'$ by Lemma 5.3.7 and $\iota_P^*\Theta \equiv 2\Xi$. This gives the assertion, since the Néron–Severi group of J is torsion-free. □

5.3.3 Topological Construction of Prym Varieties

In Theorem 5.3.9 we saw that there are two types of double coverings determining Prym varieties: namely those ramified in none or two points. In this section we study these coverings from the topological point of view. This gives a second proof of the fact that they determine Prym varieties.

Let $f: C \to C'$ be a double covering of smooth projective curves, étale or ramified in two points. As in the last section denote by P the abelian subvariety of the Jacobian $J = J(C)$ complementary to the abelian subvariety $A = \operatorname{Im} f^*$. Let $\iota: C \to C$ be the involution corresponding to the double covering f. It extends to an involution $\tilde{\iota}$ on J. In terms of $\tilde{\iota}$ the norm-endomorphisms of the abelian subvarieties A and P can be described as follows.

Lemma 5.3.11 $N_A = 1 + \tilde{\iota}$ *and* $N_P = 1 - \tilde{\iota}$.

In particular, using Proposition 5.3.3 we get

$$A = \operatorname{Im}(1 + \tilde{\iota}) = \operatorname{Ker}(1 - \tilde{\iota})^0 \quad \text{and} \quad P = \operatorname{Im}(1 - \tilde{\iota}) = \ker(1 + \tilde{\iota})^0. \tag{5.18}$$

Proof Proposition 5.3.8 gives $N_A = f^* N_f$, so for any $x = \mathcal{O}_C(\sum r_\nu p_\nu) \in J$:

$$N_A(x) = f^* N_f\Big(\mathcal{O}_C\big(\sum r_\nu p_\nu\big)\Big) = \mathcal{O}_C\Big(\sum r_\nu (p_\nu + \iota p_\nu)\Big) = x + \tilde{\iota} x = (1 + \tilde{\iota})(x).$$

Consequently $N_P = 2 - N_A = 1 - \tilde{\iota}$. □

Recall that $J = H^0(\omega_C)^*/H_1(C, \mathbb{Z})$. In these terms the induced action of ι on $H^0(\omega_C)^*$, respectively $H_1(C, \mathbb{Z})$, is just the analytic respectively rational representation of $\tilde{\iota}$. Denote by $H^0(\omega_C)^-$ and $H_1(C, \mathbb{Z})^-$ the (-1)-eigenspaces in $H^0(\omega_C)$ and $H_1(C, \mathbb{Z})$ with respect to the action of the involution ι. An immediate consequence of (5.18) is:

Proposition 5.3.12 $P = \big(H^0(\omega_C)^-\big)^* / H_1(C, \mathbb{Z})^-$.

Suppose first that f is an étale covering. Setting $g = \dim P$, the curves C' and C are of genus $g+1$ and $2g+1$. Choose a symplectic basis $\lambda_0, \lambda_1, \dots, \lambda_g, \mu_0, \mu_1, \dots, \mu_g$ of $H_1(C', \mathbb{Z})$ (see the proof of Proposition 4.1.2 and the picture before that). From the topological point of view $f: C \to C'$ is a connected degree 2 covering of the topological space C' and as such determined by a nonzero element of $H_1(C', \mathbb{Z}/2\mathbb{Z})$. We may assume that this element is the image of the cycle λ_0 in $H_1(C', \mathbb{Z}/2\mathbb{Z})$.

Topologically f can be realized as follows: cut the surface C' along μ_0 and glue two copies of it with upper and lower boundary of μ_0 reversed, so that the orientations fit together.

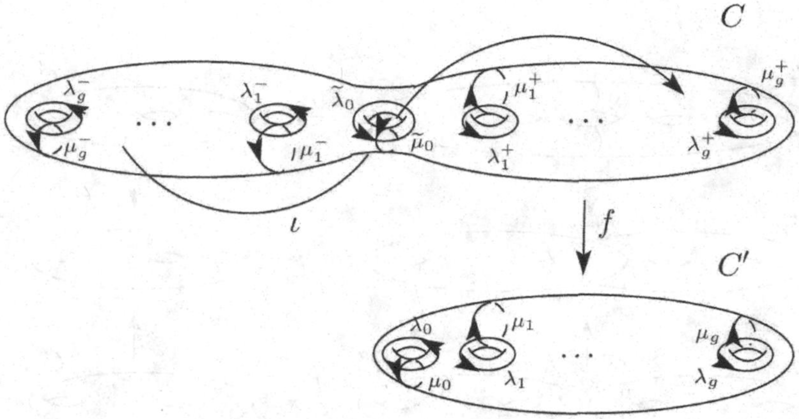

We obtain cycles $\widetilde{\lambda}_0, \lambda_i^+, \lambda_i^-, \widetilde{\mu}_0, \mu_i^+, \mu_i^-$ which obviously form a symplectic basis of $H_1(C,\mathbb{Z})$. The involution ι corresponding to the covering is just the map interchanging the copies. It is clear from the picture how ι acts on the lattice $H_1(C,\mathbb{Z})$. Obviously a basis of the (-1)-eigenspace $H_1(C,\mathbb{Z})^-$ is given by the skew symmetric cycles

$$\alpha_i := \lambda_i^+ - \lambda_i^- \,, \quad \beta_i := \mu_i^+ - \mu_i^- \,, \qquad i = 1, \dots, g.$$

Let $E\colon H_1(C,\mathbb{Z}) \times H_1(C,\mathbb{Z}) \to \mathbb{Z}$ denote the alternating form associated to the canonical polarization on $J(C)$. In order to show that P is a Prym variety, we compute its restriction to the basis α_i, β_i:

$$E(\alpha_i, \beta_j) = 2\delta_{ij}\,, \quad E(\alpha_i, \alpha_j) = E(\beta_i, \beta_j) = 0\,, \quad 1 \le i, j \le g.$$

So the induced polarization is twice a principal polarization and P is a Prym variety.

Remark 5.3.13 *Every Jacobian (J, Θ) of a smooth projective curve C of genus g is a limit of a family of Prym varieties (P_t, Ξ_t) of dimension g associated to étale double coverings.*

This was shown in Wirtinger [147] (see also Beauville [13]). We want to sketch the argument: choose distinct points p and q of C and identify them to obtain a singular curve C_0 with one double point whose normalization is C. Now choose a one-parameter family C_t of smooth genus $g+1$ curves degenerating to C_0. There is a family of 1-cycles μ_t on C_t shrinking to the singular point in C_0. The 1-cycle μ_t determines an étale double covering $f_t\colon \widetilde{C}_t \to C_t$ as described above. The curves $\widetilde{C}_t$ degenerate to a singular curve $\widetilde{C}_0$. The two singular points of $\widetilde{C}_0$ arise from shrinking the vanishing cycles μ_t' and μ_t'' of $\widetilde{C}_t$. So $\widetilde{C}_0$ is a reducible curve obtained by identifying the distinguished points $p_1 = q_2$ and $q_1 = p_2$ of two copies C' and C'' of C. The curve $\widetilde{C} = C' \cup C''$, the normalization of $\widetilde{C}_0$, is the induced double covering of C.

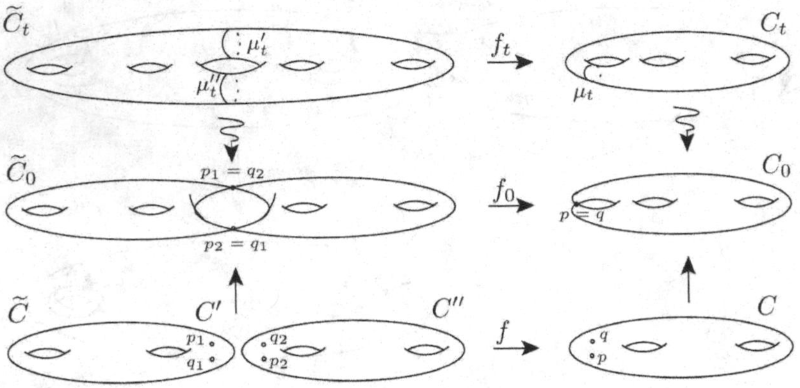

Recall that the Jacobian of $\widetilde{C}$ is the product of the Jacobians of its components; that is, $J(\widetilde{C}) = J \times J$. For the reducible double covering $f: \widetilde{C} \to C$ one can define a Prym variety P in the same way as in the irreducible case: the involution $\tilde{\iota}$ on $J(\widetilde{C}) = J \times J$ induced by the involution corresponding to the double covering f is obviously given by $\tilde{\iota}(x_1, x_2) = (x_2, x_1)$. Analogously as above, the Prym variety P for the covering f is

$$P = \operatorname{Im}(1 - \tilde{\iota}) = \{(x, -x) \mid x \in J\}$$

(see also equation (5.18)). So P is isomorphic to the Jacobian J of C. On the other hand, one can show, considering explicitly the degenerations of differential 1-forms and 1-cycles on C_t, that the family of Prym varieties (P_t) associated to the coverings $f_t: \widetilde{C}_t \to C_t$ degenerates to P.

Finally, suppose that $f: C \to C'$ is a double covering ramified over two points p_0 and q_0 of C'. Setting $g = \dim P$, the curves C' and C are of genus g and $2g$. As before, choose a symplectic basis $\lambda_1, \dots, \lambda_g, \mu_1, \dots, \mu_g$ of $H_1(C', \mathbb{Z})$.

Topologically the covering can be realized as follows: let γ be a path joining p_0 and q_0 which does not intersect any of the cycles λ_i, μ_i. Cut the surface C' along γ and glue two copies of it with upper and lower boundaries reversed, such that the orientations fit together.

We obtain cycles $\lambda_i^+, \lambda_i^-, \mu_i^+, \mu_i^-$, which obviously form a symplectic basis of $H_1(C, \mathbb{Z})$. As in the étale case one sees that the cycles

$$\alpha_i := \lambda_i^+ - \lambda_i^-, \quad \beta_i := \mu_i^+ - \mu_i^-, \qquad i = 1, \dots, g$$

form a basis of the (-1)-eigenspace $H_1(C, \mathbb{Z})^-$ of the involution ι. The restriction to $H_1(C, \mathbb{Z})^-$ of the alternating form $E: H_1(C, \mathbb{Z}) \times H_1(C, \mathbb{Z}) \to \mathbb{Z}$ of the canonical polarization on $J(C)$ is given by

$$E(\alpha_i, \beta_j) = 2\delta_{ij}\,, \quad E(\alpha_i, \alpha_j) = E(\beta_i, \beta_j) = 0\,, \quad 1 \le i, j \le g.$$

So the induced polarization on P is twice a principal polarization and P is a Prym variety.

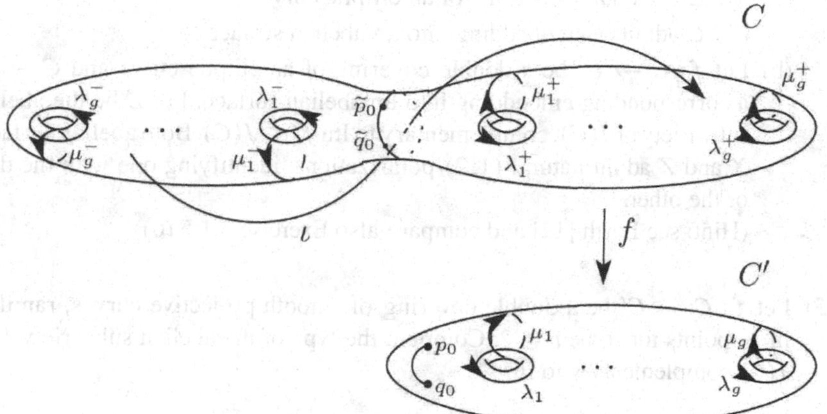

5.3.4 Exercises and Further Results

(1) Give an example of a pair of complementary abelian subvarieties of a non-principally polarized abelian variety for which Corollary 5.3.2 is not valid. (Hint: Consider a product of polarized abelian varieties with polarizations of suitable different types.)

(2) Suppose (X, L) is a polarized abelian variety of dimension g.

(a) Show that for any positive integer n there are only finitely many abelian subvarieties of X of exponent less than or equal to n.

(b) Conclude that for any smooth projective curve C and any positive integer n there are up to isomorphism only finitely many morphisms of degree less than or equal to n of C onto curves of genus ≥ 1.

A special case of (b) is the *Theorem of de Franchis* [44]: any smooth projective curve C admits up to isomorphism only finitely many morphisms onto curves of genus ≥ 2.

(3) Suppose $f\colon (X, L) \to (X', L')$ is an isogeny of polarized abelian varieties. Let Y be an abelian subvariety of X and d the exponent of the finite group $\operatorname{Ker} f|_Y$. Show that

$$\frac{1}{d^2} e(Y) \le e\big(f(Y)\big) \le e(Y).$$

(4) Let C be a smooth projective curve of genus 3.

(a) The following conditions are equivalent:
  (i) C is a double covering of an elliptic curve;
  (ii) C admits an embedding into an abelian surface X.

(b) Let $f\colon C \to C'$ be a double covering of an elliptic curve and $C \hookrightarrow X$ a corresponding embedding into an abelian surface. Let Z be the abelian subvariety of $J(C)$, complementary to $\operatorname{Im} f^* \subset J(C)$. Both abelian surfaces X and Z admit natural $(1, 2)$-polarizations, identifying one with the dual of the other.
(Hint: see Barth [12] and compare also Exercise 5.1.5 (6)).

(5) Let $f\colon C \to C'$ be a double covering of smooth projective curves, ramified in $2n$ points for some $n \ge 2$. Compute the type of the abelian subvariety P of $J(C)$ complementary to $\operatorname{Im} f^*$.

(6) Let $f\colon C \to C'$ be a finite morphism of smooth projective curves. Consider the abelian subvariety $A = \operatorname{Im} f^*$ of $J(C)$, and let k denote the largest integer such that $\operatorname{Ker}(f^* N_f)$ contains the group $J(C)_k$ of k-division points of $J(C)$.

(a) Show that $e(A) = \frac{\deg f}{k}$.
(b) Give an example of a morphism f with $e(A) < \deg f$.

(7) Let $f\colon C \to C'$ be an étale double covering of smooth projective curves. Recall that $\operatorname{Ker} N_f$ consists of two connected components, the Prym variety P associated to f and P_1. Show that $P = \big\{\sum_{\nu=1}^{2N} \mathcal{O}_C(p_\nu - \iota p_\nu) \,\big|\, N \ge 0,\ p_\nu \in C\big\}$ and $P_1 = \big\{\sum_{\nu=1}^{2N+1} \mathcal{O}_C(p_\nu - \iota p_\nu) \,\big|\, N \ge 0,\ p_\nu \in C\big\}$. A set of generators for P is given by $\{\mathcal{O}_C(p - \iota p)^2 \mid p \in C\}$. Conclude

$$P = (1 - \hat{\iota}) \operatorname{Pic}^0(C) \quad \text{and} \quad P_1 = (1 - \hat{\iota}) \operatorname{Pic}^1(C).$$

(8) Let C' be a smooth hyperelliptic curve of genus ≥ 2 and $B \subset \mathbb{P}^1$ the set of branch points of the hyperelliptic double covering.

(a) A connected étale double covering $f\colon C \to C'$ corresponds uniquely to a decomposition of B into two non-empty disjoint subsets $B = B_1 \cup B_2$, both of even cardinality.
(b) Let $B = B_1 \cup B_2$ be a decomposition inducing $f\colon C \to C'$ and $C_i \to \mathbb{P}^1$ the double covering branched over B_i. The Prym variety (P, Ξ) associated to f is isomorphic to the product $\big(J(C_1), \Theta_1\big) \times \big(J(C_2), \Theta_2\big)$.

(c) Conclude that the Jacobian of any hyperelliptic curve is the Prym variety associated to some étale double covering.
(See Mumford [98].)

(9) Let X be a smooth projective curve of genus 5 neither hyperelliptic nor trigonal.

(a) The canonical model of X is the complete intersection of three quadrics $\{Q_0 = 0\} \cap \{Q_1 = 0\} \cap \{Q_2 = 0\}$ in $\mathbb{P}^4$. The discriminant curve

$$C' = \{(x_0 : x_1 : x_2) \in \mathbb{P}^2 \mid \det(x_0Q_0 + x_1Q_1 + x_2Q_2) = 0\}$$

is a plane quintic depending only on X but not on the choice of the quadrics Q_i.

(b) C' is smooth if and only if $\mathrm{rk}(x_0Q_0 + x_1Q_1 + x_2Q_2) = 4$ for all $(x_0 : x_1 : x_2) \in C'$.

(c) Suppose C' is smooth. Then every quadric $x_0Q_0+x_1Q_1+x_2Q_2$ corresponding to a point of C' has two different rulings. The two rulings define an étale double covering $f\colon C \to C'$.

(d) With the notation of (b) the Prym variety associated to f is isomorphic to the Jacobian of X.

(e) Let $\eta \in \mathrm{Pic}^0(C)$ be the 2-division point associated to f. Show that $h^0(\mathcal{O}_{C'}(1) \otimes \eta) = 0$.
(See Masiewicki [89].)

(10) Let C be a smooth projective curve of genus $\neq 4$ and $f : \widetilde{C} \to C$ an étale double covering corresponding to a 2-division point $\eta \in \mathrm{Pic}^0(C')$.
Show that, if the Prym variety associated to f is a Jacobian, then C is either hyperelliptic or trigonal or a plane quintic with $h^0(\mathcal{O}_{C'}(1) \otimes \eta) = 0$.
(See Shokurov [129].)

The next seven exercises concern the Abel–Prym map. For this we use the following notation: Let $f : C \to C'$ be a double covering of smooth projective curves with C' of genus $g \geq 1$, étale or ramified in two points p_0 and q_0 in C. Let ι denote the corresponding involution of C.

Let (P, Ξ) denote the Prym variety of f with canonical embedding $\iota_P : P \hookrightarrow J = J(C)$. Identifying J and P with their duals via the canonical principal polarizations, the norm-endomorphism N_P is given by

$$N_P = \iota_P \widehat{\iota_P}.$$

Considering the Abel–Jacobi map $\alpha_c : C \to J$ with respect to a point $c \in C$, the composed map

$$\pi = \pi_c : C \overset{\alpha_c}{\to} J \overset{\widehat{\iota_P}}{\to} P$$

is called the *Abel–Prym map* of P (with respect to the point c).

(11) (*Universal Property of the Abel–Prym map*) For any morphism $\varphi : C \to X$ into an abelian variety X satisfying $\varphi\iota = -\varphi$, there exists a unique homomorphism $\psi : P \to X$ such that for any $c \in C$ the following diagram commutes

$$\begin{array}{ccc} C & \xrightarrow{\varphi} & X \\ {\scriptstyle \pi_c}\downarrow & & \downarrow{\scriptstyle t_{-\varphi(c)}} \\ P & \xrightarrow{\psi} & X. \end{array}$$

(See Masiewicki [89].)

(12) (a) Suppose C is not hyperelliptic. Then $\pi(p) = \pi(q)$ for points $p \neq q$ in C if and only if f is ramified in p and q; that is, $p = p_0$ and $q = q_0$.
(b) Suppose C is hyperelliptic. Then $\pi : C \to P$ is of degree 2 onto its image and $\pi(p) = \pi(q)$ for distinct points $p, q \in C$ if and only if $p + \iota(q)$ is in the unique linear system g^1_2 of C.

(13) (a) The double covering $f : C \to C'$ above can be described as follows: there is a non-trivial $\eta \in \mathrm{Pic}(C')$ with $\eta^2 \simeq \mathcal{O}_{C'}$ if f is étale and $\eta^2 \simeq \mathcal{O}_{C'}(f(p_0) + f(q_0)))$ if f is ramified in $p_0 \neq q_0$. This isomorphism defines an algebra structure on the sheaf $\mathcal{O}_{C'} \oplus \eta$ and C can be defined as the spectrum of this algebra.
(See Hartshorne [61][Ex. IV.2.7].)
(b) Show that

$$f_*\mathcal{O}_C = \mathcal{O}_{C'} \oplus \eta^{-1} \qquad \text{and} \qquad \omega_C = f^*(\omega_{C'} \otimes \eta).$$

(14) Show that the projectivized differential of the Abel–Prym map $\pi : C \to P$ is the composed map $\varphi_{\omega_{C'} \otimes \eta} f : C \to C' \to \mathbb{P}(H^0(\omega_{C'} \otimes \eta)^*)$.

(15) The differential $d\pi_p$ at the point $p \in C$ is injective unless one of the following cases holds:

(i) f is étale and C' is hyperelliptic, $\eta = \mathcal{O}_C(f(p) - q')$ for some $q' \in C'$ and $f(p)$ and q' are distinct branch points of the hyperelliptic covering;
(ii) f is ramified in $p_0, q_0 \in C$, the curve C' is hyperelliptic and $\eta = \mathcal{O}_C(f(p))$ with $2f(p) \sim f(p_0) + f(q_0)$.

(16) Suppose C is not hyperelliptic.

(i) If $f : C \to C'$ is étale, the Abel–Prym map $\pi : C \to P$ is an embedding.
(ii) If f is ramified in p_0 and q_0 in C, then $\pi : C \to P$ embeds $C - \{p_0, q_0\}$ and $\pi(p_0) = \pi(q_0)$ is an ordinary double point unless C' is hyperelliptic and $\eta = \mathcal{O}_C(p')$ for some $p' \in C'$ with $2p' \sim f(p_0 + q_0)$.

(17) Suppose C is hyperelliptic. Then we have for both types of coverings $f : C \to C'$: The curve $D = \pi(C)$ is a smooth hyperelliptic and the Prym variety (P, Ξ) is the Jacobian of D.

The next five exercises concern the singularities of the theta divisor given by an étale double covering $f : C \to C'$ with Prym variety P. As in Exercise (7) above let P_1 denote the other component of $\operatorname{Ker} N_f$. Here it turns out to be more convenient to assume that C' is of genus $g+1$ (so that P is of dimension g). Recall that a theta characteristic on a curve is a line bundle whose square is the canonical line bundle.

(18) Show that there is a theta characteristic κ on C which is the pullback of a theta characteristic on C'.

We fix such a κ in the sequel.

(19) (a) $P = \{L \in \operatorname{Ker} N_f \subset \operatorname{Pic}^0(C) \mid h^0(L \otimes \kappa) \equiv 0 \bmod 2\}$.
(b) $P_1 = \{L \in \operatorname{Ker} N_f \subset \operatorname{Pic}^0(C) \mid h^0(L \otimes \kappa) \equiv 1 \bmod 2\}$.
(Hint: Use Exercise (7) above.)

(20) Consider the theta divisor $\Theta := W_{g(C)-1} - \kappa$ in $\operatorname{Pic}^0(J)$. Then
(a) there is a theta divisor Ξ defining the principal polarization of P such that $\iota_P^*\Theta = 2\Xi$ (as divisors); and
(b) $P_1 \subset \Theta$.

(21) (a) $\dim \operatorname{sing} \Xi \geq g - 6$.
(b) If C' is a general curve of genus $g+1$, then $\operatorname{sing} \Xi$ is irreducible of dimension $g-6$ for $g \geq 7$, finite for $g = 6$ and empty for $g \leq 5$.
(See Welters [144] and Debarre [35].)

A singular point x of Ξ is called *stable* if the multiplicity of Θ is x is ≥ 4. It is called *exceptional* otherwise; that is, if it is equal to 2. The importance of exceptional singularities lies in the fact that they provide special line bundles of low degree on C'.

(22) A singularity L of Ξ (considered as a line bundle on C) is exceptional if and only if

$$L \otimes \kappa = f^*M \otimes \mathcal{O}(B)$$

with $M \in \operatorname{Pic}(C')$ such that $h^0(M) = 2$ and B is an effective divisor on C.

(23) Let $f : C \to C'$ be an étale double covering of smooth projective curves and g^r_d a complete base-point free linear system on C'. Define $V = {f^{(d)}}^{-1}(g^r_d)$, the scheme-theoretic preimage of g^r_d under $f^{(d)} : C^{(d)} \to C'^{(d)}$. As a set V consists of the divisors of degree d on C, which push down to divisors of the g^r_d.
(a) V consists of two connected components V_1 and V_2.
(b) If d is odd, the involution ι, corresponding to the double covering f, induces an isomorphism $V_1 \simeq V_2$.
(c) Suppose the following conditions are fulfilled:
(I) every fibre of the morphism $h : C' \to \mathbb{P}^r$ associated to the g^r_d contains at most one ramification point and this is of index ≤ 3;

(ii) if $r > 1$, the map k is birational onto its image.
Then V_1 and V_2 are normal and irreducible varieties.

(See Welters [143] and Beauville [14].)

(24) Let the notation be as in the previous exercise. Consider the map $\alpha_D : C^{(d)} \longrightarrow J(C)$ corresponding to the divisor $D \in C^{(d)}$. After suitable translations we may assume that $\alpha_D(V_1)$ and $\alpha_D(V_2)$ are contained in the Prym variety P associated to f. If $d > 2r$, then $\alpha_D(V_1)$ and $\alpha_D(V_2)$ have the same class in $H^{2(g-r)}(P, \mathbb{Z})$. To be more precise

$$[\alpha_D(V_i)] = \frac{2^{d-2r-1}}{(g-r)!} \bigwedge^{g-r} [\Xi] ,$$

where Ξ is a theta divisor on P and $g = \dim P$.
(See Beauville [14].)

(25) Let $\mathcal{R}_g$ denote the moduli space of non-trivial étale double coverings of curves of genus g. It is a finite covering of the moduli space $\mathcal{M}_g$ of curves of genus g of degree $2^{2g} - 1$. By associating to the étale double covering $f : C \to C'$ of $\mathcal{R}_{g+1}$ its Prym variety $P \in \mathcal{A}_{1_g}$ we get a morphism

$$p_g : \mathcal{R}_{g+1} \to \mathcal{A}_{1_g} ,$$

called the *Prym map*.

(a) The Prym map p_g is not everywhere injective.
(See Donagi's tetragonal construction [39]. For another case in which p_g is not injective, see Verra [135]. Examples of high Clifford index at which p_g is not injective were given by Izadi and Lange in [72].)
(b) (*Torelli Theorem for the Prym map*) The Prym map p_g is generically injective for $g \geq 6$. This implies that the image of p_g is of dimension $3g$ for $g \geq 6$.
(See Friedmann–Smith [45], Kanev [74], Welters [145] and Debarre [35].)
(c) The general fibre of p_5 consists of 27 elements. The Galois group of the field extension $\mathbb{C}(\mathcal{R}_6)|\mathbb{C}(\mathcal{A}_{1_5})$ is isomorphic to the Galois group of the 27 lines on a cubic surface.
(See Donagi–Smith [41].)
(d) The general fibre of p_4 is a double covering of a Fano surface.
(See Donagi [39].)
(e) For a general $(X, \Theta) \in \mathcal{A}_3$ the fibre $p_3^{-1}(X)$ is isomorphic to the Kummer variety of X.

A principally polarized abelian variety (P, Ξ) is called a *Prym–Tyurin variety for the curve C* if P is an abelian subvariety of the Jacobian (J, Θ) with canonical embedding $\iota_P : P \to J$, such that

$$\iota_P^* \Theta \equiv e\Xi$$

for some integer e, called the *exponent* of the Prym–Tyurin variety.

(26) Let σ be a correspondence of C into itself with associated endomorphism of J (see Theorem 4.6.1) denoted by the same symbol such that the endomorphism satisfies the equation

$$\sigma^2 + (e-2)\sigma - (e-1) = 0.$$

Show that $P := \mathrm{Im}(\sigma - 1)$ is a Prym–Tyurin variety for C of exponent e.

(27) (a) A Prym–Tyurin variety of exponent 2 is a usual Prym variety.

(b) Every principally polarized abelian variety of dimension g is a Prym–Tyurin variety of exponent $2^{g-1}(g-1)!$ for some curve C.
(Hint for (b): Use Bertini's Theorem to get C.)

(28) (*Universal Property of the Prym–Tyurin variety*) Let (P, Ξ) be a Prym–Tyurin variety for the curve C and $\varphi \colon C \to X$ a morphism into an abelian variety X. Assume that the induced homomorphism $\widetilde{\varphi} \colon J \to X$ satisfies

$$\widetilde{\varphi} N_A = 0.$$

(Here N_A is the norm-endomorphism of the complementary abelian subvariety A of P in J.) Then there is a unique homomorphism $\psi \colon P \to X$ such that for any $c \in C$ the following diagram commutes

$$\begin{array}{ccc} C & \xrightarrow{\varphi} & X \\ {\scriptstyle \pi_c}\downarrow & & \downarrow{\scriptstyle t_{-\varphi(c)}} \\ P & \xrightarrow{\psi} & X, \end{array}$$

where π_c denotes the composition $\pi_c : C \overset{\alpha_c}{\to} J \overset{\widehat{\iota_P}}{\to} P$.

5.4 Intermediate Jacobians

In Chapter 4 we associated to every smooth projective curve C an abelian variety which reflects its geometry, the Jacobian $J = J(C)$. If M is a smooth projective variety of dimension $n > 1$, there are several ways to associate to M an abelian variety with some geometric relevance: In Section 5.2 we defined the Picard variety $J^1 := \mathrm{Pic}^0(M)$ using the first cohomology of M and the Albanese variety $J^n := \mathrm{Alb}(M)$ using the $(2n-1)$-th cohomology of M (see Section 5.4.2 below), which is dual to J^1. So for $n = 1$, $J = J^1$ is self-dual. For $n > 2$, every cohomology of odd weight yields an abelian variety. These are the *Intermediate Jacobians* $J^p(M)$. They are in general highly decomposable complex tori. This implies that there are many different possibilities for their definition.

Weil introduced in [138] an abelian variety $(J^p_W(M), H^p_W)$ for every $1 \leq p \leq n$, called the *p-th Weil Intermediate Jacobian* (see Section 5.4.3). It has the disadvantage that it does not vary holomorphically in families in general. Due to this fact, Griffiths introduced in [54] for all $1 \leq p \leq n$ a Jacobian $(J^p_G(M), H^p_G)$, called the *p-th Griffiths Intermediate Jacobian* (Section 5.4.2). It contains the same geometric information as $(J^p_W(M), H^p_W)$ and moreover varies holomorphically with the variety M. This has the advantage that one can study properties of M via deformation. However, it is not an abelian variety in general. The hermitian form H^p_G is non-degenerate of some index which can be easily computed.

By changing the complex structure one can associate an abelian variety $(\widetilde{J}^p_G(M), \widetilde{H}_G)$ in a canonical way. In order to see what it is, at least for $p = \frac{n+1}{2}$, we introduce in Section 5.4.4 the *Lazzeri intermediate Jacobian* and show in Section 5.4.5 that both abelian varieties coincide.

In this section we follow parts of Birkenhake–Lange [22, Chapter 4].

5.4.1 Primitive Cohomology

For the definition of the Intermediate Jacobians we need some cohomological facts of Kähler theory, which we collect here, referring for the proofs to Weil [140] and Wells [142].

Let M be a smooth projective variety of dimension n. As usual let Ω^p_M denote the sheaf of holomorphic p-forms on M and $H^{p,q}(M) = H^q(M, \Omega^p_M)$. The decomposition of differential forms on M into forms of type (p, q) yields the *Hodge decomposition* (see Section 1.1.5),

$$H^r(M, \mathbb{C}) = \bigoplus_{p+q=r} H^{p,q}(M) \quad \text{with} \quad H^{p,q}(M) = \overline{H^{q,p}(M)}. \tag{5.19}$$

Note that in general the Hodge decomposition does not vary holomorphically in families. However, as shown by Griffiths in [54], the Hodge filtration, defined as follows, does vary holomorphically with M: For $p = 0, \dots, r$ define

$$F^p H^r := F^p H^r(M, \mathbb{C}) := \bigoplus_{s=p}^{r} H^{s,r-s}(M).$$

This gives the *Hodge filtration*

$$0 =: F^{r+1} H^r \subseteq F^r H^r \subseteq \cdots \subseteq F^1 H^r \subseteq F^0 H^r = H^r(M, \mathbb{C}). \tag{5.20}$$

Note that (5.19) implies for every p, $1 \leq p \leq r$,

$$H^r(M, \mathbb{C}) = F^p H^r \oplus \overline{F^{r-p+1} H^r}. \tag{5.21}$$

Now let $\omega \in H^{1,1}(M) \cap H^2(M,\mathbb{Z})$ denote the Kähler form induced by a hyperplane section of M. The cup product with ω defines the *Lefschetz operator*

$$L: H^r(M,R) \to H^{r+2}(M,R), \qquad \varphi \mapsto \omega \wedge \varphi$$

for $R = \mathbb{Z}, \mathbb{R}$ or $\mathbb{C}$. For the proof of the following theorem see Griffiths–Harris [55, p. 122].

Theorem 5.4.1 (Hard Lefschetz theorem) *The map*

$$L^{n-r}: H^r(M,\mathbb{C}) \to H^{2n-r}(M,\mathbb{C})$$

is an isomorphism for $0 \le r \le n$.

For $r \le n$ the *primitive cohomology group* $H^r_{\mathrm{pr}}(M,\mathbb{C})$ is defined by

$$H^r_{\mathrm{pr}}(M,\mathbb{C}) := \operatorname{Ker} L^{n-r+1}: H^r(M,\mathbb{C}) \to H^{2n-r+2}(M,\mathbb{C}).$$

Theorem 5.4.2 (Lefschetz Decomposition) *The Kähler form* ω *determines an isomorphism*

$$H^r(M,\mathbb{C}) = \bigoplus_s L^s H^{r-2s}_{\mathrm{pr}}(M,\mathbb{C}).$$

Here the sum runs over all s *between* $\max(0, r-n)$ *and* $\left[\frac{r}{2}\right]$.

For the proof, see Wells, [142, Corollary 4.13].

Denote by

$$H^r_{\mathbb{Z}}(M) := H^r(M,\mathbb{Z})/\text{torsion} = \operatorname{Im}\{H^r(M,\mathbb{Z}) \to H^r(M,\mathbb{C})\}$$

the torsion-free part of the integral cohomology of M and

$$H^r_{\mathrm{pr}}(M,\mathbb{Z}) := H^r_{\mathrm{pr}}(M,\mathbb{C}) \cap H^r_{\mathbb{Z}}(M).$$

Since the Lefschetz operator L is defined over $\mathbb{Z}$, the Lefschetz decomposition restricts to a decomposition of the torsion-free part of the integral cohomology

$$H^r_{\mathbb{Z}}(M) = \oplus_s L^s H^{r-2s}_{\mathrm{pr}}(M,\mathbb{Z}). \tag{5.22}$$

The Lefschetz decomposition is compatible with the Hodge decomposition (5.19), since L is the cup product with a (1,1)-form. To be more precise, defining

$$H^{p,q}_{\mathrm{pr}}(M) := H^{p+q}_{\mathrm{pr}}(M,\mathbb{C}) \cap H^{p,q}(M),$$

the Lefschetz decomposition gives

$$H^{p,q}(M) = \bigoplus_s L^s H^{p-s,q-s}_{\mathrm{pr}}(M). \tag{5.23}$$

For the definition of the Weil Intermediate Jacobian we need the $\mathbb{C}$-bilinear form

$$A : H^r_{\mathrm{pr}}(M,\mathbb{C}) \times H^r_{\mathrm{pr}}(M,\mathbb{C}) \to \mathbb{C}, \quad (\varphi,\psi) \mapsto (-1)^{\frac{r(r+1)}{2}} \int_M \omega^{n-r} \wedge \varphi \wedge \psi.$$

Using the Lefschetz decomposition, the form A extends $\mathbb{C}$-linearly to a form on $H^r(M,\mathbb{C})$, also denoted by A:

$$A : H^r(M,\mathbb{C}) \times H^r(M,\mathbb{C}) \to \mathbb{C}.$$

Hence, if $\varphi = \sum_s L^s\varphi_s$ and $\psi = \sum_s L^s\psi_s$ are the Lefschetz decompositions of $\varphi,\psi \in H^r(M,\mathbb{C})$,

$$A(\varphi,\psi) := \sum_s A(\varphi_s,\psi_s) = \sum_s (-1)^{\frac{r(r+1)}{2}+s} \int_M \omega^{n-r+2s} \wedge \varphi_s \wedge \psi_s. \tag{5.24}$$

Notice that A maps real, respectively integral, classes into $\mathbb{R}$, respectively $\mathbb{Z}$, since ω is integral.

The complex structure of M induces the $\mathbb{C}$-linear operator

$$C := \sum_{p+q=r} i^{p-q} p_{p,q} : H^r(M,\mathbb{C}) \to H^r(M,\mathbb{C}).$$

Here $p_{p,q} : H^r(M,\mathbb{C}) \to H^{p,q}$ is the natural projection map. With this notation we have the following theorem, for the proof of which we refer to Exercise 5.4.6 (1).

Lemma 5.4.3 *For all $\varphi,\psi \in H^r(M,\mathbb{C})$, $\varphi \neq 0$ we have*

(a) $A(\varphi,\psi) = (-1)^r A(\psi,\varphi)$,
(b) $A(C\varphi, C\psi) = A(\varphi,\psi)$,
(c) $A(\varphi, C\psi) = A(\psi, C\varphi)$,
(d) $A(\varphi, C\overline{\varphi}) > 0$.

5.4.2 The Griffiths Intermediate Jacobians

In this section we introduce the Griffiths Intermediate Jacobian and a polarization of some index on it, which will be computed.

Let M be a smooth complex projective variety of dimension n and p an integer between 1 and n. Consider the subvector space F^pH^{2p-1} of $H^{2p-1}(M,\mathbb{C})$. In this case (5.21) is

$$H^{2p-1}(M,\mathbb{C}) = F^pH^{2p-1} \oplus \overline{F^pH^{2p-1}}. \tag{5.25}$$

This equation induces the composed map

$$\iota : H^{2p-1}_{\mathbb{Z}}(M) \hookrightarrow H^{2p-1}(M,\mathbb{C}) \to H^{2p-1}(M,\mathbb{C})/F^pH^{2p-1} \xrightarrow{\simeq} \overline{F^pH^{2p-1}}.$$

Proposition 5.4.4 *The quotient*

$$\begin{aligned} J_G^p(M) &:= H^{2p-1}(M,\mathbb{C})/(F^pH^{2p-1} + H_{\mathbb{Z}}^{2p-1}(M)) \\ &= \overline{F^pH^{2p-1}}/\iota(H_{\mathbb{Z}}^{2p-1}(M)) \end{aligned}$$

is a complex torus of dimension $\sum_{s=p}^{2p-1} \dim H^{s,2p-1-s}$.

Proof First we claim that the map ι is injective. To see this, note that according to (5.25) any integral form $\varphi \in H_{\mathbb{Z}}^{2p-1}(M)$ can be written as $\varphi = \alpha + \overline{\alpha}$ with $\alpha \in F^pH^{2p-1}$. So the map ι is given by $\varphi = \alpha + \overline{\alpha} \mapsto \overline{\alpha}$, which implies the assertion.

Hence $J_G^p(M)$ is a complex torus, since F^pH^{2p-1} is of dimension $\sum_{s=p}^{2p-1} \dim H^{s,2p-1-s}$ and $\operatorname{rk} \iota(H_{\mathbb{Z}}^{2p-1}(M)) = \dim H^{2p-1}(M\mathbb{C}) = 2\dim F^pH^{2p-1}$. □

Proposition 5.4.5 *The cup product pairing induces an isomorphism of complex tori*

$$\widehat{J_G^p(M)} \simeq J_G^{n-p+1}(M).$$

Proof Consider the sesquilinear form

$$S : H^{2n-2p+1}(M,\mathbb{C}) \times H^{2p-1}(M,\mathbb{C}) \to \mathbb{C}, \qquad (\varphi,\psi) \mapsto 2i \int_M \varphi \wedge \overline{\psi}.$$

First we claim that its restriction to $\overline{F^{n-p+1}H^{2n-2p+1}} \times \overline{F^pH^{2p-1}}$ is non-degenerate.

To see this, note that $S(\varphi,\psi) = 2i\langle \varphi, \overline{\psi}\rangle$, where $\langle\, ,\rangle$ is the cup product pairing and $\overline{F^{n-p+1}H^{2n-2p+1}}$ and F^pH^{2p-1} are dual with respect to this pairing.

The form S induces the $\mathbb{C}$-linear isomorphism

$$S : \overline{F^{n-p+1}H^{2n-2p+1}} \to \operatorname{Hom}_{\overline{\mathbb{C}}}(\overline{F^pH^{2p-1}}, \mathbb{C}), \qquad \varphi \mapsto S(\varphi,\cdot),$$

where $\operatorname{Hom}_{\overline{\mathbb{C}}}(V,\mathbb{C})$ denotes the $\mathbb{C}$-antilinear forms on a $\mathbb{C}$-vector space V.

For the proof of the proposition it remains to show that S induces an isomorphism between the lattice $\iota(H_{\mathbb{Z}}^{2n-2p+1}(M))$ defining $J_G^{n-p+1}(M)$ and the dual lattice $\iota(H_{\mathbb{Z}}^{2p-1}(M))\widehat{}$ defining $\widehat{J_G^p(M)}$. For this recall from Section 1.4.1 that

$$\iota(H_{\mathbb{Z}}^{2p-1}(M))\widehat{} = \{\ell \in \operatorname{Hom}_{\overline{\mathbb{C}}}(\overline{F^pH^{2p-1}},\mathbb{C}) \mid \operatorname{Im} \ell(\iota(H_{\mathbb{Z}}^{2p-1}(M))) \subseteq \mathbb{Z}\}.$$

Applying S^{-1} this gives

$$S^{-1}\big(\iota(H_{\mathbb{Z}}^{2p-1}(M))\widehat{}\,\big) = \{\lambda \in \overline{F^{n-p+1}H^{2n-2p+1}} \mid \operatorname{Im} S(\lambda, \iota(H_{\mathbb{Z}}^{2p-1}(M))) \subseteq \mathbb{Z}\}.$$

But $\alpha \in \iota(H^{2p-1}_{\mathbb{Z}}(M))$ if and only if $\alpha + \overline{\alpha} \in H^{2p-1}_{\mathbb{Z}}(M)$. So

$$\begin{aligned}\operatorname{Im} S(\lambda, \alpha) &= \frac{1}{2i}[S(\lambda,\alpha) - \overline{S(\lambda,\alpha)}]\\ &= \int_M \lambda \wedge \overline{\alpha} + \int_M \overline{\lambda} \wedge \alpha = \int_M (\lambda + \overline{\lambda}) \wedge (\alpha + \overline{\alpha})\end{aligned}$$

is in $\mathbb{Z}$ if and only if $\lambda + \overline{\lambda} \in H^{2n-2p+1}_{\mathbb{Z}}(M)$; that is, $\lambda \in \iota(H^{2n-2p+1}_{\mathbb{Z}}(M))$. This completes the proof. □

Recall that we consider the first Chern class H of a line bundle on a complex torus $X = V/\Lambda$ as a hermitian form on V with integer values on Λ or equivalently an alternating integer-valued form E on Λ with $E(i\lambda, i\mu) = E(\lambda, \mu)$ for all $\lambda, \mu \in \Lambda$. If H is non-degenerate of index k, we call it a *polarization of index* k on X.

Note that according to Proposition 1.2.9 any such H (respectively E) is the first Chern class of a line bundle. By abuse of notation we also denote the line bundle as a polarization of index k. With this notation, the usual polarization on an abelian variety is a polarization of index 0.

Next we will define a polarization of some index k on the complex torus $J^p_G(M)$. According to Proposition 5.4.5 it suffices to do this for $p \leq \frac{n+1}{2}$. The remaining intermediate Jacobians will be endowed with the dual polarization.

Recall that ω denotes the Kähler form induced by a hyperplane section of M. For $p \leq \frac{n+1}{2}$ and $R = \mathbb{Z}, \mathbb{R}$ or $\mathbb{C}$, consider the alternating R-bilinear form

$$E : H^{2p-1}(M, R) \times H^{2p-1}(M, R) \to R, \qquad (\varphi, \psi) \mapsto (-1)^p \int_M \omega^{n-2p+1} \wedge \varphi \wedge \psi. \tag{5.26}$$

Its restriction to $\overline{F^p H^{2p-1}}$ leads to a hermitian form

$$H^p_G : \overline{F^p H^{2p-1}} \times \overline{F^p H^{2p-1}} \to \mathbb{C}, \qquad (\varphi, \psi) \mapsto 2iE(\varphi, \overline{\psi}).$$

If we define $h^{s,t} := \dim H^{s,t}$ for $s, t \geq 0$ and $h^{s,t} = 0$ for s or $t = -1$, we have the following theorem.

Theorem 5.4.6 *Let M be a smooth projective variety of dimension n and $1 \leq p \leq n$. Then $(J^p_G(M), H^p_G)$ is a non-degenerate complex torus of dimension $\sum_{r=p}^{2p-1} h^{r,2p-1-r}$ and index*

$$\operatorname{ind} H^p_G = \begin{cases} i(p) & \text{if } p \leq \frac{n+1}{2},\\ i(n-p+1) & \text{if } p > \frac{n+1}{2},\end{cases}$$

where for $p \leq \frac{p+1}{2}$, $i(p)$ is defined by

$$i(p) := \sum_{t=0}^{[\frac{p-2}{2}]} \sum_{s=0}^{1+2t} \left(h^{p-2-2t, p+1-2s+2t} - h^{p-3-2t, p-2s+2t}\right).$$

The non-degenerate complex torus $(J^p_G(M), H^p_G)$ is called the *p-th Griffiths Intermediate Jacobian* of M.

Proof **Step I**: *For $p \leq \frac{n+1}{2}$ the hermitian form H^p_G is the first Chern class of a line bundle on M.*

For the proof, according to Proposition 1.2.9, it suffices to show that the imaginary part of H^p_G takes integral values on the lattice $\iota(H^{2p-1}_{\mathbb{Z}}(M))$. Suppose $\alpha, \beta \in \iota(H^{2p-1}_{\mathbb{Z}}(M))$. Since $\alpha + \overline{\alpha}$ and $\beta + \overline{\beta} \in H^{2p-1}_{\mathbb{Z}}(M)$ we have

$$\begin{aligned}\operatorname{Im} H^p_G(\alpha, \beta) &= \frac{1}{2i}[H^p_G(\alpha, \beta) - H^p_G(\beta, \alpha)] \\ &= E(\alpha, \overline{\beta}) - E(\beta, \overline{\alpha}) = E(\alpha + \overline{\alpha}, \beta + \overline{\beta}) \in \mathbb{Z},\end{aligned}$$

where the last equation uses that $F^p H^{2p-1}$ and $\overline{F^p H^{2p-1}}$ are isotropic for E.

Step II: *The alternating forms A of Section 5.4.1 and E are related by*

$$E = (-1)^s A \text{ on } L^s H^{2p-1-2s}_{\mathrm{pr}}(M, \mathbb{C}) \subseteq H^{2p-1}(M, \mathbb{C}). \tag{5.27}$$

We leave the proof of this computation to the reader (see Exercise 5.4.6 (2)).

Step III: *The subvector spaces $L^s H^{p-1-t-s, p+t-s}_{\mathrm{pr}}(M)$ of $\overline{F^p H^{2p-1}}$ are mutually orthogonal with respect to H^p_G.*

For the proof note first that the subvector spaces of $L^s H^{p-1-t-s,p+t-s}_{\mathrm{pr}}$ of $\overline{F^p H^{2p-1}}$ can be illustrated by the following *Hodge–Lefschetz triangle*: $\overline{F^p H^{2p-1}}$ is the direct sum of the subvector spaces

$$\begin{array}{llllll}
H^{p-1,p}_{\mathrm{pr}} & H^{p-2,p+1}_{\mathrm{pr}} & H^{p-3,p+2}_{\mathrm{pr}} & \cdots & H^{1,2p-2}_{\mathrm{pr}} & H^{0,2p-1}_{\mathrm{pr}} \\
LH^{p-2,p-1}_{\mathrm{pr}} & LH^{p-3,p}_{\mathrm{pr}} & & \cdots & LH^{0,2p-3}_{\mathrm{pr}} & \\
L^2 H^{p-3,p-2}_{\mathrm{pr}} & & & \cdots & & \\
\vdots & \vdots & & & & \\
L^{p-2} H^{1,2}_{\mathrm{pr}} & L^{p-2} H^{0,3}_{\mathrm{pr}} & & & & \\
L^{p-1} H^{0,1}_{\mathrm{pr}} & & & & &
\end{array}$$

In this figure the subvector spaces $L^s H^{p-1-t,p+t-2s}_{\mathrm{pr}}$ of $\overline{F^p H^{2p-1}}$ are arranged horizontally according to their Hodge type and vertically according to their degree s in the Lefschetz decomposition.

Now suppose that $\varphi \in H^{p-1-t,p+t}(M)$ and $\psi \in H^{p-1-r,p+r}(M)$ with $r, t \geq 0$. If $r \neq t$,

$$H^p_G(\varphi, \psi) = 2i(-1)^p \int_M \omega^{n-2p+1} \wedge \varphi \wedge \overline{\psi} = 0,$$

since $\omega^{n-2p+1} \wedge \varphi \wedge \overline{\psi}$ is of type $(n - t + r, n + t - r) \neq (n, n)$. This shows that the columns in the triangle are pairwise orthogonal.

Hence it suffices to show that subspaces in different rows are pairwise orthogonal. So suppose $\omega^s \wedge \varphi \in L^s H_{\mathrm{pr}}^{2p-1-2s}(M,\mathbb{C})$ and $\omega^t \wedge \psi \in L^t H_{\mathrm{pr}}^{2p-1-2t}(M,\mathbb{C})$. We may assume that $t \geq s+1$. Then

$$\begin{aligned} H_G^p(\omega^s \wedge \varphi, \omega^t \wedge \psi) &= 2i(-1)^p \int_M \omega^{n-2p+1} \wedge \omega^s \wedge \varphi \wedge \omega^t \wedge \overline{\psi} \\ &= 2i(-1)^p \int_M \omega^{n-2p+1+s+t} \wedge \varphi \wedge \overline{\psi} = 0, \end{aligned}$$

since $H_{\mathrm{pr}}^{2p-1-2s}(M,\mathbb{C})$ is in the kernel of $L^{n-2p+2s+2} : H^{2p-1-2s}(M,\mathbb{C}) \to H^{2n-2p+2s+3}(M,\mathbb{C})$. This completes the proof of Step III.

Step IV: (a) $H_G^p > 0$ *on* $L^s H_{\mathrm{pr}}^{p-1-2t,p+2t-2s}(M)$ *for* $0 \leq t \leq [\frac{p-1}{2}]$, $0 \leq s \leq 2t+1$.

(b): $H_G^p < 0$ *on* $L^s H_{\mathrm{pr}}^{p-2-2t,p+1+2t-2s}(M)$ *for* $0 \leq t \leq [\frac{p-2}{2}]$, $0 \leq s \leq 2t+1$.

In particular, H_G^p *is non-degenerate.*

To see this, one checks immediately when H_G^p is positive or negative on the vector spaces $L^s H_{\mathrm{pr}}^{p-1-t,p+t-2s}$. Using the same arrangement as in the Hodge–Lefschetz triangle, the signs + or − in the following triangle indicate whether it is positive or negative on the corresponding subvector space:

$$\begin{array}{cccccc} + & - & + & \cdots & (-1)^{p-2} & (-1)^{p-1} \\ - & + & - & \cdots & (-1)^{p-1} & \\ + & - & \cdots & & & \\ \vdots & \vdots & & & & \\ (-1)^{p-2} & (-1)^{p-1} & & & & \\ (-1)^{p-1} & & & & & \end{array}$$

Note that H_G^p is always positive on $H_{\mathrm{pr}}^{p-1,p}$. The sign alternates along every row and every column.

Let $0 \neq \varphi \in L^s H_{\mathrm{pr}}^{p-1-t,p+t-2s}(M)$ with $t, s \geq 0$. By Step II and Lemma 5.4.3 we get

$$\begin{aligned} H_G^p(\varphi,\varphi) &= 2i(-1)^s A(\varphi,\overline{\varphi}) \\ &= 2(-1)^t A(\varphi,(-1)^{s+t} i\overline{\varphi}) \\ &= 2(-1)^t A(\varphi, C\varphi) = \begin{cases} > 0 & t \equiv 0 \bmod 2, \\ < 0 & t \equiv 1 \bmod 2. \end{cases} \end{aligned}$$

Step V: To complete the proof of the theorem, it remains to compute the index of H_G^p. Suppose first that $p \leq \frac{n+1}{2}$. Then the hard Lefschetz Theorem 5.4.1 implies that $L^s : H_{\mathrm{pr}}^{p-1-2s} \to H^{2p-1}(M,\mathbb{C})$ is injective for all $s \geq 0$, since $2p-1 \leq n$. Hence

$$\dim L^s H_{\mathrm{pr}}^{p-2-2t,p+1+2t-2s} = \dim H_{\mathrm{pr}}^{p-2-2t,p+1+2t-2s}. \tag{5.28}$$

We claim that for all $r, s \geq 0, r + s \leq n$,

$$\dim H_{\mathrm{pr}}^{r,s} = h^{r,s} - h^{r-1,s-1}. \tag{5.29}$$

For the proof of this equation, see Exercise 5.4.6 (3).

Using this, according to Step IV (b), the index of H_{pr}^{p} is just the sum of the dimensions of the vector spaces $L^s H_{\mathrm{pr}}^{p-1-2t,p+2t-2s}$, which completes the proof of Step V in the case $p \leq \frac{n+1}{2}$. Finally, in the case $p > \frac{n+1}{2}$ we apply this result to the dual polarization of J_G^{n-p+1}, which is of the same index, and use Proposition 5.4.5. This completes the proof of the theorem. □

Remark 5.4.7 Suppose $\pi : \mathcal{M} \to T$ is a flat family of smooth projective varieties. Let $M_t = \pi^{-1}(t)$ for $t \in T$. Recall from Section 5.4.1 (or to be more precise [54]) that the Hodge filtration varies holomorphically with t. This implies that the associated Griffiths intermediate Jacobians $(J_G^p(M_t), H_G^p)$ vary holomorphically with t. To be more precise, the relative intermediate Jacobian is by definition

$$\pi_{\mathcal{J}} : \mathcal{J}^p(\pi) := R\pi_*^{2-1}\mathbb{C}/(F^p R^{2p-1}\pi_*\mathbb{C} + R^{2p-1}\pi_*\mathbb{Z}) \longrightarrow T.$$

The fibre $\pi_{\mathcal{J}}^{-1}(t)$ is exactly the Griffiths intermediate Jacobian $J_G^p(M_t)$. The above mentioned holomorphic variation of the $J_G^p(M_t)$ means that $\mathcal{J}^p(\pi)$ carries the structure of a complex manifold and $\pi_{\mathcal{J}}$ is holomorphic.

Proposition 5.4.8 *Let M be a smooth projective variety of dimension n.*

(a) *The Griffiths intermediate Jacobian $J_G^1(M)$ is canonically isomorphic to the Picard variety* $\mathrm{Pic}^0(M)$.
(b) *The Griffiths intermediate Jacobian $J_G^n(M)$ is canonically isomorphic to the Albanese variety* $\mathrm{Alb}(M)$.

For the proof, see Exercise 5.4.6 (4).

Consider the special case of a three-dimensional variety M. According to Proposition 5.4.8 the only new Griffiths intermediate Jacobian is $J_G^2(M)$, for which we have:

Proposition 5.4.9 *Let M be a smooth projective threefold. Then $(H_G^2(M), H_G^2)$ is a non-degenerate complex torus of dimension* $\dim H_G^2(M) = h^{2,1} + h^{3,0}$ *and index* $i(2) = h^{0,3} + h^{0,1}$.

Proof In this case the Hodge–Lefschetz triangle reduces to

$$\begin{matrix} H_{\mathrm{pr}}^{1,2} & H_{\mathrm{pr}}^{0,3} \\ LH_{\mathrm{pr}}^{0,1} & \end{matrix}$$

and the hermitian form H_G^2 is positive definite on $H_{\mathrm{pr}}^{1,2}$ and negative definite on $H_{\mathrm{pr}}^{0,3} \oplus LH_{\mathrm{pr}}^{0,1} = H^{0,3} \oplus LH^{0,1}$. Hence, according to Theorem 5.4.6 H_G^2 is of dimension $h^{2,1} + h^{3,0}$ and index $i(2) = h^{0,3} + h^{0,1}$. □

Example 5.4.10

(i) Let M be an abelian threefold, say $M = V/\Lambda$, and z_1, z_2, z_3 be coordinate functions on V. Then $H^{0,3}$ is generated by $d\bar{z}_1 \wedge d\bar{z}_2 \wedge d\bar{z}_3$ and hence is 1-dimensional, and $H^{0,1}$ is generated by $d\bar{z}_1, d\bar{z}_2$ and $d\bar{z}_3$. This implies that $(J_G^2(M), H_G^2)$ is a non-degenerate 10-dimensional complex torus of index 4. So the Griffiths intermediate Jacobian $J_G^2(M)$ of an abelian threefold is not an abelian variety.

(ii) Let M be a smooth cubic threefold in $\mathbb{P}^4$. One checks that $h^{0,3} = h^{0,1} = 0$ and $h_{\text{pr}}^{1,2} = h^{1,2} = 5$. So $J_G^2(M)$ is a 5-dimensional abelian variety.

5.4.3 The Weil Intermediate Jacobian

In this section we introduce the Weil Intermediate Jacobian, which is always an abelian variety of the same dimension as the corresponding Griffiths Intermediate Jacobian.

Let M be a smooth complex projective variety of dimension n and $\omega \in H^{1,1}(M) \cap H^2(M, \mathbb{C})$ the Kähler form induced by a hyperplane section. According to Exercise 5.4.6 (5), for any integer p, $1 \leq p \leq n$, the operator $C : H^{2p-1}(M, \mathbb{C}) \to H^{2p-1}(M, \mathbb{C})$ of Section 5.4.1 defines a complex structure on the real subspace $H^{2p-1}(M, \mathbb{R})$. So

$$J_W^p(M) := (H^{2p-1}(M, \mathbb{R}), -C)/H_{\mathbb{Z}}^{2p-1}(M)$$

is a complex torus.

In order to introduce a polarization on $J_W^p(M)$, suppose first $p \leq \frac{n+1}{2}$. Recall from equation (5.24) the alternating form

$$\begin{aligned} A : H^{2p-1}(M, \mathbb{R}) \times H^{2p-1}(M, \mathbb{R}) &\to \mathbb{R}, \\ (\varphi, \psi) &\mapsto \sum_s (-1)^{p+s} \int_M \omega^{n-2p+1+2s} \wedge \varphi_s \wedge \psi_s, \end{aligned}$$

where φ_s and ψ_s are given by the Lefschetz decompositions; that is, $\varphi = \sum_s L^s \varphi_s$ and $\psi = \sum_s L^s \psi_s$. According to Lemma 5.4.3 the map

$$H_W^p : H^{2p-1}(M, \mathbb{R}) \times H^{2p-1}(M, \mathbb{R}) \to \mathbb{C}, \qquad (\varphi, \psi) \mapsto -A(C\varphi, \psi) + iA(\varphi, \psi)$$

is a hermitian form on the $\mathbb{C}$-vector space $(H^{2p-1}(M, \mathbb{R}), -C)$.

Proposition 5.4.11 *For $1 \leq p \leq \frac{n+1}{2}$ the pair $(J_W^p(M), H_W^p)$ is a polarized abelian variety.*

Proof According to Lemma 5.4.3 (d) the hermitian form H_W^p is positive definite on the $\mathbb{C}$-vector space $(H^{2p-1}(M, \mathbb{R}), -C)$ and its imaginary part A is integral valued on the lattice $H_{\mathbb{Z}}^{2p-1}(M)$, since $\omega \in H^2(M, \mathbb{Z})$. □

In order to define a polarization also on $J^p_W(M)$ for $p > \frac{n+1}{2}$, we show:

Proposition 5.4.12 *Let $p \leq \frac{n+1}{2}$. The cup product pairing induces an isomorphism*

$$\overline{J^p_W(M)} \simeq J^{n-p+1}_W(M).$$

Proof Note first that according to Lemma 5.4.3 (b) the cup product pairing

$$\langle , \rangle : H^{2n-2p+1}(M,\mathbb{R}) \times H^{2p-1}(M,\mathbb{R}) \to \mathbb{R}, \qquad (\varphi,\psi) \mapsto \int_M \varphi \wedge \psi$$

is invariant with respect to the complex structure $-C$ on both spaces; that is,

$$\langle -C\varphi, -C\psi \rangle = \langle \varphi, \psi \rangle$$

for all $\varphi \in H^{2n-2p+1}(M,\mathbb{R})$ and $\psi \in H^{2p-1}(M,\mathbb{R})$. Hence

$$B : (H^{2n-2p+1}(M,\mathbb{R}), -\mathbb{C}) \times (H^{2n-2p+1}(M,\mathbb{R}), -\mathbb{C}) \to \mathbb{C},$$
$$(\varphi,\psi) \mapsto \langle \varphi, C\psi \rangle + i\langle \varphi, \psi \rangle$$

is a non-degenerate sesquilinear form, $\mathbb{C}$-linear in the first and $\mathbb{C}$-antilinear in the second argument. The assignment

$$\varphi \mapsto B(\varphi, \cdot)$$

defines a $\mathbb{C}$-linear isomorphism, which by abuse of notation is denoted by the same letter:

$$B : (H^{2n-2p+1}(M,\mathbb{R}), -\mathbb{C}) \to \mathrm{Hom}_{\overline{\mathbb{C}}}((H^{2p-1}(M,\mathbb{R}), -C), \mathbb{C}).$$

It remains to show that B restricts to an isomorphism between $H^{2n-2p+1}_{\mathbb{Z}}(M)$ and the lattice $\overline{H^{2p-1}_{\mathbb{Z}}(M)}$ of $\overline{J^p_W(M)}$: For this recall that

$$\overline{H^{2p-1}_{\mathbb{Z}}(M)} = \{\ell \in \mathrm{Hom}_{\overline{\mathbb{C}}}((H^{2p-1}(M,\mathbb{R}), -C), \mathbb{C}) \mid \mathrm{Im}\,\ell(H^{2p-1}_{\mathbb{Z}}(M)) \subseteq \mathbb{Z}\}.$$

Applying B^{-1} this gives

$$\begin{aligned} B^{-1}\overline{H^{2p-1}_{\mathbb{Z}}(M)} &= \{\lambda \in H^{2n-2p+1}(M,\mathbb{R}) \mid \mathrm{Im}\,B(\lambda, H^{2p-1}_{\mathbb{Z}}(M)) \subseteq \mathbb{Z}\} \\ &= \{\lambda \in H^{2n-2p+1}(M,\mathbb{R}) \mid \langle \lambda, H^{2p-1}_{\mathbb{Z}}(M)\rangle \subseteq \mathbb{Z}\}. \end{aligned}$$

But this is $H^{2n-2p+1}_{\mathbb{Z}}(M) = \mathrm{Im}\{H^{2n-2p+1}(M,\mathbb{Z}) \to H^{2n-2p+1}(M,\mathbb{R})\}$, since the cup product pairing is unimodular on integral cohomology. This completes the proof. □

For $p > \frac{n+1}{2}$ denote by H_W^p the polarization on $J_W^p(M) \simeq \overline{J_W^{n-p+1}(M)}$ dual to the polarization H_W^{n-p+1} on $J_W^{n-p+1}(M)$. Then we have proved the following theorem, the assertion on the dimension being obvious.

Theorem 5.4.13 *For $p = 1, \ldots, n$, $(J_W^p(M), H_W^p)$ is an abelian variety of dimension*

$$\dim J_W^p = \frac{1}{2}\dim H^{2p-1}(M, \mathbb{C}) = \sum_{r \geq p} h^{r,2p-1-r}.$$

It is called the *p-th Weil intermediate Jacobian*. Note that $\dim J_G^p(M) = \dim J_W^p(M)$.

The Weil intermediate Jacobian $J_W^p(M)$ can also be described in terms of the complex cohomology: Consider the $\mathbb{C}$-subvector space

$$V := \bigoplus_{1-p \leq 2\nu \leq p} H^{p-1+2\nu, p-2\nu} \subset H^{2p-1}(M, \mathbb{C}).$$

From $H^{r,s} = \overline{H^{s,r}}$ and equations (5.19) and (5.21) we get

$$V \oplus \overline{V} = H^{2p-1}(M, \mathbb{C}) \quad \text{and} \quad V \cap \overline{V} = \{0\}.$$

Proposition 5.4.14 *There is a canonical isomorphism*

$$J_W^p(M) \simeq V / p_V(H_{\mathbb{Z}}^{2p-1}(M)),$$

where $p_V : H^{2p-1}(M\mathbb{C}) \to V$ denotes the natural projection.

Proof It suffices to show that the composed map

$$\Psi : H^{2p-1}(M, \mathbb{R}) \hookrightarrow H^{2p-1}(M, \mathbb{C}) \overset{p_V}{\to} V$$

induces an isomorphism of $\mathbb{C}$-vector spaces $(H^{2p-1}(M, \mathbb{R}), -C) \to V$. But clearly Ψ is an isomorphism of $\mathbb{R}$-vector spaces. Hence the assertion follows from

$$C|_V = -i\mathbf{1}_V, \qquad C|_{\overline{V}} = i\mathbf{1}_{\overline{V}},$$

which follows immediately from the definition. □

5.4.4 The Lazzeri Intermediate Jacobian

The Lazzeri Intermediate Jacobian was studied by E. Rubei in her thesis [113]. It can be defined more generally for every oriented Riemannian manifold M whose (real) dimension is twice an odd prime number.

Let M denote a compact oriented Riemannian manifold of even dimension $n = 2m$. The star operator

$$* : H^m(M, \mathbb{R}) \to H^m(M, \mathbb{R})$$

is defined as follows: According to Hodge's Theorem (see Griffiths–Harris [55, page 84]), $H^m(M, \mathbb{R})$ is canonically isomorphic to the vector space of harmonic m-forms $\mathcal{H}^m(M)$ on M. The Riemann metric induces a metric on $\mathcal{H}^m(M)$, which we denote by $\langle\, , \rangle$. Let vol denote the volume form on M. Then $*$ is defined by

$$\alpha \wedge *\beta = \langle \alpha, \beta \rangle \mathrm{vol} \qquad \text{for all} \quad \alpha, \beta \in \mathcal{H}^m(M). \tag{5.30}$$

Lemma 5.4.15 *With M as above of dimension $2m$ with $m = 2p - 1$, the operator $(-1)^{p-1}* : H^m(M, \mathbb{R}) \to H^m(M, \mathbb{R})$ defines a complex structure on $H^m(M, \mathbb{R})$.*

Proof It suffices to show that

$$** = (-1)^m.$$

But this is an easy computation (see Exercise 5.4.6 (8)). □

As above let $H^m_{\mathbb{Z}}(M)$ denote the image of $H^m(M, \mathbb{Z})$ in $H^m(M, \mathbb{R})$. Then

$$J_L(M) := (H^m(M, \mathbb{R}), (-1)^{p-1}*)/H^m_{\mathbb{Z}}(M)$$

is a complex torus of dimension $\frac{1}{2}h^m$, called the *Lazzeri Intermediate Jacobian* of M.

Recall the alternating form $E : H^m(M, \mathbb{R}) \times H^m(M, \mathbb{R}) \to \mathbb{R}$ defined in equation (5.26).

Proposition 5.4.16 *Let M be a compact oriented Riemannian manifold of dimension $2m$ with $m = 2p - 1$. The hermitian form*

$$H_L : H^m(M, \mathbb{R}) \times H^m(M, \mathbb{R}) \to \mathbb{R}, \qquad (\varphi, \psi) \mapsto E((-1)^{p-1} * \varphi, \psi) + iE(\varphi, \psi)$$

defines a principal polarization on $J_L(M)$. In particular, $J_L(M)$ is an abelian variety.

Proof H_L is positive definite, since for any $0 \neq \varphi \in H^m(M, \mathbb{R})$,

$$H_L(\varphi, \varphi) = (-1)^{p-1} E(*\varphi, \varphi) = -\int_M *\varphi \wedge \varphi = \int_M \varphi \wedge *\varphi > 0$$

by equation 5.30. The assertion follows from the fact that the cup product is unimodular on $H^m_{\mathbb{Z}}(M)$. □

If M is a smooth complex projective variety of dimension $m = 2p - 1$, then it is a compact oriented Riemann manifold of dimension $2m$. Hence the Lazzeri intermediate Jacobian is well defined. For the comparison of the complex structure $(-1)^{p-1}*$ with the complex structure C induced by the complex structure of M, see Exercise 5.4.6 (9).

Example 5.4.17 Let M be a smooth complex projective threefold. So

$$J_L(M) = (H^3(M,\mathbb{R}), -*)/H^3_{\mathbb{Z}}(M)$$

with polarization defined by the alternating form E on $H^3(M,\mathbb{R})$, whereas

$$H^2_W(M) = (H^3(M,\mathbb{R}), -C)/H^3_{\mathbb{Z}}(M)$$

with polarization defined by A. These are two polarized abelian varieties associated to M. They do not coincide, as this example shows.

The table below compares the complex structures $*$ and C as well as the alternating forms E and A on the Lefschetz decomposition of $H^3(M,\mathbb{R})$.

$H^3_{\text{pr}}(M,\mathbb{R})$	$LH^1(M,\mathbb{R})$
$-* = -C$	$* = -C$
$E = A$	$E = -A.$

For the proof use equations (5.27) and Exercise 5.4.6 (9).

The Lefschetz decomposition induces decompositions of the polarized abelian varieties $(J_L(M), E)$ and $(J_W(M), A)$: For this consider the complex tori

$$X := (H^3_{\text{pr}}(M,\mathbb{R}), -C)/H^3_{\text{pr}}(M,\mathbb{Z}) \text{ and}$$
$$Y := (LH^1(M,\mathbb{R}), -C)/LH^1_{\mathbb{Z}}(M).$$

Obviously X and Y are complex subtori of $J^2_W(M)$ and the addition map

$$\mu : (X, A|_X) \times (Y, A|_Y) \to (J^2_W(M), A) \tag{5.31}$$

is an isogeny of polarized abelian varieties. According to the above table X and

$$\overline{Y} = (LH^1(M,\mathbb{R}), C)/LH^1_{\mathbb{Z}}(M)$$

are complex subtori if $J_L(M)$ and the addition map

$$\mu : (X, A|_X) \times (\overline{Y}, -A|_{\overline{Y}}) \to (J_L(M), E) \tag{5.32}$$

is an isogeny of polarized abelian varieties. It is easy to see that in general the complex conjugate abelian varieties $(Y, A|_Y)$ and $(\overline{Y}, -A_{\overline{Y}})$ are not isogenous (see Exercise 5.4.6 (11)).

5.4.5 The Abelian Variety Associated to the Griffiths Intermediate Jacobian

Let M be a smooth projective variety of dimension n. In this section we show that to every Griffiths intermediate Jacobian one can associate an abelian variety in a canonical way. If moreover $n = 2p - 1$ this abelian variety coincides with the Lazzeri Intermediate Jacobian.

Consider first an arbitrary non-degenerate complex torus (X, L) of index $k > 0$. Write $X = V/\Lambda$ and denote by $\mathrm{Gr}_k(V)$ the Grassmannian of k-dimensional subvector spaces of V. The space

$$\mathrm{Gr}_k^-(L) := \{V_- \in \mathrm{Gr}_k(V) \mid H = c_1(L) \text{ is negative definite on } V_-\}$$

is an open non-empty subset of $\mathrm{Gr}_k(V)$.

Lemma 5.4.18 *To every $V_- \in Gr^-(L)$ one can associate a polarized abelian variety (X_{V_-}, L_{V_-}) of the same type in a canonical way such that the underlying real tori coincide.*

Proof Suppose $L = L(H, \chi)$. Denote by V_+ the orthogonal complement of V_- with respect to H. Let W denote the underlying real vector space of V and $J : W \to W$ the complex structure defining V. Certainly the direct sum decomposition $V = V_- \oplus V_+$ induces a direct sum decomposition

$$W = W_- \oplus W_+$$

over $\mathbb{R}$. Define a new complex structure I_{V_-} on W by

$$I_{V_-}|_{W_-} := -J|_{W_-}, \qquad I_{V_+}|_{W_+} := J|_{W_-}. \tag{5.33}$$

The lattice Λ in V does not depend on the complex structure, so it is also a lattice in the vector space (W, I_{V_-}) and thus

$$X_{V_-} := (W, I_{V_-})/\Lambda$$

is a complex torus. Consider $E = \mathrm{Im}\, H$ as an alternating form on the real vector space W. Recall that $H(v, W) = E(Jv, W) + iE(v, w)$ for all $v, w \in V = (W, J)$. Now

$$H_{V_-}(v, w) := E(I_{V_-}v, w) + iE(v, w) \qquad \text{for all} \quad v, w \in W$$

is a hermitian form on the complex vector space (W, I_{V_-}). This hermitian form is positive definite with integral-valued imaginary part on Λ. Moreover, the semicharacter χ of L is also a semicharacter for H_{V_-}, since $\mathrm{Im}\, H_{V_-} = \mathrm{Im}\, H$. According to the Appell–Humbert Theorem, $L_{V_-} := L(H_{V_-}, \chi)$ is a non-degenerate line bundle on X_{V_-} of index 0. The two types coincide, since the lattices and the alternating forms are the same. □

Now let M be a smooth projective variety of dimension n and $1 \leq p \leq n$. The Griffiths Intermediate Jacobian $(J_G^p(M), H_G^p)$ is a non-degenerate complex torus of index $i(p)$ given in Theorem 5.4.6. For every $V_- \in Gr_{i(p)}^-(V)$, Lemma 5.4.18 associated an abelian variety (X_{V_-}, L_{V_-}). But here the Hodge–Lefschetz decomposition provides a canonical choice for V_-, namely

$$V_- := \bigoplus_{t=0}^{[\frac{p-2}{2}]} \bigoplus_{s=0}^{2t+1} L^s H_{\mathrm{pr}}^{p-2-2t, p+1+2t-2s}(M)$$

(see Steps IV and V in the proof of Theorem 5.4.6). Denote by

$$(\widetilde{J}_G^p(M), \widetilde{H}_G) := (J_G^p(M)_{V_-}, (H_G^p)_{V_-})$$

the associated polarized abelian variety, which we call the *p-th algebraic Griffiths Intermediate Jacobian*.

In the special when that the dimension n is odd, we have:

Theorem 5.4.19 *For a smooth projective variety of dimension $n = 2p - 1$ the p-th algebraic Griffiths Intermediate Jacobian coincides with the Lazzeri Intermediate Jacobian:*

$$(\widetilde{J}_G^p(M), \widetilde{H}_G) = (J_L(M), H_L).$$

Proof Denote by V_+ the orthogonal complement of V_- with respect to the hermitian form H_G^p. According to Step V in the proof of Theorem 5.4.6) we have

$$V_+ = \bigoplus_{t=0}^{[\frac{p-1}{2}]} \bigoplus_{s=0}^{2t+1} L^s H_{\mathrm{pr}}^{p-1-2t, p+2t-2s}(M).$$

By definition, the algebraic Griffiths intermediate Jacobian $\widetilde{J}_G^p(M)$ is the complex torus

$$\widetilde{J}_G^p(M) = (H^n(M, \mathbb{R}), I_{V_-}) / H_{\mathbb{Z}}(M),$$

where the complex structure I_{V_-} is defined by equation (5.33). But by Exercise 5.4.6 (10) this is just the complex structure defining the Lazzeri Intermediate Jacobian. So $\widetilde{J}_G^p(M) = J_L(M)$. As for the polarizations, it suffices to note that the alternating form E is the imaginary part of the hermitian forms in both cases. □

5.4.6 Exercises

(1) Show that with A defined in (5.24) we have for all $\varphi, \psi \in H^r(M, \mathbb{C}), \varphi \neq 0$:

(a) $A(\varphi, \psi) = (-1)^r A(\psi, \varphi)$;
(b) $A(C\varphi, C\psi) = A(\varphi, \psi)$;
(c) $A(\varphi, C\psi) = A(\psi, C\varphi)$;
(d) $A(\varphi, C\overline{\varphi}) > 0$.

(2) Show that the alternating A of Section 5.4.1 and E of Section 5.4.2 are related by

$$E = (-1)^s A \quad \text{on} \quad L^s H^{2p-1-2s}_{\text{pr}}(M, \mathbb{C}) \subseteq H^{2p-1}(M, \mathbb{C}).$$

(3) Show that with the notation of the proof of Theorem 5.4.6,

$$\dim H^{r,s}_{\text{pr}} = h^{r,s} - h^{r-1,s-1}$$

for all $r, s \geq 0, r + s \leq n$.
(Hint: Use (5.28) and the Lefschetz decomposition.)

(4) Let M be a smooth projective variety of dimension n.

(a) The Griffiths Intermediate Jacobian $J^1_G(M)$ is canonically isomorphic to the Picard variety $\text{Pic}^0(M)$.
(b) The Griffiths Intermediate Jacobian $J^n_G(M)$ is canonically isomorphic to the Albanese variety $\text{Alb}(M)$.

(Hint: Use the Dolbeault isomorphism and Serre duality.)

(5) Show that for any integer p, $1 \leq p \leq n$, the $\mathbb{C}$-linear operator $C : H^{2p-1}(M, \mathbb{C}) \to H^{2p-1}(M, \mathbb{C})$ of Section 5.4.1 defines a complex structure on the real subvector space $H^{2p-1}(M, \mathbb{R})$.

(6) Show that for $p \leq [\frac{n+1}{2}]$ the polarization of $J^p_W(M) = V/p_V(H^{2p-1}_{\mathbb{Z}}(M))$ (with V as in Proposition 5.4.14) is the hermitian form

$$V \times V \to \mathbb{C}, \qquad (\varphi, \psi) \mapsto 2A(\varphi, C\overline{\psi}) = -2iA(\varphi, \overline{\psi}).$$

(7) Show that

$$J^1_W(M) = J^1_G(M) \qquad \text{and} \qquad J^n_W(M) = J^n_G(M).$$

(Hint: Use the description of Proposition 5.4.14 and the previous exercise of the Weil Intermediate Jacobian.)

(8) Let M be a compact oriented Riemannian manifold of dimension $2m$ with $m = 2p - 1$ and $*$ the Star operator on $H^m(M, \mathbb{R})$. Show that $** = (-1)^m$.
(Hint: The usual proof uses the local description of $*$ as given in Griffiths–Harris [55, p. 82].)

(9) Let M be a smooth projective variety of dimension $2p-1$. Show that on $H^{2p-1}(M,\mathbb{R}) = \oplus_s L^s H^{2p-1-2s}_{\mathrm{pr}}(M,\mathbb{R})$ we have

$$* = \sum_s (-1)^{p+s} C pr_{L^s H^{2p-1-2s}_{\mathrm{pr}}}.$$

(Hint: This is a special case of Wells [142, Theorem V 3.16].)

(10) Consider the complex structure $C' = \sum_{r+s=2p-1} i^{\frac{r-s}{|r-s|}} p_{r,s}$ on $H^{2p-1}(M,\mathbb{C})$. For the $\mathbb{C}$-linear extension of $*$ to $H^{2p-1}(M,\mathbb{C})$ we have

(a) $(-1)^{p-1} * = -C'$ on $L^s H^{p-1-2t,p+2t-2s}_{\mathrm{pr}} \subseteq \overline{F^p H^{2p-1}}$ for $0 \leq t \leq [\frac{p-1}{2}]$ and $0 \leq s \leq 2t+1$;

(b) $(-1)^{p-1} * = C'$ on $L^s H^{p-2-2t,p+1+2t-2s}_{\mathrm{pr}} \subseteq \overline{F^p H^{2p-1}}$ for $0 \leq t \leq [\frac{p-2}{2}]$ and $0 \leq s \leq 2t+1$.

(Hint: This is an easy but tedious computation using the previous exercise.)

(11) Give a proof of the assertion in Example 5.4.17 that in general the abelian varieties $(Y, A|_Y)$ and $(\overline{Y}, -A_{\overline{Y}})$ occurring in (5.31) and (5.32) are not isogenous. (Hint: Use the fact that Y is isogenous to the Picard variety of M and any abelian variety is the Picard variety of a 3-dimensional variety.)

(12) Show that for a compact Riemann surface M we have

$$J_L(M) = J^1_W(M) = J^1_G(M) = \text{usual Jacobian variety of } M.$$

Note that the Lazzeri intermediate Jacobian is also defined for compact Riemann surfaces.

Chapter 6
The Fourier Transform for Sheaves and Cycles

In the theory of algebraic cycles it was quite common to use the Poincaré bundle to transfer cycles on an abelian variety X to cycles on the dual abelian variety (see for example Weil [137]). To be more precise: If a is a cycle class on X, $\mathcal{P}_X$ denotes the Poincaré bundle on $X \times \widehat{X}$, and p_1 and p_2 are the projections of $X \times \widehat{X}$, then

$$F_X(a) := p_{2*}\big(c_1[\mathcal{P}] \cdot p_1^* a\big)$$

is a cycle class on $\widehat{X}$.

It was the fundamental idea of Mukai [95] to apply the same construction for sheaves. Similarly, if $\mathcal{F}$ is a coherent sheaf on X, then

$$F_X(\mathcal{F}) := p_{2*}(\mathcal{P}_X \otimes p_1^* \mathcal{F})$$

is a coherent sheaf on $\widehat{X}$. In general this sheaf is not very useful. However, if $\mathcal{F}$ is a WIT-sheaf, meaning that $R^j p_{2*}(\mathcal{P}_X \otimes p_1^* \mathcal{F}) = 0$ for all $j \neq i$. then the sheaf

$$\widehat{\mathcal{F}} := R^i p_{2*}(\mathcal{P}_X \otimes p_1^* \mathcal{F})$$

is an important coherent sheaf on $\widehat{X}$, called the *Fourier–Mukai transform* of $\mathcal{F}$. Mukai defines the transform more generally for all complexes of $\mathcal{O}_X$-modules with bounded coherent cohomology in the corresponding derived category. For this, see Remark 6.1.18. The first section gives details on WIT-sheaves and some consequences on the theory of vector bundles on abelian varieties.

For abelian varieties one knows a little more about algebraic cycles than for most other classes of smooth projective varieties. This is mainly due to the fact that the Chow group $\mathrm{Ch}(X)$ admits two ring structures, one is induced by the intersection product and the other by the Pontryagin product. The Fourier transform exchanges both ring structures. In the second section we give an introduction into the theory of algebraic cycles on abelian varieties. In order to make this as self-contained as possible we introduce the Chow ring $\mathrm{Ch}^\bullet(X)$ in Section 6.2.1 and prove the main properties of correspondences which are needed in Section 6.2.2.

H. Lange, *Abelian Varieties over the Complex Numbers*, Grundlehren Text Editions,
https://doi.org/10.1007/978-3-031-25570-0_6

Finally, Section 6.3 contains some special results on the Chow ring of an abelian variety, due to Beauville, Deninger–Murre and Künnemann.

6.1 The Fourier–Mukai Transform for WIT-sheaves

6.1.1 Some Properties of the Poincaré Bundle

The Poincaré bundle was introduced in Section 1.4.4 for any complex torus. In this chapter we need some of its properties in the case of an abelian variety.

Let $X = V/\Lambda$ be an abelian variety of dimension g. Recall from Section 1.4.4 that the Poincaré bundle $\mathcal{P} = \mathcal{P}_X$ is a holomorphic line bundle on $X \times \widehat{X}$ uniquely determined by the two properties

(i) $\mathcal{P}_X|_{X\times\{\widehat{x}\}} = P_{\widehat{x}}$ for all $\widehat{x} \in \widehat{X}$, and
(ii) $\mathcal{P}_X|_{\{0\}\times\widehat{X}}$ is trivial.

Here $P_{\widehat{x}}$ denotes the line bundle in $\mathrm{Pic}^0(X)$ corresponding to the point $\widehat{x} \in \widehat{X}$ via the identification $\widehat{X} = \mathrm{Pic}^0(X)$. Similarly denote by P_x the line bundle in $\mathrm{Pic}^0(\widehat{X})$ corresponding to $x \in X$. Denote by $s : X \times \widehat{X} \to \widehat{X} \times X$ the isomorphism exchanging factors $s(x, \widehat{x}) = (\widehat{x}, x)$.

Lemma 6.1.1 *Identifying* $\widehat{\widehat{X}} = X$, *the homomorphism*

$$\phi_{\mathcal{P}_X} : X \times \widehat{X} \to (X \times \widehat{X})\widehat{\ } = \widehat{X} \times X, \qquad z \mapsto t_z^*\mathcal{P}_X \otimes \mathcal{P}_X^{-1}$$

coincides with s. *In particular*

$$\phi_{\mathcal{P}_X}^*\mathcal{P}_{\widehat{X}} = \mathcal{P}_X.$$

Proof Recall from the proof of Theorem 1.4.10 that the hermitian form $H = c_1(\mathcal{P}_X)$ is the map

$$H : (V \times \overline{\Omega}) \times (V \times \overline{\Omega}) \to \mathbb{C}, \qquad ((v_1, l_1), (v_2, l_2)) \mapsto \overline{l_2(v_1)} + l_1(v_2).$$

Double duality on the level of the vector spaces identifies $\mathrm{Hom}_{\overline{\mathbb{C}}}(V \times \overline{\Omega}, \mathbb{C}) = \overline{\Omega} \times V$. In these terms $H((v, l), \cdot) = (l, v)$ for all $(v, l) \in V \times \overline{\Omega}$. By Lemma 1.4.5 the left-hand side of this equation is the analytic representation ϕ_H of $\phi_{\mathcal{P}_X}$. This implies $\phi_{\mathcal{P}_X} = s$. Finally, notice that $s^*\mathcal{P}_{\widehat{X}} = \mathcal{P}_X$ by properties (i) and (ii). □

We obtain that

$$P_x = \mathcal{P}_X|_{\{x\}\times\widehat{X}} = \mathcal{P}_{\widehat{X}}|_{\widehat{X}\times\{x\}} \quad \text{for all } x \in X, \text{ and}$$
$$P_{\widehat{x}} = \mathcal{P}_X|_{X\times\{\widehat{x}\}} = \mathcal{P}_{\widehat{X}}|_{\{\widehat{x}\}\times X} \quad \text{for all } \widehat{x} \in \widehat{X}.$$

Moreover we have:

Lemma 6.1.2 $((-1)_X \times \mathbf{1}_{\widehat{X}})^* \mathcal{P}_X \simeq (\mathbf{1}_X \times (-\mathbf{1})_{\widehat{X}})^* \mathcal{P}_X \simeq \mathcal{P}_X^{-1}$.

Proof Consider first the second assertion. By the seesaw principle, Corollary 1.4.9, it suffices to show that both line bundles coincide when restricted to $X \times \{\widehat{x}\}$, for all $\widehat{x} \in \widehat{X}$, and $\{0\} \times X$. But

$$(\mathbf{1}_X \times (-\mathbf{1})_{\widehat{X}})^* \mathcal{P}_X|_{X\times\{\widehat{x}\}} = \mathcal{P}_X|_{X\times\{-\widehat{x}\}} = P_{-\widehat{x}} = P_{\widehat{x}}^{-1} = \mathcal{P}_X^{-1}|_{X\times\{\widehat{x}\}}$$

and the restrictions to $\{0\} \times X$ are both trivial by Property (ii). This implies $(\mathbf{1}_X \times (-\mathbf{1})_{\widehat{X}})^* \mathcal{P}_X \simeq \mathcal{P}_X^{-1}$. The proof of the other assertion is analogous. □

Lemma 6.1.3 *$t_{(x,\widehat{x})}^* \mathcal{P}_X \simeq \mathcal{P}_X \otimes p_1^* P_{\widehat{x}} \otimes p_2^* P_x$ for all $(x, \widehat{x}) \in X \times \widehat{X}$.*

Here p_1 and p_2 denote the projections of $X \times \widehat{X}$ onto its factors.

Proof Note first that for all $\widehat{x} \in \widehat{X}$ we have

$$t_{(0,\widehat{x})}^* \mathcal{P}_X|_{\{0\}\times\widehat{X}} = \mathcal{O}_{\widehat{X}} \quad \text{and} \quad \mathcal{P}_X \otimes p_1^* P_{\widehat{x}}|_{\{0\}\times\widehat{X}} = \mathcal{O}_{\widehat{X}} \otimes q_{\widehat{X}}^*(P_{\widehat{x}}|_{\{0\}}) = \mathcal{O}_{\widehat{X}},$$

where $q_{\widehat{X}} : \widehat{X} \to \{0\}$ is the zero map. Moreover for $\widehat{x}, \widehat{y} \in \widehat{X}$,

$$\begin{aligned} t_{(0,\widehat{y})}^* \mathcal{P}_X|_{X\times\{\widehat{x}\}} &= \mathcal{P}_X|_{X\times\{\widehat{y}+\widehat{x}\}} = P_{\widehat{y}+\widehat{x}} \\ &= P_{\widehat{y}} \otimes P_{\widehat{x}} = \left(\mathcal{P}_X|_{X\times\{\widehat{y}\}}\right) \otimes \left(p_1^* P_{\widehat{x}}|_{X\times\{\widehat{y}\}}\right). \end{aligned}$$

So $t_{(0,\widehat{x})}^* \mathcal{P}_X = \mathcal{P}_X \otimes p_1^* P_{\widehat{x}}$ by the seesaw principle, Corollary 1.4.9.
Moreover by symmetry, or more explicitly, by applying what we have shown so far to $\mathcal{P}_{\widehat{X}}$,

$$t_{(x,0)}^* \mathcal{P}_X = t_{(x,0)}^* s^* \mathcal{P}_{\widehat{X}} = s^* t_{(0,x)}^* \mathcal{P}_{\widehat{X}} = s^* \mathcal{P}_{\widehat{X}} \otimes s^* p_1^* P_x = \mathcal{P}_X \otimes p_2^* P_x.$$

Combining both statements gives the assertion. □

Proposition 6.1.4 *The Poincaré bundle $\mathcal{P}_X$ is a symmetric non-degenerate line bundle on $X \times \widehat{X}$ of type $(1, \ldots, 1)$ and index $i(\mathcal{P}_X) = g$.*

Proof By Lemma 6.1.2 we have

$$(-1)_{X\times\widehat{X}}^* \mathcal{P}_X \simeq (1_X \times (-1)_{\widehat{X}})^* \mathcal{P}_X^{-1} \simeq \mathcal{P}_X,$$

so $\mathcal{P}_X$ is symmetric. By Lemma 6.1.3, $t_{(x,\widehat{x})}^* \mathcal{P}_X \simeq \mathcal{P}_X$ if and only if $x = \widehat{x} = 0$, so $K(\mathcal{P}_X) = 0$ and thus $\mathcal{P}$ is non-degenerate of type $(1, \ldots, 1)$ by Proposition 1.4.7.

As for the index, recall that $i(\mathcal{P}_X)$ is the number of negative eigenvalues of the hermitian form $c_1(\mathcal{P}_X)$ on $V \times \overline{\Omega}$. By Lemma 6.1.2, $\left((-1)_V \times 1_{\overline{\Omega}}\right)^* c_1(\mathcal{P}_X) = c_1(\mathcal{P}_X^{-1}) = -c_1(\mathcal{P}_X)$. Since it is non-degenerate it must have $g = \frac{1}{2}\dim(V \times \overline{\Omega})$ negative eigenvalues. This completes the proof. □

Denote by $\mathbb{C}_0$ the skyscraper sheaf on X, respectively $\widehat{X}$, with support 0 and fibre $\mathbb{C}$.

Corollary 6.1.5 $R^j p_{i*}\mathcal{P}_X = \begin{cases} \mathbb{C}_0 & \textit{if } j = g \textit{ for } i = 1,2, \\ 0 & \textit{if } j \neq g \textit{ for } i = 1,2. \end{cases}$

Proof According to the base change theorem (see Hartshorne [61, Thm. III 12, 11(a)]), if the natural map

$$\varphi^j(\widehat{x}) : R^j p_{2*}\mathcal{P}_X \otimes \mathbb{C}(\widehat{x}) \to H^j(X \times \{\widehat{x}\}, \mathcal{P}_X|_{X\times\{\widehat{x}\}})$$

is surjective, then it is an isomorphism and the same is true for all y in a suitable neighborhood of $\widehat{x}$ in $\widehat{X}$. But

$$h^j(\mathcal{P}_X|_{X\times\{\widehat{x}\}}) = h^j(X, P_{\widehat{x}}) \begin{cases} = 0 & \text{if } \widehat{x} \neq 0, \\ \neq 0 & \text{if } \widehat{x} = 0, \end{cases}$$

hence the support of $R^j p_{2*}\mathcal{P}_X$ is contained in $\{0\} \subset \widehat{X}$. Now we apply the Leray spectral sequence for p_2:

$$E_2^{p,q} = H^p(\widehat{X}, R^q p_{2*}\mathcal{P}_X) \Rightarrow E^{p+q} = H^{p+q}(X \times \widehat{X}, \mathcal{P}_X).$$

Since $E_2^{p,q} = 0$ for $p > 0$, the spectral sequence degenerates. Thus

$$H^0(\widehat{X}, R^j p_{2*}\mathcal{P}_X) = H^j(X \times \widehat{X}, \mathcal{P}_X) = \begin{cases} \mathbb{C} & \text{if } j = g, \\ 0 & \text{if } j \neq g. \end{cases} \tag{6.1}$$

Here the last equation is a direct consequence of Proposition 6.1.4 and Theorem 1.6.8. This gives the assertion for p_2. By symmetry we obtain the assertion for p_1.□

Denote by p_{ij} the projection of $X \times \widehat{X} \times X$ onto the i-th times the j-th factor and by $\varphi : X \times \widehat{X} \times X \to X \times \widehat{X}$ the homomorphism $\varphi(x, \widehat{x}, y) = (x + y, \widehat{x})$. This notation will be used in the following lemma.

Lemma 6.1.6 $p_{12}^*\mathcal{P}_X \otimes p_{23}^*\mathcal{P}_{\widehat{X}} \simeq \varphi^*\mathcal{P}_X$.

Proof Again we use the seesaw principle, Corollary 1.4.9, restricting to $\{0\} \times \widehat{X} \times X$ and $X \times \{(\widehat{x}, y)\}$ for all $(\widehat{x}, y) \in \widehat{X} \times X$. Denote by $\iota_0 : \widehat{X} \times X \to X \times \widehat{X} \times X$ the inclusion $\iota_0(\widehat{x}, y) = (0, \widehat{x}, y)$. Then $p_{12}\iota_0(\widehat{x}, y) = (0, \widehat{x})$ and hence

$$p_{12}^*\mathcal{P}_X|_{\{0\}\times\widehat{X}\times X} = (p_{12}\iota_0)^*\mathcal{P}_X = \mathcal{P}_X|_{\{0\}\times\widehat{X}} = \mathcal{O}_{\widehat{X}}.$$

Since $p_{23}\iota_0 = \mathbf{1}_{\widehat{X}\times X}$ and $\varphi\iota_0 = s : \widehat{X} \times X \to X \times \widehat{X}$, the exchange map, we get $p_{23}^*\mathcal{P}_{\widehat{X}}|_{\{0\}\times\widehat{X}\times X} = \mathcal{P}_{\widehat{X}}$ and $\varphi^*\mathcal{P}_X|_{\{0\}\times\widehat{X}\times X} = s^*\mathcal{P}_X = \mathcal{P}_{\widehat{X}}$. This shows that the restriction of both sides to $\{0\} \times \widehat{X} \times X$ coincide.

As for the other restrictions, note first that

$$p_{12}^*\mathcal{P}_X\,|_{X\times\{(\widehat{x},y)\}} = \mathcal{P}_X\,|_{X\times\{\widehat{x}\}} = P_{\widehat{x}}.$$

Defining $\iota_{\widehat{x}} = X \hookrightarrow X\times\widehat{X}$, $\iota_{\widehat{x}}(x) = (x,\widehat{x})$ and $\iota_y : \widehat{X} \hookrightarrow \widehat{X}\times X$, $\iota_y(\widehat{x}) = (\widehat{x},y)$ we see that

$$\varphi(x,\widehat{x},y) = t_{(y,0)}\iota_{\widehat{x}}(x) \quad \text{and} \quad p_{23}(x,\widehat{x},y) = \iota_y p_2 \iota_{\widehat{x}}(x).$$

Hence

$$p_{23}^*\mathcal{P}_{\widehat{X}}\,|_{X\times\{(\widehat{x},y)\}} = \iota_{\widehat{x}}^* p_2^* \iota_y^* \mathcal{P}_{\widehat{X}} = \iota_{\widehat{x}}^* p_2^*(\mathcal{P}_{\widehat{X}}\,|_{\widehat{X}\times\{y\}}) = (p_2^* P_y)\,|_{X\times\{\widehat{x}\}}$$

and using Lemma 6.1.3

$$\begin{aligned}\varphi^*\mathcal{P}_X\,|_{X\times\{(\widehat{x},y)\}} &= t_{(y,0)}^*\mathcal{P}_X\,|_{X\times\{\widehat{x}\}}\\ &= (\mathcal{P}_X \otimes p_2^* P_y)\,|_{X\times\{\widehat{x}\}}\\ &= \left(p_{12}^*\mathcal{P}_X \otimes p_{23}^*\mathcal{P}_{\widehat{X}}\right)\Big|_{X\times\{(\widehat{x},y)\}}.\end{aligned}$$

This implies the assertion. □

The first Chern class $c_1(\mathcal{P}_X)$ is an element of $H^2(X\times\widehat{X},\mathbb{Z})$. The Künneth decomposition gives

$$\begin{aligned}&H^2(X\times\widehat{X},\mathbb{Z})\\ \simeq&\left(H^2(X,\mathbb{Z})\otimes H^0(\widehat{X},\mathbb{Z})\right)\oplus\left(H^1(X,\mathbb{Z})\otimes H^1(\widehat{X},\mathbb{Z})\right)\oplus\left(H^0(X,\mathbb{Z})\otimes H^2(\widehat{X},\mathbb{Z})\right).\end{aligned}$$

The following lemma shows that $c_1(\mathcal{P}_X)$ is actually contained in the middle term.

Lemma 6.1.7 $c_1(\mathcal{P}_X) \in H^1(X,\mathbb{Z}) \otimes H^1(\widehat{X},\mathbb{Z})$.

Proof By Lemma 6.1.2 we have $-c_1(\mathcal{P}_X) = \left((-\mathbf{1})_X \otimes \mathbf{1}_{\widehat{X}}\right)^* c_1(\mathcal{P}_X)$. But $(-\mathbf{1})_X \times \mathbf{1}_{\widehat{X}}$ induces the identity on $H^2(X,\mathbb{Z})\otimes H^0(\widehat{X},\mathbb{Z})$ as well as on $H^0(X,\mathbb{Z})\otimes H^2(\widehat{X},\mathbb{Z})$. This implies the assertion.

Note that this can also be seen by observing that as a hermitian form $c_1(\mathcal{P}_X) = H : (V\times\overline{\Omega})\times(V\times\overline{\Omega}) \to \mathbb{C}$ is given by $H\left((v_1,l_1),\ (v_2,l_2)\right) = \overline{l_2(v_1)} + l_1(v_2)$. □

Using the canonical isomorphism

$$H^1(X,\mathbb{Z})\otimes H^1(\widehat{X},\mathbb{Z}) \simeq \mathrm{Hom}_{\mathbb{Z}}\left(H^1(X,\mathbb{Z})^*, H^1(\widehat{X},\mathbb{Z})\right)$$

and the fact that $\mathcal{P}_X$ is non-degenerate, we may consider $c_1(\mathcal{P}_X)$ as an isomorphism

$$c_1(\mathcal{P}_X) : H^1(X,\mathbb{Z})^* \to H^1(\widehat{X},\mathbb{Z}).$$

Choose a basis $e_1,\ldots,e_{2g}$ of $H^1(X,\mathbb{Z})$ and let $e_1^*,\ldots,e_{2g}^* \in H^1(X,\mathbb{Z})^*$ be the dual basis. Defining $f_i := c_1(\mathcal{P}_X)(e_i^*)$ we have:

Lemma 6.1.8 $c_1(\mathcal{P}_X) = \sum_{i=1}^{2g} e_i \otimes f_i$.

Proof As an element of $H^1(X, \mathbb{Z}) \otimes H^1(\widehat{X}, \mathbb{Z})$, the first Chern class $c_1(\mathcal{P}_X)$ is of the form $c_1(\mathcal{P}_X) = \sum c_{ij} e_i \otimes f_j$, with $c_{ij} \in \mathbb{Z}$. But then

$$f_k = c_1(\mathcal{P}_X)(e_k^*) = \sum c_{ij}(e_i \otimes f_j)(e_k^*) = \sum c_{ij} e_k^*(e_i) f_j = \sum c_{kj} f_j.$$

So $c_{kj} = \delta_{kj}$, which implies the assertion. □

6.1.2 WIT-sheaves

Let X be an abelian variety. Recall that the index $i(L)$ of a non-degenerate line bundle L on X is the number of negative eigenvalues of its associated hermitian form. Mumford's Index Theorem 1.6.5 says: For any non-degenerate $L \in \mathrm{Pic}(X)$,

$$H^j(L \otimes P) = 0 \quad \text{for all} \quad P \in \mathrm{Pic}^0(X) \quad \text{and} \quad j \neq i(L).$$

Induced by this, Mukai called more generally any coherent sheaf $\mathcal{F}$ on X an *IT-sheaf of index i* (IT stands for Index Theorem) if

$$H^j(\mathcal{F} \otimes P) = 0 \quad \text{for all} \quad P \in \mathrm{Pic}^0(X) \quad \text{and} \quad j \neq i.$$

The following lemma might be called the *Weak Index Lemma*.

Lemma 6.1.9 *Let $\mathcal{F}$ be an IT-sheaf of index i on X. Then*

(a) $R^j p_{2*}(\mathcal{P}_X \otimes p_1^* \mathcal{F}) = 0$ *for* $j \neq i$,
(b) $R^i p_{2*}(\mathcal{P}_X \otimes p_1^* \mathcal{F})$ *is locally free of finite rank on* $\widehat{X}$.

Here p_1 and p_2 denote again the projections of $X \times \widehat{X}$.

Proof Note first that $\mathcal{P}_X \otimes p_1^* \mathcal{F} \,|_{X \times \{\widehat{x}\}} = P_{\widehat{x}} \otimes \mathcal{F}$, implying that

$$H^j\left(X \times \{\widehat{x}\}, \mathcal{P}_X \otimes p_1^* \mathcal{F}|_{X \times \{\widehat{x}\}}\right) = 0$$

for $j \neq i$. Now the assertion follows from the Base Change Theorem (see Hartshorne [61, III, 12.11]) and the coherence of the direct image sheaves. □

The lemma motivates the following definition. A coherent sheaf $\mathcal{F}$ on X is called a *WIT-sheaf of index i* (WIT stands for Weak Index Theorem) if

$$R^j p_{2*}(\mathcal{P}_X \otimes p_1^* \mathcal{F}) = 0 \quad \text{for all} \quad j \neq i.$$

In this case the coherent sheaf

$$\widehat{\mathcal{F}} := R^i p_{2*}(\mathcal{P}_X \otimes p_1^* \mathcal{F})$$

is called the *Fourier* or *Fourier–Mukai transform* of $\mathcal{F}$. With this terminology Lemma 6.1.9 implies that every IT-sheaf of index i is a WIT-sheaf of the same index. In particular, a non-degenerate line bundle L of index i is a WIT-sheaf of index i and its Fourier transform $\widehat{L}$ is a vector bundle on $\widehat{X}$.

Theorem 6.1.10 (Inversion Theorem) *If $\mathcal{F}$ is a* WIT*-sheaf of index i on an abelian variety X of dimension g, then its Fourier–Mukai transform $\widehat{\mathcal{F}}$ is a* WIT*-sheaf of index $g-i$ on $\widehat{X}$, and there is a canonical isomorphism*

$$\widehat{\widehat{\mathcal{F}}} \simeq (-1)_X^*\mathcal{F}.$$

Proof Denote by q_i, respectively π_i, the projections of $\widehat{X} \times X$, respectively $X \times X$, for $i = 1, 2$ and by p_{ij} the projections of $X \times \widehat{X} \times X$ for $i, j \in \{1, 2, 3\}$. Note that by Lemma 6.1.6,

$$\mathcal{E} := p_{23}^*\mathcal{P}_{\widehat{X}} \otimes p_{12}^*(\mathcal{P}_X \otimes p_1^*\mathcal{F}) = \varphi^*\mathcal{P}_X \otimes p_{13}^*\pi_1^*\mathcal{F}.$$

Using the projection formula, flat base change with q_1, and the fact that $\mathcal{F}$ is a WIT-sheaf of index i, we have,

$$\begin{aligned} R^q p_{23_*}\mathcal{E} &= \mathcal{P}_{\widehat{X}} \otimes R^q p_{23*}p_{12}^*(\mathcal{P}_X \otimes p_1^*\mathcal{F}) \\ &= \mathcal{P}_{\widehat{X}} \otimes q_1^* R^q p_{2*}(\mathcal{P}_X \otimes p_1^*\mathcal{F}) \\ &= \begin{cases} \mathcal{P}_{\widehat{X}} \otimes q_1^*\widehat{\mathcal{F}} & \text{if } \ q = i, \\ 0 & \text{if } \ q \neq i. \end{cases} \end{aligned}$$

As for every composition of morphisms of algebraic varieties there is a spectral sequence (see for example Gelfand–Manin [50, Theorem 3.7.1])

$$E_2^{p,q} = R^p q_{2*} R^q p_{23_*}\mathcal{E} \;\Rightarrow\; E^{p+q} = R^{p+q}(q_2 p_{23})_*\mathcal{E}.$$

The above equation for $R^q p_{23_*}\mathcal{E}$ implies that the spectral sequence degenerates. In particular,

$$E^n = E_2^{n-i,i}.$$

Similarly, using the projection formula, $p_1\varphi = (\pi_1 + \pi_2)p_{13}$, flat base change with $\pi_1 + \pi_2$, and Corollary 6.1.5 we get

$$\begin{aligned} R^q p_{13_*}\mathcal{E} &= R^q p_{13_*}(\varphi^*\mathcal{P}_X) \otimes \pi_1^*\mathcal{F} \\ &= (\pi_1 + \pi_2)^* R^q p_{1_*}\mathcal{P}_X \otimes \pi_1^*\mathcal{F} \\ &= \begin{cases} (\pi_1 + \pi_2)^*\mathbb{C}_0 \otimes \pi_1^*\mathcal{F} & \text{if } \ q = g, \\ 0 & \text{if } \ q \neq g. \end{cases} \end{aligned}$$

Hence the spectral sequence for the composition $\pi_2 p_{1,3}$,

$$(E_2')^{p,q} = R^p \pi_{2*} R^q p_{13*} \mathcal{E} \Rightarrow (E')^{p+q} = R^{p+q}(\pi_2 p_{13})_* \mathcal{E},$$

also degenerates. Since $\pi_2 p_{13} = q_2 p_{23}$, we have in particular,

$$E^n = (E')^n = (E_2')^{n-g,g} \simeq R^{n-g}\pi_{2*}\Big((\pi_1 + \pi_2)^* \mathbb{C}_0 \otimes \pi_1^* \mathcal{F}\Big)$$

for all n. Identifying $(\pi_1 + \pi_2)^*(0) = \{(x, -x) \mid x \in X\}$ with X, then $(\pi_1 + \pi_2)^*\mathbb{C}_0 \otimes \pi_1^*\mathcal{F} \simeq \mathcal{F}$ and the restriction $\pi_2 \mid_{(\pi_1+\pi_2)^*(0)}$ coincides with the automorphism $(-\mathbf{1})_X$ of X. Hence $E^n = 0$ for $n \neq g$ and

$$E^g = \pi_{2*}\big((\pi_1 + \pi_2)^*\mathbb{C}_0 \otimes \pi_1^*\mathcal{F}\ \big) = (-\mathbf{1})^*\mathcal{F}.$$

Now flat base change with q_1 and the projection formula give

$$\begin{aligned} R^j q_{2*}(\mathcal{P}_{\widehat{X}} \otimes q_1^* \widehat{\mathcal{F}}) &= R^j q_{2*}\Big(\mathcal{P}_{\widehat{X}} \otimes q_1^*\big(R^i p_{2*}(\mathcal{P}_X \otimes p_1^*\mathcal{F})\big)\Big) \\ &= R^j q_{2*}\Big(\mathcal{P}_{\widehat{X}} \otimes R^i p_{23*}\big(p_{12}^*(\mathcal{P}_X \otimes p_1^*\mathcal{F})\big)\Big) \\ &= R^j q_{2*} R^i p_{23*} \mathcal{E} = E_2^{j,i} \\ &= E^{j+i} = \begin{cases} (-\mathbf{1})^*\mathcal{F} & \text{if } \ j = g - i, \\ 0 & \text{if } \ j \neq g - i. \end{cases} \end{aligned}$$

We conclude that $\widehat{\mathcal{F}}$ is a WIT-sheaf of index $g - i$ and $\widehat{\widehat{\mathcal{F}}} \simeq (-1)^*\mathcal{F}$. □

Corollary 6.1.11 *Let $\mathcal{F}$ and $\mathcal{G}$ be WIT-sheaves of index i on X and $f \in$* Hom$(\mathcal{F}, \mathcal{G})$. *Define a homomorphism $\widehat{f} : \widehat{\mathcal{F}} \to \widehat{\mathcal{G}}$ by*

$$\widehat{f} := R^i p_{2*}(\mathbf{1}_{\mathcal{P}_X} \otimes p_1^* f).$$

This makes $\widehat{\ }$ into a fully faithful functor from the category of WIT-sheaves of index i on X into the category of WIT-sheaves of index $g - i$ on $\widehat{X}$.

Proof It remains to show that $\widehat{\ }$: Hom$(\mathcal{F}, \mathcal{G}) \to$ Hom$(\widehat{\mathcal{F}}, \widehat{\mathcal{G}})$ is an isomorphism. But this follows immediately from the functoriality of the isomorphism of the Inversion Theorem 6.1.10. □

6.1.3 Some Properties of the Fourier–Mukai Transform

In this section we compile some properties of the Fourier–Mukai transform of WIT-sheaves. Most of them are consequences of the Inversion Theorem 6.1.10.

Proposition 6.1.12 *Let* $0 \to \mathcal{F} \to \mathcal{G} \to \mathcal{H} \to 0$ *be an exact sequence of coherent sheaves on* X *with WIT-sheaves* $\mathcal{F}$ *and* $\mathcal{H}$ *of index* i*. Then* $\mathcal{G}$ *is also a WIT-sheaf of index* i *and the sequence* $0 \to \widehat{\mathcal{F}} \to \widehat{\mathcal{G}} \to \widehat{\mathcal{H}} \to 0$ *is exact.*

In particular the functor $\widehat{\ }$ on the category of WIT-sheaves of index i is exact.

Proof With $0 \to \mathcal{F} \to \mathcal{G} \to \mathcal{H} \to 0$ also the sequence $0 \to \mathcal{P}_X \otimes p_1^*\mathcal{F} \to \mathcal{P}_X \otimes p_1^*\mathcal{G} \to \mathcal{P}_X \otimes p_1^*\mathcal{H} \to 0$ is exact, p_1 being flat. Now the long exact cohomology sequence for p_{2*} gives the assertion. □

Example 6.1.13 Let $\mathbb{C}_x$ denote the skyscraper sheaf on X with support $x \in X$ and fibre $\mathbb{C}$. This is an IT-sheaf of index 0, since $H^j(X, \mathbb{C}_x \otimes P) = 0$ for all $j > 0$ and all $P \in \mathrm{Pic}^0(\widehat{X})$. Its Fourier–Mukai transform is

$$\widehat{\mathbb{C}_x} = p_{2*}(\mathcal{P}_X \otimes p_1^*\mathbb{C}_x) = \mathcal{P}_X|_{\{x\}\times\widehat{X}} = P_x \in \mathrm{Pic}^0(\widehat{X}).$$

The Inversion Theorem 6.1.10 implies that every $P \in \mathrm{Pic}^0(\widehat{X})$ is a WIT-sheaf of index g. However P is not an IT-sheaf, since $H^0(P \otimes P^{-1}) = h^0(\mathcal{O}_X) \neq 0$.

More generally we have:

Proposition 6.1.14 *Let* $\mathcal{F}$ *be a coherent sheaf on* X *with 0-dimensional support. Then* $\mathcal{F}$ *is an IT-sheaf of index 0 and its Fourier–Mukai transform* $\widehat{\mathcal{F}}$ *is a vector bundle.*

Proof We apply induction on the length of $\mathcal{F}$. The case $\mathrm{length}(\mathcal{F}) = 1$ is covered by Example 6.1.13. If $\mathrm{length}(\mathcal{F}) > 1$ there is an exact sequence $0 \to \mathcal{F}' \to \mathcal{F} \to \mathbb{C}_x \to 0$. Since $\mathrm{length}(\mathcal{F}') < \mathrm{length}(\mathcal{F})$, the induction hypothesis, Proposition 6.1.12 and Lemma 6.1.9 give the assertion. □

Another class of WIT-sheaves are unipotent vector bundles. Recall that a vector bundle U on X is called *unipotent* if it admits a filtration

$$0 = U_0 \subset U_1 \subset \cdots \subset U_{r-1} \subset U_r = U$$

such that $U_i/U_{i-1} \simeq \mathcal{O}_X$ for $i = 1, \dots, r$.

Proposition 6.1.15 *A vector bundle* U *on* X *is unipotent if and only if* U *is a WIT-sheaf of index* g *and the support of its Fourier–Mukai transform satisfies* $\mathrm{supp}(\widehat{U}) = \{0\} \subset \widehat{X}$.

Proof Suppose U is unipotent of rank r. If $r = 1$, then $U = O_X$ and the assertion follows from Example 6.1.13. Suppose $r > 1$ and the assertion holds for unipotent vector bundles of rank $< r$. Then there is an exact sequence $0 \to U_{r-1} \to U_r \to O_X \to 0$ and the assertion follows from the long exact cohomology sequence for p_{2*}.

Conversely, suppose U is a WIT-sheaf of index g on X with $\operatorname{supp}(\widehat{U}) = \{0\}$. Apply induction on $n = \operatorname{length}(\widehat{U})$: For $n = 1$, the Inversion Theorem 6.1.10 and Example 6.1.13 give $(-1)^*U = \widehat{\widehat{U}} = O_X$. If $n > 1$ there is an exact sequence $0 \to V \to \widehat{U} \to \mathbb{C}_0 \to 0$. By Proposition 6.1.14, $V = \widehat{U}_{r-1}$ with a vector bundle U_{r-1}. By the induction hypothesis U_{r-1} is unipotent. Now the long exact cohomology sequence for p_{2*} of $0 \to \mathcal{P}_X \otimes p_1^*V \to \mathcal{P}_X \otimes p_1^*\widehat{U} \to \mathcal{P}_X \otimes p_1^*\mathbb{C}_0 \to 0$ implies that $U = (-1)^*\widehat{\widehat{U}}$ is unipotent. □

Let $\mathcal{P} = \mathcal{P}_X$, the Poincaré bundle of X.

Proposition 6.1.16 *Suppose $\mathcal{F}$ is a WIT-sheaf of index i on X, $x \in X$ and $\widehat{x} \in \widehat{X}$. Then $\mathcal{F} \otimes P_{\widehat{x}}$ and $t_x^*\mathcal{F}$ are WIT-sheaves of index i with Fourier–Mukai transforms*

(a) $(\mathcal{F} \otimes P_{\widehat{x}})\widehat{\ } \simeq t_{\widehat{x}}^*\widehat{\mathcal{F}}$,

(b) $(t_x^*\mathcal{F})\widehat{\ } \simeq \widehat{\mathcal{F}} \otimes P_{-x}$.

Proof According to the Inversion Theorem 6.1.10 it suffices to prove the assertion for $\mathcal{F} \otimes P_{\widehat{x}}$. Using Lemma 6.1.3 and flat base change with

$$\begin{array}{ccc} X \times \widehat{X} & \xrightarrow{t_{(0,\widehat{x})}} & X \times \widehat{X} \\ \downarrow{\scriptstyle p_2} & & \downarrow{\scriptstyle p_2} \\ \widehat{X} & \xrightarrow{t_{\widehat{x}}} & \widehat{X} \end{array}$$

we have

$$\begin{aligned} R^j p_{2*}(\mathcal{P} \otimes p_1^*(\mathcal{F} \otimes P_{\widehat{x}})) &= R^j p_{2*}(t_{(0,\widehat{x})}^*\mathcal{P} \otimes p_1^*\mathcal{F}) \\ &= t_{\widehat{x}}^* R^j p_{2*}(\mathcal{P} \otimes t_{(0,-\widehat{x})}^* p_1^*\mathcal{F}) \\ &= t_{\widehat{x}}^* R^j p_{2*}(\mathcal{P} \otimes p_1^*\mathcal{F}) = \begin{cases} t_{\widehat{x}}^*\widehat{\mathcal{F}} & \text{if } j = i, \\ 0 & \text{if } j \neq i\,. \end{cases} \end{aligned}$$

□

Let $f : Y \to X$ be an isogeny of abelian varieties, and $\widehat{f} : \widehat{X} \to \widehat{Y}$ the dual isogeny. The next proposition computes the Fourier–Mukai transform of the direct image, respectively pull back, via f of WIT-sheaves on Y, respectively on X.

Proposition 6.1.17

(a) *If $\mathcal{F}$ is a WIT-sheaf on Y of index i, then $f_*\mathcal{F}$ is a WIT-sheaf on X of index i with Fourier–Mukai transform*

$$(f_*\mathcal{F})\widehat{\ } = \widehat{f}^*\widehat{\mathcal{F}}.$$

(b) *If $\mathcal{G}$ is a WIT-sheaf of index i on X, then $f^*\mathcal{G}$ is a WIT-sheaf of index i on Y with Fourier–Mukai transform*

$$(f^*\mathcal{G})\widehat{\ } = \widehat{f}_*\widehat{\mathcal{G}}.$$

Proof (a) Let q_i denote the projections of $Y \times \widehat{X}$. By abuse of notation we denote the projections of $X \times \widehat{X}$ and $Y \times \widehat{Y}$ both by p_1 and p_2. Since $f \times \mathbf{1}_{\widehat{X}} : Y \times \widehat{X} \to X \times \widehat{X}$ is an isogeny, the spectral sequence for the composition of maps $p_2(f \times \mathbf{1}_{\widehat{X}}) = q_2$ (see Gelfand–Manin [50, Theorem 3.7.1]) degenerates; that is,

$$R^j p_{2*}(f \times \mathbf{1}_{\widehat{X}})_*(\,\cdot\,) = R^j q_{2*}(\,\cdot\,) \tag{6.2}$$

for any coherent sheaf on $Y \times \widehat{X}$. Moreover, note that

$$(f \times \mathbf{1}_{\widehat{X}})^*\mathcal{P}_X \simeq (\mathbf{1}_Y \times \widehat{f})^*\mathcal{P}_Y \tag{6.3}$$

by the universal property of the Poincaré bundle. Using flat base change for $f q_1 = p_1(f \times \mathbf{1}_{\widehat{X}})$, the projection formula, equations (6.2) and (6.3), and flat base change for $p_2(\mathbf{1}_Y \times \widehat{f}) = \widehat{f} q_2$, we get

$$\begin{aligned} R^j p_{2*}(\mathcal{P}_X \otimes p_1^*(f_*\mathcal{F})) &\simeq R^j p_{2*}(\mathcal{P}_X \otimes (f \times \mathbf{1}_{\widehat{X}})_* q_1^*\mathcal{F}) \\ &\simeq R^j p_{2*}(f \times \mathbf{1}_{\widehat{X}})_*((f \times \mathbf{1}_{\widehat{X}})^*\mathcal{P}_X \otimes q_1^*\mathcal{F}) \\ &\simeq R^j q_{2*}((f \times \mathbf{1}_{\widehat{X}})^*\mathcal{P}_X \otimes q_1^*\mathcal{F}) \\ &\simeq R^j q_{2*}((\mathbf{1}_Y \times \widehat{f})^*\mathcal{P}_Y \otimes (\mathbf{1}_Y \times \widehat{f})^* p_1^*\mathcal{F}) \\ &\simeq \widehat{f}^* R^j p_{2*}(\mathcal{P}_Y \otimes p_1^*\mathcal{F}) \\ &\simeq \begin{cases} \widehat{f}^*\widehat{\mathcal{F}} & \text{if } j = i, \\ 0 & \text{if } j \neq i. \end{cases} \end{aligned}$$

This completes the proof of assertion (a). Assertion (b) follows from (a) and the Inversion Theorem 6.1.10. □

Remark 6.1.18 There is a generalization of the Fourier–Mukai functor to all coherent sheaves of an abelian variety, also due to Mukai [95]. Here we only give the definition and refer for more details to the books by Huybrechts [69] and Polishchuk [105]. For a short introduction, see Birkenhake–Lange [24].

Let X be an abelian variety of dimension g. Denote as above by p_1 and p_2 the projections of $X \times \widehat{X}$ and by $\mathcal{P}_X$ the Poincaré bundle on $X \times \widehat{X}$. The functor from the category of $\mathcal{O}_X$-modules into the category of $\mathcal{O}_{\widehat{X}}$-modules

$$S : \mathcal{F} \mapsto p_{2*}(\mathcal{P}_X \otimes p_1^*\mathcal{F})$$

is left exact. If q_1 and q_2 denote the projection of $\widehat{X} \times X$ we have similarly the functor

$$\widehat{S} : \mathcal{G} \mapsto q_{2*}(\mathcal{P}_{\widehat{X}} \otimes q_1^* \mathcal{G}) = p_{1*}(\mathcal{P}_X \otimes p_2^* \mathcal{G})$$

from the category of $\mathcal{O}_{\widehat{X}}$-modules into the category of $\mathcal{O}_X$-modules.

Let $\mathcal{D}^b(X)$, respectively $\mathcal{D}^b(\widehat{X})$, denote the derived category of complexes of $\mathcal{O}_X$-modules, respectively $\mathcal{O}_{\widehat{X}}$-modules, with bounded coherent cohomology. Then the derived functors RS of S and $R\widehat{S}$ of $\widehat{S}$ exist. The Fourier–Mukai transform is then given by

$$RS : \mathcal{D}^b(X) \to \mathcal{D}^b(\widehat{X}), \qquad \mathcal{F}^\bullet \mapsto Rp_{2*}(\mathcal{P}_X \otimes p_1^* \mathcal{F}^\bullet),$$
$$R\widehat{S} : \mathcal{D}^b(\widehat{X}) \to \mathcal{D}^b(X), \qquad \mathcal{G}^\bullet \mapsto Rq_{2*}(\mathcal{P}_{\widehat{X}} \otimes q_1^* \mathcal{G}^\bullet).$$

The Inversion Theorem 6.1.10 then generalizes to a canonical isomorphism of functors

$$R\widehat{S} \circ RS \simeq (-1)_X^*[-g],$$

where $[-g]$ denotes the shift of a complex by g places to the right. The proof is formally very similar to the proof of the Inversion Theorem 6.1.10 (see [24, Theorem 14.7.2]). It follows from the Inversion Theorem that RS induces an equivalence of categories $\mathcal{D}^b(X) \simeq \mathcal{D}^b(\widehat{X})$.

6.1.4 Exercises

(1) Let X be an abelian variety and q_{ij} be the projection onto the ith times the jth factor of $X \times X \times \widehat{X}$ and Δ the diagonal in $X \times X$. Then

$$R^j q_{12*}(q_{13}^* \mathcal{P}_X \otimes q_{23}^* \mathcal{P}_X^{-1}) \simeq \begin{cases} \mathcal{O}_\Delta & \text{if } j = g, \\ 0 & \text{if } j \neq g. \end{cases}$$

(Hint: The proof is analogous to that of Lemma 6.1.6.)

(2) Let $\mathcal{F}$ be a WIT-vector bundle of index i on the abelian variety X Then the dual vector bundle $\mathcal{F}^*$ is a WIT-vector bundle of index $g - i$ with Fourier transform

$$(\mathcal{F}^*)\widehat{\ } = (-1)_{\widehat{X}}(\widehat{\mathcal{F}})^*.$$

Define the *Pontryagin product of sheaves* $\mathcal{F}$ and $\mathcal{G}$ on X by

$$\mathcal{F} \star \mathcal{G} := \mu_*(\pi_1^* \mathcal{F} \otimes \pi_2^* \mathcal{G}),$$

where as above $\pi_i : X \times X \to X$ for $i = 1, 2$ are the projections, and as usual

$\mu : X \times X \to X$ is the addition map. If $\mathcal{F}$ is a vector bundle, the functor

$$\mathcal{F}\star : \mathcal{G} \mapsto \mathcal{F} \star \mathcal{G}$$

is left exact on the category of coherent sheaves on X. Hence its derived functors $R^p(\mathcal{F}\star)$ are well-defined.

In the case of a non-degenerate line bundle L the following proposition expresses the functors $R^p(L\star)$ in terms of the Fourier transform.

(3) Let L be a non-degenerate line bundle and $\mathcal{F}$ a coherent sheaf on X. If $L \otimes (-1)^*\mathcal{F}$ is a WIT-sheaf of index i, then

$$R^j(L\star)\mathcal{F} \simeq R^j\mu_*(\pi_1^*L \otimes \pi_2^*\mathcal{F}) \simeq \begin{cases} L \otimes \phi_L^*\left(\widehat{(L \otimes (-1)^*\mathcal{F})}\right), & \text{if } j = i, \\ 0 & \text{if } j \neq i. \end{cases}$$

(Hint: Use the spectral sequence for the composite functor $\mu_* \circ (\pi_1^*L \otimes \pi_2^*)(\cdot)$).

(4) For a coherent sheaf $\mathcal{F}$ on X and $\widehat{x} \in \widehat{X}$

$$R^j(P_{\widehat{x}}\star)\mathcal{F} \simeq H^j(\mathcal{F} \otimes P_{-\widehat{x}}) \otimes P_{\widehat{x}}.$$

(Hint: The proof is similar to the proof of the previous exercise.)

(5) If L is a non-degenerate line bundle of index i on X, then

(a) $\phi_L^*\widehat{L} \simeq H^i(L) \otimes L^{-1}$,
(b) $\phi_{L*}L^{-1} \simeq H^i(L) \otimes \widehat{L}$.

(Hint: For (a) apply Exercise (3) above to $\mathcal{F} = \mathcal{O}_X$; for (b) apply (a) and Serre duality.)

(6) Deduce from the previous exercise that $\widehat{\mathcal{P}_X} = \mathcal{P}_{\widehat{X}}^{-1}$.

(7) Let L be an ample line bundle of type $(d_1, \ldots, d_g)$ on the abelian variety X. According to Section 2.5.1 there is the line bundle L_δ on $\widehat{X}$ defining the dual polarization. Show that L_δ and the Fourier–Mukai transform $\widehat{L}$ are related as follows:

$$\det(\widehat{L})^{-1} \equiv L_\delta^{d_2 \cdots d_g - 1}.$$

(Hint: Use Exercise (5) above and Proposition 1.4.12.)

(8) Let L be a non-degenerate line bundle on X.

(a) Show that there is an isogeny $f : X \to Y$ and a line bundle N of type $(1, \ldots, 1)$ such that $L = f^*N$.
(b) With the notation of (a) show that

$$\widehat{L} \simeq (\widehat{f}\phi_N)_* N^{-1}.$$

(Hint: For (a) use Corollary 1.4.4. For (b) use Proposition 6.1.17 and Exercise (5) above.)

(9) The Fourier transform $\widehat{L}$ of a non-degenerate line bundle L on X is a μ-semistable vector bundle of rank $|\chi(L)|$ with respect to *any* polarization H.
(Hint: Use Exercise (5) above.)

(10) Let X be an elliptic curve and let E be a vector bundle on X.

(a) If E is semistable on X with $\deg E < 0$, then E is IT of index 1 and $\widehat{E}$ is a semistable vector bundle with $\deg \widehat{E} = \operatorname{rk} E$ and $\operatorname{rk} \widehat{E} = -\deg E$.
(b) If E is semistable on X with $\deg E > 0$, then E is IT of index 0 and $\widehat{E}$ is a semistable vector bundle with $\deg \widehat{E} = \operatorname{rk} E$ and $\operatorname{rk} \widehat{E} = \deg E$.
(c) Any semistable vector bundle of degree 0 on X is homogeneous; that is $t_x^* E \simeq E$ for all $x \in X$. Show that E is WIT of index 1 and the Fourier–Mukai transform induces a bijection between the set of semistable bundles of degree 0 and the set of coherent sheaves with finite support on $\widehat{X} = X$.

6.2 The Fourier Transform on the Chow and Cohomology Rings

There is an analogue of the Fourier–Mukai transform on Chow rings and cohomology rings which is introduced in this section. The first section contains the definition of the Chow group of a smooth projective variety and some of its properties. The second section gives a generalization of correspondences between two curves, as defined in Section 4.6.1, to arbitrary smooth projective varieties. After that we are in a position to define and study the Fourier transform on the Chow ring and the cohomology ring of a smooth projective variety.

6.2.1 Chow Groups

In this section we compile some generalities about algebraic cycles and Chow groups. For more details and proofs we refer to the standard books on intersection theory (preferably Fulton [47]).

Let X be a smooth projective variety of dimension g over the field of complex numbers. Denote by $\mathcal{Z}_k(X)$ the free abelian group generated by all subvarieties of X of dimension k. Its elements are finite sums $\mathrm{V} = \sum_{i=1}^{s} n_i \mathrm{V}_i$, with $n_i \in \mathbb{Z}$ and subvarieties $\mathrm{V}_i \subset X$ of dimension k. They are called *algebraic cycles* of dimension k. Denote by $\mathcal{Z}^p(X) := \mathcal{Z}_{g-p}(X)$ the group of algebraic cycles of codimension p and $\mathcal{Z}(X) = \bigoplus_{k=0}^{g} \mathcal{Z}_k(X)$ the graded group of all cycles on X.

Let $f : X \longrightarrow Y$ be a morphism of smooth projective varieties. If f is proper, the *push forward homomorphism* $f_* : \mathcal{Z}_k(X) \longrightarrow \mathcal{Z}_k(Y)$ is the homomorphism of groups, induced by

$$f_*\mathrm{V} := \begin{cases} \deg(f|_{\mathrm{V}}) \cdot f(\mathrm{V}) & \text{if } \deg(f|_{\mathrm{V}}) < \infty, \\ 0 & \text{if otherwise} \end{cases}$$

for any subvariety $\mathrm{V} \subset X$ of dimension k. If f is flat, the *pull back homomorphism* $f^* : \mathcal{Z}^p(Y) \to \mathcal{Z}^p(X)$ is the homomorphism of groups induced by

$$f^*\mathrm{W} = f^{-1}\mathrm{W}$$

for any subvariety $\mathrm{W} \subset Y$ of codimension p. Let V and W be subvarieties of X of codimension p and q. Let $U_1, \ldots, U_k$ be the irreducible components of $\mathrm{V} \cap \mathrm{W}$. Recall that V and W *intersect properly* if $\operatorname{codim} U_i = p + q$ for $i = 1, \ldots, k$. If this is the case the *intersection product* is defined by

$$\mathrm{V} \cdot \mathrm{W} = \sum_{i=1}^{k} \operatorname{mult}_{U_i}(\mathrm{V}, \mathrm{W})\, U_i,$$

where $\operatorname{mult}_{U_i}(\mathrm{V}, \mathrm{W})$ is the local intersection multiplicity of V and W along U_i.

Two cycles $\mathrm{V} = \sum n_i \mathrm{V}_i$ and $\mathrm{W} = \sum m_j \mathrm{W}_j$ on X intersect properly if V_i and W_j intersect properly whenever $n_i \neq 0 \neq m_j$. If this is the case the *intersection product* of V with W is

$$\mathrm{V} \cdot \mathrm{W} := \sum_{i,j} n_i m_j \mathrm{V}_i \cdot \mathrm{W}_j.$$

As above let $f : X \to Y$ be a morphism of smooth projective varieties. Let p_1 and p_2 denote the projections of $Y \times X$ and $Z \in \mathcal{Z}(Y \times X)$. For a cycle $\mathrm{V} \in \mathcal{Z}(Y)$ such that Z and $p_1^*\mathrm{V}$ intersect properly,

$$Z(\mathrm{V}) := p_{2*}(Z \cdot p_1^*\mathrm{V}) \tag{6.4}$$

is a cycle on X.

With this definition we are in position to define rational equivalence: Let

$$\mathcal{Z}^p_{\mathrm{rat}}(X) = \mathcal{Z}^{\mathrm{rat}}_{g-p}(X)$$

denote the subgroup of $\mathcal{Z}^p(X)$ generated by all cycles of the form $Z(0) - Z(\infty)$, where $Z \in \mathcal{Z}^p(\mathbb{P}^1 \times X)$ such that Z intersects $\{t\} \times X$ properly for all t in an open dense subset $U \subset \mathbb{P}^1$ containing 0 and ∞. Two cycles V and $\mathrm{W} \in \mathcal{Z}^p(X)$ are called *rationally equivalent*, in notation $\mathrm{V} \sim_{\mathrm{rat}} \mathrm{W}$, if $\mathrm{V} - \mathrm{W} \in \mathcal{Z}^p_{\mathrm{rat}}(X)$. Obviously this is an equivalence relation on $\mathcal{Z}^p(X)$ for all p and thus on $\mathcal{Z}(X)$.

Remark 6.2.1 An *adequate equivalence relation* is an equivalence relation "$\sim$" on $\mathcal{Z}(\,\cdot\,)$ satisfying the following conditions

(i) $\{\mathrm{V} \in \mathcal{Z}(X) \mid \mathrm{V} \sim 0\}$ is a graded subgroup of $\mathcal{Z}(X)$.

(ii) For any $V, V_1, \ldots, V_r \in \mathcal{Z}(X)$ there is a $W \in \mathcal{Z}(X)$ with $W \sim V$ such that W intersects V_i properly for all $i = 1, \ldots, r$.
(iii) Let $Z \in \mathcal{Z}(Y \times X), V \in \mathcal{Z}(Y)$ with $V \sim 0$. If Z intersects $p_1^* V$ properly, then $Z(V) \sim 0$.

Remark 6.2.2 Rational equivalence is an *adequate equivalence relation* (see Exercise 6.2.5 (1)).

Denote by

$$\mathrm{Ch}_k(X) := \mathrm{Ch}^{g-k}(X) := \mathcal{Z}_k(X)/\mathcal{Z}_k^{\mathrm{rat}}(X)$$

the group of algebraic cycles of dimension k, respectively codimension $g-k$, on X modulo rational equivalence. Moreover, denote by $\mathrm{Ch}(X)$ the group of all algebraic cycles on X modulo rational equivalence. For any cycle $V \in \mathcal{Z}(X)$ we denote its image in $\mathrm{Ch}(X)$ by the same symbol. According to property (i) of Remark 6.2.1 dimension and codimension of cycles define gradings on the group $\mathrm{Ch}(X)$. If it is necessary to emphasize the grading, we also use the notation

$$\mathrm{Ch}_{\bullet}(X) = \bigoplus_{k=0}^{g} \mathrm{Ch}_k(X) \quad \text{and} \quad \mathrm{Ch}^{\bullet}(X) = \bigoplus_{p=0}^{g} \mathrm{Ch}^p(X).$$

Thanks to property (ii) of Remark 6.2.1 the intersection product induces a product

$$\mathrm{Ch}^p(X) \times \mathrm{Ch}^q(X) \to \mathrm{Ch}^{p+q}(X) \tag{6.5}$$

which is again called the *intersection product*. This makes $\mathrm{Ch}^{\bullet}(X)$ into a commutative associative graded ring with identity X, the *Chow ring* of X. $\mathrm{Ch}_k(X)$ and $\mathrm{Ch}^p(X)$ are called the *Chow groups of dimension* k, respectively *codimension* p*-cycles on* X.

Remark 6.2.3 For every smooth projective variety X there is a canonical isomorphism $\mathrm{Ch}^1(X) \simeq \mathrm{Pic}(X)$ (see Exercise 6.2.5 (2)). Accordingly, in this chapter we consider line bundles L as elements of $\mathrm{Ch}^1(X)$. We denote the ν-th self-intersection product $L \cdot \ldots \cdot L$ by $L^{\cdot \nu}$, in order to distinguish it from the ν-th tensor power $L^{\nu} = L \otimes \cdots \otimes L$. So $L^{\cdot \nu}$ is an element of $\mathrm{Ch}^{\nu}(X)$. Note that the intersection product used in former chapters of this book always was the intersection product of cohomology classes.

The following formulas will be applied very often. For the proofs, see for example Fulton [47].

Theorem 6.2.4 (Projection Formula) *Let $f : Y \to X$ be a proper morphism of smooth projective varieties. Then for all $\alpha \in \mathrm{Ch}(Y)$ and $\beta \in \mathrm{Ch}(X)$*

$$f_*(\alpha \cdot f^*\beta) = f_*\alpha \cdot \beta.$$

Theorem 6.2.5 (Base Change Formula) *Suppose*

$$\begin{array}{ccc} Y' & \xrightarrow{g'} & Y \\ {\scriptstyle f'}\downarrow & & \downarrow{\scriptstyle f} \\ X' & \xrightarrow{g} & X \end{array}$$

is a cartesian diagram of smooth projective varieties with f proper and g flat. Then for all $\alpha \in \mathrm{Ch}(Y)$:

$$f'_* g'^* \alpha = g^* f_* \alpha.$$

Finally, recall that for every subvariety $V \subset X$ of codimension p the fundamental class $[V]$ is an element of $H^{2p}(X,\mathbb{Z})$. This defines a map $\mathcal{Z}^p(X) \to H^{2p}(X,\mathbb{Z})$. Its kernel is denoted by $\mathcal{Z}^p_{\mathrm{hom}}(X)$. Two cycles V and $W \in \mathcal{Z}^p(X)$ are called *homologically equivalent* if $V - W \in \mathcal{Z}^p_{\mathrm{hom}}(X)$. This is again an adequate equivalence relation (see Exercise 6.2.5 (1)).

Lemma 6.2.6 *For any smooth projective variety X, the map $\mathcal{Z}^p(X) \to H^{2p}(X,\mathbb{Z})$ factorizes via the Chow group $\mathrm{Ch}^p(X)$.*

Proof It suffices to show that $\mathcal{Z}^p_{\mathrm{rat}}(X) \subset \mathcal{Z}^p_{\mathrm{hom}}(X)$, which is easy to see, the group $H^{2p}(X,\mathbb{Z})$ being discrete. □

The induced map

$$\mathrm{cl} : \mathrm{Ch}^p(X) \to H^{2p}(X,\mathbb{Z}).$$

is called the *cycle map*. Its extension to $\mathrm{Ch}^p(X)_{\mathbb{Q}} := \mathrm{Ch}^p(X) \otimes_{\mathbb{Z}} \mathbb{Q}$ is denoted by the same symbol and also called the *cycle map*.

Lemma 6.2.7 *The image of the cycle map $\mathrm{cl} : \mathrm{Ch}^p(X)_{\mathbb{Q}} \to H^{2p}(X,\mathbb{Q})$ is contained in $H^{2p}(X,\mathbb{Q}) \cap H^{p,p}(X)$.*

For the proof, see Exercise 6.2.5 (8).

6.2.2 Correspondences

In Section 4.6.1 we defined a correspondence between two smooth projective curves C_1 and C_2 to be a line bundle on $C_1 \times C_2$. According to Theorem 4.6.1 any such correspondence induces a homomorphism $\mathrm{Pic}^0(C_1) \to \mathrm{Pic}^0(C_2)$. By Remark 6.2.3 we have $\mathrm{Pic}(C_\nu) = \mathrm{Ch}^1(C_\nu)$. More generally, let X_1 and X_2 be any smooth projective varieties. In this section we will see that any cycle on $X_1 \times X_2$ induces homomorphisms between the Chow groups of X_1 and X_2.

A *correspondence (of codimension) p* between X_1 and X_2 is by definition a cycle $Z \in \mathcal{Z}^p(X_1 \times X_2)$ or a cycle class $Z \in \mathrm{Ch}^p(X_1 \times X_2)$. Let $p_i : X_1 \times X_2 \to X_i$ be the projections. According to Remark 6.2.1 (iii) the assignment

$$\mathcal{Z}_q(X_1) \to \mathcal{Z}^{p-q}(X_2), \qquad V \mapsto p_{2*}(Z \cdot p_1^* V)$$

induces a map

$$\mathrm{Ch}_q(X_1) \to \mathrm{Ch}^{p-q}(X_2) \qquad \alpha \mapsto p_{2*}(Z \cdot p_1^* \alpha).$$

For the proof that the map is well defined, that is, depends only on the cycle class and not on the cycle itself, see Exercise 6.2.5 (6). By abuse of notation we denote this map also by Z, that is,

$$Z(\alpha) := p_{2*}(Z \cdot p_1^* \alpha).$$

This gives a homomorphism of groups

$$\mathrm{Ch}^p(X_1 \times X_2) \to \bigoplus_q \mathrm{Hom}\left(\mathrm{Ch}_q(X_1), \mathrm{Ch}^{p-q}(X_2)\right)$$

respectively

$$\mathrm{Ch}^\bullet(X_1 \times X_2) \to \mathrm{Hom}\left(\mathrm{Ch}_\bullet(X_1), \mathrm{Ch}^\bullet(X_2)\right). \tag{6.6}$$

Let X_3 be a third smooth projective variety and $p_{ij} : X_1 \times X_2 \times X_3 \to X_i \times X_j$ the projections. For correspondences $Z_1 \in \mathrm{Ch}(X_1 \times X_2)$ and $Z_2 \in \mathrm{Ch}(X_2 \times X_3)$ define the composition $Z_2 \circ Z_1 \in \mathrm{Ch}(X_1 \times X_3)$ by

$$Z_2 \circ Z_1 := p_{13*}(p_{23}^* Z_2 \cdot p_{12}^* Z_1).$$

The following lemma shows that the composition of correspondences is compatible with the composition of the associated homomorphisms.

Lemma 6.2.8 *For $\alpha \in \mathrm{Ch}(X_1)$ we have*

$$Z_2 \circ Z_1(\alpha) = Z_2\big(Z_1(\alpha)\big).$$

Proof In this proof denote by p_i^{ij} and p_j^{ij} the projections of $X_i \times X_j$, and by p_i the corresponding projections $X_1 \times X_2 \times X_3$. Then using the Base Change Formula 6.2.5 with $p_2^{12} \circ p_{12} = p_2^{23} \circ p_{23}$ and the Projection Formula 6.2.4,

$$\begin{aligned}
Z_2\big(Z_1(\alpha)\big) &= p_{3*}^{23}\Big(Z_2 \cdot p_2^{23*}\big(p_{2*}^{12}(Z_1 \cdot p_1^{12*}\alpha)\big)\Big) \\
&= p_{3*}^{23}\big(Z_2 \cdot p_{23*}(p_{12}^* Z_1 \cdot p_1^* \alpha)\big) \\
&= p_{3*}^{23} p_{23*}(p_{23}^* Z_2 \cdot p_{12}^* Z_1 \cdot p_1^* \alpha) \\
&= p_{3*}^{13} p_{13*}(p_{23}^* Z_2 \cdot p_{12}^* Z_1 \cdot p_{13}^* p_1^{13^*} \alpha) \\
&\qquad (\text{using } p_3^{23} \circ p_{23} = p_3^{13} \circ p_{13} \text{ and } p_1 = p_1^{13} \circ p_{13}) \\
&= p_{3*}^{13}\big(p_{13*}(p_{23}^* Z_2 \cdot p_{12}^* Z_1) \cdot p_1^{13^*} \alpha\big) \\
&= Z_2 \circ Z_1(\alpha).
\end{aligned}$$

□

Let $s := X_1 \times X_2 \to X_2 \times X_1$ be the exchange morphism $s(x_1,x_2) = (x_2,x_1)$. It induces an isomorphism

$$\mathrm{Ch}(X_2 \times X_1) \to \mathrm{Ch}(X_1 \times X_2), \quad Z \mapsto {}^tZ := s^*Z.$$

For a proper morphism $f : X_1 \to X_2$ the *graph* of f is defined to be the correspondence

$$\Gamma_f := (\mathbf{1}_{X_1}, f)_*(X_1) \in \mathrm{Ch}(X_1 \times X_2).$$

Proposition 6.2.9

(a) $\Gamma_f(\alpha) = f_*(\alpha)$ *for all* $\alpha \in \mathrm{Ch}(X_1)$.
(b) ${}^t\Gamma_f(\beta) = f^*(\beta)$ *for all* $\beta \in \mathrm{Ch}(X_2)$.

Proof Using the Projection Formula 6.2.4 we have

$$\Gamma_f(\alpha) = p_{2*}((\mathbf{1}_{X_1}, f)_*(X_1) \cdot p_1^*\alpha) = p_{2*}(\mathbf{1}_{X_1}, f)_*((\mathbf{1}_{X_1}, f)^* p_1^*\alpha) = f_*(\alpha),$$

since $p_1 \circ (\mathbf{1}_{X_1}, f) = \mathbf{1}_{X_1}$ and $p_2 \circ (\mathbf{1}_{X_1}, f) = f$. The proof of (b) is similar (see Exercise 6.2.5 (7)). □

Proposition 6.2.10 *Let* $f_i : X_i' \to X_i$, *for* $i = 1, 2$, *be proper morphisms of smooth projective varieties. Then*
(a) $(f_1 \times f_2)^*Z = {}^t\Gamma_{f_2} \circ Z \circ \Gamma_{f_1}$ *for all* $Z \in \mathrm{Ch}(X_1 \times X_2)$,
(b) $(f_1 \times f_2)_*Z' = \Gamma_{f_2} \circ Z \circ {}^t\Gamma_{f_1}$ *for all* $Z' \in \mathrm{Ch}(X_1' \times X_2')$.

Combining Lemma 6.2.8 and Propositions 6.2.9 and 6.2.10, this immediately gives

$$((f_1 \times f_2)^*Z)(\alpha) = f_2^*\Big(Z\big(f_{1*}(\alpha)\big)\Big) \quad \text{for all} \quad \alpha \in \mathrm{Ch}(X_1') \tag{6.7}$$

and

$$((f_1 \times f_2)_*Z')(\beta) = f_{2*}\Big(Z'\big(f_1^*(\beta)\big)\Big) \quad \text{for all} \quad \beta \in \mathrm{Ch}(X_1). \tag{6.8}$$

Proof We give a proof for (a), the proof of (b) is similar.

For the proof of (a) it suffices to show that $(f_1 \times \mathbf{1}_{X_2})^*Z = Z \circ \Gamma_{f_1}$ and $(\mathbf{1}_{X_1} \times f_2)^*Z = {}^t\Gamma_{f_2} \circ Z$. Denote by p_{ij} the projections of $X_1' \times X_1 \times X_2$ and by q_1 the projection $X_1' \times X_2 \to X_1'$. Then using the Base Change Formula 6.2.5 with $(\mathbf{1}_{X_1'}, f_1) \circ q_1 = p_{12} \circ ((\mathbf{1}_{X_1'}, f_1) \times \mathbf{1}_{X_2})$ and the Projection Formula 6.2.4,

$$\begin{aligned}
Z \circ \Gamma_{f_1} &= p_{13*}(p_{23}^*Z \cdot p_{12}^*(\mathbf{1}_{X_1'}, f_1)_*(X_1')) \\
&= p_{13*}\Big(p_{23}^*Z \cdot ((\mathbf{1}_{X_1'}, f_1) \times \mathbf{1}_{X_2})_* q_1^*(X_1')\Big) \\
&= p_{13*}((\mathbf{1}_{X_1'}, f_1) \times \mathbf{1}_{X_2})_*((\mathbf{1}_{X_1'}, f_1) \times \mathbf{1}_{X_2})^* p_{23}^*Z \cdot q_1^*(X_1')) \\
&= p_{13*}((\mathbf{1}_{X_1'}, f_1) \times \mathbf{1}_{X_2})_*((\mathbf{1}_{X_1'}, f_1) \times \mathbf{1}_{X_2})^* p_{23}^*Z) \\
&= (f_1 \times \mathbf{1}_{X_2})^*Z,
\end{aligned}$$

where we used $p_{13}\circ((\mathbf{1}_{X_1'},f_1)\times\mathbf{1}_{X_2}) = \mathbf{1}_{X_1'\times X_2}$ and $p_{23}\circ((\mathbf{1}_{X_1'},f_1)\times\mathbf{1}_{X_2}) = f_1\times\mathbf{1}_{X_2}$. Using this we obtain

$$(\mathbf{1}_{X_1}\times f_2)^*Z = {}^t((f_2\times\mathbf{1}_{X_1})^{*t}Z) = {}^t({}^tZ\circ\Gamma_{f_2}) = {}^t\Gamma_{f_2}\circ Z. \qquad \square$$

Denote by Δ_{X_i} the class of the diagonal in $\mathrm{Ch}(X_i\times X_i)$ for $i = 1, 2$. Since Δ_{X_i} is the graph of the identity $\mathbf{1}_{X_i}$, we obtain as a consequence

Corollary 6.2.11 *For any correspondence $Z \in \mathrm{Ch}(X_1\times X_2)$ we have*

$$Z = Z\circ\Delta_{X_1} = \Delta_{X_2}\circ Z.$$

As an immediate consequence of Remark 6.2.1 (iii) we get:

Remark 6.2.12 For a correspondence $Z \in \mathrm{Ch}(X_1\times X_2)$, any adequate equivalence relation $\sim$ and $\alpha \in \mathrm{Ch}^{\sim}(X_1)$ we have $Z(\alpha) \in \mathrm{Ch}^{\sim}(X_2)$.

6.2.3 The Fourier Transform on the Chow Ring

The Fourier transform on the level of cycles has been thoroughly investigated by Beauville in [15]. In this section we give the definition of the Fourier transform on the Chow ring and derive some properties. In particular it exchanges, up to sign, the intersection product by the Pontryagin product.

Let X be an abelian variety of dimension g. The Fourier functor $F = F_X$ is defined on the Chow ring with $\mathbb{Q}$-coefficients:

$$\mathrm{Ch}^p(X)_{\mathbb{Q}} := \mathrm{Ch}^p(X)\otimes_{\mathbb{Z}}\mathbb{Q} \quad \text{and} \quad \mathrm{Ch}(X)_{\mathbb{Q}} := \mathrm{Ch}(X)\otimes_{\mathbb{Z}}\mathbb{Q}.$$

All definitions and properties of $\mathrm{Ch}(X)$ of Sections 6.2.1 and 6.2.2 extend to $\mathrm{Ch}(X)_{\mathbb{Q}}$ in an obvious way.

Consider the Poincaré bundle $\mathcal{P} = \mathcal{P}_X \in \mathrm{Pic}(X\times\widehat{X}) = \mathrm{Ch}^1(X\times\widehat{X})$ (see Sections 1.4.4 and 6.1.1). We denote its image in $\mathrm{Ch}^1(X\times\widehat{X})_{\mathbb{Q}}$ by the same letter. The correspondence

$$\mathbf{e}^{\mathcal{P}} := \sum_{\nu\geq 0}\tfrac{1}{\nu!}\mathcal{P}^{\cdot\nu} \in \mathrm{Ch}(X\times\widehat{X})_{\mathbb{Q}}$$

is well-defined, the sum being finite. The correspondence $\mathbf{e}^{\mathcal{P}}$ defines a homomorphism of groups

$$F = F_X : \mathrm{Ch}(X)_{\mathbb{Q}} \to \mathrm{Ch}(\widehat{X})_{\mathbb{Q}}, \quad \alpha\mapsto \mathbf{e}^{\mathcal{P}}(\alpha) = p_{2*}(\mathbf{e}^{\mathcal{P}}\cdot p_1^*\alpha),$$

called the *Fourier transform* on $\mathrm{Ch}(X)_{\mathbb{Q}}$.

Remark 6.2.13 Note that according to Fulton [47], $\mathbf{e}^{\mathcal{P}}$ coincides with the Chern character $\mathrm{ch}(\mathcal{P})$ of the line bundle $\mathcal{P}$. In Section 6.1.2 we discussed the Fourier–Mukai transform of a WIT-sheaf, associating to a WIT-sheaf on X a WIT-sheaf on the dual abelian variety $\widehat{X}$. Applying the Chern character, this construction gives the above Fourier transform on $\mathrm{Ch}(X)_{\mathbb{Q}}$.

More generally, the Chern character extends to the derived category of complexes $\mathcal{D}^b(X)$ of $\mathcal{O}_X$-modules on X (which we do not study in this book, but mentioned briefly in Remark 6.1.18) and the above Fourier transform on $\mathrm{Ch}(X)_{\mathbb{Q}}$ is related to the Fourier–Mukai transform $RF : \mathcal{D}^b(X) \to \mathcal{D}^b(\widehat{X})$ via the Chern character $\mathcal{D}^b(\cdot) \to \mathrm{Ch}_{\mathbb{Q}}(\cdot)$ by an obvious commutative diagram.

Lemma 6.2.14 *Let* $(0)_{\widehat{X}}$ *denote the zero-cycle given by the element* $0 \in \widehat{X}$*. Then*

$$F_X(X) = (-1)^g (0)_{\widehat{X}}.$$

Proof This is a consequence of the Grothendieck–Riemann–Roch Theorem applied twice, to the projection $p_2 : X \times \widehat{X} \to \widehat{X}$ and the canonical embedding $i_{\widehat{X}} : 0 \hookrightarrow \widehat{X}$, for which we refer to Fulton [47, Section 15.2].

Recall that $\mathrm{ch}(\mathcal{P}_X) = \mathbf{e}^{\mathcal{P}_X}$ and that $Rp_{2*}\mathcal{P}_X = \mathbb{C}_0[-g]$, the complex with $\mathbb{C}_0$ at the g-th place and zero elsewhere. The last equation holds by Corollary 6.1.5, where $\mathbb{C}_0$ denotes the skyscraper sheaf with fibre $\mathbb{C}$ at $0 \in \widehat{X}$.

Now if we use the fact that for homomorphisms of abelian varieties the relative tangent bundle is trivial and thus its Todd class is 1, Grothendieck–Riemann–Roch applied to the projection $p_2 : X \times \widehat{X} \to \widehat{X}$ and the WIT-sheaf $\mathcal{P}_X$ gives

$$F_X(X) = p_{2*}\,\mathbf{e}^{\mathcal{P}_X} = p_{2*}\mathrm{ch}(\mathcal{P}_X) = \mathrm{ch}(Rp_{2*}\mathcal{P}_X) = (-1)^g \mathrm{ch}(\mathbb{C}_0).$$

Applying Grothendieck–Riemann–Roch to the canonical embedding $i_{\widehat{X}} : 0 \hookrightarrow \widehat{X}$ yields similarly

$$\mathrm{ch}(\mathbb{C}_0) = \mathrm{ch}(R\, i_{\widehat{X}*}\mathcal{O}_{\mathrm{Spec}\,\mathbb{C}}) = i_{\widehat{X}*}\mathrm{ch}(\mathcal{O}_{\mathrm{Spec}\,\mathbb{C}}) = i_{\widehat{X}*}(0) = (0)_{\widehat{X}}.$$

Combining both equations gives the assertion. □

Theorem 6.2.15 (Inversion Theorem)

$$F_{\widehat{X}}F_X = (-1)^g(-1)_X^* : \mathrm{Ch}(X)_{\mathbb{Q}} \to \mathrm{Ch}(X)_{\mathbb{Q}}.$$

Proof Let q_{ij} be the projections of $X \times \widehat{X} \times X$, define $\varphi : X \times \widehat{X} \times X \to X \times \widehat{X}$ by $\varphi(x, \widehat{x}, y) = (x + y, \widehat{x})$ and let as usual $\mu : X \times X \to X$ denote the addition map. Applying Lemma 6.2.8 we get

$$
\begin{aligned}
F_{\widehat{X}} \circ F_X &= q_{13*}(q_{23}^* \, \mathbf{e}^{\mathcal{P}_{\widehat{X}}} \cdot q_{12}^* \, \mathbf{e}^{\mathcal{P}_X}) \\
&= q_{13*}\varphi^* \, \mathbf{e}^{\mathcal{P}_X} && \text{(by Lemma 6.1.6)} \\
&= \mu^* p_{1*} \, \mathbf{e}^{\mathcal{P}_X} && \text{(using base change for } p_1 \circ \varphi = \mu \circ p_{13}) \\
&= \mu^* F_{\widehat{X}}(\widehat{X}) && \text{(by definition of } F_{\widehat{X}}) \\
&= (-1)^g \mu^*(0)_X && \text{(by Lemma 6.2.14)} \\
&= (-1)^g \, {}^t\Gamma_{(-1)_X}.
\end{aligned}
$$

Now Proposition 6.2.9 implies the assertion. □

An immediate consequence of the theorem and its proof is

Corollary 6.2.16 *The Fourier transform* $F_X : \mathrm{Ch}(X)_\mathbb{Q} \to \mathrm{Ch}(\widehat{X})_\mathbb{Q}$ *is bijective with* $F_X(X) = (-1)^g (0)_{\widehat{X}}$ *and* $F_X\big((0)_X\big) = \widehat{X}$.

The Fourier transform behaves well with respect to isogenies. In fact:

Proposition 6.2.17 *Let* $f : Y \to X$ *be an isogeny of abelian varieties. Then for all* $\alpha \in \mathrm{Ch}(Y)_\mathbb{Q}$ *and* $\beta \in \mathrm{Ch}(X)_\mathbb{Q}$:

(a) $F_X f_*(\alpha) = \widehat{f}^* F_Y(\alpha)$;
(b) $F_Y f^*(\beta) = \widehat{f}_* F_X(\beta)$.

Proof The Universal Property of the Poincaré bundle implies

$$(f \times \mathbf{1}_{\widehat{X}})^* \mathcal{P}_X = (\mathbf{1}_Y \times \widehat{f})^* \mathcal{P}_Y.$$

Using this and equation (6.7) twice we get

$$
\begin{aligned}
F_X f_*(\alpha) &= \mathbf{e}^{\mathcal{P}_X}(f_*(\alpha)) \\
&= \Big((f \times \mathbf{1}_{\widehat{X}})^* \, \mathbf{e}^{\mathcal{P}_X} \Big)(\alpha) \\
&= \Big((\mathbf{1}_Y \times \widehat{f})^* \, \mathbf{e}^{\mathcal{P}_Y} \Big)(\alpha) \\
&= \widehat{f}^* (\mathbf{e}^{\mathcal{P}_Y}(\alpha)) = \widehat{f}^* F_Y(\alpha),
\end{aligned}
$$

which proves (a).

(b): Using (a) applied to $\widehat{f} : \widehat{X} \to \widehat{Y}$ and the Inversion Theorem 6.2.15 twice we obtain:

$$F_Y f^* = (-1)^g (-1)_{\widehat{Y}}^* F_Y f^* F_{\widehat{X}} F_X = (-1)^g (-1)_{\widehat{Y}}^* F_Y F_{\widehat{Y}} \widehat{f}_* F_X = \widehat{f}_* F_X. \qquad \square$$

The *Pontryagin product* on the Chow groups is defined in the same way as for homology groups (see Section 2.5.3), namely

$$\star : \mathrm{Ch}_p(X) \times \mathrm{Ch}_q(X) \xrightarrow{\times} \mathrm{Ch}_{p+q}(X \times X) \xrightarrow{\mu_*} \mathrm{Ch}_{p+q}(X),$$

where μ is the addition map. The Pontryagin product makes $\mathrm{Ch}_\bullet(X)$ and $\mathrm{Ch}_\bullet(X)_\mathbb{Q}$ into commutative associative graded rings with identity $(0)_X \in \mathrm{Ch}_0(X)$. Together with the intersection product we thus have two ring structures on $\mathrm{Ch}(X)$. The following proposition shows that the Fourier transform interchanges both (up to a sign).

Proposition 6.2.18 *For all $\alpha, \beta \in \mathrm{Ch}^\bullet_\mathbb{Q}(X)$:*

(a) $F(\alpha \star \beta) = F(\alpha) \cdot F(\beta)$;
(b) $F(\alpha \cdot \beta) = (-1)^g F(\alpha) \star F(\beta)$.

Proof (b) follows from (a) by the Inversion Theorem 6.2.15.

(a): Denote by q_i and q_{ij} the projections of $X \times X \times \widehat{X}$ and by p_i the projections of $X \times \widehat{X}$. Then, using the Base Change Formula 6.2.5 and the Projection Formula 6.2.4,

$$\begin{aligned} F(\alpha \star \beta) &= p_{2*}\Big(\mathbf{e}^{\mathcal{P}_X} \cdot p_1^*(\mu_*(\alpha \times \beta))\Big) \\ &= p_{2*}\big(\mathbf{e}^{\mathcal{P}_X} \cdot (\mu \times \mathbf{1}_{\widehat{X}})_* q_{12}^*(\alpha \times \beta)\big) \\ &= p_{2*}\big(\mathbf{e}^{\mathcal{P}_X} \cdot (\mu \times \mathbf{1}_{\widehat{X}})_*(q_1^*\alpha \cdot q_2^*\beta)\big) \\ &= p_{2*}(\mu \times \mathbf{1}_{\widehat{X}})_*\big((\mu \times \mathbf{1}_{\widehat{X}})^* \mathbf{e}^{\mathcal{P}_X} \cdot q_1^*\alpha \cdot q_2^*\beta\big). \end{aligned}$$

An immediate modification of Lemma 6.1.6 gives

$$(\mu \times \mathbf{1}_{\widehat{X}})^* \mathbf{e}^{\mathcal{P}_X} = q_{13}^* \mathbf{e}^{\mathcal{P}_X} \cdot q_{23}^* \mathbf{e}^{\mathcal{P}_X}.$$

Using $p_2 \circ (\mu \times 1_{\widehat{X}}) = q_3 = p_2 \circ q_{13}$, the computation continues as follows:

$$\begin{aligned} F(\alpha \star \beta) &= q_{3*}(q_{13}^* \mathbf{e}^{\mathcal{P}_X} \cdot q_1^*\alpha \cdot q_{23}^* \mathbf{e}^{\mathcal{P}_X} \cdot q_2^*\beta) \\ &= p_{2*} \circ q_{13*}\big(q_{13}^*(\mathbf{e}^{\mathcal{P}_X} \cdot p_1^*\alpha) \cdot q_{23}^*(\mathbf{e}^{\mathcal{P}_X} \cdot p_1^*\beta)\big) \\ &= p_{2*}\big(\mathbf{e}^{\mathcal{P}_X} \cdot p_1^*\alpha \cdot q_{13*}q_{23}^*(\mathbf{e}^{\mathcal{P}_X} \cdot p_1^*\beta)\big) \\ &= p_{2*}\big(\mathbf{e}^{\mathcal{P}_X} \cdot p_1^*\alpha \cdot p_2^*p_{2*}(\mathbf{e}^{\mathcal{P}_X} \cdot p_1^*\beta)\big) \\ &= p_{2*}(\mathbf{e}^{\mathcal{P}_X} \cdot p_1^*\alpha) \cdot p_{2*}(\mathbf{e}^{\mathcal{P}_X} \cdot p_1^*\beta) = F(\alpha) \cdot F(\beta). \end{aligned}$$ □

Recall from Section 6.1.1 the notation $P_x = \mathcal{P}_X|_{\{x\}\times\widehat{X}}$ respectively $P_{\widehat{x}} = \mathcal{P}_X|_{X\times\{\widehat{x}\}}$ considered as line bundles on $\widehat{X}$, respectively X.

Proposition 6.2.19 *For any $x \in X$ and $\widehat{x} \in \widehat{X}$:*

(a) $F(x) = \mathbf{e}^{P_x}$;
(b) $F(t_x^*\alpha) = \mathbf{e}^{-P_x} \cdot F(\alpha)$ *for all* $\alpha \in \mathrm{Ch}^\bullet(X)_\mathbb{Q}$;
(c) $(-1)^g F(P_{\widehat{x}}) = \sum_{\nu=1}^{g} \frac{1}{\nu}\big((0)_{\widehat{X}} - (\widehat{x})\big)^{\star\nu}$.

Proof (a): Consider the embedding $\iota : \widehat{X} \hookrightarrow X \times \widehat{X}$, $\widehat{x} \mapsto (x, \widehat{x})$. Then

$$F(x) = p_{2*}(\mathbf{e}^{\mathcal{P}_X} \cdot p_1^*(x)) = p_{2*}(\iota_*\iota^* \mathbf{e}^{\mathcal{P}_X}) = \mathbf{e}^{\iota^*\mathcal{P}_X} = \mathbf{e}^{P_x}.$$

(b): Note that by definition of the Pontryagin product

$$t_x^*\alpha = t_{-x*}\alpha = (-x) \star \alpha$$

in $\mathrm{Ch}^\bullet(X)_{\mathbb{Q}}$. So the assertion follows from Proposition 6.2.18 and (a).

(c): For any cycle $\alpha \in \mathrm{Ch}^\bullet_{\mathbb{Q}}(X)$:

$$\alpha = -\log\Big((X) - ((X) - \mathbf{e}^{-\alpha})\Big) = \sum_{\nu=1}^{g} \tfrac{1}{\nu}((X) - \mathbf{e}^{-\alpha})^{\cdot\nu}. \tag{6.9}$$

So we get

$$\begin{aligned}
\sum_{\nu=1}^{g} \tfrac{1}{\nu}((0)_{\widehat{X}} - (\widehat{x}))^{\star\nu} &= (-1)^g F_X F_{\widehat{X}} (-1)^*_{\widehat{X}} \sum_{\nu=1}^{g} \tfrac{1}{\nu}((0)_{\widehat{X}} - (\widehat{x}))^{\star\nu} \\
&\qquad \text{(by the Inversion Theorem 6.2.15)} \\
&= (-1)^g F_X \sum_{\nu=1}^{g} \tfrac{1}{\nu}\Big(F_{\widehat{X}}((0)_{\widehat{X}} - (-\widehat{x}))\Big)^{\cdot\nu} \\
&\qquad \text{(by Proposition 6.2.18)} \\
&= (-1)^g F_X \sum_{\nu=1}^{g} \tfrac{1}{\nu}((X) - \mathbf{e}^{-P_{\widehat{x}}})^{\cdot\nu} \\
&\qquad \text{(by Corollary 6.2.16 and (a))} \\
&= (-1)^g F_X(P_{\widehat{x}}). \qquad \text{(by equation (6.9))} \ \square
\end{aligned}$$

6.2.4 The Fourier Transform on the Cohomology Ring

Let X be an abelian variety of dimension g. The Fourier transform $F : \mathrm{Ch}(X)_{\mathbb{Q}} \to \mathrm{Ch}(\widehat{X})_{\mathbb{Q}}$ induces via the cycle map $\mathrm{cl} : \mathrm{Ch}^\bullet(X)_{\mathbb{Q}} \to H^{2\bullet}(X, \mathbb{Q})$ a homomorphism $F_H : H^\bullet(X, \mathbb{Q}) \to H^\bullet(\widehat{X}, \mathbb{Q})$. In this section we show that one can express F_H in terms of Poincaré duality.

The element

$$\mathrm{cl}(\mathbf{e}^{\mathcal{P}_X}) = \mathbf{e}^{\mathrm{cl}(\mathcal{P}_X)} = \sum_{\nu \geq 0} \tfrac{1}{\nu!}\mathrm{cl}(\mathcal{P}_X)^{\wedge\nu} \in H^\bullet(X \times \widehat{X}, \mathbb{Q})$$

defines a homomorphism of groups

$$F = F_H : H^\bullet(X, \mathbb{Q}) \to H^\bullet(\widehat{X}, \mathbb{Q}), \quad F_H(\alpha) = p_{2*}(\mathrm{cl}(\mathbf{e}^{\mathcal{P}_X}) \cdot p_1^*\alpha),$$

called the *Fourier transform (on the cohomology ring)*. By definition we have

$$F \circ \mathrm{cl} = \mathrm{cl} \circ F.$$

Using the canonical isomorphism

$$H^1(X,\mathbb{Z}) \otimes H^1(\widehat{X},\mathbb{Z}) \to \mathrm{Hom}_{\mathbb{Z}}(H^1(X,\mathbb{Z})^*, H^1(\widehat{X},\mathbb{Z})),$$

Lemmas 6.1.7 and 6.1.8 imply that we may consider the first Chern class $c_1(\mathcal{P}_X) = \mathrm{cl}(\mathcal{P}_X)$ as an isomorphism

$$c_1(\mathcal{P}_X) : H^1(X,\mathbb{Z})^* \to H^1(\widehat{X},\mathbb{Z}).$$

On the other hand, the cup product pairing

$$H^p(X,\mathbb{Z}) \otimes H^{2g-p}(X,\mathbb{Z}) \to H^{2g}(X,\mathbb{Z}) \simeq \mathbb{Z}$$

yields the Poincaré duality

$$H^p(X,\mathbb{Z}) \xrightarrow{\sim} H^{2g-p}(X,\mathbb{Z})^*.$$

Combining both we get an isomorphism

$$\alpha_p : H^p(X,\mathbb{Z}) \xrightarrow{\sim} H^{2g-p}(X,\mathbb{Z})^* \xrightarrow{\bigwedge^{2g-p} c_1(\mathcal{P}_X)} H^{2g-p}(\widehat{X},\mathbb{Z}).$$

The restriction of the Fourier transform F to $H^p(X,\mathbb{Z})$ is related to the isomorphism α_p as follows:

Proposition 6.2.20 *The Fourier transform F is an isomorphism with*

$$F|_{H^p(X,\mathbb{Z})} = (-1)^{g+\frac{1}{2}p(p+1)}\alpha_p : H^p(X,\mathbb{Z}) \to H^{2g-p}(\widehat{X},\mathbb{Z}).$$

Proof Choose a basis $e_1,\dots,e_{2g}$ of $H^1(X,\mathbb{Z})$ and let $f_1,\dots,f_{2g} \in H^1(\widehat{X},\mathbb{Z})$ be the image of the dual basis under the isomorphism $c_1(\mathcal{P}_X) : H^1(X,\mathbb{Z})^* \to H^1(\widehat{X},\mathbb{Z})$ (so we have the same notation as in Lemma 6.1.8). Denote by d the composed map

$$d : H^\bullet(X,\mathbb{Z}) \xrightarrow{\mathrm{proj}_{2g}} H^{2g}(X,\mathbb{Z}) \xrightarrow{\sim} \mathbb{Z}$$

and identify

$$H^\bullet(X \times \widehat{X},\mathbb{Z}) = H^\bullet(X,\mathbb{Z}) \otimes H^\bullet(\widehat{X},\mathbb{Z}) \tag{6.10}$$

under the Künneth isomorphism. In these terms $p_1^*x = x \otimes 1$, with $1 = \mathrm{cl}(\widehat{X}) \in H^{2g}(\widehat{X},\mathbb{Z})$, and $p_{2*}(x \otimes y) = d(x)y$ for $x \in H^\bullet(X,\mathbb{Z})$ and $y \in H^\bullet(\widehat{X},\mathbb{Z})$.

If $I = (i_1 < \cdots < i_p)$ is a multi-index in $\{1, \ldots, 2g\}$ we denote as usual $e_I = e_{i_1} \wedge \cdots \wedge e_{i_p}$. Moreover, denote by I° the complementary ordered multi-index. Then by definition of α_p and d we have

$$\alpha_p(e_I) = d(e_I \wedge e_{I^\circ}) f_{I^\circ}.$$

By Lemma 6.1.8, using the fact that the product in $H^\bullet(X, \mathbb{Z})$ is the cup product, we get,

$$\begin{aligned}
\mathbf{e}^{\mathrm{cl}(\mathcal{P}_X)} = \mathbf{e}^{\sum_{i=1}^{2g} e_i \otimes f_i} &= \bigwedge_{i=1}^{2g} \sum_{\nu \geq 0} \frac{1}{\nu!} \bigwedge^{\nu} (e_i \otimes f_i) = \bigwedge_{i=1}^{2g} (1 + e_i \otimes f_i) \\
&= \sum_{q=0}^{2g} \sum_{J=(j_1 < \cdots < j_q)} (e_{j_1} \otimes f_{j_1}) \wedge \ldots \wedge (e_{j_q} \otimes f_{j_q}) \\
&= \sum_{q=0}^{2g} \sum_{J=(j_1 < \cdots < j_q)} (-1)^{\frac{1}{2}q(q-1)} e_J \otimes f_J.
\end{aligned}$$

(using, that "$\otimes$" is alternating, due to the isomorphism (6.10).)

Combining everything, we get

$$\begin{aligned}
F(e_I) &= p_{2*}\big(\mathbf{e}^{\mathrm{cl}(\mathcal{P}_X)} \wedge p_1^* e_I\big) \\
&= \sum_{q=0}^{2g} \sum_{J=(j_1 < \ldots < j_q)} (-1)^{\frac{1}{2}q(q-1)} p_{2*}\big((e_J \otimes f_J) \wedge (e_I \otimes 1)\big) \\
&= \sum_{q=0}^{2g} \sum_{J=(j_1 < \ldots < j_q)} (-1)^{\frac{1}{2}q(q-1)} p_{2*}\big((e_I \wedge e_J) \otimes f_J\big) \\
&= \sum_{q=0}^{2g} \sum_{J=(j_1 < \ldots < j_q)} (-1)^{\frac{1}{2}q(q-1)} d(e_I \wedge e_J) f_J \\
&= (-1)^{\frac{1}{2}(2g-p)(2g-p-1)} d(e_I \wedge e_{I^\circ}) f_{I^\circ} \\
&= (-1)^{g + \frac{1}{2}p(p+1)} \alpha_p(e_I).
\end{aligned}$$

□

Extending the Fourier transform $\mathbb{C}$-linearly to $H^\bullet(X, \mathbb{C})$, the following proposition shows that F behaves well with respect to the Hodge decomposition.

Proposition 6.2.21 $F\big(H^{r,s}(X)\big) = H^{g-s, g-r}(\widehat{X})$.

Proof Notice that $\mathrm{cl}(\mathcal{P}_X) = c_1(\mathcal{P}_X) \in H^{1,1}(X \times \widehat{X}, \mathbb{Z})$. So for any $x \in H^p(X, \mathbb{C})$ we may write

$$F(x) = \sum_{\nu \geq 0} \frac{1}{\nu!} p_{2*}\Big(p_1^* x \wedge \bigwedge^{\nu} \mathrm{cl}(\mathcal{P}_X)\Big) = \frac{1}{(2g-p)!} p_{2*}\Big(p_1^* x \wedge \bigwedge^{2g-p} \mathrm{cl}(\mathcal{P}_X)\Big). \tag{6.11}$$

Moreover, if x is of Hodge type (r, s) (with $p = r + s$), then

$$p_1^* x \wedge \bigwedge^{2g-p} \mathrm{cl}(\mathcal{P}_X) \in H^{2g-p+r,2g-p+s}(X \times \widehat{X}, \mathbb{C}) = H^{2g-s,2g-r}(X \times \widehat{X}, \mathbb{C}).$$

The graph of the projection $p_2 : X \times \widehat{X} \to \widehat{X}$ is $X \times \Delta_{\widehat{X}} \in \mathrm{Ch}^g(X \times \widehat{X} \times \widehat{X})$, where $\Delta_{\widehat{X}} \in \mathrm{Ch}^g(\widehat{X} \times \widehat{X})$ denotes the diagonal. Note that for all p the map $p_{2*} : H^p(X \times \widehat{X}) \to H^{p-2g}(\widehat{X})$ factorizes as follows

$$H^p(X \times \widehat{X}) \xrightarrow{q_{12}^*} H^p(X \times \widehat{X} \times \widehat{X}) \xrightarrow{\wedge \mathrm{cl}(X \times \Delta_{\widehat{X}})} H^{p+2g}(X \times \widehat{X} \times \widehat{X})$$
$$\xrightarrow{p_{(4g,p-2g)}} H^{4g}(X \times \widehat{X}) \otimes H^{p-2g}(\widehat{X}) \simeq H^{p-2g}(\widehat{X}),$$

where $q_{12} : X \times \widehat{X} \times \widehat{X} \to X \times \widehat{X}$ is the projection onto the first and second factor and $p_{(4g,p-2g)}$ is the projection onto the $(4g, p-2g)$-th Künneth component. Since all these maps respect Hodge decompositions, so does p_{2*}; that is:

$$p_{2*}(H^{i,j}(X \times \widehat{X})) \subset H^{i-g,j-g}(\widehat{X})$$

for all i, j. So $F(x) \in H^{g-s,g-r}(\widehat{X})$. This implies the assertion, since F is an isomorphism by Proposition 6.2.20. □

6.2.5 Exercises

Let $\mathcal{Z}^p_{\mathrm{alg}}(X)$ denote the subgroup of $\mathcal{Z}^p(X)$ generated by all cycles of the form $Z(p_1) - Z(p_2)$, where $Z \in \mathcal{Z}^p(C \times X)$ with a smooth projective curve C, and $p_1, p_2 \in C$ such that Z intersects $\{p\} \times X$ properly for all p in an open dense subset $U \subset C$ containing p_1 and p_2. Two cycles V and W $\in \mathcal{Z}^p(X)$ are called *algebraically equivalent*, $V \sim_{\mathrm{alg}} W$, if $V - W \in \mathcal{Z}^p_{\mathrm{alg}}(X)$.

(1) Show that rational, algebraic and homological equivalence are adequate equivalence relations such that $\mathcal{Z}_{\mathrm{rat}}(X) \subseteq \mathcal{Z}_{\mathrm{alg}}(X) \subseteq \mathcal{Z}_{\mathrm{hom}}(X)$. The quotient $\mathrm{Ch}_{\mathrm{alg}}(X) := \mathcal{Z}_{\mathrm{alg}}(X)/\mathcal{Z}_{\mathrm{rat}}(X)$ is a graded subring of the Chow ring $\mathrm{Ch}^\bullet(X)$.

(2) Show that for every smooth projective variety X there is a canonical isomorphism $\mathrm{Ch}^1(X) \simeq \mathrm{Pic}(X)$.

(3) Show that

(a) $\mathrm{Ch}_0^{\mathrm{alg}}(X) = \ker \deg$, where the degree map deg is defined by $\deg : \mathrm{Ch}_0(X) \to \mathbb{Z}$, $\sum n_i(x_i) \mapsto \sum n_i$.

(b) $\mathrm{Ch}^1_{\mathrm{alg}}(X) = \mathrm{Pic}^0(X)$.

(4) Show that $\mathrm{Ch}^p_{\mathrm{alg}}(X)$ is a divisible group for all p.

(5) Show that with $\mathrm{Ch}_0^{\mathrm{alg}}(X)$ as in Exercise (3) above,

$$\mathrm{Ch}_0^{\mathrm{alg}}(X) \star \mathrm{Ch}^1(X) = \mathrm{Pic}^0(X).$$

(6) Let Z be a correspondence of codimension p between X_1 and X_2. Show that

(a) if $V \in \mathcal{Z}^q(X_1)$, then $p_{2*}(Z \cdot p_1^* V) \in Z^{p-q}(Z_2)$;

(b) the map

$$\mathrm{Ch}_q(X_1) \to \mathrm{Ch}^{p-q}(X_2), \qquad \alpha \mapsto p_{2*}(Z \cdot p_1^* \alpha)$$

depends only on the rational equivalence class of Z and not on Z itself.

(Hint: Use Remark 6.2.1 (iii))

(7) Let $f : X_1 \to X_2$ be a proper morphism of algebraic varieties. Show that ${}^t\Gamma_f(\beta) = f^*(\beta)$ for all $\beta \in \mathrm{Ch}(X_2)$.

(8) Show that the image of the cycle map $\mathrm{cl} : \mathrm{Ch}^p(X)_{\mathbb{Q}} \to H^{2p}(X, \mathbb{Q})$ is contained in $H^{2p}(X, \mathbb{Q}) \cap H^{p,p}(X)$.

(9) Let (X, L) be a polarized abelian variety of type $(d_1, \ldots, d_g)$ and $(\widehat{X}, L_\delta)$ its dual polarized abelian variety (see Section 2.5.1). Let C be a smooth curve in the class $L^{\cdot(g-1)} \in \mathrm{Ch}_1(X)$, (J, Θ) its Jacobian, and $\tau : \widehat{X} \to \widehat{J} = J$ the dual of the canonical homomorphism $J \to X$. Show that

$$\tau^*\Theta \equiv \left(L_\delta\right)^{(g-1)!d_2 \cdots d_{g-1}}.$$

(10) Two algebraic cycles on an abelian variety are homologically equivalent if and only if they are numerically equivalent.
(See Liebermann [85].)

6.3 Some Results on the Chow Ring of an Abelian Variety

In the first section we prove an eigenspace decomposition of the Chow ring of an abelian variety, due to Beauville [16]. A consequence is a generalization of the Poincaré Formula 4.2.2 to an arbitrary polarized abelian variety, given in Section 6.3.2. Finally, in the last two sections some results on the Künneth decomposition of the Chow ring of $X \times X$ are given, mainly due to Deninger–Murre [38] and Künnemann [78]. The most important result is the decomposition of the diagonal Δ in $\mathrm{Ch}^g(X \times X)_{\mathbb{Q}}$.

6.3.1 An Eigenspace Decomposition of $\mathrm{Ch}(X)_{\mathbb{Q}}$

Let X be an abelian variety of dimension g. Recall that for every $p = 0, \dots, g$ there is the cycle map $\mathrm{cl} : \mathrm{Ch}^p(X)_{\mathbb{Q}} \to H^{2p}(X, \mathbb{Q})$. Considering elements of $H^{2p}(X, \mathbb{Q})$ as differential forms on X, it is easy to see that for every $n \in \mathbb{Z}$ the map $n_X^* : H^{2p}(X, \mathbb{Q}) \to H^{2p}(X, \mathbb{Q})$ is multiplication by n^{2p}.

The situation is different for the Chow group $\mathrm{Ch}^p(X)_{\mathbb{Q}}$. Consider for example the case $p = 1$. There is an obvious decomposition

$$\mathrm{Ch}^1(X)_{\mathbb{Q}} = \mathrm{NS}(X)_{\mathbb{Q}} \oplus \mathrm{Pic}^0(X)_{\mathbb{Q}}$$

where the elements of $\mathrm{NS}(X)_{\mathbb{Q}}$ can be interpreted as symmetric line bundles. Now n_X^* acts on $\mathrm{NS}(X)_{\mathbb{Q}}$ by multiplication by n^2 (see Corollary 1.3.8), whereas on $\mathrm{Pic}^0(X)_{\mathbb{Q}} = \widehat{X} \otimes \mathbb{Q}$ we have $n_X^* = \widehat{n_X} = n_{\widehat{X}}$, which is multiplication by n.

There is a similar eigenspace decomposition of $\mathrm{Ch}^p(X)_{\mathbb{Q}}$ for every p. Define

$$\mathrm{Ch}^p(X)^s_{\mathbb{Q}} := \left\{\alpha \in \mathrm{Ch}^p(X)_{\mathbb{Q}} \mid n_X^*\alpha = n^{2p-s}\alpha \quad \text{for all} \quad n \in \mathbb{Z}\right\}.$$

The following theorem is due to Beauville [16].

Theorem 6.3.1 $\mathrm{Ch}^p(X)_{\mathbb{Q}} = \bigoplus_{s=p-g}^{p} \mathrm{Ch}^p(X)^s_{\mathbb{Q}}$.

Remark 6.3.2 As mentioned above, for $p = 1$ we have,

$$\mathrm{Ch}^1(X)^0_{\mathbb{Q}} = \mathrm{NS}(X)_{\mathbb{Q}} \quad \text{and} \quad \mathrm{Ch}^1(X)^1_{\mathbb{Q}} = \mathrm{Pic}^0(X)_{\mathbb{Q}}$$

and $\mathrm{Ch}^1(X)^s_{\mathbb{Q}} = 0$ for $s < 0$. In [15] and [16] Beauville conjectures that $\mathrm{Ch}^p(X)^s_{\mathbb{Q}} = 0$ for all p whenever $s < 0$.

The main tool of the proof is the Fourier transform $F : \mathrm{Ch}^\bullet(X)_{\mathbb{Q}} \to \mathrm{Ch}^\bullet(\widehat{X})_{\mathbb{Q}}$ introduced in Section 6.2.3. First we need:

Lemma 6.3.3 *Suppose $\alpha \in \mathrm{Ch}^p(X)_{\mathbb{Q}}$ and $F(\alpha) = \sum_{q=0}^{g} \beta_q$ with $\beta_q \in \mathrm{Ch}^q(\widehat{X})_{\mathbb{Q}}$. Then for all $n \in \mathbb{Z}$*

$$n_{\widehat{X}}^*\beta_q = n^{g-p+q}\beta_q.$$

Proof By definition of F,

$$\beta_q = \frac{1}{(g+q-p)!} p_{2*}(\mathcal{P}_X^{\cdot(g+q-p)} \cdot p_1^*\alpha) \in \mathrm{Ch}^q(\widehat{X})_{\mathbb{Q}}.$$

Hence using flat base change with $n_{\widehat{X}} \circ p_2 = p_2 \circ (\mathbf{1}_X \times n_{\widehat{X}})$ and the fact that $(\mathbf{1}_X \times n_{\widehat{X}})^*\mathcal{P}_X = n\mathcal{P}_X$ (which follows by an immediate computation using for example the hermitian form of $\mathcal{P}_X$ given in the proof of Theorem 1.4.10) we have

$$\begin{aligned} n_{\widehat{X}}^* \beta_q &= \frac{1}{(g+q-p)!} n_{\widehat{X}}^* p_{2*}(\mathcal{P}_X^{\cdot(g+q-p)} \cdot p_1^* \alpha) \\ &= \frac{1}{(g+q-p)!} p_{2*}((\mathbf{1}_X \times n_{\widehat{X}})^* \mathcal{P}_X^{\cdot(g+q-p)} \cdot p_1^* \alpha) \\ &= \frac{n^{g+q-p}}{(g+q-p)!} p_{2*}(\mathcal{P}_X^{\cdot(g+q-p)} \cdot p_1^* \alpha) = n^{g+q-p} \cdot \beta_q. \end{aligned}$$ □

Proposition 6.3.4 *For $\alpha \in \mathrm{Ch}^p(X)_{\mathbb{Q}}$ and $n \in \mathbb{Z} \setminus \{-1, 0, +1\}$ the following statements are equivalent:*

(i) $\alpha \in \mathrm{Ch}^p(X)_{\mathbb{Q}}^s$*; that is,* $m^* \alpha = m^{2p-s} \alpha$ *for all* $m \in \mathbb{Z}$*;*
(ii) $n_X^* \alpha = n^{2p-s} \alpha$*;*
(iii) $n_{X*} \alpha = n^{2g-2p+s} \alpha$*;*
(iv) $F(\alpha) \in \mathrm{Ch}^{g-p+s}(\widehat{X})_{\mathbb{Q}}$*;*
(v) $F(\alpha) \in \mathrm{Ch}^{g-p+s}(\widehat{X})_{\mathbb{Q}}^s$.

Proof (i) ⇒ (ii) is trivial.

(ii) ⇒ (iii): this follows from $n_{X_*} \alpha = n^{-2p+s} n_{X_*} n_X^* \alpha = n^{2g-2p+s} \alpha$ using $n_{X_*} n_X^* = \deg n_X \cdot \mathbf{1}_X = n^{2g} \mathbf{1}_X$.

(iii) ⇒ (iv) and (v): write $F(\alpha) = \sum_q \beta_q$ with $\beta_q \in \mathrm{Ch}^q(\widehat{X})_{\mathbb{Q}}$. Then using Proposition 6.2.17 and Lemma 6.3.3,

$$\begin{aligned} \sum_{q=0}^{g} \beta_q = F(\alpha) &= \frac{1}{n^{2g-2p+s}} F(n_{X_*} \alpha) = \frac{1}{n^{2g-2p+s}} n_{\widehat{X}}^* F(\alpha) \\ &= \frac{1}{n^{2g-2p+s}} \sum_{q=0}^{g} n_{\widehat{X}}^* \beta_q = \sum_{q=0}^{g} n^{p+q-g-s} \beta_q. \end{aligned}$$

Comparing coefficients this implies $F(\alpha) = \beta_{g-p+s} \in \mathrm{Ch}^{g-p+s}(\widehat{X})_{\mathbb{Q}}$. So we get (iv) and according to Lemma 6.3.3 also (v).

(v) ⇒ (i): we have for every $m \in \mathbb{Z}$,

$$\begin{aligned} m_X^* \alpha &= m_X^* (-1)^g (-1)_X^* F_{\widehat{X}} F_X \alpha && \text{(by the Inversion Theorem 6.2.15)} \\ &= (-1)^g (-1)_X^* F_{\widehat{X}} m_{\widehat{X}*} F_X(\alpha) && \text{(by Proposition 6.2.17)} \\ &= (-1)^g (-1)_X^* F_{\widehat{X}} m_{\widehat{X}*} \beta_{g-p+s} && \text{(by statement (v))} \\ &= m^{2p-2g-s} (-1)^g (-1)_X^* F_{\widehat{X}} m_{\widehat{X}*} m_{\widehat{X}}^* \beta_{g-p+s} && \text{(by Lemma 6.3.3)} \\ &= m^{2p-s} (-1)^g (-1)_X^* F_{\widehat{X}} \beta_{g-p+s} = m^{2p-s} (-1)^g (-1)_X^* F_{\widehat{X}} F_X(\alpha) = m^{2p-s} \alpha. \end{aligned}$$

It remains to show that (iv) ⇒ (v), but this follows from the equivalence of (i) and (ii), applied to $F_* \alpha$. □

Proof (of Theorem 6.3.1) Suppose $\alpha \in \mathrm{Ch}^p(X)_{\mathbb{Q}}$ and write as above $F_X(\alpha) = \sum_{q=0}^{g} \beta_q$ with $\beta_q \in \mathrm{Ch}^q(\widehat{X})_{\mathbb{Q}}$. According to Lemma 6.3.3, $\beta_q \in \mathrm{Ch}^q(X)_{\mathbb{Q}}^{p+q-g}$. So Proposition 6.3.4, (i) $\Rightarrow$ (v) applied to β_q implies $F_{\widehat{X}}(\beta_q) \in \mathrm{Ch}^p(X)_{\mathbb{Q}}^{p+q-g}$. Now using the Inversion Theorem 6.2.15,

$$\begin{aligned}\alpha &= (-1)^g(-1)^* F_{\widehat{X}} F_X(\alpha)\\ &= (-1)^g(-1)^* \sum_{q=0}^{g} F_{\widehat{X}}(\beta_q) \in \bigoplus_{q=0}^{g} \mathrm{Ch}^p(X)_{\mathbb{Q}}^{p+q-g}.\end{aligned}$$

This implies the assertion. □

6.3.2 Poincaré's Formula for Polarized Abelian Varieties

Let $(J(C), \Theta)$ be a canonically polarized Jacobian of dimension g with C embedded in $J(C)$ via an Abel–Jacobi map. In Theorem 4.2.2 we proved Poincaré's Formula relating the classes of C and Θ in $H^\bullet(J(C), \mathbb{Z})$. Using the Pontryagin product $\star$ and replacing the wedge product $\wedge$ by $\cdot$ for the intersection product, it reads as follows: for $1 \leq p \leq g$,

$$\frac{1}{p!}[\Theta]^{\cdot p} = \frac{1}{(g-p)!}[C]^{\star(g-p)}.$$

Here we use the decomposition of Theorem 6.3.1 to prove an analogous formula, due to Beauville, for an arbitrary polarized abelian variety in the Chow ring $\mathrm{Ch}(X)_{\mathbb{Q}}$. The formula is a consequence of the following theorem.

Theorem 6.3.5 *Let L be a symmetric ample line bundle on an abelian variety X of dimension g and $d = h^0(L)$. Then the following equation holds in $\mathrm{Ch}(\widehat{X})_{\mathbb{Q}}$ for $0 \leq p \leq g$:*

$$F\left(\frac{L^{\cdot p}}{p!}\right) = \frac{(-1)^{g-p}}{d} \phi_{L*}\left(\frac{L^{\cdot(g-p)}}{(g-p)!}\right).$$

Proof First we claim

$$F(\mathbf{e}^L) = \tfrac{1}{d}\phi_{L*}\, \mathbf{e}^{-L}. \tag{6.12}$$

To see this, denote by $q_i : X \times X \to X$ the projections and by $\mu : X \times X \to X$ the addition map. Then comparing dimensions we get

$$q_{2*}\mu^* \mathbf{e}^L = \sum_{\nu \geq 0} \tfrac{1}{\nu!} q_{2*}\mu^* L^{\cdot \nu} = \tfrac{1}{g!} q_{2*}\mu^* L^{\cdot g} = d\,[X]. \tag{6.13}$$

Hence

$$
\begin{aligned}
F(\mathbf{e}^L) &= p_{2*}\,\mathbf{e}^{\mathcal{P}+p_1^*L} \\
&= \tfrac{1}{d^2}\, p_{2*}(\mathbf{1}_X \times \phi_L)_*(\mathbf{1}_X \times \phi_L)^*\,\mathbf{e}^{\mathcal{P}+p_1^*L} && (\text{using } \deg \phi_L = d^2) \\
&= \tfrac{1}{d^2}\, \phi_{L*} q_{2*}(\mathbf{1}_X \times \phi_L)^*\,\mathbf{e}^{\mathcal{P}+p_1^*L} \\
&= \tfrac{1}{d^2}\, \phi_{L*} q_{2*}\,\mathbf{e}^{\mu^* L - q_2^* L} && (\text{with Exercise 1.4.5 (10)}) \\
&= \tfrac{1}{d}\, \phi_{L*}\,\mathbf{e}^{-L} && (\text{by the projection formula and equation (6.13)})
\end{aligned}
$$

which proves (6.12). Now

$$\sum_{\nu \geq 0} F\Big(\frac{L^{\cdot \nu}}{\nu!}\Big) = F(\mathbf{e}^L) = \frac{1}{d}\,\phi_{L*}\,\mathbf{e}^{-L} = \sum_{\mu \geq 0} \frac{(-1)^\mu}{d}\phi_{L*}\Big(\frac{L^{\cdot \mu}}{\mu!}\Big).$$

But $L \in \mathrm{NS}(X)_{\mathbb{Q}} = \mathrm{Ch}^1(X)^0_{\mathbb{Q}}$, L being symmetric and ample. So $\phi_{L*}(L^{\cdot\mu}) \in \mathrm{Ch}^\mu(\widehat{X})^0_{\mathbb{Q}}$ and by Proposition 6.3.4, $F(L^{\cdot\nu}) \in \mathrm{Ch}^{g-\nu}(\widehat{X})^0_{\mathbb{Q}}$. Now Theorem 6.3.1 implies the assertion. □

Define

$$c_L := \frac{L^{\cdot(g-1)}}{d\,(g-1)!} \in \mathrm{Ch}_1(X)_{\mathbb{Q}}.$$

Notice that, for a Jacobian $(J(C), \Theta)$, by Poincaré's formula, Theorem 4.2.2, the image of the cycle c_Θ in the cohomology ring is just the fundamental class of the curve C. More generally we have more precisely in $\mathrm{Ch}^p(X)_{\mathbb{Q}}$:

Theorem 6.3.6 (Poincaré's Formula) *For any polarized abelian variety (X, L) with $d = h^0(L)$ and $0 \leq p \leq g$,*

$$\frac{L^{\cdot p}}{p!} = d\,\frac{c_L^{\star(g-p)}}{(g-p)!} \quad \textit{in } \mathrm{Ch}^p(X)_{\mathbb{Q}}.$$

Proof Using Theorem 6.3.5 twice we have

$$
\begin{aligned}
d\,\phi_{L*}\Big(\frac{c_L^{\star(g-p)}}{(g-p)!}\Big) &= \frac{d}{(g-p)!}\phi_{L*}\Big(\frac{L^{\cdot(g-1)}}{d(g-1)!}\Big)^{\star(g-p)} \\
&= \frac{(-1)^{(g-1)(g-p)}d}{(g-p)!}\,F(L)^{\star(g-p)} \\
&= (-1)^p\, d\, F\Big(\frac{L^{\cdot(g-p)}}{(g-p)!}\Big) && (\text{by Proposition 6.2.18 (b)}) \\
&= \phi_{L*}\Big(\frac{L^{\cdot p}}{p!}\Big).
\end{aligned}
$$

This implies the assertion, since ϕ_{L*} is bijective on $\mathrm{Ch}(X)_{\mathbb{Q}}$. □

The special case $p = g$ is worth mentioning:

Corollary 6.3.7 $\frac{L^{\cdot g}}{g!} = d\,(0)$ *in* $\mathrm{Ch}_0(X)_{\mathbb{Q}}$.

Recall that (0) is the unit element of the ring $(\mathrm{Ch}(X)_{\mathbb{Q}}, +, \star)$.

6.3.3 The Künneth Decomposition of $\mathrm{Ch}^p(X \times X)_{\mathbb{Q}}$

From Algebraic Topology one knows the Künneth decomposition

$$H^n(X \times X, \mathbb{Q}) = \bigoplus_{p+q=n} H^p(X, \mathbb{Q}) \otimes H^q(X, \mathbb{Q}) \tag{6.14}$$

for any variety X. In this section we will see that in the case of an abelian variety X there is a decomposition of $\mathrm{Ch}^p(X \times X)_{\mathbb{Q}}$ compatible with (6.14) via the cycle map. The results of this section are due to Deninger–Murre [38] and Künnemann [78], where they were proven in the language of motives.

Consider the abelian variety $X \times X$ as a family of abelian varieties over X via the first projection $p_1 : X \times X \to X$. Then

$$n_{X \times X/X} = \mathbf{1}_X \times n_X : X \times X \to X \times X$$

is the multiplication by n on every fibre of $X \times X/X$. Similarly as in the case of an abelian variety (see Section 6.3.1) define the eigenspaces

$$\mathrm{Ch}^p(X{\times}X)^s_{\mathbb{Q}} := \left\{\alpha \in \mathrm{Ch}^p(X{\times}X)_{\mathbb{Q}} \mid (\mathbf{1}_X{\times}n_X)^*\alpha = n^{2p-s}\alpha \text{ for all } n \in \mathbb{Z}\right\}. \tag{6.15}$$

It is easy to see (see Exercise 6.3.5 (2)) that the proof of Theorem 6.3.1 generalizes directly to give for any $0 \le p \le 2g$ a decomposition

$$\mathrm{Ch}^p(X \times X)_{\mathbb{Q}} = \bigoplus_{s=\max(p-g,2p-2g)}^{\min(2p,p+g)} \mathrm{Ch}^p(X \times X)^s_{\mathbb{Q}}. \tag{6.16}$$

The following proposition shows that this decomposition is compatible with the Künneth decomposition of $H^{2p}(X \times X, \mathbb{Q})$ via the cycle map. Hence it makes sense to call (6.16) the *Künneth decomposition* of $\mathrm{Ch}^p(X \times X)_{\mathbb{Q}}$.

Proposition 6.3.8 $\mathrm{cl}\big(\mathrm{Ch}^p(X \times X)^s_{\mathbb{Q}}\big) \subset H^s(X, \mathbb{Q}) \otimes H^{2p-s}(X, \mathbb{Q})$.

Since $H^q(X, \mathbb{Q}) = 0$ for $q < 0$ and $q > 2g$, this shows that, if $s < 0$ or $s > 2g$, the eigenspace $\mathrm{Ch}^p(X \times X)^s_{\mathbb{Q}}$ is in the kernel of the cycle map.

Proof The statement is a consequence of the commutativity of the following diagrams

$$\begin{array}{ccc}
\mathrm{Ch}^p(X\times X)^s_{\mathbb{Q}} & \xrightarrow{\cdot n^{2p-s}} & \mathrm{Ch}^p(X\times X)^s_{\mathbb{Q}} \\
\downarrow \mathrm{cl} & & \downarrow \mathrm{cl} \\
H^{2p}(X\times X,\mathbb{Q}) & \xrightarrow{(1_X\times n_X)^*} & H^{2p}(X\times X,\mathbb{Q}) \\
\downarrow \mathrm{proj} & & \downarrow \mathrm{proj} \\
H^t(X,\mathbb{Q})\otimes H^{2p-t}(X,\mathbb{Q}) & \xrightarrow{1\otimes \cdot n^{2p-t}} & H^t(X,\mathbb{Q})\otimes H^{2p-t}(X,\mathbb{Q})
\end{array}$$

for any $0 \le t \le 2p$. □

6.3.4 The Künneth Decomposition of the Diagonal

Denote by $\Delta = \Delta_X$ the class of the diagonal of $X \times X$ in $\mathrm{Ch}^g(X\times X)_{\mathbb{Q}}$. If $\pi_{i,X} = \pi_i$ denotes the component of Δ in $\mathrm{Ch}^g(X\times X)^{2g-i}_{\mathbb{Q}}$, then the Künneth decomposition of Δ is

$$\Delta = \sum_{i=0}^{2g} \pi_i.$$

Proposition 6.3.9 *The correspondences π_i are orthogonal idempotents commuting with ${}^t\Gamma_f$ for every homomorphism of abelian varieties $f : X \to Y$. To be more precise,*

(a) $\pi_i \circ \pi_j = \delta_{ij}\pi_i$ *for all* $0 \le i, j \le 2g$,
(b) ${}^t\Gamma_f \circ \pi_{i,Y} = \pi_{i,X} \circ {}^t\Gamma_f$.

Proof (a): For all $n \in \mathbb{Z}$ we have by Proposition 6.2.10 and the definition of π_i:

$$ {}^t\Gamma_{n_X} \circ \pi_i = (\mathbf{1}_X \times n_X)^*\pi_i = n^i\,\pi_i. \tag{6.17}$$

This implies

$$ {}^t\Gamma_{n_X} = {}^t\Gamma_{n_X} \circ \Delta = \textstyle\sum_{i=0}^{2g} {}^t\Gamma_{n_X} \circ \pi_i = \sum_{i=0}^{2g} n^i\,\pi_i. \tag{6.18}$$

Combining both equations we get $n^i\,\pi_i = {}^t\Gamma_{n_X} \circ \pi_i = \sum_{j=0}^{2g} n^j\,\pi_j \circ \pi_i$. But this is equivalent to

$$\textstyle\sum_{j=0, j\neq i}^{2g} n^j\,\pi_j \circ \pi_i + n^i\,(\pi_i \circ \pi_i - \pi_i) = 0,$$

which implies (a), the last equation being valid for all $n \in \mathbb{Z}$.

(b): Using (a) and equation (6.18) we get,

$$\pi_i \circ {}^t\Gamma_{n_X} = \textstyle\sum_{j=0}^{2g} n^j \pi_i \circ \pi_j = n^i\,\pi_i. \tag{6.19}$$

Equations (6.17) and (6.19) prove (b) for the homomorphisms n_X with $n \in \mathbb{Z}$.

For an arbitrary homomorphism of abelian varieties $f : X \to Y$ it suffices to show that

$$\pi_{j,X} \circ {}^t\Gamma_f \circ \pi_{i,Y} = 0 \quad \text{for } i \neq j,$$

since then

$$\begin{aligned} {}^t\Gamma_f \circ \pi_{i,Y} &= \Delta_X \circ {}^t\Gamma_f \circ \pi_{i,Y} = \textstyle\sum_j \pi_{j,X} \circ {}^t\Gamma_f \circ \pi_{i,Y} = \pi_{i,X} \circ {}^t\Gamma_f \circ \pi_{i,Y} \\ &= \textstyle\sum_j \pi_{i,X} \circ {}^t\Gamma_f \circ \pi_{j,Y} = \pi_{i,X} \circ {}^t\Gamma_f \circ \Delta_Y = \pi_{i,X} \circ {}^t\Gamma_f. \end{aligned}$$

But using equations (6.17) and (6.19) and the fact that every homomorphism commutes with multiplication by n we have for all $n \in \mathbb{Z}$,

$$\begin{aligned} n^i\, \pi_{j,X} \circ {}^t\Gamma_f \circ \pi_{i,Y} &= \pi_{j,X} \circ {}^t\Gamma_f \circ {}^t\Gamma_{n_Y} \circ \pi_{i,Y} \\ &= \pi_{j,X} \circ {}^t\Gamma_{n_X} \circ {}^t\Gamma_f \circ \pi_{i,Y} \\ &= n^j\, \pi_{j,X} \circ {}^t\Gamma_f \circ \pi_{i,Y}, \end{aligned}$$

which implies the assertion. □

Proposition 6.3.10 ${}^t\pi_i = \pi_{2g-i}$ *for* $i = 0, \ldots, 2g$.

Proof Since $\sum_{i=0}^{2g} {}^t\pi_i$ is also a decomposition of $\Delta = {}^t\Delta$, it suffices to show that ${}^t\pi_i \in \mathrm{Ch}^g(X \times X)^i_{\mathbb{Q}}$, or equivalently that $(\mathbf{1}_X \times n_X)^{*t}\pi_i = n^{2g-i} \cdot {}^t\pi_i$. For this note first that

$$(n_X \times \mathbf{1}_X)^{*t}\pi_i = {}^t\big((\mathbf{1}_X \times n_X)^*\pi_i\big) = n^i \cdot {}^t\pi_i$$

and hence using Proposition 6.2.10 (b),

$${}^t\pi_i \circ {}^t\Gamma_{n_X} = (n_X \times \mathbf{1}_X)_*{}^t\pi_i = n^{-i}\,(n_X \times \mathbf{1}_X)_*(n_X \times \mathbf{1}_X)^{*t}\pi_i = n^{2g-i} \cdot {}^t\pi_i.$$

As in the previous proof, this yields

$${}^t\Gamma_{n_X} = {}^t\Delta \circ {}^t\Gamma_{n_X} = \textstyle\sum_{i=0}^{2g} {}^t\pi_i \circ {}^t\Gamma_{n_X} = \sum_{i=0}^{2g} n^{2g-i} \cdot {}^t\pi_i,$$

and one concludes with Proposition 6.2.10 (a),

$$(\mathbf{1}_X \times n_X)^{*\,t}\pi_i = {}^t\Gamma_{n_X} \circ {}^t\pi_i = \sum_{j=0}^{2g} n^{2g-j} \cdot {}^t\pi_j \circ {}^t\pi_i = n^{2g-i} \cdot {}^t\pi_i.$$ □

The correspondences π_i are projectors of $\mathrm{Ch}^g(X \times X)_{\mathbb{Q}}$. The following proposition shows that they also induce projectors of $\mathrm{Ch}^p(X \times X)_{\mathbb{Q}}$.

Proposition 6.3.11 *For every* $p = 1, \ldots, 2g$ *the map*

$$\mathrm{Ch}^p(X \times X)_{\mathbb{Q}} \longrightarrow \mathrm{Ch}^p(X \times X)^{2p-i}_{\mathbb{Q}}, \quad \alpha \mapsto \pi_i \circ \alpha$$

is a projection.

Proof First note that for any $\alpha \in \mathrm{Ch}^p(X \times X)_{\mathbb{Q}}$ the image $\pi_i \circ \alpha$ is a member of $\mathrm{Ch}^p(X \times X)_{\mathbb{Q}}^{2p-i}$, since by Proposition 6.2.10,

$$(\mathbf{1}_X \times n_X)^*(\pi_i \circ \alpha) = {}^t\Gamma_{n_X} \circ \pi_i \circ \alpha = \left((\mathbf{1}_X \times n_X)^* \pi_i\right) \circ \alpha = n^i \pi_i \circ \alpha.$$

It remains to show that $\pi_i \circ \alpha = \alpha$ for any $\alpha \in \mathrm{Ch}^p(X \times X)_{\mathbb{Q}}^{2p-i}$. But for any such α,

$$\begin{aligned} n^i \alpha &= (\mathbf{1}_X \times n_X)^* \alpha = (\mathbf{1}_X \times n_X)^*(\Delta \circ \alpha) \\ &= \textstyle\sum_{j=0}^{2g} (\mathbf{1}_X \times n_X)^*(\pi_j \circ \alpha) = \sum_{j=0}^{2g} n^j \pi_j \circ \alpha \end{aligned}$$

for all $n \in \mathbb{Z}$. Considering this as a polynomial in n implies the assertion. □

The following theorem, due to Künnemann [78], gives an explicit description of the correspondences π_i. For this we need some notation: Identify the fibre product $(X \times X) \times_X (X \times X) \longrightarrow X$ of the family of abelian varieties $p_1 : X \times X \longrightarrow X$ with itself, with the projection $q_1 : X \times X \times X \longrightarrow X$. Then the relative addition map of the family $p_1 : X \times X \longrightarrow X$ is given by

$$\mathbf{1}_X \times \mu : X \times X \times X \longrightarrow X \times X.$$

The *relative Pontryagin product*

$$\star : \mathrm{Ch}(X \times X)_{\mathbb{Q}} \otimes \mathrm{Ch}(X \times X)_{\mathbb{Q}} \longrightarrow \mathrm{Ch}(X \times X)_{\mathbb{Q}}$$

is defined by

$$\alpha \star \beta := (\mathbf{1}_X \times \mu)_*(\alpha \times_X \beta).$$

As in the absolute case (see Section 6.2.3), $(\mathrm{Ch}(X \times X)_{\mathbb{Q}}, +, \star)$ is a commutative ring with unit element $X \times (0)$. For every $\alpha \in \mathrm{Ch}(X \times X)_{\mathbb{Q}},\ \alpha \neq 0$ the correspondence

$$\log \alpha := \sum_{\nu \geq 1} \frac{(-1)^{\nu-1}}{\nu} \left(\alpha - X \times (0)\right)^{\star \nu}$$

is well defined, the sum being finite.

Theorem 6.3.12 $\pi_i = \frac{1}{(2g-i)!} (\log \Delta)^{\star(2g-i)}$.

For the proof we need the following lemma.

Lemma 6.3.13 $\Delta^{\star n} = \Gamma_{n_X}$.

Proof The n-fold fibre product of the family of abelian varieties $p_1 : X \times X \longrightarrow X$ with itself identifies with the first projection $q_1 : X \times X^n \longrightarrow X$. The corresponding n-fold relative addition map is $\mathbf{1}_X \times \mu^n : X \times X^n \longrightarrow X \times X$, where $\mu^n(x_1, \dots, x_n) := x_1 + \dots + x_n$. Denote by $\delta^{1+n} : X \longrightarrow X \times X^n$ the diagonal map $\delta^{1+n}(x) := (x, \dots, x)$. Then we have

$$\Delta^{\star n} = (\mathbf{1}_X \times \mu^n)_*(\Delta \times_X \dots \times_X \Delta) = (\mathbf{1}_X \times \mu^n)_* \delta_*^{1+n}(X) = (\mathbf{1}_X, n_X)_*(X) = \Gamma_{n_X}. \quad \square$$

Proof (of Theorem 6.3.12) Define $\tilde{\pi}_i := \frac{1}{(2g-i)!}(\log \Delta)^{\star(2g-i)}$. We have to show that $\tilde{\pi}_i = \pi_i$ or equivalently ${}^t\tilde{\pi}_i = {}^t\pi_i = \pi_{2g-i}$. First note that

$$\sum_{i=0}^{2g} {}^t\tilde{\pi}_i = \sum_{i=0}^{2g} \frac{1}{i!}(\log {}^t\Delta)^{\star i} = \sum_{i \geq 0} \frac{1}{i!}(\log \Delta)^{\star i} = \Delta.$$

Hence it suffices to show that ${}^t\tilde{\pi}_i \in \mathrm{Ch}^g(X \times X)^i_{\mathbb{Q}}$. For all $n \in \mathbb{Z}$ we have, using Proposition 6.2.10 and Lemma 6.3.13,

$$\begin{aligned}(n_X \times \mathbf{1}_X)^*\tilde{\pi}_i &= \frac{1}{(2g-i)!}(\log(n_X \times \mathbf{1}_X)^*\Delta)^{\star(2g-i)} \\ &= \frac{1}{(2g-i)!}(\log(\Delta \circ \Gamma_{n_X}))^{\star(2g-i)} \\ &= \frac{1}{(2g-i)!}(\log(\Gamma_{n_X}))^{\star(2g-i)}] = \frac{1}{(2g-i)!}(\log(\Delta^{\star n}))^{\star(2g-i)} \\ &= \frac{1}{(2g-i)!}(n \log \Delta)^{\star(2g-i)} = n^{2g-i}\tilde{\pi}_i.\end{aligned}$$

Hence

$$(\mathbf{1}_X \times n_X)^{*t}\tilde{\pi}_i = {}^t\big((n_X \times \mathbf{1}_X)^*\tilde{\pi}_i\big) = n^{2g-i} \cdot {}^t\tilde{\pi}_i;$$

that is, ${}^t\tilde{\pi}_i \in \mathrm{Ch}^g(X \times X)^i_{\mathbb{Q}}$. □

For the special cases $i = 0$ and $2g$ the theorem gives:

Corollary 6.3.14 $\pi_{2g} = X \times (0)$ *and* $\pi_0 = (0) \times X$.

Proof This follows, since $\pi_{2g} = (\log \Delta)^{\star 0} = X \times (0)$ is the unit element of the ring $(\mathrm{Ch}(X \times X)_{\mathbb{Q}}, +, \star)$, and $\pi_0 = {}^t\pi_{2g}$ by Proposition 6.3.10. □

6.3.5 Exercises and Further Results

(1) Let (X, L) be a polarized abelian variety of dimension g. Show that for all $p, q \geq 0$ we have in $\mathrm{Ch}^{p+q-g}(X)_{\mathbb{Q}}$,

$$\frac{L^{\cdot p}}{p!} \star \frac{L^{\cdot q}}{q!} = h^0(L)\binom{2g-p-q}{g-p}\frac{L^{\cdot(p+q-g)}}{(p+q-g)!}.$$

(Hint: Use Poincaré's Formula 6.3.6.)

(2) Let $p_1 : X \times X \to X$ be a family of abelian varieties of dimension g over X via the first projection and consider the spaces $\mathrm{Ch}^p(X \times X)^s_{\mathbb{Q}}$ as in equation (6.15). Show that for any p, $0 \le p \le 2g$, there is a decomposition

$$\mathrm{Ch}^p(X \times X)_{\mathbb{Q}} = \bigoplus_{s=\max(p-g,2p-2g)}^{\min(2p,p+g)} \mathrm{Ch}^p(X \times X)^s_{\mathbb{Q}}.$$

(Hint: Generalize the proof of Theorem 6.3.1.)

For the following results let C be a smooth projective curve of genus g with Jacobian variety (J, Θ). Denote by $W_i \in \mathrm{Ch}_i(J)$, $1 \le i \le g-1$, the cycle class of the image of the i-th symmetric self-product $C^{(i)}$ via the Abel–Jacobi map $\alpha_{\mathcal{O}_C(ic)}$ with respect to some fixed point c. Moreover let W_0 denote the class of 0 in $\mathrm{Ch}_0(J)_{\mathbb{Q}}$.

(3) Let $W_i^* := (-1)_* W_i + \kappa$, where $\kappa = \omega_C((-2g+2)c)$ denotes the canonical point of J. Then the following relations hold in the Chow ring $\mathrm{Ch}(J)$:

$$\begin{aligned} W_{g-1} - W^*_{g-1} &= 0 \\ W_{g-2} - W_{g-1}W^*_{g-1} + W^*_{g-2} &= 0 \\ &\vdots \\ W_0 - W_1 W^*_{g-1} + \cdots + (-1)^g W^*_0 &= 0. \end{aligned}$$

(See Mattuck [92].)

In particular, W_{g-1} is algebraically equivalent to $(-1)_* W_{g-1}$. If C is a hyperelliptic curve, the hyperelliptic involution induces the homomorphism $(-1)_J$ on J. This implies $W_i = (-1)_* W_i$ for all i. Moreover for any smooth projective curve and any i the cycles W_i and $(-1)_* W_i$ are homologically equivalent, since $(-1)_J$ induces the identity on $H^*(J, \mathbb{Z})$. On the other hand one has:

(4) Let C be a general curve of genus $g \ge 3$. Then W_i is not algebraically equivalent to $(-1)_* W_i$ for $1 \le i \le g-2$.
(See Ceresa [31].)

Let

$$A(J) := \mathrm{Ch}(J)_{\mathbb{Q}} / \mathrm{Ch}^{\mathrm{alg}}(J)_{\mathbb{Q}}$$

denote the $\mathbb{Q}$-algebra of algebraic cycles on J modulo algebraic equivalence. For $i = 1, \ldots, g-1$ denote by w_i the image of W_i in $A(J)$, let Z_i denote the $\mathbb{Q}$-subvector space of $A(J)$ generated by the classes $n_* w_i$ for all $n \in \mathbb{Z}$. Moreover, consider the Künneth components π_i of Δ in $\mathrm{Ch}^g(J \times J)^{2g-i}_{\mathbb{Q}}$ as in Section 6.3.4 and for $C \subset J$ define $C^- := (-1)_{J_*}(C)$. For the next three results see Colombo–van Geemen [33].

(5) (a) If $\pi_{2g-i} W_{g-n} \neq 0$, then $2(g-n) \leq i \leq 2g-n$ and $\pi_{2n} W_{g-n} = \frac{1}{n!}\Theta^{\cdot n}$ in $\mathrm{Ch}^n(J)_{\mathbb{Q}}$.
(b) $\dim_{\mathbb{Q}} Z_{g-n} \leq n$ for $1 \leq n \leq g-1$.

(6) Suppose C is a generic d-gonal curve.

(a) If $\pi_{2g-i} C \nsim_{\mathrm{alg}} 0$, then $2 \leq i \leq d$.
(b) The cycles $\pi_{2g-2}C, \ldots \pi_{2g-d}C$ span Z_1 and thus $\dim_{\mathbb{Q}} Z_1 \leq d-1$.

(7) (a) If $\pi_i C \sim_{\mathrm{alg}} 0$ for $i \leq 2g-4$ (for example, if C is trigonal), then
$\pi_{2g-2}C \sim_{\mathrm{alg}} \frac{1}{2}(C + C^-)$,
$\pi_{2g-3}C \sim_{\mathrm{alg}} \frac{1}{2}(C - C^-)$, and
$n_* C \sim_{\mathrm{alg}} \frac{n^3+n^2}{2} C - \frac{n^3-n^2}{2} C^-$.
(b) If $\pi_i C \sim_{\mathrm{alg}} 0$ for $i \leq 2g-5$ (for example, if C is four-gonal), then
$\pi_{2g-2}C \sim_{\mathrm{alg}} \frac{1}{12}(-2_*C + 12C + 4C^-)$,
$\pi_{2g-3}C \sim_{\mathrm{alg}} \frac{1}{2}(C - C^-)$, and
$\pi_{2g-4}C \sim_{\mathrm{alg}} \frac{1}{12}(2_*C - 6C + 2C^-)$.

(8) Define the tautological subring R of $A(J)$ to be the smallest $\mathbb{Q}$-subvector space of $A(J)$ containing the class of C, which is stable under intersection- and Pontryagin products and under n^* and n_* for all $n \in \mathbb{Z}$.

(a) R is the $\mathbb{Q}$-subalgebra of $A(J)$ generated by $w_1, \ldots, w_{g-1}$.
(b) If C is d-gonal, then R is generated by $w_{g-d+1}, \ldots, w_{g-1}$.

(See Beauville [17].)

(9) Let M be a smooth projective variety with Albanese map $\alpha_{p_0} : M \to \mathrm{Alb}(M)$ with respect to a point $p_0 \in M$.

(a) Show that the Albanese map α_{p_0} factors via a group homomorphism $\mathrm{Ch}_0(M) \to \mathrm{Alb}(M)(\mathbb{C})$ into the group of points of $\mathrm{Alb}(M)$, also called the Albanese map.
(b) (Roitman's Theorem) The Albanese map $\mathrm{Ch}_0(M) \to \mathrm{Alb}(M)(\mathbb{C})$ induces an isomorphism on the ℓ-torsion points for every prime ℓ.

(See Roitman [112].)

Chapter 7
Introduction to the Hodge Conjecture for Abelian Varieties

The original version of the conjecture was first formulated by Hodge in his 1950 ICM congress address [64]. In 1961 Atiyah and Hirzebruch formulated a modified version of it in [11], which since then has been considered as the Hodge conjecture.

Although the Hodge conjecture is formulated for arbitrary smooth complex projective varieties, abelian varieties are considered as a test for its validity. In this chapter we give an introduction to this subject.

There are numerous examples of abelian varieties for which the Hodge conjecture has been proven. Here we outline the proof only in the two simplest cases, namely for general polarized abelian varieties and for general Jacobian varieties. For these the proof is particularly simple, and moreover uses the method which is applied in almost all other cases. For a few more examples, see Sections 7.2.4 and 7.3.3.

A class in $H^{2p}(X,\mathbb{Q})$ is called *algebraic* if it is in the image of the cycle map $\mathrm{cl} : CH^p(X)_{\mathbb{Q}} \to H^{2p}(X,\mathbb{Q})$ of Chapter 6. On the other hand, the elements of the subvector space

$$H^{2p}_{\mathrm{Hodge}}(X) := H^{2p}(X,\mathbb{Q}) \cap H^{p,p}(X)$$

are called *Hodge classes*. According to Lemma 6.2.7 every algebraic class is a Hodge class. The *Hodge (p,p)-conjecture* asserts the converse, i.e.,

$$\mathrm{cl}\big(\mathrm{Ch}^p(X)_{\mathbb{Q}}\big) = H^{2p}_{\mathrm{Hodge}}(X).$$

The proof of the Hodge (p,p)-conjecture for a variety X consists of two steps:

(1) Determine the vector subspace of Hodge classes $H^{2p}_{\mathrm{Hodge}}(X)$.

(2) Verify that every element of $H^{2p}_{\mathrm{Hodge}}(X)$ is algebraic.

Both steps are difficult in general. In the case of abelian varieties Mumford had for step (1) the fundamental idea to realize $H^{2p}_{\mathrm{Hodge}}(X)$ as the space of invariants under a certain reductive group, the Hodge-group, sometimes also called the special Mumford–Tate group. This reduces step (1) to classical invariant theory. For general

H. Lange, *Abelian Varieties over the Complex Numbers*, Grundlehren Text Editions,
https://doi.org/10.1007/978-3-031-25570-0_7

polarized abelian varieties and general Jacobians step (2) is particularly easy, due to the fact that for these varieties the ring of Hodge classes $H^{2\bullet}_{\mathrm{Hodge}}(X)$ is generated by classes of divisors.

In the first section we introduce Hodge structures and study their relation to complex structures. The aim is to express the Siegel upper half space $\mathfrak{H}_g$ in terms of symplectic complex structures. In Section 7.2 the Hodge group is introduced and shown to be reductive. The third section contains the proof of a theorem of Mattuck, saying that for a general polarized abelian variety (X, L) the Hodge ring $H^{2\bullet}(X)$ is generated by the class of L, which easily implies the Hodge conjecture. The proof of the Hodge conjecture for general Jacobian varieties is a special case of a more general result of Biswas–Narasimhan [25], who showed an analogous conjecture for certain moduli spaces of stable parabolic vector bundles. Finally, in the last section some more special results on the conjecture are quoted in order to induce the reader to further studies.

In this chapter we apply some basic facts for algebraic and Lie groups, mainly in the case of the symplectic group. Moreover, we use the following notation: If V is a K-vector space and A a K-algebra we denote $V \otimes_K A$ by V_A. If G is an algebraic group defined over K, $G(A)$ denotes the group of A-valued points of the algebraic variety G.

7.1 Complex Structures

In Section 7.1.1 we study the relation of Hodge structures and complex structures and use this in Section 7.1.2 to express the Siegel upper half plane $\mathcal{H}_g$ in terms of symplectic complex structures.

7.1.1 Hodge Structures and Complex Structures

Let V be a real vector space of finite dimension and n a nonzero integer. A *Hodge structure of weight n* on V is by definition a decomposition

$$V_{\mathbb{C}} = \bigoplus_{p+q=n} V^{p,q}$$

with complex subvector spaces $V^{p,q}$ satisfying

$$\overline{V^{p,q}} = V^{q,p}$$

for all p, q with $p+q = n$. In the same way one defines Hodge structures on $\mathbb{Q}$-vector spaces, which sometimes are also called *rational Hodge structures*. In this section we consider Hodge structures of weight -1. In particular we will see that Hodge structures of weight -1 are equivalent to complex structures. So Hodge structures may be thought of as generalizations of complex structures.

A *complex structure* on the real vector space V is an element J of the general linear group $\mathrm{GL}(V)$ with $J^2 = -\mathbf{1}_V$. If a complex structure exists on V, then necessarily the dimension of V is even, say $\dim V = 2g$, and the pair (V, J) is a complex vector space of dimension g with respect to the scalar multiplication $\mathbb{C} \times V \longrightarrow V$, $(x+iy, v) \mapsto xv + yJ(v)$. In this chapter we denote by $\mathbb{S}^1$ the circle group

$$\mathbb{S}^1 := \{z \in \mathbb{C} \,\big|\, |z| = 1\},$$

in order to stress the fact that it is a real Lie group. (Elsewhere we denoted it by $\mathbb{C}_1$.) Recall that a *weight* of a representation $h : \mathbb{S}^1 \longrightarrow \mathrm{GL}(V)$ is by definition a character $\chi : \mathbb{S}^1 \to \mathbb{C}^*$ such that the *weight space*

$$V_\chi := \{v \in V_\mathbb{C} \mid h(z)(v) = \chi(z) \cdot v\}$$

is different from zero. The *multiplicity* of the weight χ is the dimension of the vector space V_χ. Every integer n defines a character $\mathbb{S}^1 \to \mathbb{C}^*$, $z \mapsto z^n$, which we simply denote by n.

Proposition 7.1.1 *Let V be a real vector space of dimension $2g$. There are canonical bijections between the sets of*

(i) *complex structures on V,*
(ii) *representations $h : \mathbb{S}^1 \longrightarrow \mathrm{GL}(V)$ with weights 1 and -1, both of multiplicity g,*
(iii) *Hodge structures of weight -1 on V.*

Proof (i)⇒ (ii): Let $J : V \longrightarrow V$ be a complex structure. The endomorphism J has eigenvalues $\pm i$, each of multiplicity g. Let $V_\pm \subseteq V_\mathbb{C}$ be the corresponding eigenspaces. Define a representation

$$h_J : \mathbb{S}^1 \longrightarrow \mathrm{GL}(V), \quad z \mapsto \cos\theta \cdot \mathbf{1}_V + \sin\theta \cdot J$$

where $z = e^{i\theta}$. Then for $v \in V_\pm$ we have

$$h_J(z)(v) = \cos\theta \cdot v \pm \sin\theta \cdot iv = z^{\pm 1} \cdot v.$$

Hence the weights of h_J are ± 1, each with multiplicity g.

(ii) ⇒ (iii): Let $h : \mathbb{S}^1 \longrightarrow \mathrm{GL}(V)$ be a representation as in (ii) with weight spaces $V_{\pm 1} \subseteq V_\mathbb{C}$. For any $v = \sum_\nu v_\nu \otimes z_\nu \in V \otimes_\mathbb{R} \mathbb{C} = V_\mathbb{C}$ let $\overline{v} = \sum_\nu v_\nu \otimes \overline{z}_\nu$. Then for $z \in \mathbb{S}^1$ and $v \in V_{\pm 1}$

$$h(z)(\overline{v}) = \overline{h(z)(v)} = \overline{z^{\pm 1} \cdot v} = z^{\mp 1} \cdot \overline{v},$$

implying $\overline{V_{+1}} = V_{-1}$. Since moreover $V_\mathbb{C} = V_{+1} \oplus V_{-1}$, this defines a Hodge structure of weight -1 on V with $V^{-1,0} = V_{+1}$ and $V^{0,-1} = V_{-1}$.

(iii) ⇒ (i): Suppose $V_{\mathbb{C}} = V^{-1,0} \oplus V^{0,-1}$ is a Hodge structure of weight -1 on V. Define a $\mathbb{C}$-linear automorphism $J : V_{\mathbb{C}} \longrightarrow V_{\mathbb{C}}$ by

$$J|_{V^{-1,0}} = i \cdot \mathbf{1}_{V^{-1,0}} \text{ and } J|_{V^{0,-1}} = -i \cdot \mathbf{1}_{V^{0,-1}}.$$

Then $J^2 = -\mathbf{1}_{V_{\mathbb{C}}}$ and it suffices to show that $J(V) \subseteq V$. But $V = \{v + \overline{v} \mid v \in V^{-1,0}\}$ and

$$J(v + \overline{v}) = iv + (-i)\overline{v} = iv + \overline{iv} \in V.$$

It is easy to see that these assignments are compatible bijections. □

Remark 7.1.2 The image of any representation h as in Proposition 7.1.1 (ii) is contained in the special linear group $\mathrm{SL}(V)$.

For the proof see Exercise 7.1.3 (1).

Proposition 7.1.3 *Let J be a complex structure on V with corresponding representation $h : \mathbb{S}^1 \to \mathrm{GL}(V)$. The differential $\mathrm{d}h_1 : T_1\mathbb{S}^1 = \mathbb{R} \longrightarrow \mathfrak{gl}(V)$ is given by*

$$\mathrm{d}h_1(\theta) = \theta \cdot J.$$

Proof It suffices to show that the following diagram commutes

$$\begin{array}{ccc} T_1\mathbb{S}^1 = \mathbb{R} & \xrightarrow{\theta \mapsto \theta J} & \mathfrak{gl}(V) \\ \Big\downarrow{\scriptstyle \mathbf{e}^i} & & \Big\downarrow{\scriptstyle \exp} \\ \mathbb{S}^1 & \xrightarrow{h} & \mathrm{GL}(V), \end{array}$$

where exp denotes the usual exponential map of Lie theory. But for every $z = \mathbf{e}^{i\theta} = \cos\theta + i\sin\theta$ we have

$$\begin{aligned} \exp(\theta \cdot J) &= \sum_{k \geq 0} \tfrac{\theta^k}{k!} J^k = \sum_{k \geq 0} (-1)^k \tfrac{\theta^{2k}}{(2k)!} \cdot \mathbf{1}_V + \sum_{k \geq 0} (-1)^k \tfrac{\theta^{2k+1}}{(2k+1)!} J \\ &= \cos\theta \cdot \mathbf{1}_V + \sin\theta \cdot J = h(z). \end{aligned}$$

□

7.1.2 Symplectic Complex Structures

In this section we give a description of the Siegel upper half space $\mathcal{H}_g$ in terms of complex structures.

As in the last section let V be a real vector space of dimension $2g$. Fix a lattice $\Lambda \subseteq V$ and a non-degenerate alternating form $E : V \times V \longrightarrow \mathbb{R}$ taking integer values on Λ. As usual denote by $\mathrm{Sp}(V, E)$ the symplectic group associated to E. A complex structure J on V contained in $\mathrm{Sp}(V, E)$ is called a *symplectic complex structure*.

Note that for every symplectic complex structure J the form $E(J\cdot,\cdot): V\times V \longrightarrow \mathbb{R}$ is a non-degenerate symmetric bilinear form. Let $C_0(\mathrm{Sp}(V,E))$ denote the set of symplectic complex structures in $\mathrm{Sp}(V,E)$ such that $E(J\cdot,\cdot)$ is positive definite.

Remark 7.1.4 Let J be a complex structure on V with associated representation $h:\mathbb{S}^1 \to \mathrm{GL}(V)$ and Hodge structure $V_{\mathbb{C}} = V^{-1,0}\oplus V^{0,-1}$ of weight -1 as in Proposition 7.1.1, then J is symplectic if and only if one of the following equivalent conditions holds

(i) h factorizes via $\mathrm{Sp}(V,E)$,
(ii) $V^{-1,0}$ and $V^{0,-1}$ are isotropic with respect to E.

For the proof, see Exercise 7.1.3 (2).

Lemma 7.1.5 *Let Λ be a lattice in the real vector space V. Any complex structure $J\in C_0(\mathrm{Sp}(V,E))$ defines a polarized abelian variety.*

Proof If V_J denotes the complex vector space (V,J), then

$$X_J := V_J/\Lambda$$

is a complex torus. Moreover E defines a polarization on X_J, since $E(\Lambda,\Lambda)\subseteq\mathbb{Z}$, $J^*E=E$ and $E(J\cdot,\cdot)$ is positive definite, since $J\in C_0(\mathrm{Sp}(V,E))$ (see Section 1.2.2). □

It is clear that the type $D=\mathrm{diag}(d_1,\dots,d_g)$ of the polarization of the lemma does not depend on J but only on the pair (Λ,E). So the pair (X_J,E) is a polarized abelian variety of type D. We will use this fact to show that the space $C_0(\mathrm{Sp}(V,E))$ admits the structure of a complex manifold, isomorphic to the Siegel upper half space $\mathfrak{H}_g$.

For this choose a symplectic basis $\lambda_1,\dots,\lambda_{2g}$ of $\Lambda\subset V$ with respect to E. So $\left(\begin{smallmatrix}0 & D\\ -D & 0\end{smallmatrix}\right)$ is the matrix of E with respect to $\lambda_1,\dots\lambda_{2g}$. The basis defines an isomorphism (see Section 3.1.4 equation (3.7))

$$\lambda: \mathrm{Sp}(V,E)\longrightarrow \mathrm{Sp}^D_{2g}(\mathbb{R}) = \left\{R\in \mathrm{M}_{2g}(\mathbb{R}) \,\middle|\, R\left(\begin{smallmatrix}0 & D\\ -D & 0\end{smallmatrix}\right){}^tR = \left(\begin{smallmatrix}0 & D\\ -D & 0\end{smallmatrix}\right)\right\}.$$

Defining

$$C_0(\mathrm{Sp}^D_{2g}(\mathbb{R})) := \left\{J\in \mathrm{Sp}^D_{2g}(\mathbb{R}) \,\middle|\, J^2=-\mathbf{1}_{2g},\ {}^tJ\left(\begin{smallmatrix}0 & D\\ -D & 0\end{smallmatrix}\right) \text{ symmetric positive definite}\right\},$$

the isomorphism λ restricts to a bijection

$$\lambda: C_0(\mathrm{Sp}(V,E))\longrightarrow C_0(\mathrm{Sp}^D_{2g}(\mathbb{R})). \tag{7.1}$$

Given $J \in C_0(\mathrm{Sp}_{2g}^D(\mathbb{R}))$ there is a uniquely defined basis of the complex vector space $V_J = (\mathbb{R}^{2g}, J)$ such that the period matrix of the abelian variety X_J is of the form

$$\pi = (Z, D)$$

for some $Z \in \mathfrak{H}_g$ (see Section 3.1.1). This defines a map

$$C_0(\mathrm{Sp}_{2g}^D(\mathbb{R})) \longrightarrow \mathfrak{H}_g, \qquad J \mapsto Z. \tag{7.2}$$

The matrices J and Z are related by the following commutative diagram

$$\begin{array}{ccc} V \simeq \mathbb{R}^{2g} & \xrightarrow{(Z,D)} & \mathbb{C}^g \simeq V_J \\ {\scriptstyle J}\downarrow & & \downarrow{\scriptstyle i1_g} \\ \mathbb{R}^{2g} & \xrightarrow{(Z,D)} & \mathbb{C}^g. \end{array} \tag{7.3}$$

This shows that the matrices J and Z determine each other, so the map (7.2) is bijective. For the proof of the following lemma see Exercise 7.1.3 (3).

Lemma 7.1.6

(1) *If* $J = \left(\begin{smallmatrix} \alpha & \beta \\ \gamma & \delta \end{smallmatrix}\right)$ *then* $Z = D\beta^{-1}\alpha + iD\beta^{-1}$.
(2) *If* $Z = X + iY$, *then* $J = \left(\begin{smallmatrix} Y^{-1}X & Y^{-1}D \\ -D^{-1}(XY^{-1}X+Y) & -D^{-1}XY^{-1}D \end{smallmatrix}\right)$.

The symplectic group $\mathrm{Sp}_{2g}^D(\mathbb{R})$ acts on $C_0(\mathrm{Sp}_{2g}^D(\mathbb{R}))$ (from the left) by

$$(R, J) \mapsto {}^tR^{-1} J\, {}^tR.$$

On the other hand, according to equation (3.9) the group $\mathrm{Sp}_{2g}^D(\mathbb{R})$ acts (also from the left) on $\mathfrak{H}_g$ by

$$(R, Z) \mapsto R\langle Z\rangle := (aZ + bD)(D^{-1}cZ + D^{-1}dD)^{-1}$$

where $R = \left(\begin{smallmatrix} a & b \\ c & d \end{smallmatrix}\right)$.

Lemma 7.1.7 *The bijection* $C_0(\mathrm{Sp}_{2g}^D(\mathbb{R})) \longrightarrow \mathfrak{H}_g$ *is* $\mathrm{Sp}_{2g}^D(\mathbb{R})$*-equivariant.*

Proof Let J and $J' \in C_0(\mathrm{Sp}_{2g}^D(\mathbb{R}))$ be the complex structures corresponding to Z and $R\langle Z\rangle$ respectively, as in diagram (7.3). We have to show that $J' = {}^tR^{-1} J\, {}^tR$, but this follows from the commutative diagram

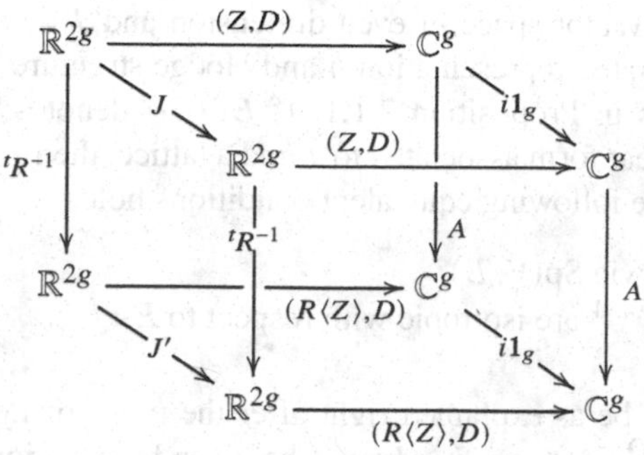

where $R = \begin{pmatrix} a & b \\ c & d \end{pmatrix}$ and $A = {}^t(D^{-1}cZ + D^{-1}dD)^{-1}$. □

Combining the maps (7.1) and (7.2) we obtain a bijection

$$C_0(\mathrm{Sp}(V,E)) \longrightarrow \mathfrak{H}_g.$$

Thus $C_0(\mathrm{Sp}(V,E))$ inherits the structure of a complex manifold.

Proposition 7.1.8 *The complex structure on $C_0(\mathrm{Sp}(V,E))$ does not depend on the choice of the symplectic basis.*

Proof Any two isomorphisms (7.1) differ by conjugation by a symplectic matrix R. So the assertion follows from Lemma 7.1.7. □

Proposition 7.1.9 *A complex structure $J \in C_0(\mathrm{Sp}(V,E))$ defines an isomorphism*

$$C_0(\mathrm{Sp}(V,E)) \simeq \mathrm{Sp}(V,E)/\big(\mathrm{Sp}(V,E) \cap GL(V_J)\big).$$

Here $\mathrm{GL}(V_J) = \{M \in \mathrm{GL}(V) \mid JM = MJ\}$.

Proof According to Proposition 3.1.7 the symplectic group acts transitively on $\mathcal{H}_g$. By Lemma 7.1.7 this implies that the map

$$\mathrm{Sp}(V,E) \longrightarrow C_0(\mathrm{Sp}(V,E)), \quad M \mapsto MJM^{-1}$$

is surjective. Obviously the stabilizer of J in $\mathrm{Sp}(V,E)$ is $\mathrm{Sp}(V,E) \cap GL(V_J)$. This implies the assertion. □

7.1.3 Exercises

(1) Let V be a real vector space of dimension $2g$ and $h : \mathbb{S}_1 \to \mathrm{GL}(V)$ representations of weight 1 and -1, both of multiplicity g. Show that the image of h is contained in the special linear group $\mathrm{SL}(V)$.

(2) Let V be a real vector space of even dimension and J be a complex structure on V with associated representation h and Hodge structure $V_{\mathbb{C}} = V^{-1,0} \oplus V^{0,-1}$ of weight -1 as in Proposition 7.1.1. If $E(J\cdot, \cdot)$ denotes the non-degenerate symmetric bilinear form associated to J and a lattice, then J is symplectic if and only if one of the following equivalent conditions holds

(i) h factorizes via $\mathrm{Sp}(V, E)$,
(ii) $V^{-1,0}$ and $V^{0,-1}$ are isotropic with respect to E.

(3) Let the notation be as explained right after the proof of Lemma 7.1.5 and the bijection $C_0(\mathrm{Sp}_{2g}^D(\mathbb{R})) \longrightarrow \mathfrak{H}_g$, $J \mapsto Z$ be given by equation (7.2). Show that J and Z are related as follows:

(a) If $J = \left(\begin{smallmatrix} \alpha & \beta \\ \gamma & \delta \end{smallmatrix}\right)$ then $Z = D\beta^{-1}\alpha + iD\beta^{-1}$.
(b) If $Z = X + iY$, then $J = \left(\begin{smallmatrix} Y^{-1}X & Y^{-1}D \\ -D^{-1}(XY^{-1}X+Y) & -D^{-1}XY^{-1}D \end{smallmatrix}\right)$.

7.2 The Hodge Group of an Abelian Variety

The section contains the definition of the Hodge group of an abelian variety, the interpretation of the Hodge classes of codimension p as invariants and finally the proof of the reductivity of the Hodge group.

7.2.1 The Hodge Group Hg(X)

Let X be an abelian variety of dimension g. Consider the $\mathbb{Q}$-vector space

$$V := H_1(X, \mathbb{Q})$$

of dimension $2g$. According to Theorem 1.1.21 its dual vector space $V^* = H^1(X, \mathbb{Q})$ admits a canonical Hodge structure of weight 1

$$V^*_{\mathbb{C}} = H^1(X, \mathbb{C}) = H^{1,0}(X) \oplus H^{0,1}(X)$$

with $H^{p,q}(X) = H^q(\Omega^p_X)$. Taking duals we obtain a Hodge structure of weight -1 on V

$$V_{\mathbb{C}} = H_1(X, \mathbb{C}) = V^{-1,0} \oplus V^{0,-1}$$

with

$$V^{-1,0} = \left\{\ell : H^1(X, \mathbb{C}) \,\middle|\, \ell|_{H^{0,1}(X)} = 0\right\} = H^{1,0}(X)^*$$

and $V^{0,-1} = H^{0,1}(X)^*$. According to Proposition 7.1.1 and Remark 7.1.2 this corresponds to a representation

$$h : \mathbb{S}^1 \longrightarrow \mathrm{SL}(V_{\mathbb{R}}).$$

The *Hodge group* $\mathrm{Hg}(X)$ (sometimes also called the *special Mumford–Tate group*) is by definition the smallest algebraic subgroup of $\mathrm{SL}(V)$ defined over $\mathbb{Q}$ satisfying

$$h(\mathbb{S}^1) \subseteq \mathrm{Hg}(X)(\mathbb{R}).$$

Lemma 7.2.1 *The Hodge group* $\mathrm{Hg}(X)$ *is a connected algebraic group.*

Proof A connected component of an algebraic group defined over $\mathbb{Q}$ is defined over $\mathbb{Q}$. On the other hand the group $\mathbb{S}^1$ is connected. □

Remark 7.2.2

(1) The subgroup $\mathrm{Hg}(X)(\mathbb{C})$ of $\mathrm{SL}(V_{\mathbb{C}})$ is generated by the conjugates $h(\mathbb{S}^1)^\sigma$, with $\sigma \in \mathrm{Aut}(\mathbb{C})$, of $h(\mathbb{S}^1)$ in $\mathrm{SL}(V_{\mathbb{C}})$.
(2) As usual let $\mathbb{G}_m$ denote the multiplicative group. The *Mumford–Tate group* of X is by definition the scalar extension $\mathbb{G}_m\cdot \mathrm{Hg}(X)$; that is, the smallest algebraic subgroup of $\mathrm{GL}(V)$ defined over $\mathbb{Q}$ and containing $h(\mathbb{S}^1)$ as well as all nonzero scalar matrices. In particular

$$(\mathbb{G}_m \cdot \mathrm{Hg}(X))(\mathbb{C}) = \{\alpha \cdot M \in \mathrm{GL}(V_{\mathbb{C}}) \,|\, \alpha \in \mathbb{C}^*,\ M \in \mathrm{Hg}(X)(\mathbb{C})\}.$$

According to Proposition 7.1.1 the representation $h : \mathbb{S}^1 \longrightarrow \mathrm{SL}(V_{\mathbb{R}})$ is equivalent to a complex structure J on $V_{\mathbb{R}} = H_1(X, \mathbb{R})$. Obviously the associated complex vector space $V_J = (V_{\mathbb{R}}, J)$ is the tangent space of the abelian variety X and thus $X = X_J = V_J/\Lambda$ (where $\Lambda = H_1(X, \mathbb{Z})$) with the notation of Section 7.1.2. Let E be a polarization on the abelian variety X. By definition E is the first Chern class of an ample line bundle on X, and as such, an element of $H^2(X, \mathbb{Z})$. As in Corollary 1.1.19 we have

$$H^2(X, \mathbb{Z}) \subset H^2(X, \mathbb{Q}) = \bigwedge^2 H^1(X, \mathbb{Q}) = \bigwedge^2 V^*.$$

Hence we may consider the polarization E as an alternating form

$$E : V \times V \longrightarrow \mathbb{Q}.$$

According to Section 2.1.5 the $\mathbb{R}$-linear extension of E satisfies the Riemann relations

(i) $E(Jv, Jw) = E(v, w)$,
(ii) $E(Jv, v) > 0$,

for all $v, w \in V_{\mathbb{R}}$ with $v \neq 0$. In particular, E is non-degenerate. The symplectic group $\mathrm{Sp}(V, E)$ associated to E is an algebraic subgroup of $\mathrm{SL}(V)$ defined over $\mathbb{Q}$. Its real points are

$$\mathrm{Sp}(V, E)(\mathbb{R}) = \{M \in \mathrm{SL}(V_{\mathbb{R}}) \,|\, E(M\,\cdot, M\,\cdot) = E(\cdot, \cdot)\},$$

which is exactly the symplectic group $\mathrm{Sp}(V_{\mathbb{R}}, E)$.

Proposition 7.2.3 *The Hodge group* $\mathrm{Hg}(X)$ *is an algebraic subgroup of* $\mathrm{Sp}(V, E)$.

Note that $\mathrm{Hg}(X) \subseteq \mathrm{Sp}(V, E)$ for *any* polarization E of X.

Proof It suffices to show that $h(\mathbb{S}^1) \subset \mathrm{Sp}(V, E)(\mathbb{R})$, since $\mathrm{Sp}(V, E)$ is defined over $\mathbb{Q}$. But for $z = e^{i\theta} \in \mathbb{S}^1$ the Riemann relations imply for all $v, w \in V_{\mathbb{R}}$,

$$\begin{aligned} E(h(z)v, h(z)w) &= E((\cos\theta\, \mathbf{1}_{V_{\mathbb{R}}} + \sin\theta J)v, (\cos\theta\, \mathbf{1}_{V_{\mathbb{R}}} + \sin\theta J)w) \\ &= (\cos^2\theta + sin^2\theta)E(v, w) = E(v, w). \end{aligned}$$ □

7.2.2 Hodge Classes as Invariants

The $\mathbb{Q}$-vector space

$$H^{2p}_{\mathrm{Hodge}}(X) := H^{2p}(X, \mathbb{Q}) \cap H^{p,p}(X)$$

is called the *vector space of Hodge classes* of codimension p. The next theorem shows that the Hodge group can be used to determine this vector space.

Consider $H^1(X, \mathbb{Q}) = V^*$ as the dual representation of the Hodge group $\mathrm{Hg}(X)$ and $H^k(X, \mathbb{Q}) = \bigwedge^k H^1(X, \mathbb{Q}) = \bigwedge^k V^*$ as its k-th exterior product. $\mathrm{Hg}(X)$ acts on $H^k(X, \mathbb{Q})$ in the usual way. Let $H^k(X, \mathbb{Q})^{\mathrm{Hg}(X)}$ denote the subspace of invariants.

Theorem 7.2.4 $H^{2p}_{\mathrm{Hodge}}(X) = H^{2p}(X, \mathbb{Q})^{\mathrm{Hg}(X)}$.

Proof **Step I:** $H^{2p}_{\mathrm{Hodge}}(X) \subseteq H^{2p}(X, \mathbb{Q})^{\mathrm{Hg}(X)}$.
Note that $H^{2p}_{\mathrm{Hodge}}(X)$ is a subvector space of $H^{2p}(X, \mathbb{Q}) = \bigwedge^{2p} V^*$. So

$$G_p := \left\{M \in \mathrm{SL}(V) \,\middle|\, \textstyle\bigwedge^{2p} M^*|_{H^{2p}_{\mathrm{Hodge}}(X)} = \mathbf{1}_V\right\}$$

is an algebraic subgroup of $\mathrm{SL}(V)$ defined over $\mathbb{Q}$. Hence it suffices to show that

$$h(\mathbb{S}^1) \subseteq G_p(\mathbb{R}),$$

since then $\mathrm{Hg}(X) \subseteq G_p$ and thus

$$H^{2p}_{\mathrm{Hodge}}(X) \subseteq H^{2p}(X, \mathbb{Q})^{G_p} \subseteq H^{2p}(X, \mathbb{Q})^{\mathrm{Hg}(X)}.$$

For this note that $H^{p,p}(X) = \bigwedge^p H^{1,0}(X) \otimes \bigwedge^p H^{0,1}(X)$ according to Theorem 1.1.21. But $H^{1,0}(X)$ and $H^{0,1}(X)$ are the weight spaces of the dual representation h^*. So for any $z \in S^1$, the exterior power $\bigwedge^{2p} h^*(z)$ acts on $\bigwedge^p H^{1,0}(X) \otimes \bigwedge^p H^{0,1}(X)$ by multiplication with $z^p \bar{z}^p = 1$. This implies the assertion.

Step II: $H^{2p}(X,\mathbb{Q})^{\mathrm{Hg}(X)} \subseteq H^{2p}_{\mathrm{Hodge}}(X)$.
We have to show that $H^{2p}(X,\mathbb{Q})^{\mathrm{Hg}(X)} \subseteq H^{p,p}(X)$. For this it suffices to show that

$$H^{2p}(X,\mathbb{R})^{h(\mathbb{S}^1)} \subseteq H^{p,p}(X),$$

since $h(\mathbb{S}^1) \subseteq \mathrm{Hg}(X)(\mathbb{R})$. But

$$H^{2p}(X,\mathbb{R}) \subseteq H^{2p}(X,\mathbb{C}) = \bigoplus\nolimits_{r+s=2p} H^{r,s}(X)$$

and the right-hand side is a decomposition of $\mathbb{S}^1$-modules: In fact, $h^*(z)$ acts on $H^{r,s}(X) = \bigwedge^r H^{1,0}(X) \otimes \bigwedge^s H^{0,1}(X)$ by multiplication with $z^r\bar{z}^s$. Hence $H^{p,p}(X)$ is the only trivial subrepresentation of $H^{2p}(X,\mathbb{C})$, which implies the assertion. □

Tensoring with $\mathbb{C}$ we obtain

$$H^{2p}_{\mathrm{Hodge}}(X) \otimes_{\mathbb{Q}} \mathbb{C} = H^{2p}(X,\mathbb{C})^{\mathrm{Hg}(X)(\mathbb{C})}.$$

Thus, in order to compute the dimension of the vector space of Hodge classes of codimension p, one can apply invariant theory for the complex group $\mathrm{Hg}(X)(\mathbb{C})$.

Proposition 7.2.5 *The rational representation induces an isomorphism of* $\mathbb{Q}$*-algebras*

$$\mathrm{End}_{\mathbb{Q}}(X) \simeq \mathrm{End}(V)^{\mathrm{Hg}(X)},$$

where $\mathrm{Hg}(X)$ *acts on* $\mathrm{End}(V)$ *by conjugation.*

Proof Recall that $V = H_1(X,\mathbb{Z}) \otimes_{\mathbb{Z}} \mathbb{Q}$. Hence we can embed $\mathrm{End}_{\mathbb{Q}}(X)$ into $\mathrm{End}(V)$ via the rational representation:

$$\mathrm{End}_{\mathbb{Q}}(X) = \{\varphi \in \mathrm{End}(V) \mid \varphi J = J\varphi \text{ on } V_{\mathbb{R}}\} = \mathrm{End}(V) \cap \mathrm{End}(V_{\mathbb{R}})^{h(\mathbb{S}^1)}.$$

Since $\mathrm{Hg}(X)$ is defined over $\mathbb{Q}$ and $h(\mathbb{S}^1) \subseteq \mathrm{Hg}(X)(\mathbb{R})$, we have

$$\begin{aligned}\mathrm{End}(V)^{\mathrm{Hg}(X)} &= \mathrm{End}(V) \cap \mathrm{End}(V_{\mathbb{R}})^{\mathrm{Hg}(X)(\mathbb{R})}\\ &\subseteq \mathrm{End}(V) \cap \mathrm{End}(V_{\mathbb{R}})^{h(\mathbb{S}^1)} = \mathrm{End}_{\mathbb{Q}}(X).\end{aligned}$$

For the converse inclusion recall that $\mathrm{SL}(V)$ acts on $\mathrm{End}(V)$ by conjugation. Consider the centralizer $G \subset \mathrm{SL}(V)$ of the subvector space $\mathrm{End}_{\mathbb{Q}}(X) \subset \mathrm{End}(V)$, in particular

$$G(\mathbb{R}) = \left\{M \in \mathrm{SL}(V_{\mathbb{R}}) \,\middle|\, M\varphi M^{-1} = \varphi \text{ for all } \varphi \in \mathrm{End}_{\mathbb{Q}}(X)\right\}.$$

Obviously J and hence $h(\mathbb{S}^1)$ is contained in $G(\mathbb{R})$. Since G is defined over $\mathbb{Q}$, this implies $\mathrm{Hg}(X) \subset G$ and thus

$$\mathrm{End}_{\mathbb{Q}}(X) \subseteq \mathrm{End}(V)^G \subseteq \mathrm{End}(V)^{\mathrm{Hg}(X)}.$$ □

7.2.3 Reductivity of the Hodge Group and a Criterion for its Commutativity

The following proposition gives a criterion for $\mathrm{Hg}(X)$ to be an abelian group.

Proposition 7.2.6 *For an abelian variety X of dimension g the following conditions are equivalent:*

(i) *the Hodge group* $\mathrm{Hg}(X)$ *is commutative;*
(ii) $\mathrm{End}_{\mathbb{Q}}(X)$ *contains a commutative semisimple $\mathbb{Q}$-algebra of dimension* $2g$.

This applies for example to abelian varieties of CM-type in the sense of Exercise 2.6.3 (4). Slightly more generally, abelian varieties satisfying condition (ii) are also sometimes said to be of *CM-type*.

Proof (i)$\Rightarrow$ (ii): Suppose $\mathrm{Hg}(X)$ is commutative. Since $\mathrm{Hg}(X)$ is connected it is contained in a maximal torus $T \subset \mathrm{GL}(V) \subset \mathrm{End}(V)$. So $T \subset \mathrm{End}(V)^{\mathrm{Hg}(X)} = \mathrm{End}_{\mathbb{Q}}(X)$ by Proposition 7.2.5. So $\mathrm{End}_{\mathbb{Q}}(X)$ satisfies condition (ii), since $[T:\mathbb{Q}] = \dim_{\mathbb{Q}} V = 2\dim X$.
(ii) $\Rightarrow$ (i): Let $T \subset \mathrm{End}_{\mathbb{Q}}(X)$ be a commutative semisimple $\mathbb{Q}$-algebra of rank $2g$ over $\mathbb{Q}$. Consider T as a maximal commutative semisimple subalgebra of $\mathrm{End}(V)$ via the rational representation. According to Proposition 7.2.5 the Hodge group $\mathrm{Hg}(X) \subset \mathrm{End}(V)$ commutes with T. By the maximality of T, the centralizer of T is T itself. Hence $\mathrm{Hg}(X)$ is contained in T and thus $\mathrm{Hg}(X)$ is also commutative. □

Recall that a linear algebraic group G defined over a field K of characteristic 0 is called *reductive* if every K-representation of G is fully reducible. Equivalent to this is that there are normal subgroups G_a and G_s of G defined over K such that G_a is a torus, G_s is semi-simple (meaning that $G_s(K)$ is semi-simple) and $G(K)$ is an almost direct product of $G_a(K)$ and $G_s(K)$ (that is, $G_a(K) \cdot G_s(K)$ is Zariski-dense in $G(K)$ and $G_a(K) \cap G_s(K)$ is finite). Note that G is reductive if and only if the scalar extension G_L is reductive for some field extension $L|K$ (see Satake [115, Section I.3]).

The following proposition, due to Mumford [99], is important for applying invariant theoretical results.

Proposition 7.2.7 $\mathrm{Hg}(X)$ *is a reductive group.*

Proof As mentioned above, the group $\mathrm{Hg}(X)$, defined over $\mathbb{Q}$, is reductive if and only if $\mathrm{Hg}(X)_L$ is reductive for some field extension L of $\mathbb{Q}$. Hence the group $\mathrm{Hg}(X)$ is reductive if and only if its complexification $\mathrm{Hg}(X)_{\mathbb{C}}$ is reductive and it suffices to show that $\mathrm{Hg}(X)_{\mathbb{C}}$ admits a reductive real form.

Consider the complex structure $J = h(i)$ as an element of $\mathrm{Hg}(X)_{\mathbb{R}}$. The inner automorphism $\mathrm{Ad}\, J$ of $\mathrm{Hg}(X)_{\mathbb{R}}$, $M \mapsto JMJ^{-1}$, is an involution. So denoting by "$^-$" complex conjugation on $\mathrm{Hg}(X)(\mathbb{C})$,

$$^{-J} := \mathrm{Ad}\, J \circ {}^{-} = {}^{-} \circ \mathrm{Ad}\, J$$

is a second complex conjugation on $\mathrm{Hg}(X)(\mathbb{C})$ and defines a second real form $\mathrm{Hg}(X)^{-J}$ of $\mathrm{Hg}(X)$. In order to check that $\mathrm{Hg}(X)^{-J}$ is reductive it suffices to show that the real Lie group $\mathrm{Hg}(X)^{-J}(\mathbb{R})$ is compact (see Satake [115, Chapter I, Prop. 3.3] or Humphreys [67, Thm 35.3]). For this consider the positive definite hermitian form

$$Q : V_{\mathbb{Q}} \times V_{\mathbb{Q}} \longrightarrow \mathbb{C}, \quad Q(v,w) = E(Jv, \overline{w}).$$

Let $M \in \mathrm{Hg}(X)^{-J}(\mathbb{R})$; that is, $M \in \mathrm{Hg}(X)(\mathbb{C})$ with $\overline{\mathrm{Ad}\,J(M)} = M$. Then we have for all $v, w \in V_{\mathbb{C}}$

$$\begin{aligned} Q(\mathrm{Ad}\,J(M)v, \mathrm{Ad}\,J(M)w) &= E(J\mathrm{Ad}\,J(M)v, \overline{\mathrm{Ad}\,J(M)w}) \\ &= E(MJv, M\overline{w}) \\ &= E(Jv, \overline{w}) = Q(v,w). \end{aligned}$$

So $\mathrm{Hg}(X)^{-J}(\mathbb{R})$ is isomorphic to a closed subgroup of the orthogonal group with respect to the form Q and as such it is compact. □

7.2.4 Exercises and Further Results

(1) If X is an abelian variety, the group $\mathrm{GL}_n(\mathbb{C})$ acts on the cohomology rings $H^\bullet(X^n, \mathbb{C})$ of all powers X^n of X in a natural way. Show that the Hodge group $\mathrm{Hg}(X)$ can be characterized as the largest algebraic subgroup of $\mathrm{GL}_n(\mathbb{C})$ which is defined over $\mathbb{Q}$ and which leaves invariant the elements of the Hodge rings $H^\bullet_{\mathrm{Hodge}}(X^n)$ for all $n \geq 1$.

(2) The Hodge group $\mathrm{Hg}(X)$ of an abelian variety X is semisimple if and only if the centre of the Mumford–Tate group of X (see Remark 7.2.2 (2)) consists only of scalars.

(3) $\mathrm{End}_{\mathbb{Q}}(X) = \mathrm{End}_{\mathrm{Hg}(X)}(H_1(X,\mathbb{Q}))$ for any abelian variety X.

(4) Let (X, L) be a polarized abelian variety. Consider the polarization as an alternating form E on the $\mathbb{Q}$-vector space $W = H_1(X,\mathbb{Q})$. The *Lefschetz group* $\mathrm{Lf}(X)$ is by definition the connected component of the centralizer of $\mathrm{End}_{\mathbb{Q}}(X)$ in $\mathrm{Sp}(W,E)$:

$$\mathrm{Lf}(X) := \{g \in \mathrm{Sp}(W,E) \mid g \circ \varphi = \varphi \circ g \text{ for all } \varphi \in \mathrm{End}_{\mathbb{Q}}(X)\}^0.$$

Show

(a) $\mathrm{Lf}(X)$ does not depend on the polarization.
(b) $\mathrm{Lf}(X)$ is an algebraic group defined over $\mathbb{Q}$ containing $\mathrm{Hg}(X)$.
(c) If there is an isogeny $X \sim X_1^{n_1} \times \cdots \times X_r^{n_r}$ with pairwise non-isogenous simple abelian varieties, then $\mathrm{Lf}(X) \simeq \mathrm{Lf}(X_1) \times \cdots \times \mathrm{Lf}(X_r)$.

For any abelian variety X let $D^\bullet(X)$ denote the subring of $H^{2\bullet}_{\text{Hodge}}(X)$ generated by $H^0_{\text{Hodge}}(X)$ and $H^2_{\text{Hodge}}(X)$ (see Section 7.3.1).

(5) Let X be an abelian variety such that $\text{End}_{\mathbb{Q}}(X)$ is a field and with $\text{Hg}(X) = \text{Lf}(X)$. Then

$$H^\bullet_{\text{Hodge}}(X^n) = D^\bullet(X^n) \qquad \text{for all} \quad n \geq 1.$$

(See Tankeev [130].)

An *abelian variety of Weil-type* of dimension $2n$ is a pair (X, K) with a $2n$-dimensional abelian variety X and an imaginary quadratic number field K admitting an embedding $K \hookrightarrow \text{End}_{\mathbb{Q}}(X)$ such that the restricted analytic representation $\rho_a|_K$ is equivalent to the representation $\rho : K \longrightarrow \text{M}_{2n}(\mathbb{C}),\ z \mapsto \begin{pmatrix} z\mathbf{1}_n & 0 \\ 0 & \bar{z}\mathbf{1}_n \end{pmatrix}$.

A *polarized abelian variety of Weil-type* is a triplet (X, K, H) with an abelian variety of Weil-type (X, K) and a polarization H on X such that complex conjugation "$^-$" on K extends to the Rosati involution "$'$" with respect to H. If $X = V/\Lambda$, we consider H as a hermitian form on V. If E is the alternating form corresponding to H, we also write (X, K, E).

(6) Let (X, K, H) be a polarized abelian variety of Weil-type with $K = \mathbb{Q}(\sqrt{-d})$. Show that the condition on H above is equivalent to

$$\rho(\sqrt{-d})^* H = dH.$$

(7) Show that every abelian variety of Weil type (X, K) admits a polarization H such that (X, K, H) is a polarized abelian variety of Weil type.

(8) For a general polarized abelian variety (X, K, E) of Weil type the Hodge group is $\text{Hg}(X) = \text{SU}(H_1(X, \mathbb{Q}), E)$.

(9) Let (X, K, H) be a polarized abelian variety of Weil type. The embedding $K \hookrightarrow \text{End}_{\mathbb{Q}}(X)$ defines K-vector space structures on $H_1(X, \mathbb{Q})$ and its dual vector space $H^1(X, \mathbb{Q})$. Show that there is a canonical embedding of vector spaces

$$\bigwedge_K^{2n} H^1(X, \mathbb{Q}) \hookrightarrow H^{2n}_{\text{Hodge}}(X).$$

The cycles in the image of this map are called *Weil–Hodge cycles*.

(10) Fix an imaginary quadratic number field $K = \mathbb{Q}(\sqrt{-d})$ and let $n > 1$. For a general $2n$-dimensional abelian variety (X, K) of Weil-type one has $H^2_{\text{Hodge}}(X) = D^1(X) = \mathbb{Q}$ (and thus $D^p(X) = \mathbb{Q}$ for all p) but $H^{2n}_{\text{Hodge}}(X) = \mathbb{Q}^3$, the direct sum of $D^n(X)$ and the space of Weil–Hodge cycles.
(See Weil [141].)

(11) Every simple abelian variety X of dimension 4 with $H^4_{\text{Hodge}}(X) \neq D^2(X)$ is of Weil-type.
(See Moonen–Zarhin [93].)

7.3 The Hodge Conjecture for General Abelian and Jacobian Varieties

Section 7.3.1 contains the proof of a theorem of Mattuck [92] of which the Hodge conjecture for a general polarized abelian variety is an easy consequence. In Section 7.3.2 a proof of the Hodge conjecture for a general Jacobian variety is given. It is a special case of a more general result of Biswas and Narasimhan [25]. Finally, in the last section, apart from a few exercises, we quote without proof several results on the Hodge conjecture for special abelian varieties.

7.3.1 The Theorem of Mattuck

Let X be an abelian variety. The images of algebraic cycles under the cycle map $\mathrm{cl} : \mathrm{Ch}^\bullet(X)_{\mathbb{Q}} \longrightarrow H^{2\bullet}(X,\mathbb{Q})$ are called *algebraic classes*. Every algebraic class is a $\mathbb{Q}$-linear combination of fundamental classes of algebraic subvarieties of X. According to Lemma 6.2.7 every algebraic class is a Hodge class, that is,

$$\mathrm{cl}(\mathrm{Ch}^\bullet(X)_{\mathbb{Q}}) \subseteq H^{2\bullet}_{\mathrm{Hodge}}(X).$$

The *Hodge conjecture* asserts that even equality holds. For fixed codimension p the statement

$$\mathrm{cl}(\mathrm{Ch}^p(X)_{\mathbb{Q}}) = H^{2p}_{\mathrm{Hodge}}(X)$$

is called the *Hodge* (p,p)*-conjecture*.

According to Remark 6.2.3 and for trivial reasons the Hodge (p,p)-conjecture holds for $p = 0$ and 1. The aim of this section is to show that for general polarized abelian varieties the Hodge (p,p)-conjecture holds for all p. Here *general* means an abelian variety outside the union of countably many proper analytic subvarieties of the moduli space of polarized abelian varieties of a fixed type (see the proof of Proposition 7.3.2).

Denote by $D^\bullet = D^\bullet(X)$ the subring of $H^{2\bullet}_{\mathrm{Hodge}}(X) = \bigoplus_p H^{2p}_{\mathrm{Hodge}}(X)$ generated by $H^0_{\mathrm{Hodge}}(X)$ and $H^2_{\mathrm{Hodge}}(X)$. By what we have said above the cycle classes in $D^\bullet$ are all algebraic. Hence the Hodge (p,p)-conjecture is true if

$$D^p = H^{2p}_{\mathrm{Hodge}}(X).$$

The following theorem of Mattuck [92] implies the Hodge (p,p)-conjecture for a general polarized abelian variety.

Theorem 7.3.1 *For a general polarized abelian variety (X,E) of dimension g and all $p = 0,\dots,g$,*

$$H^{2p}_{\mathrm{Hodge}}(X) = D^p \simeq \mathbb{Q}.$$

For the proof we first compute the Hodge group of X. According to Proposition 7.2.3, $\mathrm{Hg}(X) \subseteq \mathrm{Sp}(V, E)$, where $V = H_1(X, \mathbb{Q})$.

Proposition 7.3.2 *For a general polarized abelian variety* (X, E)

$$\mathrm{Hg}(X) = \mathrm{Sp}(V, E).$$

Proof Suppose $J \in C_0(\mathrm{Sp}(V, E))$ is the complex structure defining (X, E) as in Section 7.1.2 and $h : \mathbb{S}^1 \to \mathrm{Sp}(V, E)$ is the corresponding representation (see Proposition 7.1.1). According to Proposition 7.1.3 the following diagram commutes

$$\begin{array}{ccc} \mathbb{R} & \xrightarrow{\theta \mapsto \theta J} & \mathfrak{sp}(V, E) \\ {\scriptstyle e^i}\big\downarrow & & \big\downarrow{\scriptstyle \exp} \\ \mathbb{S}^1 & \xrightarrow{\ h\ } & \mathrm{Sp}(V, E). \end{array}$$

In particular $J = 1 \cdot J \in \mathfrak{sp}(V, E)$ and hence $C_0(\mathrm{Sp}(V, E))$ is a subset of $\mathfrak{sp}(V, E)$. In fact, $C_0(\mathrm{Sp}(V, E))$ is even a submanifold of $\mathfrak{sp}(V, E)$, since $C_0(\mathrm{Sp}(V, E)) \subset \mathfrak{sp}(V, E)$ is the orbit of $J \in \mathfrak{sp}(V, E)$ with respect to the adjoint representation $\mathrm{Ad} : \mathrm{Sp}(V, E) \longrightarrow \mathrm{GL}(\mathfrak{sp}(V, E))$, $M \mapsto MJM^{-1}$ (see proof of Proposition 7.1.9). Moreover, $C_0(\mathrm{Sp}(V, E))$ is not contained in a proper linear subspace of $\mathfrak{sp}(V, E)$, the adjoint representation being irreducible. (If Ad were reducible, then $\mathfrak{sp}(V, E)$ would admit a non-trivial ideal, contradicting the fact that $\mathfrak{sp}(V, E)$ is simple.)

Suppose for some $J \in C_0(\mathrm{Sp}(V, E))$ the Hodge group $\mathrm{Hg}(X_J) \neq \mathrm{Sp}(V, E)$. Then its Lie algebra $\mathfrak{hg}(X_J)$ is a proper Lie subalgebra of $\mathfrak{sp}(V, E)$ defined over $\mathbb{Q}$, containing J, and $\mathfrak{hg}(X_J) \subset C_0(\mathrm{Sp}(V, E))$ is a lower-dimensional analytic subvariety of $C_0(\mathrm{Sp}(V, E))$. But there are only countably many such Lie subalgebras. Now $C_0(\mathrm{Sp}(V, E))$ is not the union of countably many lower-dimensional subvarieties. So for a general $J \in C_0(\mathrm{Sp}(V, E))$, that is, a general polarized abelian variety (X_J, E), we have for the Hodge group $\mathrm{Hg}(X_J) = \mathrm{Sp}(V, E)$. □

Now Theorem 7.3.1 is a consequence of the following proposition.

Proposition 7.3.3 *Let* (X, E) *be a polarized abelian variety with* $\mathrm{Hg}(X) = \mathrm{Sp}(V, E)$. *Then*

$$H^{2p}_{\mathrm{Hodge}}(X) = D^p(X) \simeq \mathbb{Q} \quad \textit{for all } p = 0, \dots, g.$$

Proof $V_{\mathbb{C}}^* = H^1(X, \mathbb{C})$ is the standard representation of

$$\mathrm{Hg}(X)(\mathbb{C}) = \mathrm{Sp}(V, E)(\mathbb{C}) \simeq \mathrm{Sp}_{2g}(\mathbb{C}).$$

According to Lemma 7.3.6 below the subspace of invariants in $\bigwedge^{2p} V_{\mathbb{C}}^* = H^{2p}(X, \mathbb{C})$ with respect to $\mathrm{Sp}_{2g}(\mathbb{C})$ is one-dimensional and is spanned by $\bigwedge^p E$, the p-fold product of the alternating form $E \in H^2(X, \mathbb{C})$. Hence the assertion follows from Theorem 7.2.4. □

7.3.2 The Hodge Conjecture for a General Jacobian

In this section we prove the Hodge (p, p)-conjecture for a general Jacobian. The notion of a *general* Jacobian is defined in the same way as the notion of a general polarized abelian variety. As we saw in Remark 4.7.5, Torelli's Theorem 4.3.1 implies that under the map $\mathcal{M}_g \to \mathcal{A}_{1_g}$ the moduli space of Jacobians of a fixed dimension maps bijectively to an irreducible locally closed subvariety of the same dimension of the corresponding moduli space of principally polarized abelian varieties. Hence a general Jacobian means a Jacobian outside countable many proper subvarieties.

The idea is to use a universal curve and its associated family of Jacobians. As in the previous section the Hodge (p, p)-conjecture is a direct consequence of the following theorem.

Theorem 7.3.4 *For a general polarized Jacobian $(J(C), \Theta)$ of dimension g and all $p = 0, \ldots, g$*

$$H^{2p}_{\mathrm{Hodge}}(J(C)) = D^p(J(C)) \simeq \mathbb{Q}.$$

As an immediate consequence we obtain:

Corollary 7.3.5 *The Néron–Severi group of a general polarized Jacobian $(J(C), \Theta)$ is isomorphic to $\mathbb{Z}$.*

For the proof of Theorem 7.3.4 we need some facts concerning representations of the symplectic group. Let V denote a $\mathbb{C}$-vector space of dimension $2g$ and $E : V \times V \longrightarrow \mathbb{C}$ a non-degenerate alternating form. Identifying $V = V^*$ via the isomorphism defined by E we consider E as an element of $\bigwedge^2 V$. By definition the element $E \in \bigwedge^2 V$ is invariant under the action of the symplectic group $\mathrm{Sp}(V, E)$. This implies that the operator

$$\mathbf{L} : \bigwedge^{\bullet} V \longrightarrow \bigwedge^{\bullet+2} V, \quad u \mapsto E \wedge u$$

is $\mathrm{Sp}(V, E)$-equivariant; that is, $\mathbf{L}$ is a homomorphism of $\mathrm{Sp}(V, E)$-representations. It is called the *Lefschetz operator*. We use it in the special case $V = H^1(X, \mathbb{C})$ and $E = c_1(L)$, with a polarized abelian variety (X, L). For $k = 0, \ldots, g$ consider the subvector space

$$P^k := \ker \mathbf{L}^{g-k+1} : \bigwedge^{k} V \longrightarrow \bigwedge^{2g-k+2} V.$$

Its elements are called *primitive*. We need the following facts, for which we refer to Bourbaki [29, § 13.3].

(1) $\mathbf{L}^{g-k} : \bigwedge^k V \longrightarrow \bigwedge^{2g-k} V$ is an isomorphism of $\mathrm{Sp}(V, E)$-representations, for $k = 0, \ldots, g$.
(2) $P^0, \ldots, P^g$ are pairwise non-isomorphic, irreducible representations of $\mathrm{Sp}(V, E)$, with P^0 the trivial representation.
(3) $\bigwedge^k V = P^k \oplus \mathbf{L}P^{k-2} \oplus \mathbf{L}^2 P^{k-4} \oplus \cdots$, for $k = 0, \ldots, g$.

Note that (3) is the decomposition of $\bigwedge^k V$ into irreducible $\mathrm{Sp}(V, E)$ subrepresentations which are pairwise non-isomorphic by (1) and (2). (In the special case $V = H^1(X, \mathbb{C})$ and $E = c_1(L)$ as above (3) is called the *Lefschetz decomposition* of $H^1(X, \mathbb{C})$ (see Section 5.4.1).)

Lemma 7.3.6 *All invariants of* $\mathrm{Sp}(V, E)$ *in* $\bigwedge^\bullet V$ *are contained in the subspace spanned by* $1, E, \bigwedge^2 E, \dots, \bigwedge^g E$.

Proof According to the decomposition (3) the only trivial subrepresentations of $\bigwedge^\bullet V = \bigoplus_{k=0}^g \bigwedge^k V$ are $\mathbf{L}^k P^0$ with $2k \leq g$, and these are spanned by $\bigwedge^p E$, since $P^0 = \bigwedge^0 V \simeq \mathbb{C}$. By (2) this implies the assertion. □

Choose a decomposition

$$V = V^+ \oplus V^-$$

into isotropic subvector spaces V^+ and V^- with respect to the alternating form E. (In the special case $V = H^1(X, \mathbb{C})$ and $E = c_1(L)$ one can take here $V^+ = H^{1,0}(X)$ and $V^- = H^{0,1}(X)$.)

Lemma 7.3.7 *Every* $\mathrm{Sp}(V, E)$*-stable subspace* W *of* $\bigwedge^p V^+ \otimes \bigwedge^p V^- \subset \bigwedge^{2p} V$ *is contained in* $\left(\bigwedge^{2p} V\right)^{\mathrm{Sp}(V,E)}$.

Proof Suppose first $2p \leq g$. It suffices to consider the case that W is an irreducible representation of $\mathrm{Sp}(V, E)$. Assuming $\mathrm{Sp}(V, E)$ acts non-trivially on W, properties (1) and (3) imply

$$W = \mathbf{L}^{p-\nu} P^{2\nu} \qquad \text{for some } 0 < \nu \leq p.$$

We claim that the operator $\mathbf{L}$ restricts to operators

$$\mathbf{L} : \bigwedge^{k_1} V^+ \otimes \bigwedge^{k_2} V^- \longrightarrow \bigwedge^{k_1+1} V^+ \otimes \bigwedge^{k_2+1} V^-.$$

For this it suffices to show that $E \in V^+ \otimes V^- \subset \bigwedge^2 V$. But this follows from the fact that E, considered as an isomorphism $V^* \longrightarrow V$, restricts to isomorphisms $\mathrm{Hom}(V^\pm, \mathbb{C}) \longrightarrow V^\mp$, the vector spaces $V^\pm$ being isotropic with respect to E.
Now $\mathbf{L}^{g-2\nu+1}\left(\bigwedge^{2\nu} V^+\right) \subset \bigwedge^{g+1} V^+ \otimes \bigwedge^{g-2\nu+1} V^- = 0$ gives $\bigwedge^{2\nu} V^+ \subset P^{2\nu} = \ker \mathbf{L}^{g-2\nu+1}|_{\bigwedge^{2\nu} V}$. But then $W = \mathbf{L}^{p-\nu} P^{2\nu}$ contains the subspace $\mathbf{L}^{p-\nu}(\bigwedge^{2\nu} V^+) = \bigwedge^{p+\nu} V^+ \otimes \bigwedge^{p-\nu} V^-$, contradicting the assumption $W \subset \bigwedge^p V^+ \otimes \bigwedge^p V^-$.

If $2p > g$, the assertion follows from property (1) and the first part of the proof. □

We also need the following lemma. Again we use the moduli space of curves of a fixed genus.

Lemma 7.3.8 *Let* $\mathcal{M}_g$ *denote the coarse moduli space of smooth projective curves of genus* g. *For* $g \geq 3$ *there is a dense Zariski open subset* U *of* $\mathcal{M}_g$ *paramatrizing curves without non-trivial automorphism. In particular, there is a universal family* $C \to U$ *of curves of genus* g *over* U.

Proof Let C be a curve of genus g admitting a non-trivial automorphism φ. We may assume that C is non-hyperelliptic. Then according to Torelli's Theorem 4.3.1 the extension $\widetilde{\varphi}$ of φ to the Jacobian is an automorphism of $J = J(C)$ of the same order as φ. But the last sentence of Section 2.4.1 implies that $\widetilde{\varphi}$ is of order $\leq d_0$, the maximal order of an element of $\mathrm{GL}_{2g}(\mathbb{F}_3)$.

If $d(\leq d_0)$ is the order of the automorphism φ of C, φ induces a Galois covering $f : C \to C'$ of degree d onto a smooth projective curve C', which necessarily is of genus $< g$. For any $0 \leq g' < g$ and $2 \leq d \leq d_0$ let $H(g, g', d) \subseteq \mathcal{M}_g$ denote the subspace parametrizing curves $C \in \mathcal{M}_g$ which admit a morphism $C \to C'$ of degree d onto a curve C' of genus g'. $H(g, g', d)$ is a locally closed subset of $\mathcal{M}_g$. Hence it suffices to show that

$$\dim H(g, g', d) < 3g - 3 = \dim \mathcal{M}_g,$$

since then

$$U := \mathcal{M}_g \setminus \bigcup_{0 \leq g' < g,\ 2 \leq d \leq d_0} \overline{H(g, g', d)}$$

is open and dense in $\mathcal{M}_g$ and satisfies the assertion. But, if $C \in H(g, g', d)$, with associated morphism $f : C \to C'$, Hurwitz's formula says

$$2g - 2 = d(2g' - 2) + \delta,$$

where δ is the degree of the ramification divisor of f. This implies that

$$\begin{aligned} \dim H(g, g', d) &\leq \dim \mathcal{M}_{g'} + \delta + \begin{cases} +0 & \text{if } g' \geq 2, \\ -1 & \text{if } g' = 1, \\ -3 & \text{if } g' = 0, \end{cases} \\ &= 3g' - 3 + \delta \\ &< \tfrac{3}{2} d(2g' - 2) + \tfrac{3}{2}\delta \\ &= 3g - 3 = \dim \mathcal{M}_g. \end{aligned}$$

The last assertion of the lemma is a consequence of the construction of the moduli space $\mathcal{M}_g$ (see Mumford et al [101]). □

Proof (of Theorem 7.3.4) According to Corollary 4.7.3 the general principally polarized abelian surface is the Jacobian of a curve of genus 2 and thus the assertion follows from Theorem 7.3.1. The same statement is true for $g \leq 1$.

Hence we may assume $g \geq 3$. According to Lemma 7.3.8 there is a Zariski open dense subset U which admits a universal family $\mathcal{C} \longrightarrow U$ of curves of genus g. Let

$$\pi : \mathcal{J} \longrightarrow U$$

be the associated family of Jacobian varieties and

$$\widetilde{\pi} : \widetilde{\mathcal{J}} \longrightarrow \widetilde{U}$$

its pull back via the universal covering $q : \widetilde{U} \longrightarrow U$ of U. The family $\pi : \mathcal{J} \longrightarrow U$ admits a line bundle $\mathcal{L}$ restricting to the principal polarization on every fibre $J_u = J(C_u) := \pi^{-1}(u)$ for $u \in U$. For any $u \in U$ define

$$E_u := c_1(\mathcal{L}|_{J_u}) \in D^1(J_u) \subseteq H^2(J_u, \mathbb{Z}).$$

Choose $u_0 \in U$ and $\alpha_0 \in H^{2p}_{\text{Hodge}}(J_u)$. Then α_0 extends to a section α of $R^{2p}\widetilde{\pi}_*\mathbb{Q}$, $\widetilde{U}$ being the universal covering of U. The set

$$A_{\alpha_0} := \left\{\tilde{u} \in \widetilde{U} \,\middle|\, \alpha(\tilde{u}) \in H^{2p}_{\text{Hodge}}(J_{q(\tilde{u})})\right\}$$

is an analytic subset of $\widetilde{U}$. To see this, note that for every $\tilde{u} \in \widetilde{U}$ the element $\alpha(\tilde{u})$ is a real cohomology class and thus the condition

$$\alpha(\tilde{u}) \in H^{2p}_{\text{Hodge}}(J_{q(\tilde{u})}) = H^{2p}_{\text{Hodge}}(\widetilde{J}_{\tilde{u}}) = H^{2p}(\widetilde{J}_{\tilde{u}}, \mathbb{Q}) \cap H^{p,p}(\widetilde{J}_{\tilde{u}})$$

is equivalent to $\alpha(\tilde{u})$ belonging to the part F^p of the Hodge filtration of $H^{2p}(\widetilde{J}_{\tilde{u}}, \mathbb{C})$. But according to a theorem of Griffiths [54], the Hodge filtration moves holomorphically in a family. We claim:

$$\text{If } \alpha_0 \notin \mathbb{Q} \cdot \bigwedge^p E_{u_0}, \text{ then } A_{\alpha_0} \neq \widetilde{U}. \tag{7.4}$$

Suppose equation (7.4) is proven, then

$$B := \bigcup_{p=1}^{g} \quad \bigcup_{\alpha_0 \in H^{2p}_{\text{Hodge}}(J_{u_0}) \setminus \mathbb{Q} \bigwedge^p E_{u_0}} \quad q(A_{\alpha_0})$$

is a countable union of proper analytic subsets of U, since $H^{2p}(J_{u_0}, \mathbb{Q})$ is countable. But a complex variety of positive dimension U is not a countable union of proper analytic subsets. So for any $u \in U \setminus B$ all Hodge classes are contained in the subspace spanned by $1, E_u, \ldots, \bigwedge^g E_u$. This completes the proof of the assertion.

It remains to verify (7.4): The fundamental group $\pi_1(U, u_0)$ acts on $V := H^1(J_{u_0}, \mathbb{C})$ in the usual way. This action respects the intersection form E_{u_0}. This implies that the representation $\pi_1(U, u_0) \longrightarrow \mathrm{GL}(V)$ factorizes via a homomorphism $\rho : \pi_1(U, u_0) \longrightarrow \mathrm{Sp}(V, E_{u_0})$. According to a classical theorem (see Magnus et al [86, Theorem N 13]) the homomorphism ρ is surjective.

Suppose now $A_{\alpha_0} = \widetilde{U}$; that is, $\alpha(\tilde{u})$ is of type (p, p) for all $\tilde{u} \in \widetilde{U}$. So the subspace Σ of $H^{2p}(J_{u_0}, \mathbb{C}) = \bigwedge^{2p} V$ generated by $\mathrm{Sp}(V, E_{u_0}) \cdot \alpha_0$ is contained in $H^{p,p}(J_{u_0})$. According to Lemma 7.3.7, Σ is invariant under the action of $\mathrm{Sp}(V, E_{u_0})$. So by Lemma 7.3.6 we have $\alpha_0 \in \mathbb{Q} \bigwedge^p E_{u_0}$, contradicting the assumption. □

Remark 7.3.9 According to a hint of one of the referees one can shorten the proof of Theorem 7.3.4 by using [37, Proposition 7.5] to conclude that the Hodge group of a general Jacobian is $\mathrm{Sp}_{2g}(\mathbb{Q})$ and then deduce the Hodge conjecture (see [7, Section 6], where this is spelled out).

7.3.3 Exercises and Further Results

Although the Hodge conjecture is true for general polarized abelian and Jacobian varieties, there are many families of abelian varieties where the conjecture is open. To the best of my knowledge we owe this insight to Weil [141]. In this section some special results on the subject are given with a reference, but without proof. The first two exercises should be solved by everybody.

(1) Let $f : X \to Y$ be an isogeny of abelian varieties. Show that

(a) f induces an isomorphism $H^{2p}_{\text{Hodge}}(Y) \xrightarrow{\simeq} H^{2p}_{\text{Hodge}}(X)$.
(b) The Hodge (p, p)-conjecture is true for X if and only if it is true for Y.

(2) Show that

(a) (Lefschetz-Hodge Theorem) The Hodge (1,1)-conjecture is true for any abelian variety.
(b) The Hodge $(n-1, n-1)$-conjecture is true for any abelian variety of dimension n.
(c) Conclude: the Hodge conjecture is true for abelian varieties of dimension ≤ 3.

(Hint: For (a) use the exponential sequence. For (b) use property (1) for the Lefschetz operator in Section 7.3.2.)

(3) For an abelian variety X which is isogenous to the n-fold power of an elliptic curve E:

(a) $H^{2p}_{\text{Hodge}}(X) = D^p(X)$ for all p, and thus the Hodge (p, p)-conjecture holds for all p.
(b) $\dim H^{2p}_{\text{Hodge}}(X) = \binom{n}{d}^2 - \binom{n}{d-1} \cdot \binom{n}{d+1} \epsilon(E)$ for all $0 \leq d \leq n$ with $\epsilon(E) = 0$ or 1 if E admits complex multiplication or not.

(See Tate [132].)

(4) Let $X = (V/\Lambda, E)$ be a general polarized abelian variety, the polarization E considered as an alternating form on V. According to Proposition 7.3.2 the Hodge group is $\text{Hg}(X) = \text{Sp}(V, E)$.
Noting that the space of invariants of $\text{Sp}(V, E)$ in $H^{2p}(X, \mathbb{Q})$ is one-dimensional (see Lemma 7.3.6), we get with Theorem 7.2.4 a second proof of Mattuck's Theorem 7.3.1.

(5) Let (X, K, E) be a polarized abelian variety of Weil type with $K = \mathbb{Q}(\sqrt{-d}$ and an alternating form E.

(a) The map $H : H_1(X.\mathbb{Q}) \times H_1(X, \mathbb{Q}) \to K,\ (x, y) \mapsto E(x, \sqrt{-d} \cdot y) + \sqrt{-d}E(x, y)$, is a non-degenerate Hermitian form on the K-vector space $H_1(X, \mathbb{Q})$; that is, K-linear in the second and K-antilinear in the first factor.
(b) The hermitian form is of signature (n, n).

(c) Let $\det H$ be the determinant of the matrix of H with respect to some K-basis of $H_1(X, \mathbb{Q})$ and let $\mathrm{Nm} : K^* \to \mathbb{Q}^*$, $a + \sqrt{-d} \mapsto a^2 + db^2$ be the usual norm homomorphism. Then the image of $\det H$ in $\mathbb{Q}^*/\mathrm{Nm}(K^*)$ does not depend on the choice of the K-basis nor on the lattice defining X. It is a discrete invariant called the *discriminant of* (X, K, E).
(See van Geemen [49].)

(6) Given K, there are infinitely many irreducible families of polarized abelian varieties of dimension $g \geq 4$ of Weil type (X, K, E), distinguished by the discriminant as defined in the previous exercise.
(See Schoen [119] and van Geemen [49].)

(7) The Hodge $(2, 2)$-conjecture is true for a general four-dimensional polarized abelian variety of Weil type

(a) for $K = \mathbb{Q}(\sqrt{-3})$ in any of the infinitely many families of the previous exercise,
(b) for $K = \mathbb{Q}(i)$ and discriminant 1.

(For (a) see Schoen [119] and for (b) see Schoen [118].)

(8) A general polarized 6-dimensional abelian variety (X, K, E) of Weil type with $K = \mathbb{Q}(\sqrt{-3})$ and discriminant 1 is obtained as the Prym variety of an étale cover $C_{10} \to C_4$.
(See Faber [42, Theorem 3.1].)

(9) The Hodge conjecture is true for a general polarized abelian variety of dimension 6 of Weil type with $K = \mathbb{Q}(\sqrt{-3})$ of the previous exercise.
(See van Geemen [49, 7.3]) and Schoen [119].)

(9) Exercise 7.2.4 (11) reduces the proof of the Hodge conjecture for abelian varieties of dimension 4 to those of Weil type.
(See Ramón Mari [108, Theorem 4.11].)

(10) Let K be any imaginary quadratic field. The Hodge conjecture is true for a general polarized abelian variety (X, K, E) of Weil type of dimension 4 of discriminant 1.
(See Markman [87].)

(11) For a simple abelian variety X whose dimension is a prime number one has $H^{2p}_{\mathrm{Hodge}}(X) = D^p(X)$ for all p, and thus the Hodge (p, p)-conjecture holds for all p.
(See Tankeev [130].)

(12) For any odd m let C_m be the smooth projective curve with affine equation $y^2 = x^m - 1$.

(a) The Hodge $(2, 2)$-conjecture is true for the Jacobian $J(C_m)$.
(b) $H^4_{\text{Hodge}}(J(C_m)) = D^2(J(C_m))$ if and only if $m \not\equiv 0 \bmod 3$.
(c) $\dim(H^4_{\text{Hodge}}(J(C_m))/D^2(J(C_m)) \geq \frac{m}{3} - 1$ if $m \equiv 0 \bmod 3$.

(See Shioda [126, (6) f)].)

References

1. Adler, M. and van Moerbeke, P. The complex geometry of the Kowalewski-Painlevé analysis. *Inv. Math.* **97**, 3–51 (1989).
2. Adler, M., van Moerbeke, P. and Vanhaecke, P. *Algebraic Integrability, Painlevé Geometry and Lie Algebras*. Springer, Ergebn. der Mathem. und ihrer Grenzgeb. **47** (2004).
3. Albert, A.A. A solution of the principal problem in the theory of Riemann matrices. *Ann. Math.* **35**, 500–515 (1934) and On the construction of Riemann matrices II. *Ann. of Math.* **36**, 376–394 (1935).
4. Albert, A.A. *Structure of Algebras*. Am. J. Math. Colloquium Publications **24** (1961).
5. Andreotti, A. On a theorem of Torelli. *Am. J. Math.* **80**, 801–828 (1958).
6. Appell, P. Sur les fontions périodiques de deux variables. *J. de Math.* Sér. IV, **7**, 157–219 (1891).
7. Arapura, D. Motivation for Hodge cycles. *Adv. in Math.* **207**, 762–781 (2006).
8. Arbarello, E., Cornalba, M., Griffiths, P.A. and Harris, J. *Geometry of Algebraic Curves*. Vol. 1. Grundlehren **267**, Springer (1985).
9. Arbarello, E., Codogni, G. and Pareschi, G. Characterizing Jacobians via the KP equation and via flexes and degenerate trisecants to the Kummer variety: an algebro-geometric approach. To appear in *Journ. reine angew. Mathematik.*
10. Ash, A., Mumford, D., Rapoport, M. and Tai, Y. Smooth Compactification of Locally Symmetric Varieties. In: *Lie Groups: History, Frontiers and Applications*, Vol 4, Math. Sci. Press (1975).
11. Atiyah, M.F. and Hirzebruch, F. Analytic cycles on complex manifolds. *Topology* 1, 25–46 (1961).
12. Barth, W. Abelian Surfaces with $(1, 2)$-Polarization. In: *Algebraic Geometry*, Adv. Studies in Pure Math. **10**, pp. 41–84, Sendai (1985).
13. Beauville, A. Prym varieties and the Schottky problem. *Invent. Math.* **41**, 149–196 (1977).
14. Beauville, A. Sous-variétés spéciales des variétés de Prym. *Comp. Math.* **45**, 357–383 (1982).
15. Beauville, A. Quelque Remarques sur la Transformation de Fourier dans L'Anneau de Chow d'une Variétié Abélienne. In: *Proc. Jap.-Fr. Conf., Tokyo and Kyoto 1982*, Springer LNM **1016**, pp. 238–260 (1983).
16. Beauville, A. Sur l'anneau de Chow d'une variété abélienne. *Math. Ann.* **273**, 647–651 (1986).
17. Beauville, A. Algebraic cycles on Jacobian varieties. *Comp. Math.* **140**, 683–688, (2004).
18. Beauville, A. and Debarre, O. Une relation entre deux approches du probleme de Schottky. *Inv. math.* **86**, 195–207 (1986).
19. Beilinson, A. and Polishchuk, A. Torelli theorem via Fourier transform. In: *Moduli of Abelian Varieties*, C. Faber et al, eds, pp. 127–132, Birkhäuser (2001).

H. Lange, *Abelian Varieties over the Complex Numbers*, Grundlehren Text Editions,
https://doi.org/10.1007/978-3-031-25570-0

20. Bertrand, D. *Minimal heights and polarizations on abelian varieties*. Math. Sciences Research Inst., Berkeley, 06220-87 (1987).
21. Birkenhake, Ch. and Lange, H. Norm-endomorphisms of abelian subvarieties. In: *Classification of Irregular Varieties, Proc. Trento 1990*, Springer LNM **1515**, pp. 21–32 (1992).
22. Birkenhake, Ch. and Lange, H. *Complex Tori*. Progress in Mathem. **177**, Birkhäuser (1999).
23. Birkenhake, Ch. and Lange, H. The dual polarization of an abelian variety. *Arch. Math.* **73**, 380–389 (1999).
24. Birkenhake, Ch. and Lange, H. *Complex Abelian Varieties*. Springer, Grundl. der Math. Wiss. **302**, second edition (2004).
25. Biswas, I. and Narasimhan, M.S. Hodge Classes of Moduli Spaces of Parabolic Bundles over the General Curve. *J. Alg. Geom.* **6**, 697–715 (1997).
26. Bolza, O. On Binary Sextics with Linear Transformations onto themselves. *Am. J. Math.* **10**, 47–70 (1888).
27. Borel, A. *Introduction aux groupes arithmétiques*. Hermann, Paris (1969).
28. Bourbaki, N. *Algèbre. Chap 9*. Hermann (1958).
29. Bourbaki, N. *Groupes et Algèbres de Lie, Chap 8*. Hermann (1975).
30. Cartan, H. Quotients of complex analytic spaces. In: *Contributions to function theory*, pp. 1–15, Bombay (1960).
31. Ceresa, G. C is not algebraically equivalent to C^- in its Jacobian. *Ann. of Math.* **117**, 285–291 (1983).
32. Collino, A. A New Proof of the Ran–Matsusaka Criterion for Jacobians. *Proc. AMS* **92**, 329–331 (1984).
33. Colombo, E. and van Geemen, B. Note on curves in a jacobian. *Compos. Math.* **88**, 333–353 (1993).
34. Comessatti, A. Sulle trasformazione hermitiane delle varietà di Jacobi. Atti. R. Accad. Sci. Torino **50**, 439–455 (1914–15).
35. Debarre, O. Sur le problème de Torelli pour les variétés de Prym. *Amer. J. Math.* **111**, 111–134 (1989).
36. Deligne, P. *La théorie de Kodaira*, Exposé 5 in Sém. Géom. Alg., Orsay (1967–1968).
37. Deligne, P. La conjecture der Weil pour les surface K3. *Inv. Math.* **15**, 206–225 (1972).
38. Deninger, Ch. and Murre, J. Motivic decomposition of abelian schemes and the Fourier transform. *J. Reine Angew. Math.* **422**, 201–219 (1991).
39. Donagi, R. The tetragonal construction. *Bull. AMS* **4**, 181–185 (1981).
40. Donagi, R. and Markman, E. Spectral covers, algebraically complete integrable, Hamiltonian systems, and moduli of bundles. In: *Integrable systems and quantum groups (expanded CIME lectures, Montecatini Terme, 1993)*, Springer Lect. Notes Math. **1620**, pp. 1-119 (1996).
41. Donagi, R. and Smith, R. The structure of the Prym map. *Acta Math.* **146**, 25–102 (1981).
42. Faber, C. Prym varieties of cyclic triple covers. *Math. Z.* **199**, 61–79 (1988).
43. Fischer, G. *Complex Analytic Geometry*. Springer LNM **538** (1976).
44. de Franchis, M. Un Teorema sulle Involuzioni irrazionali. *Rend. Cir. Mat. Palermo* **36**, 368 (1913).
45. Friedman, R. and Smith, R. The generic Torelli theorem for the Prym map. *Invent. Math.* **67**, 473–490 (1982).
46. Frobenius, F.G. Über die Grundlagen der Theorie der Jacobischen Functionen. *Journal Reine Angew. Math.* **97**, 16–48, 188–224 (1884).
47. Fulton, R. *Intersection Theory*. Erg. der Math. **2**, Springer (1984).
48. van Geemen, B. Siegel modular forms vanishing on the moduli space of curves. *Inv. Math.* **78**, 329–349 (1984).
49. van Geemen, B. *An introduction to the Hodge conjecture for abelian varieties*. Springer LNM **1594**, 233–252 (1994).
50. Gelfand, S.I. and Manin, J. *Methods of Homological Algebra*. Springer (1991).
51. Grauert, H. *Ein Theorem der Analytischen Garbentheorie und die Modulräume komplexer Strukturen*. Publ. Math. IHES **5** (1960).

52. Grauert, H. and Remmert, R. *Coherent Sheaves*. Grundlehren **265**, Springer (1984).
53. Greenberg, M.J. and Harper, J.R. *Algebraic Topology, A First Course*. Benjamin (1981).
54. Griffiths, Ph. Periods of integrals on algebraic manifolds I, II. Amer. J. Math. **90**, 568–626, 805–865 (1968).
55. Griffiths, Ph. and Harris, J. *Principles of Algebraic Geometry*. John Wiley & Sons (1978).
56. Grothendieck, A. *Fondements de la Géométrie Algébrique*, Sém. Bourbaki 1957–1962 (1962).
57. Grothendieck, A. Hodge's general conjecture is false for trivial reasons. *Topology* **8**, 299–303 (1969).
58. Grothendieck, A. and Dieudonné, J. *Eléments de Géometrie Algébrique I*. Grundlehren **166**, Springer (1971).
59. Grushevsky, S. The Schottky problem. In: *Current Developments in Alg. Geom.*, MSRI Publications **59**, pp. 129–164 (2011).
60. Harris, J. *Algebraic Geometry*. Grad. Texts Math. **133**, Springer (1992).
61. Hartshorne, R. *Algebraic Geometry*. Grad. Texts Math. **52**, Springer (1977).
62. Hirzebruch, F. *Topological Methods in Algebraic Geometry*. Grundlehren **131**, Springer (1974).
63. Hochschild, G. *The structure of Lie groups*. Holden Day (1965).
64. Hodge, W.V.D. The topological invariants of algebraic varieties. In: *Proc. Int. Congr. Math.*, pp. 182–192 (1950).
65. Hulek, K. and Lange, H. Examples of abelian surfaces in $\mathbb{P}^4$. *Journal Reine Angew. Math.* **363**, 200–216 (1985).
66. Humbert, G. Théorie générale des surfaces hyperelliptiques. *J. de Math.* Sér. IV, **9**, 29–170 and 361–475 (1893).
67. Humphreys, J.E. *Linear Algebraic Groups*. Springer (1975).
68. Hurwitz, A. Über algebraische Korrespondenzen und das verallgemeinerte Korrespondenzprinzip. *Math. Ann.* **28**, 561–585 (1887).
69. Huybrechts, D, *Fourier–Mukai transform in algebraic geometry*. Clarendon Press, Oxford (2006).
70. Igusa, J. *Theta Functions*. Grundlehren **194**, Springer (1972).
71. Igusa, J. On the irreducibility of Schottky's divisor. *J. Fac. Sci Tokyo* **28**, 531–545 (1981).
72. Izadi, E. and Lange, H. Counter-examples of high Clifford index to Prym–Torelli. *J. Alg. Geom.* **21**, 769–787 (2012).
73. Jacobson, N. *Basic Algebra* I, II. Freeman (1980, 1985).
74. Kanev, V. The global Torelli theorem for Prym varieties at a generic point. *Math. USSR-Izv.* **20**, 235–258 (1983).
75. Kempf, G. Appendix to: Varieties defined by quadratic equations, by D. Mumford, in *Questions on Algebraic Varieties* CIME, Roma, pp. 95–100 (1970).
76. Koizumi, S. The Ring of Algebraic Correspondences on a Generic Curve of Genus g. *Nagoya Math. J.* **60**, 173–180 (1976).
77. Krazer, A. *Lehrbuch der Thetafunktionen*. Teubner, Leipzig (1903).
78. Künnemann, K. A Lefschetz decomposition for the Chow motiv of abelian schemes. *Invent. Math.* **113**, 85–102 (1993).
79. Krichever, I. Characterizing Jacobians via trisecants of the Kummer variety. *Ann. Math.* **172**, 485–516 (2010).
80. Lange, H. Abelian varieties with several principal polarizations. *Duke Math. J.* **55**, 617–628 (1988).
81. Lange, H. Produkte elliptischer Kurven. *Nachr. Ak. Wiss. Göttingen, math.-phys. Kl.* **II**, 95–108 (1975).
82. Lange, H. and Narasimhan, M.S. Squares of ample line bundles on abelian varieties. *Expos. Math.* **7**, 275–287 (1989).
83. Lange, H. and Rodriguez, R.E. *Decomposition of Jacobians by Prym varieties*. Springer LNM **2310** (2022).
84. Lefschetz, S. On certain numerical invariants of algebraic varieties with application to abelian varieties. *Trans. AMS* **22**, 327–482 (1921).

85. Liebermann, D. Numerical and homological equivalence on Hodge manifolds. *Am. J. Math.* **90**, 366–374 (1968).
86. Magnus, W., Karras, K. and Solitar, D. *Combinatorical Group Theory, Presentations of Groups in Terms of Generators and Relations*. Interscience Publ. (1966).
87. Markman, E. The monodromy of generalized Kummer varieties and algebraic cycles on their intermediate Jacobians. *J. Eur. Math. Soc.* (2022), published online first.
88. Martens, H.H. A new proof of Torelli's theorem. *Ann. of Math.* **78**, 107–111 (1963).
89. Masiewicki, L. Universal properties of Prym varieties with an application to algebraic curves of genus five. *Trans. AMS* **222**, 221–240 (1976).
90. Matsusaka, T. On a theorem of Torelli. *Am. J. Math.* **80**, 784–800 (1958).
91. Matsusaka, T. On a characterization of a Jacobian variety. *Mem. Coll. Sc. Kyoto* **32**, 1–19 (1959).
92. Mattuck, A. Symmetric products and Jacobians. *Am. J. Math.* **83**, 189–206 (1961).
93. Moonen, B.J.J. and Zarhin, Y.G. Weil classes on abelian varieties. *J. Reine Angew. Math.* **496**, 83–92 (1998).
94. Morikawa, H. Cycles and Endomorphisms of Abelian Varieties. *Nagoya Math. J.* **7**, 95–102 (1954).
95. Mukai, S. Duality between $D(X)$ and $D(\widehat{X})$ with its application to Picard sheaves. *Nagoya Math. J.* **81**, 153–175 (1981).
96. Mumford, D. On the equations defining Abelian varieties I, II, III. *Inv. Math.* **1**, 287–354 (1966), **3**, 75–135 (1967), **3**, 215–244 (1967).
97. Mumford, D. *Abelian varieties*. Oxford Univ. Press (1970).
98. Mumford, D. Prym varieties I. In: *Contr. to Analysis*, Academic Press, pp. 325–350 (1974).
99. Mumford, D. A Note on Shimura's Paper "Discontinuous Groups and Abelian Varieties". *Math. Ann.* **181**, 345–351 (1969).
100. Mumford, D. *Curves and their Jacobians*. Univ. of Michigan Press (1975).
101. Mumford, D. Fogarty, J. and Kirwan, F. *Geometric Invariant Theory*. Erg. der Math. **34**, Springer (1994).
102. Narasimhan, M.S. and Nori, M.V. Polarizations on an abelian variety. *Proc. Indian Ac. Sc., Math. Sci.* **90**, 125–128 (1981).
103. Ohbuchi, A. Some remarks on simple line bundles on abelian varieties. *Manuscr. Math.* **57**, 225–238 (1987).
104. Poincaré, H. and Picard, E. Sur un théorème de Riemann relatif aux fonctions de n variables indépendentes admettant $2n$ systèmes de périodes. *Comptes Rendus* **97**, 1284–1287 (1883).
105. Polishchuk, A. *Abelian Varieties, Theta Functions and the Fourier Transform*. Cambridge Univ. Press (2003).
106. Pontryagin, L. Homologies in Compact Lie Groups. *Mat. Sbornik, Novaya Seriya* **6**(3), 389–422 (1939).
107. Ramanan, S. Ample divisors on abelian surfaces. *Proc. London Math. Soc.* **51**, 231–245 (1985).
108. Ramón Mari, J. On the Hodge conjecture for products of certain surfaces. *Collect. Math.* **59**, 1–26 (2008).
109. Ran, Z. On Subvarieties of Abelian Varieties. *Inv. Math.* **62**, 459–479 (1981).
110. Ran, Z. The Structure of Gauss-like Maps. *Comp. Math.* **52**, 171–177 (1984).
111. Reider, I. Vector bundles of rank 2 and linear systems on algebraic surfaces. *Ann. of Math.* **127**, 309–316 (1988).
112. Roitman, A.A. The torsion of the group of 0-cycles modulo rational equivalence. *Ann. Math.* **111**, 553--569 (1980).
113. Rubei E. *Abelian varieties: their projective geometry and applications to Riemannian manifolds*. Tesi di perfezionamente, Scuola Norm. Sup. Pisa (1998).
114. Ruppert, W. When is an Abelian Surface Isomorphic or Isogenous to a Product of Elliptic Curves? *Math. Z.* **203**, 293–299 (1990).

115. Satake, I. *Algebraic Structures of Symmetric Domains*. Iwanami Shoten (1980).
116. Schindler, B. *Period Matrices of Hyperelliptic Curves*. Thesis Erlangen (1991). For part of it see: *Manuscr. Math.* **78**, 369–380 (1993).
117. Schoen, C. Produkte abelscher Varietäten und Moduln über Ordnungen. *J. Reine Angew. Math.* **429**, 115–123 (1992).
118. Schoen, C. Hodge classes on self-products of a variety with an automorphism. *Comp. Math.* **65**, 3–32 (1988).
119. Schoen, C. Addendum to: Hodge classes on self-products of a variety with an automorphism. *Comp. Math.* **114**, 329–336 (1998).
120. Schottky, F. *Zur Theorie der Abelschen Funktionen von vier Variablen*. *J. Reine Angew. Math.* 102, 304–352 (1888).
121. Schottky, F. and Jung, H. Neue Sätze über Symmetralfunktionen und Abelsche Funtionen der Riemann'schen Theorie. *Sitzungsber. Preuss. Akad. Wiss. Berlin, Phys. Math. Kl.* **1**, 282–297 and 732–750 (1909).
122. Serre, J.-P. Géométrie algébrique et géométrie analytique. *Ann. Inst. Fourier* **6**, 1–42 (1955).
123. Serre, J.-P. *Groupes algébriques et corps de classes*. Hermann (1959).
124. Shafarevich, I.R. *Basic Algebraic Geometry*. Grundlehren **213**, Springer (1974).
125. Shioda, T. The period map of abelian surfaces. *J. Fac. Sci. Univ. Tokyo* **25**, 47–59 (1978).
126. Shioda, T. What is known about the Hodge Conjecture? In: *Algebraic Varieties and Analytic Varieties*, Adv. Studies in Pure Mathematics **1**, pp. 55–68 (1983).
127. Shioda, T. and Mitani, N. Singular abelian surfaces and binary quadratic forms. In: *Classification of Algebraic Varieties and Compact Complex Manifolds. Proceedings 1974.* Springer LNM **412**, pp. 255–287 (1974).
128. Shiota, T. Characterization of Jacobian varieties in terms of soliton equations. *Inv. math.* **83**, 365–377 (1986).
129. Shokurov, V.V. Distinguishing Prymians from Jacobians. *Invent. Math.* **65**, 209–219 (1981).
130. Tankeev, S.G. On algebraic cycles on surfaces and abelian varieties. *Math. USSR Isv.* **18**, 349–380 (1982).
131. Tankeev, S.G. Cycles on simple abelian varieties of prime dimension. *Isv. Ross. Akad. Nauk SSSR* **57**, 192–206 (1993).
132. Tate, J. Algebraic cycles and poles of zeta functions. In: *Arithmetic and Algebraic Geometry*, Harper and Row, pp. 93–110 (1965).
133. Torelli, R. Sulle varietà di Jacobi. *Rend. R. Accad. Lincei* **22**, 98–103 (1913).
134. Umemura, H. Some results in the theory of vector bundles. *Nagoya Math. J.* **52**, 97–128 (1973).
135. Verra, A. The Prym map has degree two on plane sextics. In: *Proceedings Fano Conference, Torino*, pp. 735–759 (2002).
136. van der Waerden, B.L. *Algebra I, II*. Springer (1971).
137. Weil, A. *Courbes algébriques et variétés abéliennes*. Hermann (1948).
138. Weil, A. On Picard varieties. *Am. J. of Math.* **74**, 865–894 (1952).
139. Weil, A. Zum Beweis des Torellischen Satzes. *Nachr. Akad. Wiss. Göttingen, Math. Phys. Kl.* **11a**, 33–53 (1957).
140. Weil, A. *Variétés Kähleriennes*. Hermann (1958).
141. Weil, A. Abelian varieties and the Hodge ring. In: *Collected Works*, vol 3, [1977c], Springer, pp. 421–429 (1979).
142. Wells, R.O. *Differential Analysis on Complex Manifolds*. Grad. Texts Math. **65**, Springer (1980).
143. Welters, G. Divisor Varieties, Prym Varieties and a Conjecture of Tyurin. *Preprint Nr.* **139**, University Utrecht (1980).
144. Welters, G. A theorem of Gieseker-Petri type for Prym varieties. *Ann. Sc. Éc. Norm. Sup.* **18**, 671–683 (1985).
145. Welters, G. Recovering the curve data from a general Prym variety. *Am. J. Math.* **109**, 165–182 (1987).
146. Weyl, H. *On generalized Riemann matrices*, *Ann. of Math.* **35**, 714–729 (1934).
147. Wirtinger, W. *Untersuchungen über Thetafunctionen*. Teubner (1895).

Notation Index

H. Lange, *Abelian Varieties over the Complex Numbers*, Grundlehren Text Editions,
https://doi.org/10.1007/978-3-031-25570-0

Subject Index

H. Lange, *Abelian Varieties over the Complex Numbers*, Grundlehren Text Editions,
https://doi.org/10.1007/978-3-031-25570-0

Printed in the United States
by Baker & Taylor Publisher Services

Printed in the United States
by Baker & Taylor Publisher Services